Lecture Notes in Computer Science 6909

Commenced Publication in 1973
Founding and Former Series Editors:
Gerhard Goos, Juris Hartmanis, and Jan van Leeuwen

T0074310

Johann Eder   Maria Bielikova
A Min Tjoa (Eds.)

Advances in Databases and Information Systems

15th International Conference, ADBIS 2011
Vienna, Austria, September 20-23, 2011
Proceedings

 Springer

Volume Editors

Johann Eder
Alpen Adria Universität Klagenfurt
Institut für Informatik-Systeme
Universitätsstr. 65, 9020 Klagenfurt, Austria
E-mail: johann.eder@aau.at

Maria Bielikova
Slovak University of Technology in Bratislava
Faculty of Informatics and Information Technologies
Ilkovicova 3, 842 16 Bratislava, Slovakia
E-mail: maria.bielikova@stuba.sk

A Min Tjoa
Technische Universität Wien
Institut für Softwaretechnik
Favoritenstr. 9-11/188, 1040 Wien, Austria
E-mail: amin@ifs.tuwien.ac.at

ISSN 0302-9743   e-ISSN 1611-3349
ISBN 978-3-642-23736-2   e-ISBN 978-3-642-23737-9
DOI 10.1007/978-3-642-23737-9
Springer Heidelberg Dordrecht London New York

Library of Congress Control Number: 2011935601

CR Subject Classification (1998): H.2, H.4, H.3, C.2, J.1, H.2.8

LNCS Sublibrary: SL 3 – Information Systems and Application, incl. Internet/Web
and HCI

Typesetting: Camera-ready by author, data conversion by Scientific Publishing Services, Chennai, India

Printed on acid-free paper

Springer is part of Springer Science+Business Media (www.springer.com)

Preface

The main objective of the ADBIS series of conferences is to provide a highly visible forum for the dissemination of research accomplishments and to promote interactions and collaborations among the database and information systems research communities from Central and Eastern European countries and with the rest of the world. The ADBIS conferences provide an international platform for the presentation of research on database theory, information systems, advanced DBMS technologies, and their advanced applications.

ADBIS 2011 was the 15th instantiation of this series. It took place in Vienna, Austria, September 20–23, 2011, organized by the Institute of Software Technology and Interactive Systems of the Vienna University of Technology. ADBIS 2011 continued the ADBIS series held in St. Petersburg (1997), Poznan (1998),Maribor (1999), Prague (2000), Vilnius (2001), Bratislava (2002), Dresden (2003), Budapest (2004), Tallinn (2005), Thessaloniki (2006), Varna (2007), Pori (2008), Riga (2009), and Novi Sad (2010).

This volume contains 30 contributed research papers selected from 105 submissions from 31 countries. The thorough reviewing process – each paper was reviewed by three to five Program Committee members – was highly competitive as the acceptance rate of 29% indicates. In addition to the contributed papers, these proceedings contain two papers documenting invited keynote talks.

Furthermore, 22 papers were accepted as short papers. These are included in the additional proceedings "ADBIS 2011 Research Communications" published by the Austrian Computer Society and in the CEUR workshop proceedings (www.ceur.org). These proceedings also contain the papers accepted for the Doctoral Consortium.

This is the place to express our gratitude to all those who made ADBIS 2011 possible by generously and voluntarily sharing their knowledge, skills and time: the local Organization Chair Amirreza "Nick" Tahamtan for providing an excellent environment for the conference, the Austrian Computer Society for their help in all organizational matters, and all other colleagues holding offices. In particular, we thank the Program Committee members as well as the additional reviewers for devoting their expertise and time to ensure the high quality of the conference in an extensive review and discussion process. And last but not least, we are grateful to all the authors who showed their appreciation of the conference by submitting their valuable work.

September 2011

Johann Eder
Maria Bielikova
A Min Tjoa

Conference Organization

General Chair

A Min Tjoa   University of Technology Vienna, Austria

Program Chairs

Johann Eder   Alps-Adria University Klagenfurt, Austria
Maria Bielikova   Slovak University of Technology in Bratislava, Slovakia

Local Organization

Amirreza "Nick" Tahamtan   University of Technology Vienna, Austria

Workshop Chairs

Klaus-Dieter Schewe   SCCH Hagenberg and Johannes Kepler University Linz, Austria
Tadeusz Morzy   Technical University Poznan, Poland
Robert Wrembel   Technical University Poznan, Poland

Doctoral Consortium Chairs

Rainer Manthey   University of Bonn, Germany
Boris Novikov   St. Petersburg University, Russia

Program Committee

Suad Alagic   University of Southern Maine, USA
Witold Andrzejewski   Poznan University of Technology, Poland
Costin Badica   University of Craiova, Romania
Ladjel Bellatreche   LISI/ENSMA Poitiers University, France
Alberto Belussi   University of Verona, Italy
Andras Benczur   ELTE, Hungary
Maria Bielikova   Slovak University of Technology in Bratislava, Slovakia
Omar Boucelma   LSIS- CNRS, France
Stephane Bressan   National University of Singapore, Singapore
Dumitru Dan Burdescu   University of Craiova, Romania

Athena Vakali   Aristotle University of Thessaloniki, Greece
Olegas Vasilecas   Vilnius Gediminas Technical University,
                               Lithuania
Panos Vassiliadis   University of Ioannina, Greece
Peter Vojtas   Charles University Prague, Czech Republic
Tatjana Welzer   University of Maribor, Slovenia
Vladimir Zadorozhny   University of Pittsburgh, USA
Jaroslav Zendulka   Brno University of Technology, Czech Republic

Additional Reviewers

Abdelhamid Bouchachia, Austria   Claudiu Cristian Musat, Romania
Dmytro Buy, Ukraine   Barbara Oliboni, Italy
Federico Cavalieri, Italy   Mohamed Outahajala, France
Andrii Cherniak, USA   Ming Ouyang, USA
Amitava Datta, Australia   John Owens, USA
Hans-Georg Fill, Austria   Konstantinos Pelechrinis, USA
Olga Gkountouna, Greece   Vlad Posea, Romania
Riadh Ben Halima, Tunisia   Gabriele Pozzani, Italy
Irfan Ul Haq, Austria   Traian Rebedea, Romania
Bingsheng He, Singapore   Tarmo Robal, Estonia
David Hoksza, Czech Republic   Margit Schwab, Austria
Vedran Hrgovcic, Austria   Yi Song, Singapore
Ying-Feng Hsu, USA   Arnaud Soulet, France
Sonja Kabicher, Austria   Wee Hyong Tok, Singapore
Martin Krulis, Czech Republic   Antoine Veillard, Singapore
Maria Leitner, Austria   Thanasis Vergoulis, Greece
John Liagouris, Greece   Alexey Vovchenko, Russia
Sebastian Link, New Zealand   Robert Wrembel, Poland
Jakub Lokoc, Czech Republic   Anna Yarygina, Russian Federation
Baljeet Malhotra, Singapore   Bechir Zalila, Tunisia
Juergen Mangler, Austria   Iyad Zikra, Sweden
Sara Migliorini, Italy

ADBIS Steering Committee Chair

Leonid Kalinichenko   Russian Academy of Science, Russia

ADBIS Steering Committee

Paolo Atzeni, Italy   Johann Eder, Austria
Andras Benczur, Hungary   Marite Kirikova, Latvia
Albertas Caplinskas, Lithuania   Hele-Mai Haav, Estonia
Barbara Catania, Italy   Mirjana Ivanovic, Serbia

Table of Contents

DB Systems

Spatial Data

Query Processing 2

Information Systems

Physical DB Design

Evolution, Integrity, Security

Data Semantics

Ontological Query Answering via Rewriting

Georg Gottlob[1], Giorgio Orsi[2], and Andreas Pieris[1]

[1]Department of Computer Science, University of Oxford, UK
[2]Institute for the Future of Computing, University of Oxford, UK
firstname.lastname@cs.ox.ac.uk

Abstract. Ontological databases extend traditional databases with ontological constraints. This technology is crucial for many applications such as semantic data publishing and integration as well as model-driven database design. For many classes of ontological constraints, query answering can be solved via query rewriting. In particular, given a conjunctive query and a set of ontological constraints, the query is compiled into a first-order query, called the perfect rewriting, that encodes the intensional knowledge implied by the constraints. Then, for every database D, the answer is obtained by directly evaluating the perfect rewriting over D. Since first-order queries can be easily translated into SQL, ontological query answering can be delegated to traditional DBMSs. This allows us to utilize all the query optimization techniques available in the underlying DBMS. This paper surveys current approaches to rewriting-based query answering of ontological databases.

1 Introduction

The adoption of ontologies and semantic technology in companies, governmental organizations, and academia is becoming nowadays more and more prominent, especially for knowledge representation and data management. Thanks to their expressive power and formal semantics, ontologies have also been adopted as high-level conceptual descriptions of the data in a database, often replacing traditional metadata and documentation such as data dictionaries, UML class-diagrams and E/R schemata.

Recently, the relationship of ontologies and databases tightened, originating a new type of data management systems where a relational database is enriched by an ontological theory that enforces expressive constraints over the database. Such constraints go far beyond traditional integrity constraints and can be used to enable complex reasoning tasks over the database instances. However, the main task in an ontological database remains that of query answering. A number of commercial data management systems provide ontological querying capabilities in their current solutions (see, e.g., [6,18]). The main problem is how to couple these two different types of technologies smoothly and efficiently.

In an ontological database, queries are evaluated against an instance in such a way that the answer takes into account the semantic consequences of the ontology. Formally, if $Q : q(\mathbf{X}) \leftarrow \phi(\mathbf{X}, \mathbf{Y})$ is a *conjunctive query (CQ)* with output variables $\mathbf{X}$, then its answer in the ontological database consists of all the tuples $\mathbf{t}$ of constants such that $D \cup \Sigma \models \exists \mathbf{u}\, \phi(\mathbf{t}, \mathbf{u})$, or, equivalently, $\mathbf{t}$ belongs to the answer of Q over I, for each instance I that contains D and satisfies Σ.

J. Eder, M. Bielikova, and A.M. Tjoa (Eds.): ADBIS 2011, LNCS 6909, pp. 1–18, 2011.

Identifying expressive fragments of ontological theories under which query answering is decidable is a challenging new problem for database research. Moreover, in order to query very large databases, also tractability w.r.t. data complexity (i.e., when the query and the ontology are fixed) is required. A semantic property guaranteeing tractability is *first-order rewritability* (hence *FO-rewritability*), introduced by Calvanese et al. [16] in the context of description logics. In particular, given a conjunctive query and an ontology, the query can be transformed into a first-order query, called the perfect rewriting, that "embeds" the constraints of the ontology. Then, for every extensional database D, the answer to the query is obtained by evaluating the rewritten query against D. Since the data complexity of evaluating first-order queries is in the highly tractable complexity class AC_0 [32], query answering under FO-rewritable classes is also highly tractable. In addition, since each first-order query can be translated into an equivalent SQL expression, query answering can be delegated to a traditional relational DBMS, thus exploiting the underlying optimizations.

Example 1. Consider the ontological theory Σ consisting of the rule

$$person(X) \rightarrow \exists Y\, father(Y, X), person(Y)$$

stating that every person has a father, who is himself a person. Let Q be the CQ $q(B) \leftarrow father(A, B)$ asking for persons who have a father. Intuitively, due to the rule in Σ, not only do we have to query $father$, but we also need to query $person$, since all the persons necessarily have a father. The perfect rewriting Q_Σ will thus be the logical union of Q and of the query $q(B) \leftarrow person(B)$.   ∎

A prominent family of languages that enjoy FO-rewritability is the *DL-Lite* family of description logics (DLs) introduced by Calvanese et al. in [16]. A further contribution to the field has been the introduction of the *Datalog$^\pm$* family [9,11], whose languages extend the well-known Datalog language (see, e.g., [1]) by allowing in rule heads existential quantifiers (in the same way as value invention in Datalog [7]), the equality predicate and the truth constant *false*. Interestingly, various languages in the Datalog$^\pm$ family are FO-rewritable.

Roadmap. The paper proceeds as follows. After a brief recall of preliminary notions given in the next section, we present FO-rewritability in Section 3. Known formalisms enjoying FO-rewritability are presented in Section 4, while existing algorithms to compute the first-order rewriting are described in Section 5. Finally, Section 6 draws some conclusions and outlines some future research directions.

2   Preliminaries

In this section we recall some basics on databases, tuple-generating dependencies, queries, and the chase procedure.

General. We define the following pairwise disjoint (possibly infinite) sets of symbols: *(i)* a set Γ of *constants* (constitute the "normal" domain of a database), *(ii)* a set Γ_N of *labelled nulls* (representing unknown values, and thus can be also seen as variables), and *(iii)* a set Γ_V of *variables* (used in queries and constraints). Different constants represent different values (*unique name assumption*), while different nulls may represent the same value. We denote by **X** sequences of variables $X_1, \ldots, X_k$, where $k \geqslant 0$. Let $[n]$ be the set $\{1, \ldots, n\}$, for any integer $n \geqslant 1$.

A *relational schema* $\mathcal{R}$ (or simply *schema*) is a set of *relational symbols* (or *predicates*), each with its associated arity. A *position* $r[i]$ (in a schema $\mathcal{R}$) is identified by a predicate $r \in \mathcal{R}$ and its i-th argument (or attribute). A *term* t is a constant, null, or variable. An *atomic formula* (or simply *atom*) has the form $r(t_1, \ldots, t_n)$, where r is an n-ary predicate, and $t_1, \ldots, t_n$ are terms. Conjunctions of atoms are often identified with the sets of their atoms. A *relational instance* (or simply *instance*) I for a schema $\mathcal{R}$ is a (possibly infinite) set of atoms of the form $r(\mathbf{t})$, where r is an n-ary predicate of $\mathcal{R}$, and $\mathbf{t} \in (\Gamma \cup \Gamma_N)^n$. A *database* for $\mathcal{R}$ is a finite instance for $\mathcal{R}$ which contains only *ground* atoms, i.e., atoms whose arguments are constants of Γ.

A *substitution* from a set of symbols S_1 to another set of symbols S_2 is a function $h : S_1 \rightarrow S_2$ defined as follows: *(i)* $\varnothing$ is a substitution (empty substitution), *(ii)* if h is a substitution, then $h \cup \{X \rightarrow Y\}$ is a substitution, where $X \in S_1$ and $Y \in S_2$, and h does not already contain some $X \rightarrow Z$ with $Y \neq Z$. If $X \rightarrow Y \in h$, then we write $h(X) = Y$. A *homomorphism* from a set of atoms A_1 to a set of atoms A_2, both over the same schema $\mathcal{R}$, is a substitution $h : \Gamma \cup \Gamma_N \cup \Gamma_V \rightarrow \Gamma \cup \Gamma_N \cup \Gamma_V$ such that: *(i)* if $t \in \Gamma$, then $h(t) = t$, and *(ii)* if $r(t_1, \ldots, t_n)$ is in A_1, then $h(r(t_1, \ldots, t_n)) = r(h(t_1), \ldots, h(t_n))$ is in A_2. The notion of homomorphism naturally extends to conjunctions of atoms.

Tuple-Generating Dependencies. A *tuple-generating dependency (TGD)* σ over a schema $\mathcal{R}$ is a first-order formula $\forall \mathbf{X} \forall \mathbf{Y} \, \varphi(\mathbf{X}, \mathbf{Y}) \rightarrow \exists \mathbf{Z} \, \psi(\mathbf{X}, \mathbf{Z})$, where $\varphi(\mathbf{X}, \mathbf{Y})$ and $\psi(\mathbf{X}, \mathbf{Z})$ are conjunctions of atoms over $\mathcal{R}$, called the *body* and the *head* of σ, denoted as $body(\sigma)$ and $head(\sigma)$, respectively. Notice that the well-known *inclusion dependencies (IDs)*, see, e.g., [1], are the simplest type of TGDs with just one body-atom and one head-atom, without repetition of variables. For example, assuming that *dept* is a binary predicate and *emp* is a ternary predicate, the ID $dept[2] \subseteq emp[1]$, which asserts that each manager is an employee, can be expressed using the TGD $\forall X \forall Y \, dept(X, Y) \rightarrow \exists Z \exists W \, emp(Y, Z, W)$. Henceforth, to avoid notational clutter, we will omit the universal quantifiers in TGDs. A TGD σ is satisfied by an instance I for $\mathcal{R}$ iff, whenever there exists a homomorphism h such that $h(body(\sigma)) \subseteq I$, there exists an extension h' of h (i.e., $h' \supseteq h$) such that $h'(head(\sigma)) \subseteq I$.

Datalog. A *Datalog rule* ρ is an expression of the form $\underline{a}_0 \leftarrow \underline{a}_1, \ldots, \underline{a}_n$, for $n \geqslant 0$, where each $\underline{a}_i$ is an atom and every variable occurring in $\underline{a}_0$ appears in at least one atom of $\{\underline{a}_1, \ldots, \underline{a}_n\}$. The atom $\underline{a}_0$ is called the *head* of ρ, denoted as $head(\rho)$, while $\underline{a}_1, \ldots, \underline{a}_n$ is called the *body* of ρ, denoted as $body(\rho)$. A *Datalog program* Π over a schema $\mathcal{R}$ is a set of Datalog rules such that, for each $\rho \in \Pi$, the predicate of $head(\rho)$ does not occur in $\mathcal{R}$. Π is *non-recursive* if its *predicate graph* G is acyclic; G has as vertices the predicates occurring in Π, and there exists an edge from p_1 to p_2 if there exists a rule $\rho \in \Pi$ such that p_1 occurs in $body(\rho)$ and p_2 occurs in $head(\rho)$. The *extensional database (EDB)* predicates are those that do not occur in the head of any rule of Π; all the other predicates are called *intensional database (IDB)* predicates. A *model* of Π is an instance over $\mathcal{R}$ that satisfies the set of TGDs obtained by considering the program Π as a set of universally quantified implications. The semantics of Π w.r.t. an EDB D for $\mathcal{R}$, denoted as $\Pi(D)$, is the minimum model of Π containing D (which is unique and always exists). $\Pi(D)$ can be computed by a least fixpoint iteration starting from the EDB D and adding at each iteration all new facts generated by a single rule application. Note that $\Pi(D)$ is always finite and all values appearing in it are from the universe of D. If there exists a constant k such that, for every EDB D, $\Pi(D)$ can be obtained after k iterations of the fixpoint operator, then Π is said *bounded*. Intuitively, if a (recursive) Datalog program is bounded it is essentially non-recursive, although it appears to be recursive syntactically.

Queries. An n-ary *Datalog query* Q over a schema $\mathcal{R}$ is a pair $\langle q, \Pi \rangle$, where Π is a Datalog program over $\mathcal{R}$ and q is an n-ary predicate which occurs in the head of some rule of Π. Q is a *non-recursive* (resp., *bounded*) *Datalog query* if Π is non-recursive (resp., bounded). Q is a *union of conjunctive queries (UCQs)* if q is the only IDB predicate in Π, and for each rule $\rho \in \Pi$, q does not occur in $body(\rho)$. Finally, Q is a *conjunctive query (CQ)* if it is a UCQs, and Π contains exactly one rule.

The *answer* to a Datalog query $Q = \langle q, \Pi \rangle$ over a database D is the set of n-tuples $\{\mathbf{t} \mid q(\mathbf{t}) \in \Pi(D)\}$, denoted as $Q(D)$. Notice that if Q is a UCQs, then the problem whether $\mathbf{t} \in Q(D)$ is equivalent to the problem whether there exists $\rho \in \Pi$ such that $body(\rho)$ is mapped by a homomorphism h to D, and $\mathbf{t} = h(\mathbf{X})$, where $\mathbf{X}$ are the variables occurring in $head(\rho)$.

Conjunctive Query Answering under TGDs. Given a database D for a schema $\mathcal{R}$, and a set Σ of TGDs over $\mathcal{R}$, the *models* of D w.r.t. Σ is the set of instances $mods(D, \Sigma) = \{I \mid I \supseteq D$ and I satisfies $\Sigma\}$. Given an n-ary UCQs $Q = \langle q, \Pi \rangle$, by Q_ρ we denote the n-ary CQ $\langle q, \{\rho\} \rangle$, where $\rho \in \Pi$. The *answer* to Q w.r.t. D and Σ, denoted as $ans(Q, D, \Sigma)$, is the set of n-tuples

$$\bigcup_{\rho \in \Pi} \{\mathbf{t} \mid \mathbf{t} \in \Gamma^n \text{ and } \mathbf{t} \in Q_\rho(I), \text{ for each } I \in mods(D, \Sigma)\}.$$

Notice that the associated decision problem is undecidable [5], even when the schema and the set of TGDs are fixed [8], or even if we consider singleton sets of TGDs [4].

TGD Chase Procedure. The *chase procedure* (or simply *chase*) is a fundamental algorithmic tool introduced for checking implication of dependencies [27], and later for checking query containment [24]. Informally, the chase is a process of repairing a database w.r.t. a set of dependencies so that the resulted instance satisfies the dependencies. We shall use the term chase interchangeably for both the procedure and its result. The chase works on an instance through the so-called TGD *chase rule*.

TGD CHASE RULE. Consider a database D for a schema $\mathcal{R}$, and a TGD σ : $\varphi(\mathbf{X}, \mathbf{Y}) \rightarrow \exists \mathbf{Z} \, \psi(\mathbf{X}, \mathbf{Z})$ over $\mathcal{R}$. If σ is *applicable* to D, i.e., there exists a homomorphism h such that $h(\varphi(\mathbf{X}, \mathbf{Y})) \subseteq D$ then: *(i)* define $h' \supseteq h$ such that $h'(Z_i) = z_i$, for each $Z_i \in \mathbf{Z}$, where $z_i \in \Gamma_N$ is a "fresh" labelled null not introduced before, and following lexicographically all those introduced so far, and *(ii)* add to D the set of atoms in $h'(\psi(\mathbf{X}, \mathbf{Z}))$.

Given a database D and set of TGDs Σ, the chase procedure for D w.r.t. Σ consists of an exhaustive application of the TGD chase rules, which leads to a (possibly infinite) instance, denoted as $chase(D, \Sigma)$. We assume that the chase algorithm is *fair*, i.e., each TGD that must be applied during the construction of the chase is eventually applied.

Example 2. Consider the set Σ constituted by the TGDs $\sigma_1 : r(X, Y), s(Y) \rightarrow \exists Z\, r(Z, X)$ and $\sigma_2 : r(X, Y) \rightarrow s(X)$, and let D be the database $\{r(a, b), s(b)\}$. During the construction of $chase(D, \Sigma)$ we first apply σ_1, and we add the atom $r(z_1, a)$, where z_1 is a "fresh" null. Moreover, σ_2 is applicable and we add the atom $s(a)$. Now, σ_1 is applicable and the atom $r(z_2, z_1)$ is obtained, where z_2 is a "fresh" null. Also, σ_2 is applicable and the atom $s(z_1)$ is generated. It is clear that there is no finite chase satisfying both σ_1 and σ_2.   ∎

The fairness assumption implies that the (possibly infinite) chase of D w.r.t. Σ is a *universal model* of D w.r.t. Σ, i.e., for each instance $I \in mods(D, \Sigma)$, there exists a homomorphism from $chase(D, \Sigma)$ to I [20,19]. This fact allows us to show that the chase is a formal algorithmic tool for query answering under TGDs. In other words, given a UCQs Q, a database D, and a set Σ of TGDs, $ans(Q, D, \Sigma)$ can be obtained by evaluating Q over $chase(D, \Sigma)$, and discarding tuples containing at least one null.

3   First-Order Rewritability

In the setting of query answering under TGDs, one is usually interested in the data complexity, i.e., the complexity calculated by considering only the data as part of the input, while the query and the set of TGDs are fixed. This is because the size of the data is predominant with that of the other inputs. In particular, to be able to work with very large databases, we need query answering to be highly tractable in data complexity and possibly feasible by relational query engines.

First-order rewritability, introduced in the context of description logics (under the name first-order logic reducibility) [16], guarantees the above desirable properties. Roughly, given a UCQs Q and a set of TGDs, a first-order query can be constructed, called the *perfect rewriting*, that takes into account the semantic consequences of the TGDs. Then, the answer to the query w.r.t. a database D and the set of TGDs is obtained by evaluating the perfect rewriting over D. An n-ary *first-order query* Q is an open first-order logic formula $\phi(\mathbf{X})$ with free variables $\mathbf{X}$, where $|\mathbf{X}| = n$. The answer to Q w.r.t. a database D is the set $Q(D)$ of n-tuples of Γ^n such that, when assigned to the free variables, make the formula ϕ *true* (see, e.g., [1]).

Definition 1. *Consider a set Σ of TGDs over a schema $\mathcal{R}$. We say that Σ is first-order rewritable if, for every UCQs Q over $\mathcal{R}$, a first-order query Q_Σ can be constructed such that $ans(Q, D, \Sigma) = Q_\Sigma(D)$, for every database D for $\mathcal{R}$.*

It is well-known that evaluation of first-order queries is in the highly tractable class AC$_0$ in data complexity [32]. This is the complexity class of recognizing words in languages defined by constant-depth Boolean circuits with (unlimited fan-in) AND and OR gates (see, e.g., [29]). Given that every first-order query can be written into an equivalent (non-recursive) SQL query, in practical terms this means that query answering under first-order rewritable sets of TGDs can be deferred to a standard DBMS, exploiting the underlying optimizations.

Example 3. Let $\mathcal{R}$ be the relational schema constituted by the predicates *runs*(Dept_Id, Project_Id), *in_area*(Project_Id, Area) and *external*(Ext_Id, Area, Project_Id). Consider the set Σ constituted by the single TGD

$$\sigma : runs(W, X), in\_area(X, Y) \rightarrow \exists Z\, external(Z, Y, X),$$

which expresses the fact that for each project there exists an external controller, special-ized on the area of the project, that works on it. The conjunctive query $Q = \langle q, \{\rho\}\rangle$, where $\rho : q(B) \leftarrow external(A, database, B)$, asks for all the projects in the database area which have an external controller. Intuitively, due to the TGD σ, not only do we have to query *external*, but we also need to look for projects in the database area, as such projects will have necessarily an external controller. The perfect rewriting will thus be the UCQs (which is a first-order query) $Q_\Sigma = \langle q, \Pi\rangle$, where Π contains the rules

$$q(B) \leftarrow external(A, database, B),$$
$$q(B) \leftarrow runs(A, B), in\_area(B, database).$$

It is not difficult to see that Q_Σ can be written in SQL as follows:

```
SELECT E.Project_Id FROM external E
WHERE E.Area = "database"
UNION
SELECT R.Project_Id FROM runs R, in_area I
WHERE R.Project_Id = I.Project_Id AND I.Area = "database"
```■

Let us clarify that the problem of identifying first-order rewritable sets of TGDs is un-decidable. This can be established easily by exploiting known results. Given a Datalog program Π (which can be seen as a set of TGDs without existentially quantified vari-ables) if it is bounded, then it is also first-order rewritable. The converse was established by Ajtai and Gurevich [2]. Therefore, the class of bounded Datalog programs coincides with the class of first-order rewritable Datalog programs. It is well-known that the prob-lem whether a given Datalog program is bounded is undecidable [21]. From the above discussion, we immediately get that indeed the problem whether a set of TGDs is first-order rewritable is undecidable.

First-order rewritable sets are strictly connected to *finite unification sets* [4]. Roughly, a set Σ of TGDs is a finite unification set if, for every UCQs Q, the perfect rewriting Q_Σ of Q obtained by backward-chaining through unification, according to the TGDs of Σ, is finite. It is immediate to see that a finite unification set is trivially a first-order rewritable set. However, it is not known whether there exists a first-order rewritable set which is not a finite unification set.

4 Concrete First-Order Rewritable Classes

As already explained, first-order rewritability is just an abstract property which is, in general, not possible to recognize. In this section we present the known syntactic first-order rewritable classes of TGDs, which are actually members of the Datalog$^\pm$ family.

Linear TGDs. A TGD is *linear* if its body contains a single atom [9]. Using linear TGDs we can assert, for example, that everyone supervising her/himself is a manager: $supervises(X, X) \rightarrow mgr(X)$; notice that this TGD does not fall into the class of inclusion dependencies (IDs) since the repetition of variables is not allowed. Linear TGDs are generalized by *multi-linear* TGDs. A TGD σ is multi-linear if each atom of $body(\sigma)$ is a *guard*, i.e., contains all the universally quantified variables of σ. Obviously, each linear TGD is trivially multi-linear since its single body-atom is automatically a guard. With multi-linear TGDs we can assert, for instance, that each employee who is also a manager supervises some other employee: $emp(X), mgr(X) \rightarrow \exists Y\ supervises(X, Y)$; clearly, the above TGD is neither an ID nor linear.

As we discuss below, (multi-)linear TGDs (combined with negative constraints and key dependencies, that is, additional features which do not increase the complexity of query answering) generalize several prominent and highly tractable formalisms for ontology reasoning, in particular, the main languages of the DL-Lite family [16].

The main weakness of linear TGDs is the fact that they do not allow for joins in TGD-bodies. Although multi-linear TGDs allow for joins (like in the above example), the joins that can be expressed are very restrictive; recall that each body-variable must occur in every body-atom. For example, joins like the one over the variable X in the body of the TGD $runs(W, X), in\_area(X, Y) \rightarrow \exists Z\ external(Z, Y, X)$ (given in Example 3) is not expressible using multi-linear TGDs. An expressive first-order rewritable class of TGDs that can cope with such cases is discussed in the following paragraph.

Sticky Sets of TGDs. The class of *sticky* sets of TGDs [11] (formally defined below) is a sufficient syntactic condition that ensures the so-called *sticky property* of the chase, which is as follows. For every database D, assume that during the chase of D w.r.t. a set Σ of TGDs, we apply a TGD $\sigma \in \Sigma$ which has a variable V appearing more than once in its body. Assume also that V maps (via a homomorphism) on the symbol z, and that by virtue of this application the atom $\underline{a}$ is generated. In this case, for each atom $\underline{b} \in body(\sigma)$, we say that $\underline{a}$ is derived from $\underline{b}$. Then, z appears in $\underline{a}$, and in all atoms resulting from some chase derivation sequence starting from $\underline{a}$, "sticking" to them (hence the name "sticky" sets of TGDs).

The definition of sticky sets of TGDs is based heavily on a variable-marking procedure called SMarking. This procedure accepts as input a set of TGDs Σ, and marks the variables that occur in the body of the TGDs of Σ. Formally, SMarking(Σ) works as follows. First, we apply the so-called *initial marking* step: for each TGD $\sigma \in \Sigma$, and for each variable V in $body(\sigma)$, if there exists an atom $\underline{a}$ in $head(\sigma)$ such that V does not appear in $\underline{a}$, then we mark each occurrence of V in $body(\sigma)$. Then, we apply exhaustively (i.e., until a fixpoint is reached) the *propagation* step: for each pair of TGDs $\langle \sigma, \sigma' \rangle \in \Sigma \times \Sigma$ (including the case $\sigma = \sigma'$), if a $\forall$-variable V occurs in $head(\sigma)$ at positions $\pi_1, \ldots, \pi_m$, for $m \geqslant 1$, and there exists an atom $\underline{a} \in body(\sigma')$ such that at each position $\pi_1, \ldots, \pi_m$ a marked variable occurs, then we mark each occurrence of V in $body(\sigma)$. The formal definition of sticky sets of TGDs follows.

Definition 2. *A set Σ of TGDs is* sticky *if there is no TGD $\sigma \in$ SMarking(Σ) such that a marked variable occurs in $body(\sigma)$ more than once.*

Example 4. Assume that after the application of SMarking on a set Σ of TGDs we obtain the following set (we mark variables with a cap, e.g., $\hat{X}$):

$$dept(\hat{V}, \hat{W}) \rightarrow \exists X \exists Y \exists Z\, emp(W, X, Y, Z),$$
$$emp(\hat{V}, \hat{W}, \hat{X}, \hat{Y}) \rightarrow \exists Z\, dept(W, Z), runs(W, Y), in\_area(Y, X),$$
$$runs(\hat{W}, X), in\_area(X, Y) \rightarrow \exists Z\, external(Z, Y, X).$$

Clearly, for each TGD $\sigma \in$ SMarking(Σ), there is no marked variable that occurs in $body(\sigma)$ more than once, and thus Σ is a sticky set of TGDs.   ∎

Despite their expressiveness, sticky sets of TGDs are not powerful enough to be able to capture simple cases such as the TGD $r(X, Y, X) \rightarrow \exists Z\, s(Y, Z)$; clearly, the variable X is marked, and thus stickiness is violated. Notice that the above TGD is linear. A first-order rewritable class which captures both sticky sets of TGDs and linear TGD, called *sticky-join* sets of TGDs, is proposed in [13]. The main disadvantage of this class is the fact that the identification problem, i.e., whether a set of TGDs is sticky-join, is computationally hard; in particular, PSPACE-hard. Notice that the identification problem under (multi-)linear TGDs and sticky sets of TGDs is feasible in PTIME.

Bounded Derivation-Depth Property. The key property, underlying the classes of TGDs presented above, which implies first-order rewritability is the so-called *bounded derivation-depth property (BDDP)* [9]. As already explained, given an n-ary UCQs Q, a database D, and a set Σ of TGDs, the problem whether an n-tuple $\mathbf{t} \in \Gamma^n$ belongs to $ans(Q, D, \Sigma)$ is equivalent to the problem whether $\mathbf{t} \in Q(chase(D, \Sigma))$. However, $chase(D, \Sigma)$ is (in general) infinite, and thus not explicitly computable. The BDDP implies that, instead of evaluating Q over $chase(D, \Sigma)$, it suffices to evaluate it over a finite part of the chase which depends only on Q and $\mathcal{R}$. Roughly, $chase(D, \Sigma)$ can be decomposed in levels, where D has level 0, and an atom has level $k + 1$ if it is obtained, during the chase, due to atoms with maximum level k. We refer to the part of the chase up to level k as $chase^k(D, \Sigma)$. For the formal definitions we refer the reader to [9,11].

Definition 3. *A set Σ of TGDs enjoys the BDDP if, for every n-ary UCQs Q over a schema $\mathcal{R}$, for every database D for $\mathcal{R}$, and for every n-tuple $\mathbf{t} \in \Gamma^n$, if $\mathbf{t} \in ans(Q, D, \Sigma)$, then $\mathbf{t} \in Q(chase^k(D, \Sigma))$, where k depends only on Q and Σ, but not on the database D.*

The proof of the fact that the BDDP is a sufficient condition for first-order rewritability, given in [9], hinges on the fact that by exploiting the initial finite part of the chase implied by the BDDP, it is possible to construct a first-order query which is a perfect rewriting (see algorithm BDDP-rewrite in Section 5).

The DL-Lite Family. Description Logics (DLs), are decidable fragments of first-order logic, based on concepts (classes of objects) and roles (relations on concepts). Several variants of them have been investigated, where a central issue is the trade-off between expressiveness and complexity. Interestingly, several *lite* DLs exist which are first-order rewritable; in particular, the members of the well-known DL-Lite family [16], with which we assume the reader is familiar. In what follows, we briefly discuss how the DL-Lite family is related to the aforementioned first-order rewritable classes of TGDs; see also [9,11].

The main DL-Lite languages, namely, DL-Lite$_\mathcal{F}$, DL-Lite$_\mathcal{R}$ and DL-Lite$_\mathcal{A}$, can be reduced to linear TGDs and sticky sets of TGDs, combined with *negative constraints (NCs)* of the form $\forall \mathbf{X}\varphi(\mathbf{X}) \rightarrow \bot$, where $\bot$ denotes the truth constant *false*, and *key dependencies (KD)*. These additional features do not increase the complexity of query answering. This holds since, apart from a preliminary check (which is equivalent to query answering under the TGDs) to verify that the NCs and the KDs are satisfied by the given database and the set of TGDs, we are allowed to ignore the NCs and the KDs [9,11]. Moreover, the DLs DL-Lite$_{\mathcal{F},\sqcap}$, DL-Lite$_{\mathcal{R},\sqcap}$ and DL-Lite$_{\mathcal{A},\sqcap}$ obtained from DL-Lite$_\mathcal{F}$, DL-Lite$_\mathcal{R}$ and DL-Lite$_\mathcal{A}$, respectively, by additionally allowing conjunction in the left-hand side of the axioms, can be reduced to multi-linear TGDs (with NCs and KDs). Furthermore, the above DLs (with binary roles) have a counterpart in the DLR-Lite family, which allows for n-ary roles, along with suitable constructs to deal with them [17]. These extended languages can be also reduced to (multi-)linear TGDs and sticky sets of TGDs (with NCs and KDs) [11].

We conclude this section by giving a simple example how a DL-Lite$_\mathcal{F}$ TBox is reduced to a set of TGDs (with NCs and KDs).

Example 5. The DL-Lite$_\mathcal{F}$ TBox $\mathcal{T}$ constituted by *person* $\sqsubseteq \exists father^-$, $\exists father \sqsubseteq person$, $\exists father \sqsubseteq \neg\exists mother$ and (funct *father$^-$*) asserts that: *(i)* each person has a father, *(ii)* who is himself a person, *(iii)* fathers and mothers are disjoint sets, and *(iv)* each person has only one father. $\mathcal{T}$ can be reduced to the following set of TGDs with NCs and KDs:

$$person(X) \rightarrow \exists Y\, father(Y, X), \quad father(X, Y), mother(X, Z) \rightarrow \bot,$$
$$father(X, Y) \rightarrow person(X), \quad father(X, Y), father(Z, Y) \rightarrow X = Z;$$

the last rule is known as *equality-generating dependency*, and is equivalent to the KD $key(father) = \{2\}$. Observe that the first two formulas are linear TGDs and also a sticky set of TGDs. ∎

The ER$_\bot^+$ Family. Other known first-order rewritable formalisms are the members of the ER$_\bot^+$ family of conceptual models proposed in [10,12]. These models extend the well-known Entity-Relationship model with is-a constraints among entities and relationships, plus functional and mandatory participation constraints. Also, disjointness and non-participation constraints (but also more general ones) can be expressed. Query answering under ER$_\bot^+$ models can be reduced to query answering under *non-conflicting conceptual dependencies*, that is, a restricted class of linear TGDs and sticky sets of TGDs with KDs and NCs. The particularity of non-conflicting CDs is that, once the given database and the set of TGDs satisfy the KDs and the NCs (and this problem is equivalent to query answering under the TGDs alone), then we can proceed by considering only the TGDs.

5   Rewriting Approaches

Several techniques for computing the perfect rewriting of a query w.r.t. a set of FO-rewritable constraints can be found in the literature.

BDDP-Rewrite. Calì et al. in [9] presented an algorithm that computes the perfect rewriting $\langle q, \Pi_\Sigma \rangle$ of an n-ary UCQs $\langle q, \Pi \rangle$ w.r.t a set of TGDs Σ over a schema $\mathcal{R}$ enjoying the bounded-derivation-depth property. By definition of BDDP, for every database D and every n-ary tuple $\mathbf{t} \in \Gamma^n$, if $\mathbf{t} \in ans(Q, D, \Sigma)$, then $\mathbf{t} \in Q(chase^k(D, \Sigma))$, where k does not depend on D. Clearly, every atom in $chase(D, \Sigma)$ is obtained by at most $b = \max_{\sigma \in \Sigma}\{|body(\sigma)|\}$ atoms. Therefore, if $\mathbf{t} \in ans(Q, D, \Sigma)$, and thus there exists $\rho \in \Pi$ of the form $q(\mathbf{X}) \leftarrow \phi(\mathbf{X}, \mathbf{Y})$ such that $\phi(\mathbf{X}, \mathbf{Y})$ is mapped by a homomorphism h to $chase^k(D, \Sigma)$ and $h(\mathbf{X}) = \mathbf{t}$, the number of ancestors of $h(body(\rho))$ at level zero (i.e., the database level) is at most $n \cdot b^k$, where $n = \max_{\rho \in \Pi}\{|body(\rho)|\}$. The desired query $\langle q, \Pi_\Sigma \rangle$ is constructed as follows:

1. Let $\{S_1, \dots, S_m\}$, for $m \geq 1$, be all the possible sets of $n \cdot b^k$ atoms over $\mathcal{R}$ having nulls of Γ_N and constants that appear in Π as arguments.

2. For each $i \in [m]$, let $C_i = chase^k(S_i, \Sigma)$ (considering S_i as a database). For every rule $q(\mathbf{X}) \leftarrow \phi(\mathbf{X}, \mathbf{Y})$ of Π, and for every homomorphism h such that $h(\phi(\mathbf{X}, \mathbf{Y})) \subseteq C_i$ and $h(\mathbf{X}) \in \Gamma^n$, let ϕ_i be the logical conjunction of the atoms obtained from S_i by replacing each distinct null of Γ_N with a distinct variable of Γ_V, and add to Π_Σ the rule $q(h(\mathbf{X})) \leftarrow \phi_i$.

Roughly, each rule of Π_Σ corresponds to some derivation of n atoms (soundness), while every derivation of n atoms in the levels of the chase up to k (i.e., all those sufficient to answer Q) corresponds to a rule of Π_Σ (completeness). However, **BDDP-rewrite** is not very well-suited for practical implementations. For this reason, several other techniques, which are reviewed in the rest of this section, have been proposed.

ID-Rewrite. A more viable way of constructing the perfect rewriting is by proceeding "backward" from the input query "towards" the atoms of the database. An early algorithm of this kind is **ID-rewrite** introduced by Calì et al. in [15]. This algorithm takes as input a UCQs $\langle q, \Pi \rangle$ and a set of inclusion dependencies Σ over a relational schema $\mathcal{R}$, and constructs a UCQs $\langle q, \Pi_\Sigma \rangle$ by iterating over two steps, namely, *rewriting* and *reduction*, until a fix-point in the construction of the rewriting is reached.

The rewriting step works as follows. An ID $\sigma \in \Sigma$ is *applicable* to a rule $\rho \in \Pi_\Sigma$ if there exists $\underline{a} \in body(\rho)$ such that *(i)* $\underline{a}$ and $head(\sigma)$ unify, and *(ii)* if a bound term in ρ (i.e., a constant of Γ or a variable that occurs more than once in ρ) appears at position π, then at position π in $head(\sigma)$ a universally-quantified variable occurs. Then, for each ID $\sigma \in \Sigma$ and for each rule $\rho \in \Pi_\Sigma$ (notice that $\Pi \subseteq \Pi_\Sigma$), if σ is applicable to ρ due to an atom $\underline{a} \in body(\rho)$, then $\gamma(\rho')$ is added to Π_Σ, where γ is the most general unifier (MGU) of $\underline{a}$ and $head(\sigma)$, and ρ' is obtained from ρ by replacing $\underline{a}$ with $body(\sigma)$. Roughly, this step uses the IDs of Σ as rewriting rules.

Example 6. Consider the query $\langle q, \{\rho\} \rangle$, where $\rho : q(A) \leftarrow p(A, B), t(A, C, D)$, and the set Σ of IDs:

$$\sigma_1 : s(X) \rightarrow \exists Y \exists Z \ t(X, Y, Z)$$
$$\sigma_2 : t(X, Y, Z) \rightarrow r(Y, Z).$$

The atom $t(A, C, D)$ unifies with $head(\sigma_1)$ through the MGU $\gamma = \{X \rightarrow A, Y \rightarrow C, Z \rightarrow D\}$, where the existential variables Y and Z map to the unbound variables C and D. ID-rewrite constructs the rule $q(A) \leftarrow p(A, B), s(A)$ and adds it to Π_Σ.   ∎

The applicability condition may prevent the generation of queries that are vital to guarantee completeness of the rewritten query, as shown by the following example.

Example 7. Consider the query $\langle q, \{\rho\} \rangle$ where $\rho : q(A) \leftarrow t(A, B), s(B)$ and the set Σ of IDs:
$$\sigma_1 : p(X) \rightarrow \exists Y\, t(X, Y)$$
$$\sigma_2 : t(X, Y) \rightarrow s(Y).$$
The only viable strategy in this case is to rewrite the atom $s(B)$ using σ_2, since the atom $t(A, B)$ is blocked by the applicability condition due to the bound variable B. The rule that we obtain is $\rho_1 : q(A) \leftarrow t(A, B), t(X, B)$, where X is a fresh variable. Notice that in ρ_1 the variable B remains bound, thus no other atoms can be rewritten and $\Pi_\Sigma = \{\rho, \rho_1\}$. Consider now the database $D = \{p(a)\}$. Clearly, $chase(D, \Sigma) = \{p(a), t(a, z_1), s(z_1)\}$ and ρ maps to $chase(D, \Sigma)$; however, none of the rules in Π_Σ maps to D and therefore the rewriting is not complete.   ∎

This problem is solved by applying the reduction step: for each rule $\rho \in \Pi_\Sigma$, if $body(\rho)$ contains two atoms $\underline{a_1}$ and $\underline{a_2}$ that unify, then the rule $\gamma(\rho)$ is added to Π_Σ, where γ is the MGU of $\underline{a_1}$ and $\underline{a_2}$.

Example 8. With reference to Example 7, the reduction step would generate the rule $\rho_2 : q(A) \leftarrow t(A, B)$ from ρ_1. The atom $t(A, B)$ can be now rewritten using σ_1 to produce the rule $\rho_3 : q(A) \leftarrow p(A)$ that makes the perfect rewriting complete.   ∎

ID-rewrite is at the basis of PerfectRef, the original algorithm used by the DL-Lite reasoner QuOnto[1]. The main disadvantage of ID-rewrite is the large number of rules generated by the reduction step. Since the unifications are done blindly, most of the generated rules are redundant and do not contribute to the actual completeness of the perfect-rewriting. In addition, the size of the constructed perfect rewriting, intended as the number of rules in Π_Σ, is intrinsically exponential in the size of the input query and of Σ since the constructed rewriting is in disjunctive normal form.

UCQ-Rewrite. Calì et al. in [14] extended ID-rewrite to cope with more general classes of TGDs than IDs; in particular, linear and sticky(-join) TGDs. The main difference is in the applicability condition that must be extended to be able to deal with the repetition of head variables in the constraints. In particular, if a bound term in the input query occurs in an atom $\underline{a}$ at positions $\pi_1, \ldots, \pi_m$, for $m \geq 2$, then either, for each $i \in [m]$, the variable at position π_i in $head(\sigma)$ occurs also in $body(\sigma)$, or at positions $\pi_1, \ldots, \pi_m$ in $head(\sigma)$ the same existentially quantified variable occurs. Apart from this extension, UCQ-rewrite works as ID-rewrite and, therefore, it inherits all its drawbacks.

TGD-Rewrite. In a subsequent work [22], Gottlob et al. proposed a rewriting technique for Datalog$^\pm$ that sensibly improves the UCQ-rewrite algorithm by avoiding the redundant rules produced by the reduction step. TGD-rewrite substitutes the reduction step

---

[1] http://www.dis.uniroma1.it/ quonto/

of UCQ-rewrite with a proper *atom factorization* procedure defined as follows. First, given a query $\langle q, \Pi \rangle$, each position π in an atom $\underline{a} \in body(\rho)$, where $\rho \in \Pi$, is called *existential* w.r.t. a set Σ of TGDs if there exists a TGD $\sigma \in \Sigma$ such that $\underline{a}$ unifies with $head(\sigma)$, and the term at position π in $head(\sigma)$ is an existentially quantified variable. A set of atoms $S \subseteq body(\rho)$ is *factorizable* w.r.t a TGD σ iff: *(i)* for each pair of atoms $\underline{a}, \underline{b}$ of S, $\underline{a}$ and $\underline{b}$ unify, and *(ii)* if a variable V appears at an existential position in an atom of S, then V does not occur in $body(\rho) \setminus S$, and also V occurs only at existential positions. Intuitively, the factorization produces only useful rules, i.e., by rewriting their atoms we obtain rules that are needed to ensure completeness.

Example 9. Consider the TGD $\sigma : s(X), r(X, Y) \rightarrow \exists Z\, t(X, Z, Z)$ and the three Boolean CQs, i.e., CQs of arity zero:

$$\rho_1 : q() \leftarrow t(A, C, B), t(A, E, C)$$
$$\rho_2 : q() \leftarrow s(C), t(A, C, B), t(A, E, C)$$
$$\rho_3 : q() \leftarrow t(A, C, B), t(C, E, B).$$

Clearly, the atoms in the body of ρ_1 unify through the MGU $\gamma = \{E \rightarrow C, B \rightarrow C\}$, and they are also factorizable since the variables C, B and E appear only at existential positions in ρ_1. The factorization results in the rule $\rho_1' : q() \leftarrow t(A, C, C)$; notice that σ is not applicable to ρ_1 but it is applicable to ρ_1'. On the contrary, despite the fact that the atoms $t(A, C, B)$ and $t(A, E, C)$ in ρ_2 unify, since the variable C appears also at position $s[1]$ which is not existential w.r.t $\{\sigma\}$, the atoms are not factorizable. The same holds for ρ_3, where the atoms $t(A, C, B)$ and $t(C, E, B)$ unify but the variable C appears at $t[1]$ which is not existential w.r.t. $\{\sigma\}$. ∎

Another optimization introduced by TGD-rewrite, consists of the removal of the atoms in the body of a rule that are logically implied (w.r.t. Σ) by other atoms in the same rule. This procedure avoids the construction of redundant rules during the rewriting process. Notice that the elimination procedure works only for linear TGDs and that the worst-case size of the perfect rewriting constructed by TGD-rewrite remains exponential in the size of the query and the set of constraints.

Requiem. An alternative resolution-based rewriting technique has been proposed by Peréz-Urbina et al. in [30]. Requiem[2] is designed to work with non-FO-rewritable description logics, but can also be applied to DL-Lite TBoxes to produce a perfect rewriting in form of UCQs. The algorithm proceeds in three steps.

Skolemization. First, the existential quantifiers in the constraints of Σ are eliminated by reducing each TGD into *Skolem normal form*. During this step, Σ is transformed into an equisatisfiable set Σ_f of rules.

Saturation. In order to compute the perfect rewriting, Requiem systematically constructs all the clauses that can be derived through resolution from the set Σ_f and the query $\langle q, \Pi \rangle$. The saturation technique adopted by Requiem is inspired by the technique introduced by Joiner [25] to decide a first-order fragment by resolution. Joiner

---

[2] http://www.comlab.ox.ac.uk/projects/requiem/home.html

established that, in order to obtain a refutation decision procedure for a first-order fragment $\mathcal{L}$ it is sufficient to: *(i)* select a sound and complete clausal calculus $\mathcal{C}$, *(ii)* identify a set of clauses $\mathcal{N}$ such that $\mathcal{N}$ is finite for a finite signature and the translation of a formula in $\mathcal{N}$ into clauses produces only clauses in $\mathcal{N}$ and, *(iii)* prove that $\mathcal{N}$ is closed under $\mathcal{C}$. Given the above setting, it is possible to construct a finite set of clauses by saturating the set $\Pi \cup \Sigma_f$ that can be used to retrieve all and only the answers to the input query. Depending on the language of constraints, **Requiem** adopts a language-specific RFS (Resolution with free selection) clausal calculus for this task. Once the saturated set has been computed, **Requiem** eliminates all the rules containing function symbols since they are not necessary for query-answering purposes. The output is a Datalog program representing the perfect rewriting $\langle q, \Pi_\Sigma \rangle$ of the input query w.r.t. Σ.

Unfolding. Since the Datalog program is possibly recursive, **Requiem** proceeds to the unfolding of the rules to construct the corresponding UCQs. This is done by iteratively expanding the rules in Π_Σ that have q as head predicate, using the other rules in Π_Σ until a UCQs is obtained. The termination condition for the expansion process depends on the language of Σ; in particular, for DL-Lite it terminates when no new rules are generated.

Example 10. Consider the set of TGDs and the query of Example 7. The Skolem normal form of the set of TGDs is

$$p(X) \rightarrow t(X, f(X))$$
$$t(X, Y) \rightarrow s(Y).$$

The saturation step produces the following set of rules:

$$p(X) \rightarrow s(f(X))$$
$$p(X), s(f(X)) \rightarrow q(X)$$
$$p(X), t(X, f(X)) \rightarrow q(X)$$
$$p(X) \rightarrow q(X).$$

The Datalog perfect rewriting $\langle q, \Pi_\Sigma \rangle$ consists therefore of the function-free rules (input query included):

$$t(X, Y), s(Y) \rightarrow q(X)$$
$$t(X, Y) \rightarrow s(Y)$$
$$p(X) \rightarrow q(X).$$

The unfolding step then rewrites the atom $s(Y)$ with $t(X, Y)$ producing the following UCQs that is the final rewriting produced by **Requiem**.

$$q(X) \leftarrow t(X, Y), s(Y)$$
$$q(X) \leftarrow t(X, Y), t(X, Z)$$
$$q(X) \leftarrow p(X).$$

∎

Comb-Rewrite. Kontchakov et al. in [26] proposed a technique to compute the perfect rewriting of a UCQs $Q = \langle q, \Pi \rangle$ w.r.t. a set Σ of DL-Lite$^{\mathcal{N}}_{horn}$ constraints, i.e., DL-Lite$_{horn}$ extended with number restrictions [3]. The constructed rewriting has the property of being of polynomial size w.r.t. the size of Q and Σ at the price of requiring a completion procedure of the input database D (*combined rewriting*).

During the completion step, the database D is extended with tuples that satisfy (i.e., witness) the constraints in Σ, in a similar way as in the chase procedure, obtaining the so-called *canonical instance* D_{comp}. Differently from the chase, the completion procedure introduces fresh nulls only when it is not possible to satisfy a constraint using the nulls already in D. This procedure ensures that D_{comp} can be constructed in polynomial time w.r.t. the size of D. On the other hand, D_{comp} is not in general a model for Σ but it can be used to "simulate" a universal model.

Since answering Q over D_{comp} might lead to unsound answers, in the rewriting step, Comb-rewrite reformulates Q in order to embed constraints that exclude unsound tuples from the answer. The reformulated query is a formula $Q_\Sigma = \langle q, \Pi_\Sigma \rangle$ where each $\rho_\Sigma \in \Pi_\Sigma$ is constructed as follows: $head(\rho_\Sigma) = head(\rho)$ and $body(\rho_\Sigma) = body(\rho) \wedge \phi_{cert} \wedge \phi_{tree}$. The formula ϕ_{cert} is a conjunction of inequalities ensuring that no variable in $head(\rho_\Sigma)$ is mapped to a null value, i.e., only tuples of constants are returned, while the formula ϕ_{tree} ensures that whenever $body(\rho_\Sigma)$, for some $\rho_\Sigma \in \Pi_\Sigma$, maps to atoms in D_{comp} containing some null value, then $body(\rho_\Sigma)$ must be homomorphically embeddable into a tree-shaped subsets of D_{comp}.

It can be proven that $ans(Q, D, \Sigma) = Q_\Sigma(D_{comp})$, however, a disadvantage of this technique is that requires updates to the input database D, while all the other presented approaches are purely intensional.

Presto. To address the problem of the exponential blowup of the size of the rewriting, Rosati and Almatelli proposed Presto [31], an algorithm that computes the perfect-rewriting of a conjunctive query w.r.t. a DL-Lite TBox as a non-recursive Datalog program instead of a UCQs. Rosati et al. noticed that one of the reasons for the exponential size of UCQs rewritings is in the joins between existential variables. Some of these joins can be systematically eliminated by leveraging on the structure of the set of constraints. Presto is based on three main ideas.

Split existential joins. Given a query $\langle q, \Pi \rangle$, each rule $\rho \in \Pi$ is split into its existential-join connected components, i.e., subsets of atoms in $body(\rho)$ connected by existential variables. Each connected component is then associated to a fresh auxiliary predicate.

Example 11. Consider the rule $\rho : q(X, Y) \leftarrow p(X, W), r(W, T), p(Y, Z), s(Z)$ and the following set Σ of TGDs:

$$t(X, Y) \rightarrow p(X, Y)$$
$$u(X, Y) \rightarrow t(X, Y)$$
$$p(X, Y) \rightarrow \exists Z\, r(Y, Z)$$
$$p(X, Y) \rightarrow s(Y).$$

Algorithms such as ID-rewrite and UCQ-rewrite generate about 70 rules, where the atoms unifying with $p(X, Y)$ are substituted with atoms of the form $t(X, Y)$ and $u(X, Y)$ in all the possible combinations as in the following:

$$q(X, Y) \leftarrow \underline{p(X, W)}, r(W, T), \underline{p(Y, Z)}, s(Z)$$
$$q(X, Y) \leftarrow \underline{t(X, W)}, r(W, T), \underline{p(Y, Z)}, s(Z)$$
$$q(X, Y) \leftarrow \underline{u(X, W)}, r(W, T), \underline{p(Y, Z)}, s(Z)$$
$$q(X, Y) \leftarrow \underline{t(X, W)}, r(W, T), \underline{t(Y, Z)}, s(Z)$$
$$\dots$$

Instead, **Presto** splits ρ into two rules $\rho_1 : q_1(X) \leftarrow p(X,W), r(W,T)$ and $\rho_2 : q_2(Y) \leftarrow p(Y,Z), s(Z)$ that are then combined by the rule $\rho_3 : q(X,Y) \leftarrow q_1(X), q_2(Y)$. Notice that ρ is equivalent to the set $\{\rho_1, \rho_2, \rho_3\}$, but now it is possible to process the two components independently.   ∎

Compress hierarchical expansions. A backward-chaining algorithm iteratively expands the atoms in the query using the TGDs in Σ. **Presto** is able to compress such expansions into Datalog rules by introducing, for each predicate r in the schema, an auxiliary predicate that "collects" all the "specializations" of r.

Example 12. Consider the set of TGDs of Example 11. The predicates t and u are specializations of the predicate p, therefore **Presto** introduces an auxiliary predicate aux_p and the set of rules:

$$aux_p(X,Y) \leftarrow p(X,Y)$$
$$aux_p(X,Y) \leftarrow t(X,Y)$$
$$aux_p(X,Y) \leftarrow u(X,Y),$$

and substitutes the predicate aux_p, whenever p, t and u occur in the rules, as follows:

$$q(X,Y) \leftarrow q_1(X), q_2(Y)$$
$$q_1(X) \leftarrow aux_p(X,W), r(W,T)$$
$$q_2(Y) \leftarrow aux_p(Y,Z), s(Z).$$ ∎

Elimination of existential joins. **Presto** uses the concept of *most-general subsumee* to eliminate the unnecessary existential joins from the rules. Roughly, **Presto** eliminates those atoms in a rule whose existence is already implied by other atoms in the query. Consider again the set Σ of Example 11. The atom $s(Z)$ of ρ is implied by the atom $p(Y,Z)$ under Σ. In the same way, the atom $r(W,T)$ is implied by $p(X,W)$. The final rewriting is therefore the following set of six rules, where the unnecessary atoms have been eliminated:

$$q(X,Y) \leftarrow q_1(X), q_2(Y)$$
$$q_1(X) \leftarrow aux_p(X,W)$$
$$q_2(Y) \leftarrow aux_p(Y,Z)$$
$$aux_p(X,Y) \leftarrow p(X,Y)$$
$$aux_p(X,Y) \leftarrow t(X,Y)$$
$$aux_p(X,Y) \leftarrow u(X,Y).$$

The size of the perfect rewriting produced by **Presto** is exponential only in the number of non-eliminable existential-join variables of the given query; such variables are a subset of the join variables of the query, and are typically less than the number of atoms in the query. However, if we restrict the language to DL-Lite$_{core}$ it is possible to construct a rewriting of polynomial size also w.r.t. the size of the query [26].

DTG-Rewrite. Orsi and Pieris presented in [28] an algorithm to compute the perfect-rewriting of a UCQs $\langle q, \Pi \rangle$ w.r.t. a set of linear TGDs Σ as a bounded Datalog program $\langle q, \Pi_\Sigma \rangle$. **DTG-rewrite** proceeds in four steps.

Skolemization. First, the existential quantifiers in the TGDs of Σ are eliminated as in **Requiem** producing an equisatisfiable set of rules Σ_f.

Rule Saturation. This step computes the so-called *saturated set* of Σ_f, written $\Pi_{r\text{-}sat}$, by applying the well-known *resolution* inference rule. A rule of $\Pi_{r\text{-}sat}$, obtained by applying k times the resolution rule, "mimics" a derivation of the chase under Σ_f which involves $k + 1$ applications of the TGD chase rule. Notice that $\Pi_{r\text{-}sat}$ is, in general, infinite. However, since linear TGDs enjoy the BDDP (see Section 2), it suffices to "mimic" the chase up to a finite level and, therefore, only a finite part of $\Pi_{r\text{-}sat}$ is constructed.

Query saturation. During this step the so-called saturated query denoted as $\Pi_{q\text{-}sat}$ of $\langle q, \Pi \rangle$ is computed. This is done by resolving the rules in Π with the non-function-free rules of $\Pi_{r\text{-}sat}$. Roughly, given a database D, if one of the atoms due to which the input query maps to $chase(D, \Sigma_f)$ was obtained by a chase derivation that involves TGDs with functional terms, then **DTG-rewrite** constructs a rule that "bypasses" this derivation. Notice that the saturated query is, in general, infinite. Nevertheless, since linear TGDs have the BDDP, as in the case of $\Pi_{r\text{-}sat}$, only a finite part is constructed.

Finalization. In this step the rewritten query is obtained by adding the function-free rules of $\Pi_{r\text{-}sat} \cup \Pi_{q\text{-}sat}$ to Π_Σ.

Differently from **Presto**, no auxiliary predicates are introduced to compress the hierarchical expansions. This is done in **DTG-rewrite** by reusing the predicates of Σ. In addition, by targeting a bounded Datalog program instead of a non-recursive Datalog program, **DTG-rewrite** is able to compress more the size of the rewriting while preserving the good computational properties of first-order queries. **DTG-rewrite** also improves **Requiem** by avoiding the expansion of the Datalog rewriting into UCQs and by avoiding the unnecessary intermediate rules during the rule saturation step.

Example 13. Consider the set of TGDs and the query of Example 10. The skolemization step of **DTG-rewrite** proceeds exactly as in **Requiem** whereas the saturation step is split in two phases. The rule-saturation step is independent on the query and produces only the rule $p(X) \rightarrow s(f(X))$. The query saturation step then expands the query using only the rules with function symbols in the saturated set where at each step the redundant atoms in the query are eliminated. The final set of rules produced by **DTG-rewrite** in this case is the following

$$q(X) \leftarrow t(X, Y)$$
$$q(X) \leftarrow p(X).$$

Notice that the rule $t(X, Y) \rightarrow s(Y)$ is not part of the rewriting since it does not contribute to obtain any tuple in the output predicate q. Redundant rules can be identified *a-priori* using the notion of predicate graph of a set of rules. ∎

Poly-Rewrite. Recently, Gottlob and Schwentick [23] proved that for every query Q, for every database D, and for every set Σ of TGDs that falls in any of the aforementioned syntactic classes of TGDs, we can obtain the answers of Q w.r.t. D and Σ by evaluating Q over a finite part of $chase(D, \Sigma)$, obtained after polynomially (w.r.t. Q and Σ) many chase steps. This implies that for all the above classes, it is possible to construct a polynomial-size (w.r.t. Q and Σ) non-recursive Datalog rewriting.

Table 1. Summary of rewriting techniques

Algorithm	language	input	output	size
ID-rewrite	IDs	UCQs	UCQs	exponential in Q and Σ
UCQ-rewrite	SJTGDs	UCQs	UCQs	exponential in Q and Σ
TGD-rewrite	SJTGDs	UCQs	UCQs	exponential in Q and Σ
Requiem	DL-Lite	UCQs	UCQs	exponential in Q and Σ
Comb-rewrite	DL-Lite$_{horn}^{\mathcal{N}}$	UCQs	UCQs + views	polynomial in Q and Σ
Presto	DL-Lite	UCQs	non-recursive Datalog	exponential in Q - polynomial in Σ
DTG-rewrite	LTGDs	UCQs	bounded Datalog	exponential in Q - polynomial in Σ
Poly-rewrite	SJTGDs	UCQs	non-recursive Datalog	polynomial in Q and Σ

Table 1 summarizes the techniques discussed above where Q represents the size of atoms in each rule of the input query and Σ is the size of the set of constraints. By SJT-GDs and LTGDs we refer to sticky(-join) sets of TGDs and linear TGDs, respectively. Recall that sticky-join sets of TGDs capture linear TGDs and sticky sets of TGDs.

6 Conclusion

This work discussed FO-rewritability and related properties with particular focus to their application to ontological query answering. We also presented existing FO-rewritable classes of languages and overviewed recent algorithms to compute the perfect rewriting of a query w.r.t. a set of ontological constraints. Future research directions aim at identifying more expressive classes of FO-rewritable languages and devising more effective rewriting optimization techniques.

Acknowledgements. This research has received funding from the European Research Council under the European Community's Seventh Framework Programme (FP7/2007–2013) / ERC grant agreement DIADEM no. 246858 and from the Oxford Martin School's grant no. LC0910-019.

References

1. Abiteboul, S., Hull, R., Vianu, V.: Foundations of Databases. Addison-Wesley, Reading (1995)
2. Ajtai, M., Gurevich, Y.: Datalog vs. first-order logic. In: Proc. of FOCS, pp. 142–147 (1989)
3. Artale, A., Calvanese, D., Kontchakov, R., Zakharyaschev, M.: The DL-Lite family and relations. J. of Artificial Intelligence Research 36, 1–69 (2009)
4. Baget, J.-F., Leclère, M., Mugnier, M.-L., Salvat, E.: On rules with existential variables: Walking the decidability line. Artif. Intell. 175(9-10), 1620–1654 (2011)
5. Beeri, C., Vardi, M.Y.: The implication problem for data dependencies. In: Even, S., Kariv, O. (eds.) ICALP 1981. LNCS, vol. 115, pp. 73–85. Springer, Heidelberg (1981)
6. Bishop, B., Kiryakov, A., Ognyanoff, D., Peikov, I., Tashev, Z., Velkov, R.: Owlim: A family of scalable semantic repositories. Semantic Web 2(1), 33–42 (2011)
7. Cabibbo, L.: The expressive power of stratified logic programs with value invention. Inf. Comput. 147(1), 22–56 (1998)
8. Calì, A., Gottlob, G., Kifer, M.: Taming the infinite chase: Query answering under expressive relational constraints. In: Proc. of KR, pp. 70–80 (2008)

9. Calì, A., Gottlob, G., Lukasiewicz, T.: A general Datalog-based framework for tractable query answering over ontologies. In: Proc. of PODS, pp. 77–86 (2009)
10. Calì, A., Gottlob, G., Pieris, A.: Tractable query answering over conceptual schemata. In: Laender, A.H.F., Castano, S., Dayal, U., Casati, F., de Oliveira, J.P.M. (eds.) ER 2009. LNCS, vol. 5829, pp. 175–190. Springer, Heidelberg (2009)
11. Calì, A., Gottlob, G., Pieris, A.: Advanced processing for ontological queries. PVLDB 3(1), 554–565 (2010)
12. Calì, A., Gottlob, G., Pieris, A.: Query answering under expressive Entity-Relationship schemata. In: Parsons, J., Saeki, M., Shoval, P., Woo, C., Wand, Y. (eds.) ER 2010. LNCS, vol. 6412, pp. 347–361. Springer, Heidelberg (2010)
13. Calì, A., Gottlob, G., Pieris, A.: Query answering under non-guarded rules in Datalog$^\pm$. In: Hitzler, P., Lukasiewicz, T. (eds.) RR 2010. LNCS, vol. 6333, pp. 1–17. Springer, Heidelberg (2010)
14. Calì, A., Gottlob, G., Pieris, A.: Query rewriting under non-guarded rules. In: Proc. AMW (2010)
15. Calì, A., Lembo, D., Rosati, R.: Query rewriting and answering under constraints in data integration systems. In: Proc. of IJCAI, pp. 16–21 (2003)
16. Calvanese, D., De Giacomo, G., Lembo, D., Lenzerini, M., Rosati, R.: Tractable reasoning and efficient query answering in description logics: The DL-lite family. J. Autom. Reasoning 39(3), 385–429 (2007)
17. Calvanese, D., Giacomo, G.D., Lembo, D., Lenzerini, M., Rosati, R.: Data complexity of query answering in description logics. In: Proc. of KR, pp. 260–270 (2006)
18. Chong, E., Das, S., Eadon, G., Srinivasan, J.: An efficient SQL-based RDF querying scheme. In: Proc. of the 31th Intl Conf. on Very Large Data Bases (VLDB), pp. 1216–1227 (2005)
19. Deutsch, A., Nash, A., Remmel, J.B.: The chase revisisted. In: Proc. of PODS, pp. 149–158 (2008)
20. Fagin, R., Kolaitis, P.G., Miller, R.J., Popa, L.: Data exchange: Semantics and query answering. Theor. Comput. Sci. 336(1), 89–124 (2005)
21. Gaifman, H., Mairson, H.G., Sagiv, Y., Vardi, M.Y.: Undecidable optimization problems for database logic programs. J. ACM 40(3), 683–713 (1993)
22. Gottlob, G., Orsi, G., Pieris, A.: Ontological queries: Rewriting and optimization. In: Proc. of ICDE, pp. 2–13 (2011)
23. Gottlob, G., Schwentick, T.: Rewriting ontological queries into small non-recursive Datalog programs. In: Proc. of DL (2011)
24. Johnson, D.S., Klug, A.C.: Testing containment of conjunctive queries under functional and inclusion dependencies. J. Comput. Syst. Sci. 28(1), 167–189 (1984)
25. Joiner, W.H.: Resolution strategies as decision procedures. J. ACM 23, 398–417 (1976)
26. Kontchakov, R., Lutz, C., Toman, D., Wolter, F., Zakharyaschev, M.: The combined approach to query answering in DL-Lite. In: Proc. of KR (2010)
27. Maier, D., Mendelzon, A.O., Sagiv, Y.: Testing implications of data dependencies. ACM Trans. Database Syst. 4(4), 455–469 (1979)
28. Orsi, G., Pieris, A.: Optimizing query answering under ontological constraints. In: PVLDB (2011) (in press)
29. Papadimitriou, C.H.: Computational complexity. Addison-Wesley, Reading (1994)
30. Pérez-Urbina, H., Motik, B., Horrocks, I.: Tractable query answering and rewriting under description logic constraints. Journal of Applied Logic 8(2), 151–232 (2009)
31. Rosati, R., Almatelli, A.: Improving query answering over DL-Lite ontologies. In: Proc. KR (2010)
32. Vardi, M.Y.: On the complexity of bounded-variable queries. In: Proc. of PODS, pp. 266–276 (1995)

On the Convergence of Data and Process Engineering

Marlon Dumas

University of Tartu, Estonia
marlon.dumas@ut.ee

Abstract. It is common practice in contemporary information systems engineering to combine data engineering methods with process engineering methods. However, these methods are applied rather independently and at different layers of an information system. This situation engenders an impedance mismatch between the process layer and the business logic and data layers in contemporary information systems. We expose some of the issues that this impedance mismatch raises by means of a concrete example. We then discuss emerging paradigms for seamlessly integrating data and process engineering.

1 Introduction

Data engineering is a well-trodden field endowed with a mature body of methods and tools. Proven data analysis and design methods allow data engineers to capture complex data requirements and to refine these requirements down to the level of database schemas in a seamless and largely standardized manner. Concomitantly, database systems and associated middleware enable the development of robust and scalable data-driven applications to support a wide spectrum of business functions. Furthermore, contemporary packaged enterprise systems support hundreds of business activities on top of shared databases, while Master Data Management (MDM) methods provide guidance for managing and governing data across application and organizational boundaries.

Eventually though, individual business functions supported by database applications need to be integrated in order to automate end-to-end business processes such as order-to-cash processes. This facet of information systems engineering falls under the realm of business process engineering.

Business process engineering is also an established discipline with its own body of methods and tools. Process analysis and design methods typically rely on process models that capture how tasks, events and decision points are inter-connected, and what data objects are consumed and produced throughout a process. These models are first captured at a high level of abstraction and then refined down to executable process models that are deployed in Business Process Management Systems (BPMS). These systems orchestrate the execution of business processes by delegating work to human actors and moving data across multiple applications, possibly across organizational boundaries.

But while data engineering and process engineering are each endowed with their own body of mature methods and tools, these methods and tools are at best loosely integrated. When it comes to accessing data, BPMS typically rely on request-response interactions with database applications or packaged enterprise systems. Typically, data fetched from these systems are copied into the "working memory" of a business process.

J. Eder, M. Bielikova, and A.M. Tjoa (Eds.): ADBIS 2011, LNCS 6909, pp. 19–26, 2011.

The collected data are then used to evaluate business rules and are distributed to other systems as required by the logic encoded in the business process.

More generally, in contemporary information systems engineering, data entities and business processes are analyzed, designed, implemented and tested separately, using fundamentally different methods. This divide between data and process engineering is driven by various factors, including the fact that data are shared across multiple processes, that data and processes evolve at different rates and according to different requirements. Notwithstanding these reasons, the "data vs. processes" divide leads to redundancy that, in the long run, hinders on the coherence and maintainability of information systems. In particular, the data vs. processes divide has the following effects:

– *Process-related and function-related data redundancy*. The BPMS maintains data about the *state* of the process, since these data are needed in order to enable the system to schedule tasks, react to events and to evaluate predicates attached to decision points in the process. On the other hand, data entities manipulated by the process are stored in the database(s). The result is that data are managed sometimes redundantly at the database layer and at the process execution layer, thereby adding development and maintenance complexity.

– *Business rules redundancy*. Some business rules are encoded at the level of the business process and also at the level of the database application (or even the database system itself in the form of triggers or integrity constraints). This rule redundancy hampers maintainability and potentially leads to inconsistencies.

Service-oriented architectures (SOAs) facilitate the inter-connection of applications and application components. Their emergence has greatly facilitated the integration of data-driven and process-driven applications. SOAs have also enabled packaged enterprise software vendors to "open the box" by providing standardized programmatic access to the vast functionality of their systems. But per se, SOAs do not address the problem of data and process integration, since data-centric services and process-centric services are still developed separately using different methods. A case in point is Thomas Erl's service-oriented design method [5], which advocates that process-centric services should be strictly layered on top of data-centric (a.k.a. entity-centric) services. Erl's approach consists of two distinct methods for designing process-centric services and entity-centric services. This same principle permeates in many other service-oriented design methods [6].

This talk will give an overview of emerging approaches that aim at addressing the shortcomings of the data and process engineering divide. In particular, the talk will discuss the emerging artifact-centric process management paradigm, and how this paradigm, in conjunction with SOA platforms, allow organizations to achieve higher levels of integration and higher responsiveness to process change.

2 Illustrative Scenario

In order to concretely illustrate the drawbacks that the "process vs. data divide" create in contemporary information systems, we consider a *Build-to-order* (BTO) process at a company herewith called *MetalWorks*. A BTO process is an order-to-delivery process where the products to be sold are manufactured on the basis of a confirmed purchase order. In other words, the manufacturer does not maintain ready-to-ship products in stock. Instead, products are manufactured on demand when the customer orders them. This approach is used in the context of customized products, such as metallurgical products, where customers often submit orders for products with very specific requirements.

The process is depicted using the BPMN notation in Figure 1. The process starts when MetalWorks receives a Purchase Order (PO) from one of its customers. This PO is called the *customer PO*. The customer PO may contain one or multiple *line items*. Each line item refers to a different product.

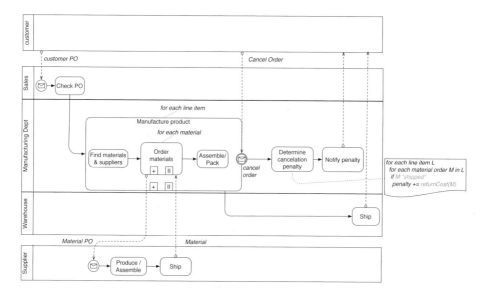

Fig. 1. BPMN model of Build-to-order (BTO) process

Upon receiving a customer PO, a Sales Officer checks the PO to determine if all the line items in the order can be produced within the timeframes indicated in the PO. As a result of this check, the Sales Officer may either confirm the customer PO or ask the customer to revise the terms of the PO. Assuming that the PO is confirmed, one *work order* is created for each line item in the customer PO. In other words, one customer PO spawns multiple work orders (one per line item).

In order to manufacture a product, raw materials are typically required. Accordingly, a Production Engineer inspects each work order in order to determine which raw materials are required in order to fulfill it. The Production Engineer annotates the work order with the list of required required raw materials. Each raw material listed in the work order is then checked by a Warehouse Manager. The Warehouse Manager determines whether the required raw material is available in stock or has to be ordered. If the material has to be ordered, the Warehouse Manager selects a suitable supplier for the raw material and sends a PO to the selected supplier. This PO for a raw material is called a *material PO*, and it is different from the customer PO. A material PO is a PO sent by MetalWorks to one of its suppliers, whereas a customer PO is a PO received by MetalWorks from one of its customers.

Once all materials required to fulfil a work order are available, the product is assembled and packed. Eventually all work orders spawned by a customer PO are ready. At that point in time, the products are shipped to the customer.

At any point during this process, the customer may send a Cancel Order message for a given PO. When this happens, the Sales Officer determines if the order can still be cancelled, and if so, whether or not the customer should pay a penalty. If the order can be cancelled without penalty, all the work related to that order is stopped and the customer is notified that the cancellation has been successful. If the customer needs to pay a penalty, the penalty is calculated and the customer is informed of the amount. The customer then needs to confirm the cancellation. It may also happen that some line items of a customer PO cannot be cancelled because the corresponding products have already been assembled and packed.

The key data entities involved in this business process are depicted in Figure 2. Importantly, the data entities in this model are distributed across multiple organizational units (indicated between brackets below the name of each entity). The customer PO is managed by the Sales department, the work orders are managed by the Manufacturing department, and the material POs are managed by the Procurement Department (not shown in the BPMN model) but also by the corresponding supplier. Each supplier offers a system that allows MetalWorks to track the material POs it submits.

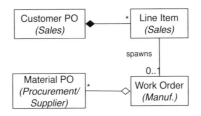

Fig. 2. High-level data model for the BTO process

The "Cancel Order" operation in this scenario raises a number of issues that illustrate the shortcomings raised by the "data vs. process divide". Specifically, in order to calculate the cancellation penalty, we need to have data about the status of the business process, like for example which instances of the "Fulfill work order" subprocess have already generated a material PO, which instances of the "Order Material" subprocess have been shipped, which instances of "Fulfill work order" are in the "assemble/-pack" phase, and which instances of "Fulfill work order" have already completed. This "process execution status" is recorded by the BPMS, typically in its own underlying database. On the other hand, we also need to have data about the supplier to which each material order is sent, the return policies of these suppliers, and the costs for returning materials back to the suppliers. These data are kept in different modules of MetalWork's enterprise system (e.g. sales, manufacturing and supplier management modules).

When it comes to implementing such a "Cancel Order" operation, information system engineers have two options:

1. Insert additional tasks in the executable process model in order to pull the data required to calculate the penalty from MetalWork's enterprise system. The aggregated data, together with the process execution status data, are then passed to a "penalty calculation service" that returns the amount of the penalty.
2. Push the process execution status data from the BPMS to MetalWork's enterprise system, so that the enterprise system has all the data required to compute the penalty. Having done this, the BPMS can then invoke a "penalty calculation service" that fetches all the required data from the enterprise system, computes the penalty and returns it back to the BPMS. The penalty calculation service can be implemented directly on top of the enterprise system.

The first approach has the drawback that the (executable) process model is polluted with low-level data collection tasks. The second approach has the drawback that the process execution status data is kept redundantly in the BPMS and in the enterprise system's underlying database. Furthermore, this also means that additional tasks need to be added in the executable process model in order to update the enterprise system every time that a step of the process is completed. In other words, the status of customer POs, work orders and material POs have to be updated after every step in the process.

In both cases, the end-result is that the business process model is "polluted" with tasks whose sole purpose is to move data between the process execution layer and the system(s) that manage the key entities of the BTO process, namely the customer PO, work order and material PO. In this setting, what SOAs achieve is that they facilitate data access. They make it easier for developers to implement the necessary interactions between the process layer and the data and business logic layer, but they do not obviate nor reduce the need for coding such interactions and weaving them into the process model. Fundamentally, the business process and the data entities that evolve as part of this process (i.e. the POs and work orders) are conceived separately and "stitched together" by means of low-level and sometimes brittle request-response interactions.

3   The Artifact-Centric Process Management Paradigm

Mainstream process modeling notations such as BPMN can be thought as being *activity-centric* in the sense that process models are structured in terms of flows of events and activities. Modularity is achieved by decomposing activities into subprocesses. Data manipulation is captured either by means of global variables defined within the scope of a process or subprocess, or by means of conceptually passive *data objects* that are created, read and/or updated by the events and activities in the process. In contrast, the database applications and/or enterprise systems on top of which these processes execute are usually structured in terms of objects that encapsulate data and/or behavior. There is thus an "impedance mismatch" between the process layer and the business logic and data layers.

In contrast, *artifact-centric* process modeling paradigms [3,4] aim at conceptually integrating the process layer and the business logic and data layers. Their main tenet is that business processes should be conceived in terms of collections of *artifacts* that encapsulate data and have an associated lifecycle. In an artifact-centric process modeling paradigm, the BTO process would be conceived not in terms of activities or subprocesses as in Figure 1, but in terms of artifacts such as *customer PO*, *work order* and *material PO*. Each of these artifacts has a lifecycle. For example, the *customer PO* goes through states such as *submitted*, *checked*, *in procurement*, *in production*, *ready to ship* and *shipped*. Transitions between these states are triggered by *events* coming from human actors, modules of an enterprise system (possibly exposed as services) and possibly other artifacts. The latter means that artifacts are inter-linked. For example, completion of all work orders spawned by a customer PO leads to the customer PO moving to the state *ready to ship*. Activities such as *Ship Products* are enabled by artifacts being in certain states and completion of these activities may cause artifacts to move from one state to another.

In an artifact-centric world, the status of customer POs and other artifacts in the BTO process are conceptually encapsulated together with the rest of the data that is relevant to that artifact. Moreover, artifacts are linked together and one artifact can query data from its associated artifacts. In the context of the BTO process, the customer PO is linked to the work orders it spawns, which themselves are linked to the material POs. Data access can seamlessly occur across these links. In particular, it is possible to write a single query to calculate the cancelation penalty for a customer PO, without having to insert additional tasks in the process in order to move data across the process and data layers for the purpose of such penalty calculation. Conceptually, all data associated to a *customer PO* are clustered together and can be seamlessly accessed.

The artifact-centric paradigm has its roots in object-orientation. In fact, artifacts are sometimes called *business objects* in the literature [7,8]. It is debatable to what extent *artifacts* are distinguishable from "coarse-grained objects". Generally, what distinguishes artifacts is that they can be directly related to business processes, which entails that artifacts have non-trivial lifecycles. With respect to the working example, a *line item* would most likely be modeled as an object, since it encapsulates data and possibly also behavior, but it is unlikely to be considered an artifact insofar as it does not have a lifecycle on its own.

4   Outlook: Artifact-Centric Interoperation Hubs

While artifacts conceptually reduce the impedance mismatch between the process layer and the business logic and data layers, it is still the case that these layers are usually separated and kept in distinct tiers in contemporary multi-tier system architectures. This imperative sometimes comes from the existence of organizational boundaries (e.g. between Sales and Manufacturing), and is even stronger when these organizational units are completely independent. For example, data about the status of a *material PO* is maintained both by MetalWorks and by its suppliers. Thus, at the lower-level of abstraction data still needs to be moved around. Service-oriented architectures provide a proven approach to construct the interfaces that facilitate such data movement, but still, someone has to design and implement the operations to enact this data movement.

The Artifact-Centric Service Interoperation (ACSI) project[1] aims at combining the artifact-centric process management paradigm with SOAs in order to achieve higher levels of abstraction during business process integration across organizational boundaries. The key principles of the ACSI project is that processes should be conceived as systems of artifacts, and these systems of artifacts are bound to services. The binding between artifacts and services specifies where should the data of the artifact be pushed to, or where it should be pulled from, and when. For example, when a supplier ships a material PO to MetalWorks, it sends an *Advance Shipment Notification* via the service-oriented system interfaces provided by MetalWorks. When MetalWorks's system receives this notification, the status of the copy of the corresponding material PO maintained at MetalWorks is updated. If a "Cancel Order" message was received by MetalWorks at that point in time, the penalty calculation would be performed using the updated status for this material PO. Importantly, process analysts do not reason in terms of request-response interactions. Instead, they reason in terms of artifacts, their lifecycles, operations and associated data. Interactions across system boundaries are conceived at a lower level of abstraction and specified in *artifact-to-service bindings*.

Several challenges lie ahead on the road to artifact-centric interoperation hubs, including:

- Artifact-centric process modeling: While in the realm of activity-centric process modeling, a consensus has been formed around BPMN as a standard modeling notation, such consensus is far from being achieved in the field of artifact-centric process modeling. Some researchers have studied the possibility of using communicating state machines as the foundation for an artifact-centric modeling [7,8], while others advocate more declarative approaches [4]. In any case, one key requirement of such a language is that process state and other artifact-related data should be viewed in a uniform manner, as opposed to activity-centric process modeling notations (including BPMN) where data access and manipulation is specified at the level of tasks and decision points, using a different language than the one used to specify the flow of control of the process.
- Verification and consistency checking: When verifying the semantic correctness of activity-centric process models, tool developers generally adopt a control-flow abstraction, meaning that they focus on the flow of control between activities. This abstraction allows process models to be statically checked using graph analysis techniques, Petri net-based techniques or more generally, finite-state model checking techniques. In the context of artifact-centric systems however, static verification ought to consider both the data and the control-flow viewpoints in an integrated manner. As soon as the data is put into the equation, verification problems become considerably harder to tackle due to potentially infinite state spaces [1,2]. In the context of artifact-centric interoperation hub, a specific verification problem is that of checking the consistency between the data and behavior specified in the artifact models and the protocols of the services to which these artifacts are bound.
- Reverse-engineering: Most of today's systems are arguably not constructed in terms of artifacts. Therefore, the emergence of artifact-centric paradigm will necessitate

---

[1] http://www.acsi-project.eu/

methods and tools to reverse-engineer artifact-centric models from existing systems. One direction to tackle this challenge is by means of process mining techniques, which aim at synthesizing process models from system execution logs. Existing process mining techniques however are geared towards activity-centric processes and it remains an open question how to extend or adapt these techniques to artifact-centric processes [9].

Initial results of the ACSI project along these and other streams can be found in the project's Web site: http://www.acsi-project.eu/.

Acknowledgments. This paper is the result of collective discussions within the ACSI project team. Thanks especially to Rick Hull for numerous discussions on this topic. The BTO process was designed jointly with Rick Hull, Lior Limonad and Marco Montali. The ACSI project is funded by the European Commission's FP7 ICT Program.

References

1. Bagheri Hariri, B., Calvanese, D., De Giacomo, G., De Masellis, R., Felli, P.: Foundations of relational artifacts verification. In: Rinderle-Ma, S., Toumani, F., Wolf, K. (eds.) BPM 2011. LNCS, vol. 6896, pp. 379–395. Springer, Heidelberg (2011)
2. Belardinelli, F., Lomuscio, A., Patrizi, F.: A computationally grounded semantics for artifact centric systems and abstraction results. In: Proceedings of the 22nd International Joint Conference on Artificial Intelligence, IJCAI (July 2011)
3. Bhattacharya, K., Caswell, N.S., Kumaran, S., Nigam, A., Wu, F.Y.: Artifact-centered operational modeling: Lessons from customer engagements. IBM Systems Journal 46(4), 703–721 (2007)
4. Cohn, D., Hull, R.: Business artifacts: A data-centric approach to modeling business operations and processes. IEEE Data Eng. Bull. 32(3), 3–9 (2009)
5. Erl, T.: Service-Oriented Architecture (SOA): Concepts, Technology, and Design. Prentice Hall, Englewood Cliffs (2005)
6. Kohlborn, T., Korthaus, A., Chan, T., Rosemann, M.: Identification and analysis of business and software services - a consolidated approach. IEEE Transactions on Services Computing 2(1), 50–64 (2009)
7. Küster, J.M., Ryndina, K., Gall, H.: Generation of business process models for object life cycle compliance. In: Alonso, G., Dadam, P., Rosemann, M. (eds.) BPM 2007. LNCS, vol. 4714, pp. 165–181. Springer, Heidelberg (2007)
8. Redding, G., Dumas, M., ter Hofstede, A.H.M., Iordachescu, A.: Generating business process models from object behavior models. IS Management 25(4), 319–331 (2008)
9. van der Aalst, W.M.P.: Process Mining: Discovery, Conformance and Enhancement of Business Processes. Springer, Heidelberg (2011)

Mixing Bottom-Up and Top-Down XPath Query Evaluation

Markus Benter, Stefan Böttcher, and Rita Hartel

University of Paderborn (Germany)
Computer Science
Fürstenallee 11
D-33102 Paderborn
{benter,stb,rst}@uni-paderborn.de

Abstract. Available XPath evaluators basically follow one of two strategies to evaluate an XPath query on hierarchical XML data: either they evaluate it top-down or they evaluate it bottom-up. In this paper, we present an approach that allows evaluating an XPath query in arbitrary directions, including a mixture of bottom-up and top-down direction. For each location step, it can be decided whether to evaluate it top-down or bottom-up, such that we can start e.g. with a location step of low selectivity and evaluate all child-axis steps top-down at the same time. As our experiments have shown, this approach allows for a very efficient XPath evaluation which is 15 times faster than the JDK1.6 XPath query evaluation (JAXP) and which is several times faster than MonetDB if the file size is ≤ 30 MB or the query to be evaluated contains at least one location step that has a low selectivity. Furthermore, our approach is applicable to most compressed XML formats too, which may prevent swapping when a large XML document does not fit into main memory but its compressed representation does.

Keywords: XML, top-down XPath evaluation, bottom-up XPath evaluation.

1 Introduction

1.1 Motivation

XML gains more and more popularity not only as a data exchange format, but also as a storage, archive or data management format and XPath is the main standard to express path queries on XML data.

Whenever XPath query evaluation is a bottleneck of an application, a fast XPath query evaluator is desired. If in addition, XML documents may become larger than the available main memory space, it may be a significant advantage when the fast XPath query evaluator can process XPath queries on compressed XML documents that can still fit into main memory. We present such a fast XPath query evaluator that relies on just a minimal set of XML navigation steps, such that it is applicable not only to plain XML data, but also to most queryable compressed XML data formats.

J. Eder, M. Bielikova, and A.M. Tjoa (Eds.): ADBIS 2011, LNCS 6909, pp. 27–41, 2011.

1.2 Contributions

"Traditional" XPath evaluators typically evaluate the hierarchical XML data either top-down or bottom-up, as both techniques provide advantages for different classes of queries. In this paper we present an approach that allows XPath evaluation in any direction and that combines the following properties:

- The approach presented in this paper supports both, bottom-up and top-down XPath query evaluation on an XPath subset that extends *core XPath* as defined in [1] by comparisons of paths to constants within predicate filters.
- Even more, our approach allows a dynamic mixture of bottom-up and top-down query evaluation, such that for each location step, it can be decided whether to evaluate it top-down or bottom-up and at which time of the query evaluation process.
- Our approach is powerful and generic as it requires only minimal support from the underlying XML format. That is, our approach can be applied to any un-compressed or compressed XML representation that provides access to XML nodes via the node's name and provides navigation via the binary axes first-child, first-child^{-1}, next-sibling, and next-sibling^{-1}, and nevertheless, our approach supports all the other XPath axes of core XPath (e.g. ancestor, descendant, following and preceding) within XPath queries.
- We have evaluated query performance on two different XML representations – one uncompressed and one compressed – that are integrated into our approach. Besides a DOM-based XML representation, we have implemented a second XML main-memory representation, that is based on Succinct compression [2] and that – if combined with an index – not only allows for an XPath evaluation as fast as the DOM-based representation, but also needs only 20% of the main memory required by a DOM representation.
- Finally, we have implemented different 'evaluation strategies' that decide, which sub-queries of a given XPath query to evaluate in which direction, i.e. top-down or bottom-up, and at which time of the evaluation process. Further-more, we have evaluated and compared these navigation strategies within a se-ries of experiments to determine which is the most efficient navigation strategy to evaluate XPath queries. Our experiments have shown that for our test queries, the mixed approach is up to 7 times faster than bottom-up evaluation and up to 56 times faster than top-down evaluation.

1.3 Query Language

The subset of XPath expressions supported by our approach extends the set of *core XPath* as defined in [1], as our approach beyond [2] additionally allows comparisons of paths to constants within predicate filters. This XPath subset supported by our approach is defined by the following EBNF grammar:

```
cxp             ::= `/' locationpath
locationpath    ::= locationstep ('/' locationstep)*
locationstep    ::= x `::' t | x `::' t `[' pred `]'
pred            ::= pred `and' pred | pred `or' pred | `not' `(' pred `)'
                    | locationpath | locationpath `=' const |`(' pred `)'
```

"cxp" is the start symbol, "x" represents an axis (self, child, parent, descendant-or-self, descendant, ancestor-or-self, ancestor, following, preceding, following-sibling, preceding-sibling), "const" represents a constant, and "t" represents a "node test" (either an XML node name test or "*", meaning "any node name").

Note that our system supports – aside from the evaluation in top-down or in bottom-up direction – using the sibling axes in XPath queries, whereas other approaches like XMLTK[3], χαοζ[4], AFilter [5], YFilter[6], XScan[7], SPEX[8], and XSQ[9] are limited to using the parent-child and the ancestor-descendant axes only.

1.4 Paper Organization

This paper is organized as follows: Section 2 summarizes the fundamental concepts used for describing our approach to evaluate XPath queries consisting of a single path and XPath queries with filters. Furthermore, this section describes the different evaluation strategies that could be used for evaluating an XPath query. The third section outlines some of the experiments that compare the different evaluation strategies of our prototype with each other and with other XPath evaluators. Section 4 gives an overview of related work and is followed by the Summary and Conclusions.

2 Our Solution

2.1 Overview of Our Solution

We follow the ideas of [1] and [10] to rewrite the given XPath queries, such that they no longer use all the core XPath axes, but only a small set of basic binary axes containing the axes first-child, first-child^{-1}, next-sibling, next-sibling^{-1}, and self. Table 1 shows how to rewrite each standard XPath axis into a regular expression using only the basic binary axes.

Table 1. Axis definition in terms of the basic binary axes

Axis	Binary expression
child	first-child, (next-sibling)*
parent	(next-sibling^{-1})*, first-child^{-1}
descendant	first-child, (first-child I next-sibling)*
ancestor	(first-child^{-1} I next-sibling^{-1})*, first-child^{-1}
following-sibling	next-sibling, (next-sibling)*
preceding-sibling	next-sibling^{-1}, (next-sibling^{-1})*
following	(((first-child^{-1} I next-sibling^{-1})*,first-child^{-1}) I self), next-sibling, (next-sibling)*, ((first-child, (first-child I next-sibling)*) I self)
preceding	(((first-child^{-1} I next-sibling^{-1})*,first-child^{-1}) I self), next-sibling^{-1}, (next-sibling^{-1})*, ((first-child, (first-child I next-sibling)*) I self)

Based on the binary XPath expressions given in Table1, we provide an atomic automaton using the binary axes for each XPath axis. For example, Figure 1 (a) shows an automaton generated for a location step child::a, and Figure 1 (b) shows an automaton generated for a location step parent::a. fc represents the first-child axis, fcR the axis first-child^{-1}, ns the next-sibling axis, nsR the axis next-sibling^{-1}, and self the self axis.

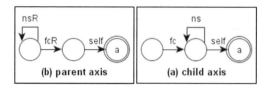

Fig. 1. Atomic automata for the location steps (a) child::a and (b) parent::a

Similar to the approach provided in[10], we translate each location step LSi of a given XPath query into a single atomic automaton BAi. The advantage of reducing all the XML axes listed in Table 1 to the basic axes (first-child, first-child^{-1}, next-sibling, next-sibling^{-1}, and self) is that we require the XML representation only to support the navigation along the basic axes together with an efficient access to all nodes that have a given node name. This requirement is met e.g. for uncompressed XML by the DOM representation or e.g. for compressed XML by the succinct representation[2]. Therefore, the presented approach can be applied to uncompressed XML as well as to compressed XML, if the XML format supports at least navigation along the basic axes and access to node names, although our approach applied to the XML format supports the much larger superset of core XPath described in Section 1.3.

The XPath query is represented as a special kind of non-deterministic finite automaton that we call a 'token automaton'. A token automaton not only contains states and transitions, but also allows for using each state in any number of tokens each of which represents an answer to a sub-query within the XML document. According to the events produced by the input XML document representation, the token-automaton fires transitions and transfers tokens, i.e. generates new tokens, along the binary axes first-child, first-child^{-1}, next-sibling, next-sibling^{-1}, and self.

The atomic token automata build a construction kit from which the final automaton representing an XPath query is built. In contrast to traditional automata, not the input – i.e., the XML document representation – controls, which transitions can be fired next, but there exist an external controlling instance – called *DecisionModule* – that decides, which transition will be fired next. In other words, the DecisionModule decides for each location step of the query whether it is evaluated top-down or bottom-up, and at which time of the query evaluation this location step is evaluated.

Each transition of the automaton can either be fired top-down, i.e., it consumes the binary axis that is denoted by the transition label and the tokens are transferred in the direction given by the transition, or it can be fired bottom-up, i.e., it consumes the inverse of the binary axis denoted by the transition label and the tokens are transferred opposite to the given direction. We assume, that the used XML compression provides – similar as it is provided by DOM – access to a list of nodes that fulfill a given node name test and supports navigation via the binary axes first-child, first-child^{-1}, next-sibling, next-sibling^{-1}, and self.

2.2 XPath Automata

Each atomic automaton contains one state that is called a stable state and that carries a node name test as label and that accepts only the tokens referring to those XML nodes which fulfill the given node name test. Stable states are marked by a double circle.

The notation of the transitions of the automaton only shows the top-down evaluation; the bottom-up evaluation can be taken by reversing the transition direction and by replacing each transition label by its reverse. The other atomic automata are built in a similar way to the child::a automaton shown in Fig. 1 according to the regular expressions provided in Table 1. If a location step LSi is followed by a location step LSj in a query Q, we concatenate the atomic automaton BAi corresponding to LSi and the atomic automaton BAj corresponding to LSj to the token automaton XPQ of query Q by drawing a self transition from the final state of BAi to the start state of BAj.

For example, Fig. 2(a) shows the automaton for the query Q = //a/b. All states have as label an ID of the form s0,…,s6, and the stable states have as an additional label the node name test that has to be fulfilled by an XML node in order to be accepted by this stable state. The root (state s0) is connected via a self-axis to the automaton for //a (states s1-s3) which is connected by another self-axis to the automaton for /b (states s4-s6).

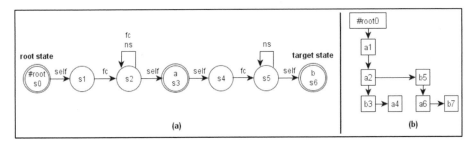

(a)

(b)

Fig. 2. (a) XPath automaton for query //a/b , and (b) a small example document where each node is represented by a node ID and the node's label

2.3 Evaluation of Filter-Less Paths

Overview: In order to evaluate an XPath query Q, first, the automaton A for Q is built as described in the previous section. Each pair (sx,sy) of stable states in A for which a path from sx to sy exists in A represents a relative XPath expression R that is a sub-sequence of location steps of Q.

Second, tokens each of which represents an answer to such a relative XPath expression R are created, transferred, joined, and deleted until all tokens that represent an answer to Q are computed. Let sx, sy be stable states in A, let R be the sub-query of Q that corresponds to the sub-automaton including all paths and states from sx to sy of the automaton A for Q, and let nv, nw be nodes in the given XML document. Then the token T=(nv/sx, nw/sy) represents an answer nw to the sub-query nv/R .

DecisionModule: A DecisionModule controls the order and the direction (bottom-up or top-down), in which the sub-queries are evaluated, i.e., it decides which of the stable states are taken as start states and for which state occurring in a token, partial sub-query evaluation is continued, i.e., which tokens are transferred next in which direction, and when tokens are joined.

Token Creation in Start States: Start states can be defined at any time during the execution. At any time, each state, none of the incoming or outgoing transitions of which had been fired, can be chosen as an additional start state. Whenever the DecisionModule declares a stable state s to be a start state, for each node n in the XML document that fulfills the node name test of s, a token (n/s, n/s) is created.

If for example the state with ID s6 of Fig. 2(a) is declared as a start state, i.e., we look for XML elements that are answers to the sub-query //self::b, tokens (3/s6, 3/s6), (5/s6, 5/s6), and (7/s6, 7/s6) are created for the nodes b3, b5, and b7 in Fig. 2(b), where b is the node label followed by the node ID (3,5,or 7).

Token Deletion: For each state s in A, except for the root state and the target state, when all transitions from s and all transitions to s have been fired and all join tokens for s have been computed as described below, all tokens containing s as the start state or as the final state are automatically deleted. Automatic token deletion can be partially switched off in order to implement a navigation cache as described in Section 2.4.

Partial Sub-query Evaluation: A sub-query R is top-down partially evaluated by firing all the transitions on a path from sx to sy. This operates on all tokens T=(nu/sz, nv/sx) that contain sx as their final state, i.e. represent an answer to a sub-query represented by paths in A ending in state sx, and it eventually generates new tokens that contain sy as their final state, i.e., for each answer nw to nv/R, this generates a new token T'=(nu/sz, nw/sy).

Similarly, a sub-query is bottom-up partially evaluated by firing all the inverted transitions of the transitions on a path from sx to sy in reversed order. This operates on all tokens T=(nw/sy, nt/sz) that contain sy as their start state, and it eventually generates new tokens that contain sx as their start state, i.e., for each answer nv to nw/R^{-1}, this generates a new token T'=(nv/sx, nt/sz).

Let si and sj be stable or non-stable states. To fire a transition with label fc (or ns or fcR or nsR respectively) that starts in state si and ends in state sj for a token T=(nu/sz, nv/si) with final state si in top-down direction means the following: to check, whether there exists a node with ID nw in the XML tree such that the node with ID nw is the first-child (or the next-sibling or fcR or previous-sibling respectively) of the node with ID nv. If such a node exists, a token T'=(nu/sz, nw/sj) is generated, otherwise no token is generated.

Correspondingly, to fire a transition with label fcR (or nsR or fc or ns respectively) that starts in state si and ends in state sj for a token T=(nw/sj, nt/su) with start state sj in bottom-up direction means the following: to check, whether there exists a node with ID nv in the XML tree such that the node with ID nw is the first-child (or the next-sibling or fcR or previous-sibling respectively) of the node with ID nv. If such a node exists, a token T'=(nv/si, nt/su) is generated, otherwise no token is generated.

A transition with a label self from the start state si to the end state sj can be fired for each token having si as final state in case of top-down evaluation and for each token having sj as start state in case of bottom-up evaluation. Firing the transition during top-down evaluation generates for each token T=(nu/sz, nv/si) another token T'=(nu/sz, nv/sj), whereas firing the transition during bottom-up evaluation generates for each token T=(nv/sj, nu/sz) another token T'=(nv/si, nu/sz).

Whenever a token with a non-stable state s' is generated during sub-query evaluation in a direction D (bottom-up or top-down), all transitions that can be fired from s' in the same direction as D are fired. Thereafter, tokens containing s' are deleted. This processing of tokens containing unstable states is repeated until all existing tokens have reached stable states again. When this happens, the DecisionModule gets the control again and decides which tokens are transferred next.

Consider for example the tokens (3/s6, 3/s6), (5/s6, 5/s6), and (7/s6, 7/s6) representing answers to the sub-query //self::b. If we fire the transitions from state s6 to state s3 in bottom-up direction, this removes and transfers the tokens having state s6 as their start state, and it will stop, when all tokens are either deleted or transferred to tokens having state s3 as their start state. In this case, the generated tokens (2/s3, 3/s6), (1/s3, 5/s6) say that the sub-query self::a/b represented by the sub-automaton between the states s3 and s6 applied to the XML nodes with IDs 2 (and 1 respectively) yields as answers the XML nodes that have the IDs 3 (and 5 respectively).

If we additionally create a token (0/s0, 0/s0) in the state s0 for the XML root node and transfer this token from s0 top-down, when token generation stops, we get the tokens (0/s0, 1/s3), (0/s0, 2/s3),(0/s0, 4/s3), and (0/s0, 6/s3) saying that the sub-query //a represented by the states of the sub-automaton between state s0 and state s3 returns the XML nodes with IDs 1, 2, 4, and 6 as answers.

Token Joining: Whenever at the end of a token transfer phase, the same stable state sy occurs as final state in tokens T1 and as start state in other tokens T2, we perform a so called 'token joining' and join those pairs (T1,T2) of tokens that relate sy to the same XML node nv with each other. A token joining of two tokens T1=(sx/nu, sy/nv) and T2=(sy/nv, sz/nw) yields a new join token T3=(sx/nu, sz/nw).

In our example, the tokens T1 ∈ { (2/s3, 3/s6) , (1/s3, 5/s6) } have the start state 3, and the tokens T2 ∈ { (0/s0, 1/s3), (0/s0, 2/s3),(0/s0, 4/s3), (0/s0, 6/s3) } have the final state 3. If we perform token joining on all pairs (T1,T2) of tokens, we get the join tokens j1=(0/s0, 3/s6) and j2=(0/s0, 5/s6). These join tokens express that the answers to the concatenated sub-query //a/self::a/b represented by the automaton between state s0 and state s6 applied to the XML node with ID 0 (the root node) returns the XML nodes with IDs 3 and 5 as answers. As all incoming and outgoing transitions of s3 have been fired, and all join tokens involving s3 have been computed, thereafter all tokens containing s3 as start state or as final state are deleted.

Token joining can also be used for finally joining the answers when query evaluation starts at an inner state and proceeds in different directions. If we declare for example state s3 as the single start state and transfer the tokens (1/s3, 1/s3), (2/s3, 2/s3), (4/s3, 4/s3), and (6/s3, 6/s3) top-down and bottom-up, 6 additional tokens are generated and the state s0 occurs as start state in the 4 tokens (0/s0, 1/s3), (0/s0, 2/s3), (0/s0, 4/s3), and (0/s0, 6/s3) and the state s6 occurs as final state in the 2 tokens (1/s3, 5/s6) and (2/s3, 3/s6). Finally, token joining calculates the final results (0/s0, 3/s6) and (0/s0, 5/s6) that express that by applying the query //a/b represented by the (sub-) automaton from the start state s0 to the final state s6 to the XML node with ID 0 (the root node) yields the XML nodes with IDs 3 and 5 as query results.

2.4   Optimization Using a Navigation Cache

If we consider the XML tree of Fig. 2(b) and the query //a//b and the nodes with ID 5 and with ID 7 and transfer the tokens bottom-up in a naïve way, similar new tokens are generated for the nodes a2, a1, and #root0, i.e., we pass this path in the tree more than once. In order to overcome this weakness, we have introduced the concept of a so called *navigation cache* that caches tokens representing sub-query evaluations of multiple paths in the XML document tree.

For example, if the token for b5 is transferred first, the token for b7 can read the cache information of node b5 and can be transferred directly to the root node without having to pass the path via b5, a2, a1, and #root0 a second time. This information is being used for bottom-up evaluation only and is not considered for top-down evaluation.

2.5 Evaluation of Queries with Filters

Whenever a location step L that is represented by a pair (sx,sy) on the main path of an XPath query Q contains one or more predicate filters, each predicate filter Fi is represented by a filter automaton Ai having a state si as its root and a final state sfi representing the final state of the main path of the filter. A filter automaton Ai has the same design and functionality as the automaton for the main path of Q as described in the previous section. As with all location steps, each location step within a filter automaton can be evaluated top-down or bottom-up.

Token transfer between the root state si of a filter automaton Ai and the final state sy representing the location step L having filter Fi, can be done either top-down, i.e. from sy to si, if tokens containing sy are generated first, or bottom-up, i.e. from si to sy, if Fi is evaluated first.

Bottom-Up Token Transfer: If the tokens are transferred bottom-up, i.e. the filter path for Fi is evaluated before tokens containing the node sy are generated, let ST={ (n1/si, nf1/sf1), ..., (nk/si, nfk/sfk) } be the set of all the tokens computed for path from si to sfi. Then the set N={n1,...nk} contains exactly those XML nodes for which Fi evaluates to true.

Then, the automaton state sy to which the filter path is connected reacts similar to a state without an attached filter with the difference, that not for each XML node that fulfills the given node name test a token is created, but only for those XML nodes contained in the set N which fulfill the given node name test.

Top-Down Token Transfer: Otherwise, i.e., if tokens are transferred top-down from sy to si, we follow an idea of [10]: Whenever a token T1=(...,nv/sy) or a token T1= (nv/sy,...) that contains the state sy is generated, this token gets a reservation that depends on whether or not the filter automaton for Fi evaluates to true for the XML node nv. At the same time, the filter automaton for Fi is switched active, i.e., a token (nv/si, nv/si) is generated which turns si into a start state of the filter automaton.

If the filter automaton Fi finally evaluates to true for the XML node nv, the reservation for Fi is deleted. We say that the execution of the filter automaton for Fi having a start state si and a final state sfi on its main path evaluates to true for the XML node nv, if and only if eventually a token (nv/si, nw/sfi) is generated for a XML node nw. Otherwise, we say that the evaluation of the filter automaton for Fi evaluates to false for the XML node nv, and the token T1 itself is deleted and considered invalid. However, if, finally, all the reservations for a filter attached to sy are deleted, the T1 token is considered valid.

The states of the filter automaton can be connected to other filter automata, such that nested filter automata for implementing nested XPath filter expressions can be evaluated by this concept as well.

2.6 Evaluation Strategies

We have implemented different types of DecisionModules that follow different evaluation strategies in order to evaluate an XPath query. The first two Decision Modules follow the 'traditional' ways to evaluate queries.

- The Top-Down-Module declares the root state as the only start state. Tokens are added to the first state of each filter automaton FA as soon as a token is added to the state to which FA is attached to. All paths, i.e. the main path of the XPath expression and all filter paths, are evaluated top-down.
- The Bottom-Up-Module declares the target state of the automaton for the main path and each target state of a filter automaton as the start states. All paths are evaluated bottom-up.
- The Minimum-Module considers the locations steps in the main path and in all filter paths and declares the stable state of that location step having the lowest selectivity of the whole query as the only start state. If the start state is part of a filter, the corresponding filter path is evaluated top-down and bottom-up starting at the start state and the result is added to the state of the main path to which the filter is attached. Then, the state of the main path behaves like a start state: From that given start state, the remaining main path of the XPath query is evaluated bottom-up to the root and top-down to the target state of the main path. Furthermore, all other filter paths are evaluated top-down.

Determining the location step having the lowest selectivity is not trivial. Currently, we are using a simple heuristics that regards that location step LS=/axis::nnt as the location step having the lowest selectivity, for which the least number of nodes exist in the document that fulfill the node name test nnt.

3 Evaluation of Our Prototype Implementation

3.1 Experimental Setup

Our test system has an Intel Core 2 Duo with 2,53 GHz (T9400) processor and 4 GB 1066 DDR 3 RAM. The prototype is implemented in Java and runs on JDK 1.6 Update 21 with an extended RAM and function stack (parameters -Xmx1300M -Xss4096k). For MonetDB [11], we have used the Oct2010-SP1 build and the measured execution time is Trans+Shred+Query. For eXist-DB (http://exist-db.org/) we have used version 1.4.0.

Our evaluation was performed on the documents generated by the XMark benchmark [12] with original XML document sizes varying from around 2 MB to 50 MB. We have evaluated our prototype on the queries A1-A7 and B2-B4 of the XPathMark [13] benchmark suite as well as on some additional, practice-oriented queries (Q1-Q6) for showing the advantages of our system (especially on queries with location steps of outstanding low selectivity). The queries that we used are shown in Table 2.

Table 2. Queries used in our prototype evaluation

A1	/site/closed_auctions/closed_auction/annotation/description/text/keyword
A2	//closed_auction//keyword
A3	/site/closed_auctions/closed_auction//keyword
A4	/site/closed_auctions/closed_auction[annotation/description/text/keyword]/date
A5	/site/closed_auctions/closed_auction[descendant::keyword]/date
A6	/site/people/person[profile/gender and profile/age]/name
A7	//keyword
B2	//keyword/ancestor::listitem/text/keyword
B3	/site/open_auctions/open_auction/bidder[following-sibling::bidder]
B4	/site/open_auctions/open_auction/bidder[preceding-sibling::bidder]
Q1	//people//age
Q2	/site/people/person[profile/age=42]
Q3	//person[.//gender='female']/name
Q4	//person[.//country='United States']/name
Q5	//person[.//country='United States' and .//gender='female']/name
Q6	//item[payment='Creditcard']

3.2 Comparison of DecisionModules

In our first series of measurements, we compared the three DecisionModules with each other. We performed all measurements on two XML representations: on succinct compression as compressed XML representation and on DOM as uncompressed XML representation. As the experiments have shown that using our prototype based on the Java DOM representation yields similar execution times, but requires 5 times more main memory than the execution based on the succinct compression, we concentrate on presenting the results received for the succinct compression in this section.

Fig. 3 compares the three DecisionModules "Top-Down(TD)", "Bottom-Up(BU)" and "Minimum(Min)". Navigation Caching is always enabled, as our evaluations have shown that this technique in general improves the performance. The evaluation shown in Fig. 3 was performed on a ~22MB document with 340,000 nodes (XMark factor 0.2), but other document sizes show the same results as the execution times scales linear with increasing document size. The overall observation is that strategy BU outperforms the execution time of the strategy TD in most cases, but that Min is the best evaluation strategy in nearly all cases (except for query Q1 and Q2).

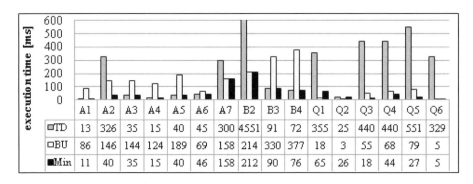

Fig. 3. Comparison of the DecisionModules TD, BU, and Min

An example for this observation is query A2: A top-down evaluation traverses the entire document due to the initial descendant axis. The bottom-up module can do better because it starts at the *keyword* nodes and evaluates the descendant location steps bottom-up. But the Min-Module performs best on A2: It has a low selectivity, i.e., it selects the 2,000 *closed_auction* nodes (compared with 14,000 *keyword* nodes) as start nodes and therefore can avoid a lot of the navigation caching overhead of the BU-Module, but only has to traverse relative small sub-trees (the *closed_aution* sub-trees) compared with the entire document the TD-Module has to traverse. A similar behavior can be observed on query A7.

In some of the XPathMark-queries (A1, A3, A4, A5, A6), the Min-Module behaves similar to the TD-Module. As these queries do not have descendant-axis location steps or the descendant-axis steps are at the end of query, top-down evaluation is nearly optimal because only relative small sub-trees are traversed.

We have added Q1-Q6 to show results of more complex queries. On query Q4, the Min-Module can profit from the filter, as only a fraction of the people are from the *United States* (around 38%). Therefore, the Min-Module starts at all *United States* text-nodes, evaluates the *person* descendant-axis location step bottom-up and the *name* location step top-down. This evaluation strategy improves the performance significantly: the Min-Module is 10 times faster than the TD-Module and 1.5 times faster than the BU-Module.

Note that the Min-Module can sometimes perform better on *more complex* queries if selectivity becomes lower: Query Q5 is evaluated faster than query Q4, because the additional filter condition *gender='female'* further reduces the number of selected *person* nodes (only 5% of the *persons* are *female* and from the *United States*). In this case, the Min-Module is 20 times faster than the TD-Module and 3 times faster than the BU-Module.

3.3 Comparison with Other Evaluators

In Fig. 4, we measured the average execution time of all queries (A1-A7,B2-B4, Q1-Q6). As we can observe, the Module "Min" of our prototype is scaling linear with increasing document size. Furthermore, it is significantly outperforming JAXP, as it is around 15 times faster than JAXP. The Module "Min" outperforms eXist-DB as well. For files up

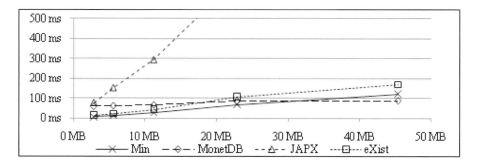

Fig. 4. Comparison of our prototype with other XPath evaluators

to ~30 MB, the Module "Min" is outperforming MonetDB as well. This is due to the fact that MonetDB needs a high overhead for query optimization but scales nearly constantly for for the document sizes tested in our evaluation. For queries with low selectivity (e.g. Q3 and Q6) our approach can outperform MonetDB also for files having a size of more than 30 MB. Furthermore, when comparing run-time, note that our prototype is a Java application, whereas MonetDB is a strongly optimized C application.

4   Related Works

There exist several different approaches to the evaluation of XPath queries on XML data. They can be divided into categories by the subset of XPath that they support. Nearly all of them are based on automata (X-scan[7], XMLTK[3], YFilter[6], [10],[13], [14], AFilter [5], XSQ [9], SPEX [8]) or parse trees ([15], [4], [16], [17]). All of them support the axes child and descendant-or-self and most of them support predicate filters and wildcards, but besides [10] and [18] none of them support the sibling-axes as our solution does.

The approach presented in [1] defines bottom-up as well as top-down semantics and presents an bottom-up and a top-down processing algorithm that both run in low-degree polynomial time for full XPath and an enhanced algorithm that runs in linear time for Core XPath that evaluates the main path top-down and the filter paths bottom-up. In contrast to this approach, we try to combine the advantages of bottom-up and top-down processing by choosing bottom-up or top-down evaluation for each location-step, such that an algorithm is developed that runs very efficient in practice. As our evaluation has shown, the mixed strategy MinimumModule performs and scales better that the pure strategies top-down or bottom-up.

For the automata-based approaches, the XML input stream is the controlling instance that is used as input for the automata representing the Query.[19] and [20] present a compressed representation for XML together with an XPath evaluator that is based on tree automata and that allows to skip irrelevant parts of the compressed XML document during the evaluation process. They allow selecting a single start point and follow the path to the root bottom-up and the path to the "leafs" of the query top-down. In contrast to[19] and [20], we allow the selection of any number of start points and the evaluation of the sub-queries in any direction.

The approach presented in[18] supports the axes self, child, descendant, following and following-sibling but does not support backward axes. It translates the queries into expressions over the binary axes first-child and next-sibling and then constructs a two-layered NFA that consumes the SAX events start-element, end-element and character. The first layer evaluates the main path of the query, whereas the second layer is responsible for the evaluation of the predicate filters. Our previous approach [10] supports all forward axes but supports backward axes only if they are rewritten to forward axes before query evaluation starts. It translates queries into an automaton that consumes the binary events first-child and next-sibling. It can evaluate streams in top-down direction only. XMLTK[3], and YFilter[6], [13], [14] and X-scan[7] are based on the lazy construction of deterministic finite automata (DFA), i.e., the DFA is not generated completely at the beginning, but additional states are added only when needed. AFilter [5] is adaptable in terms of the memory requirement, i.e., it needs a base memory that is linear in query and data size. If more memory is provided

to AFilter, AFilter uses the remaining main memory for a caching approach to evaluate queries faster than with only the base memory. XSQ [9] and SPEX [8] use a
hierarchical arrangement or network of transducers, i.e., automata extended by actions
attached to the states, extended by a buffer to evaluate XPath queries.

Parse trees – in contrast to automata – take the control of the evaluation process
themselves, i.e., they decide which node of the parse tree will be processed next and
check with the XML input document, whether this node can be processed. The approach presented in [21] translates the input query into a set of parse trees. Whenever
a matching of a leaf node of a parse tree is found within the data stream, the relevant
data is stored in form of a tuple that is afterwards evaluated to check whether predi-
cate- and join conditions are fulfilled. χαoζ[4] and [15] build a parse tree as well
(plus a parse-dag in [4], as they support the parent and the ancestor axis in addition).
This parse tree is used for 'predicting' the next matching nodes and the level in which
they have to occur. The approach discussed in [16] collapses the parse tree into a
prefix trie by combining common prefix sequences of child-axis location steps of
different queries into a leaner single path of the prefix trie. The approach presented in
[17] uses a parse tree that stores XML nodes that are solutions to the parse tree node's
sub-query within a stack that is attached to each node.

The authors of [22] show that queries containing joins on attribute values can be
computed in time linear of the XML document but exponentially of the query size.
They evaluate one path to the join attribute top-down and the path to the second join
attribute bottom-up. They require a special index on the attribute values and a pointer
structure representation of the XML document, such that the idea is not applicable to
arbitrary XML representations as e.g. compressed XML.

ROX [23] is a run-time optimizer for XQuery that is used as a MonetDB extension.
It is based on an indexed representation of the XML document that is stored in form
of relational data. It consists of a relational query optimizer for the 'relational parts' of
an XQuery and an XML query optimizer that is intertwined with the query execution,
i.e., that adapts the query execution plan during the query execution. In contrast to our
approach, ROX can be applied to the indexed XML document in form of a relational
representation only and cannot be applied to compressed XML.

In comparison to all these approaches, we additionally support the 'sibling'-axes
following and following-sibling. Furthermore, beyond [21] and [9], our approach is
capable to parse streams of recursive XML, i.e., data in which the same element
names do occur repeatedly along a root-to-leaf path. In comparison to [10] and [18],
we have used an extended automata model which supports also bottom-up evaluation
and mixed evaluation strategies.

5 Summary and Conclusions

Whenever XPath query evaluation is the bottleneck of an application, and main memory is small in comparison to memory requirements for fast query evaluation, a fast
in-memory XPath evaluator that works also on compressed XML structures may be a
significant improvement towards a better run-time.

In this paper, we have presented an XPath query processor that can evaluate XPath
queries on each XML representation that supports a small number of basic binary

axes (first-child, first-child^{-1}, next-sibling, next-sibling^{-1}, and self), like e.g. DOM or the compressed XML representation 'Succinct' [2]. Our query processor decomposes and normalizes each XPath query, such that the resulting path queries contain only the basic binary axes, and then converts them into lean token automata. A DecisionModule decides for each location step which evaluation strategy to follow, i.e., which location step to evaluate when and in which direction.

Our tests have shown, that our query processor is very efficient and outperforms other approaches like JAXP provided by JDK 1.6 and yields results faster than MonetDB – a database that allows the native storage of XML files and that uses an index on this data to speed up the query evaluation – for files up to ~30 MB in general or for queries with at least one location step that has a low selectivity.

As XPath is being used as data access standard in XSLT and XQuery, we are optimistic that the technology proposed in this paper can be used within XSLT processors or XQuery processors too.

References

1. Gottlob, G., Koch, C., Pichler, R.: Efficient algorithms for processing XPath queries. ACM Trans. Database Syst. 30, 444–491 (2005)
2. Böttcher, S., Hartel, R., Heinzemann, C.: BSBC: Towards a Succinct Data Format for XML Streams. In: WEBIST 2008, Funchal, Madeira, Portugal, pp.13–21 (2008)
3. Avila-Campillo, I., Green, T., Gupta, A., Onizuka, M., Raven, D., Suciu, D.: XMLTK: An XML toolkit for scalable XML stream processing. In: Proceedings of PLANX (2002)
4. Barton, C., Charles, P., Goyal, D., Raghavachari, M., Fontoura, M., Josifovski, V.: Streaming XPath Processing with Forward and Backward Axes. In: ICDE, Bangalore, India, pp. 455–466 (2003)
5. Candan, K., Hsiung, W.-P., Chen, S., Tatemura, J., Agrawal, D.: AFilter: Adaptable XML Filtering with Prefix-Caching and Suffix-Clustering. In: VLDB, Seoul, Korea (2006)
6. Diao, Y., Rizvi, S., Franklin, M.: Towards an Internet-Scale XML Dissemination Service. In: VLDB, Toronto, Canada, pp. 612–623 (2004)
7. Ives, Z., Halevy, A., Weld, D.: An XML query engine for network-bound data. The VLDB Journal 11(1), 380–402 (2002)
8. Olteanu, D., Kiesling, T., Bry, F.: An Evaluation of Regular Path Expressions with Qualifiers against XML Streams. In: ICDE, Bangalore, India, pp. 702–704 (2003)
9. Peng, F., Chawathe, S.: XPath Queries on Streaming Data. In: ACM SIGMOD, San Diego, California, USA, pp.431–442 (2003)
10. Böttcher, S., Steinmetz, R.: Evaluating XPath Queries on XML Data Streams. In: Cooper, R., Kennedy, J. (eds.) BNCOD 2007. LNCS, vol. 4587, pp. 101–113. Springer, Heidelberg (2007)
11. Boncz, P., Grust, T., Keulen, M., Manegold, S., Rittinger, J., Teubner, J.: MonetDB/XQuery: a fast XQuery processor powered by a relational engine (2006)
12. Schmidt, A., Waas, F., Kersten, M., Carey, M., Manolescu, I., Busse, R.: XMark: A Benchmark for XML Data Management. In: VLDB, Hong Kong, pp.974–985 (2002)
13. Green, T., Gupta, A., Miklau, G., Onizuka, M., Suciu, D.: Processing XML streams with deterministic automata and stream indexes. ACM Trans. Database Syst. 29 (2004)
14. Gupta, A., Suciu, D.: Stream Processing of XPath Queries with Predicates. In: ACM SIGMOD, San Diego, California, USA, pp.419–430 (2003)

15. Bar-Yossef, Z., Fontoura, M., Josifovski, V.: On the memory requirements of XPath evaluation over XML streams. J. Comput. Syst. Sci. 73(3), 391–441 (2007)
16. Chan, C., Felber, P., Garofalakis, M., Rastogi, R.: Efficient Filtering of XML Documents with XPath Expressions. In: ICDE, San Jose, CA, USA, pp.235–244 (2002)
17. Chen, Y., Davidson, S., Zheng, Y.: An Efficient XPath Query Processor for XML Streams. In: ICDE 2006, Atlanta, GA, USA, p.79 (2006)
18. Onizuka, M.: Processing XPath queries with forward and downward axes over XML streams. In: EDBT 2010, Lausanne, Switzerland, pp.27–38 (2010)
19. Arroyuelo, D., Claude, F., Maneth, S., Mäkinen, V., Navarro, G., Nguyen, K., Siren, J., Välimäki, N.: Fast in-memory XPath search using compressed indexes. In: ICDE 2010, Long Beach, California, USA, pp. 417–428 (2010)
20. Maneth, S., Nguyen, K.: XPath Whole Query Optimization. PVLDB 3(1), 882–893 (2010)
21. Josifovski, V., Fontoura, M., Barta, A.: Querying XML streams. VLDB Journal 14, 197–210 (2005)
22. Bojanczyk, M., Parys, P.: XPath evaluation in linear time, pp. 241–250 (2008)
23. Kader, R., Boncz, P., Manegold, S., Keulen, M.: ROX: run-time optimization of XQueries, pp. 615–626 (2009)

Querying Versioned Software Repositories

Dietrich Christopeit, Michael Böhlen,
Carl-Christian Kanne, and Arturas Mazeika

{christo,boehlen}@ifi.uzh.ch,
kanne@informatik.uni-mannheim.de, amazeika@mpi-inf.mpg.de

Abstract. Large parts of today's data is stored in text documents that undergo a series of changes during their lifetime. For instance during the development of a software product the source code changes frequently. Currently, managing such data relies on version control systems (VCSs). Extracting information from large documents and their different versions is a manual and tedious process. We present QVESTOR, a system that allows to declaratively query documents. It leverages information about the structure of a document that is available as a context-free grammar and allows to declaratively query document versions through a grammar annotated with relational algebra expressions. We define and illustrate the annotation of grammars with relational algebra expressions and show how to translate the annotations to easy to use SQL views.

1 Introduction

Modern software engineering tools process large repositories of source code to assist software developers and analysts with the code retrieval as well as with the computation of various metrics over the source code. Frequently, such tools use handcrafted custom code to extract information and compute metrics. This is tedious, error-prone and brittle. Some approaches offer efficient but very specialized and limited querying capabilities (retrieve a given version of a file), other approaches offer general but hard to formulate and inefficient querying capabilities (extract lines of code from all files satisfying a regular expressions), or yet other easy to formulate, efficient, but only predefined querying capabilities (extract all names of functions from the versions of the source).

In this paper we propose QVESTOR (<u>q</u>uerying <u>v</u>ersioned <u>s</u>oftware reposi<u>tor</u>ies), a prototype implementation of a software query and analysis tool that (i) offers a general querying interface, (ii) allows a natural and easy way to formulate queries, and (iii) answers queries efficiently. QVESTOR (i) parses the source of the documents using context-free grammars, (ii) allows to formulate queries declaratively using the components of the grammar (i.e., both the semantics and specifics of the code), and (iii) uses database query optimization techniques. This yields a general (applies for any data that adheres to a predefined grammar), elegant, concise (expresses in relational algebra operators), yet efficient (allows easy optimizations) approach to extract structured data from repositories. This enables a higher degree of reuse, and leverages database query processing techniques to the analysis of source code.

J. Eder, M. Bielikova, and A.M. Tjoa (Eds.): ADBIS 2011, LNCS 6909, pp. 42–55, 2011.

The development of QVESTOR is subtle and requires to integrate technologies and concepts from both compiler theory (to extract and store the data in relations) and database theory (to define a query language, capabilities, and show how to answer queries over extracted data). We use annotations from compiler theory as a means to formulate and execute queries. Similar to compiler theory, our annotations are program codes that are assigned to the alternatives of the rules of the grammar that are executed once the alternative is selected. In contrast to compiler theory, we associate relations with the alternatives and express annotations declaratively. Query execution then iteratively executes these queries starting with the bottom annotations (i.e., annotations over terminal relations) and finishing with the top annotations in the grammar.

The paper is organized as follows: In Section 2 we discuss the related approaches. In Section 3 we introduce our running example and discuss compiler and database essentials in the context of our approach. In Section 4 we set the foundation and present the building blocks for our system. Then declarative querying (Section 4) with the help of annotations of the leaves and propagation and combination of results in inner nodes is given. In Section 5 we outline the architecture of QVESTOR and sketch the algorithm to transform a list of grammar annotations into SQL view definitions. Finally, we conclude and offer future work in Section 6.

2   Related Work

Krishnamurthy et al. developed SystemT [7]. This system uses a declarative query language AQL to extract information from blog-entries in natural language. With AQL it is possible to express grammar rules in an SQL-like style that describe what the user wants to retrieve (from the blogs). The authors found out that parsing a document is costly and therefore applied rewrite methods to reduce execution costs. In our system we also try to avoid the parsing of data if not needed. In contrast to SystemT we want to focus in QVESTOR on querying versioned data.

Fischer et al. [4] retrieve information from version control data and Bugzilla Bug reports to analyze software evolution. By making this information available in an SQL database simple code evolution queries are possible. However, these queries merely use regular expressions rather than being able to query the code itself. In [6] Kemerer et al. use information from change logs to compute statistical information about software changes. The approach does not query over changes but merely computes the basic statistics. In our system such queries can be formulated declaratively much easier. In addition, our system is not limited to the basic statistics but also allows any sophisticated queries expressible through annotations and the grammar of the source code. Solutions for flexible querying of the source code are proposed by Paul et al. [9]. Our system differs from their SCA algebra and ESCAPE system in the way that for each queryable source code component there must be an object definition for the (OO) data store. Thus the possible queries are to some extend limited by the specification of the objects of the data store. Furthermore, our system aims to reuse as much existing query facilities as possible. Therefore, we chose to use relational algebra expressions that are attached to the grammar specification of the data to be queried. While in ESCAPE new object definitions would have to be specified if the granularity of the queries changes we would just have to change grammar annotations.

Chen et al. [3] describe a system for C code analysis using relational databases. Unlike our proposed system the CIA system does not facilitate a tight coupling between the database, declarative querying, and versions of the source code. For CIA several tools must preprocess the files and store the processed information in a database for declarative querying. Query capabilities are limited by the preprocessing part. In our system tight coupling enables us to query syntax and semantic of source code files directly in the database and does not limit results. Other systems like Rigi [8] or SHriMP [10] mainly focus on visualization of program source dependencies.

Abiteboul et al. [1] describe how semi-structured documents can be queried in an OO-DBMS using a grammar for the document's structure. While queries that involve structural elements are mostly rewritten and parts of the query are pushed into the grammar automatically as annotations we aim for an approach where the user can manage the annotations of the parse tree. In [2] querying of XML documents is described on the native XML-DBMS Natix[5]. While we want to support a wide range of semi-structured versioned data describable by grammars we also want to translate our grammar annotations to generate views that can easily be used in queries.

3   Running Example, Grammar, Parse Trees, and DB Schema

3.1   Example

Our running example consists of three versions of the C-like source code. We start (version one) with a very basic function, which evolves (version two) into another function, which, in turn, evolves into an even more complicated version (version three) of nested loops.

The example (see Figure 1) represents key aspects of evolving code. It consists of evolving signature of functions (version one), change in function calls (version two), additions and deletions of new functions (version three). In this paper we show how to declaratively query the evolving code for all such constructs.

3.2   Grammar

The grammar establishes a structure for the (otherwise unstructured) versioned documents. In addition to general substring queries over unstructured versioned documents, this allows to formulate queries related to concepts of the code. For C code, for example, one can formulate queries involving variables, functions, and specific statements of the language. Every rule ($r_i ::= u_{i,0} \mid u_{i,1} \mid \ldots \mid u_{i,m}$)

```
int a(int v) {
    printf("3");
    return 0;
}
```

```
int a(float v) {
    return x(50);
}

int x(int q) {
    return q*2;
}
```

```
float a(int v) {
    loop
        int j;
        loop
            loop
            ...
            end
        end
    end
    return v/3;
}
```

Version 1   Version 2   Version 3

Fig. 1. Three versions of C-style like source code

consists of the left hand side (abbreviated LHS; e.g. r_i), the right hand side (abbreviated RHS; e.g. $u_{i,0} \mid u_{i,1} \mid \ldots \mid u_{i,m}$) and the assignment symbol (::=) that divides the rule into the LHS and RHS. The LHS introduces a new identifier r_i; The RHS defines the rule for the new identifier. In the most general form, the RHS consists of alternatives $(u_{i,j})$ separated by delimiter $\mid$. Every alternative $u_{i,j}$, in turn, consists of components $u_{i,j} = \$1_{i,j} \ldots \$n_{i,j}$. We use the simplified notation for the components $u = u_{i,j} = \$1 \ldots \$n$ whenever it is clear from the context which components we are referring to.

The components of the alternatives may be either terminal symbols, identifiers defined by other rules or the current rule (in this case the rule is called recursive), or be the empty symbol ε.

To use a specific attribute of grammar component k we use the dot (.) to access the attribute of the component. For example, $\$k$.C refers to the content of the component and $\$k$.P refers to the parent of the component.

Consider, for example, rule

$$\text{expr} ::= \text{expr ' } * \text{' expr} \mid \text{expr '} / \text{' expr} \mid \text{IDENT} \mid \text{fnCall} \mid \text{const}$$
$$= u_{i,0} \mid u_{i,1} \mid u_{i,2} \mid u_{i,3} \mid u_{i,4}.$$

The rule has five alternatives; the components of the 0th alternative are $u_{i,0} = \$1 \$2 \$3$, where $\$1 = \text{expr}, \$3 = \text{expr}$ and $\$2 = ' * '$.

Table 1 summarizes the grammar used in the running examples of the simplified C code .

3.3   Parse Trees

The parse tree of the software code represents the syntactic structure and elements of the source code. The parser builds the parse tree by recursively applying the grammar over the source code. The parser recursively tries to match rules by starting to match sequences of terminals. If such a sequence matches a rule it creates a node representing the LHS of the rule and attaches as leaves the nodes representing the terminals. The same is done if a rule matches a sequence of RHS nodes that are used in another rule. Step by step the parser builds a so-called parse tree bottom-up.

Table 1. The grammar of the simplified C code used throughout the paper

Rule No	Rule
0	start :: = fnDefLst
1	fnDefLst :: = fnDef fnDefLst | fnDef
2	fnDef :: = type IDENT '(' varDecl ')' '{' stmtLst 'return' expr ';' '}'
3	type :: = INT | FLOAT
4	varDecl :: = type IDENT';' | type IDENT '=' const';'
5	stmtLst :: = stmt ';' stmtLst | stmt ';' | ε
6	stmt :: = fnCall | varDecl | loopStmt
7	fnCall :: = IDENT '(' const ')'
8	const :: = INTCONST | STRCONST
9	expr :: = expr '*' expr | expr '/' expr | IDENT | fnCall | const
10	loopStmt :: = 'loop' stmtLst 'end'

Consider, for example, the source code of Version 1 in Figure 1 and grammar in Table 1. The parser starts with matching 'int' to the terminal INT in rule 3, creates a 'type' node and proceeds with 'a' as 'IDENT'. Now the parser has two choices. It can create a node 'varDecl' if the next symbol is ';' and attach 'type' and 'IDENT' as children. On the other hand, if the next symbol is '(' it can proceed with matching rule 2 – eventually creating a node 'fnDef'. This is the case here. The process is continued in bottom-up manner until all the source code is processed. The resulting parse tree is shown in Figure 2(a) (the complete parse trees for all versions are shown in Figure 2).

3.4   Terminal Relations

With every grammar component (like IDENT, INT, FLOAT or expr) in the grammar we associate a relation and store the tuples related to the component in the associated relation. The schema of the relations consists of the following attributes : ID (uniquely identifies the node), V (code version), L (line number in the document; each word is in a separate line), C (content of terminal symbols), N (name of rule), S (next right sibling, and next right component in grammar rule) and P (parent ID of the node). This allows to describe and fully reconstruct the version trees from the table. We reference the relations in the following way. Let $\$k_{i,j}$ refer to the kth component of the jth alternative of the ith rule. If the component is a terminal symbol then the associated relation is referred by $\tau_{k,i,j}$.

For example consider relation $\tau_{1,7,0}$ (i.e. relation for $\$1_{7,0}$ component). For our running example the result is given in Table 2.

If the component is a non-terminal then the associated relation is referred by $t_{i,j}$.

Table 2. Relation of the components of the alternatives of the grammar rules

ID	V	L	C	N	S	P
7	1	8	printf	IDENT	-	10
8	2	9	x	IDENT	-	11

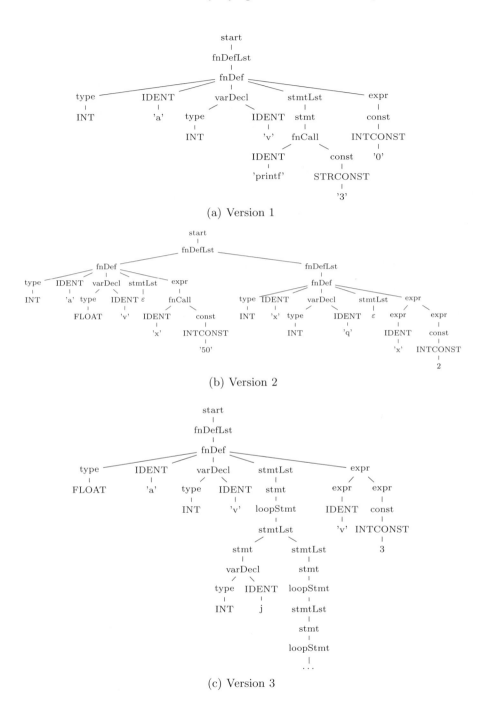

(a) Version 1

(b) Version 2

(c) Version 3

Fig. 2. Parse trees

4   Declarative Querying

4.1   Data at Leaves and Data at Inner Nodes

In this paper we query data that is associated with nodes in the parse tree. Conceptually we distinguish between two types:

Definition 1 (Node Data). *While the parser builds up the parse tree certain information is collected for every node – leaf node or inner node:*

- *ID (uniquely identifies the node)*
- *V (code version)*
- *L (line number in document; one word per line)*
- *C (content of terminal symbols)*
- *N (name of rule)*
- *S (next right sibling, next right component in grammar rule)*
- *P (parent ID of the node)*

*The set of all **node data** of a given parse tree is denoted by $\mathcal{ND}$.*

To formulate declarative queries in our system we use the node data of different nodes, combine it with relational algebra operators to get **composed data**.

Definition 2 (Composed Data). *Let k be a node in the parse tree that has $n \in \mathbb{N}_0$ children $c_0, \ldots, c_n$. Let node k further be a node that is the RHS of rule k option j. Let $d_k \in \mathcal{ND}$ be the **node data** of node k and $d_{c_0}, \ldots, d_{c_n}, d_{c_i} \in \mathcal{ND}, 0 \le i \le n$ be the node data of the children of n. Let $expr_i, 0 \le i \le n$ and $expr_k$ be relational algebra expressions.*

*Without loss of generality let $c_0, \ldots, c_n$ be leaves. The **Composed Data** is calculated for one version like the following for:*

- $c_i\colon expr_i(d_{c_i}) = \tau_{i,k,j}$
- $k\colon expr_k(expr_0(d_{c_0}), expr_1(d_{c_1}), \ldots, expr_n(d_{c_n}), d_k) = t_{k,j}$

If there exist $m, m \in \mathbb{N}$ versions then:

- $\tau_{i,k,j} = \bigcup\limits_{m} expr_i(d_{c_i})$

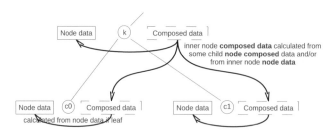

Fig. 3. Data attached to Nodes in Parse Tree

$$- \; t_{k,j} = \bigcup_m expr_k(expr_0(d_{c_0}), expr_1(d_{c_1}), \dots, expr_n(d_{c_n}), d_k)$$

$\tau_{i,k,j}$ and $t_{k,j}$ are the union of the $\tau_{i,k,j}$ and $t_{k,j}$ of the individual versions.
In Figure 3 the situation is shown for node k with two children c_0 and c_1.

4.2 Annotations

In compiler theory, annotations are program codes that are assigned to the alternatives $(u_{i,j})$ of the rules. Once the alternative is selected during build up of the parse tree, the corresponding annotation is executed, a result is calculated and/or output. Therefore, annotations can be viewed as ways to both formulate and answer queries over the source code. In this paper we focus on the declarative capabilities of the query formulations of annotations. Since we use database operations including selection, projection, and join, existing database techniques are applied to optimize and answer such complicated queries.

All our annotations are expressed in terms of the node data (see Definition 1) and composed data (see Definition 2). For example to access all IDENT nodes in expressions (expr) and print their content (C attribute) we need to formulate and execute this query:

$$\Pi_{\tau_{1,9,2}.C}(\tau_{1,9,2})$$

This example accesses the properties of only one node and does not require any joins. Consider an example now, when we want to select all variable names which get assigned a constant 5. These are the names of all IDENT nodes (cf. alternative 1, rule 4 in Table 1) and all CONST nodes (cf. alternative 1, rule 4 in Table 1) such IDENT and CONST have the same LHS varDecl.

$$\Pi_{\tau_{2,4,1}.C}(\tau_{2,4,1} \bowtie_{\tau_{2,4,1}.P=\tau_{4,4,1}.P} (\sigma_{\tau_{4,4,1}.C=5}(\tau_{4,4,1})))$$

In general, our annotations allow the following operators and predicates:

- Selection σ_P, projection π_P, join $\bowtie_P$, cartesian product $\times$, renaming ρ_V, set operators $(\cup, \cap, -)$, and aggregation ϑ.
- Schema identifier $\mathbf{S}(A)$: denotes the schema of a relation A
- Schema modification: addition of an attribute $\circ$ e.g. $\mathbf{S}(A) \circ C$; and removal of the attribute $-$ e.g. $\mathbf{S}(A) - C$
- Predicates: $<, >, =, \leq, \geq, \neq$

We describe the declarative querying and formulation annotations in turn. First we show how to formulate annotations over the leaf nodes, then we generalize it for all nodes, and eventually we explain how to use relational algebra operators in the annotations.

4.3 Annotating the Leaves

An annotation over a leaf node applies the given relational algebra operators over the associated relation and returns the relation. The schema of the returned relation solely depends on the operators applied over the source relation. Since the source relation is always (ID, V, L, C, N, S, P) (see Section 3.4 and Definition 1) the selection over the relation returns the relation of the same schema. Similarly set union, difference, and intersection based on the relation does not change the schema. In contrast, other relational algebra operations change the schema (return larger, smaller, or renamed schemata). For example, the projection operator (usually) reduces the number of attributes in the schema, while (self-) join and Cartesian product doubles the number of attributes in the schema. The general form of the annotation over the leaf node is of the form:

$$t_{i,j} = op_1(\dots op_n(\tau_{k,i,j})\dots), \tag{1}$$

where $op_1, \dots, op_n$ are relational algebra operators, and $\tau_{k,i,j}$ is the terminal relation of the alternative of the rule; relational operators may use only the attributes of the current terminal relation. The resulting relation is called $t_{i,j}$ and associated with the alternative $u(i,j)$.

Consider, for example, the following annotation:

$$t_{7,0} = \Pi_V(\sigma_{C='printf'}(\tau_{1,7,0})).$$

Then the annotation selects all versions of the source code that calls function printf. The following table is returned as the answer to this query:

$t_{7,0}$
V
1

4.4 Annotating the Non-leaves

Very similar reasoning applies for the annotations of the non-leaves including the relational algebra operators and the schema of the returned result. The key difference is that now the relational algebra operators include the relations and information from one level (in terms of parse tree/components of the grammar rules) below the relation it is formulated at. For example, consider the query that selects all the versions of the functions that have return parameter of type integer (INT). This results in the following query:

- $t_{2,0} = \Pi_{\tau_{2,2,0}.C}(t_{3,0} \bowtie_{t_{3,0}.P=\tau_{2,2,0}.P} \tau_{2,2,0})$
- $t_{3,0} = \tau_{1,3,0}$

The answer of the query is the following relation:

$t_{2,0}$
V
1
3

4.5   General Queries of Software Repositories

A general query over software repository consists of a set of annotations such that (i) every annotation is a proper annotation either of leaves or non-leaves (see Sections 4.3 and 4.4) and (ii) if a node is annotated then there must exist a path to a leaf such that every node on a path is annotated.

4.6   Use Case: Find All Versions of the Software That Have Loops of Depth Three or Higher

Identifying the software versions that have feature nested loops of depth of at least 3 is interesting both conceptually and technically. Conceptually, this can indicate versions that have performance issues: nested loops of depth of at least 3 mean at least cubic complexity and may call for attention. Technically, this is a challenge, because in principal, annotations employ context free grammars and expressing such constructs show the power of the language.

Table 3. Annotation query (along with the grammar rules)

Rule No	Rule	Annotation
0	start :: = fnDefLst	
1	fnDefLst :: = fnDef fnDefLst | fnDef	
2	fnDef :: = type IDENT '(' varDecl ')' '{' stmtLst 'return' expr ';' '}'	$\pi_{t_{5,0}.V}(\sigma_{t_{5,0}.D \geqslant 3}(t_{5,0}))$
3	type :: = INT | FLOAT	
4	varDecl :: = type IDENT';'	$\rho_{S(\tau_{2,4,0}) \circ D}(\pi_{S(\tau_{2,4,0})} \circ 0(\tau_{2,4,0}))$
	| type IDENT '=' CONST';'	$\rho_{S(\tau_{2,4,1}) \circ D}(\pi_{S(\tau_{2,4,1})} \circ 0(\tau_{2,4,1}))$
5	stmtLst :: = stmt ';' stmtLst	$(t_{6,0} \cup t_{6,1} \cup t_{6,2}) \cup t_{5,0}$
	| stmt';'	$t_{6,0} \cup t_{6,1} \cup t_{6,2}$
	|ε	
6	stmt :: = fnCall	$t_{7,0}$
	| varDecl	$t_{4,0} \cup t_{4,1}$
	| loopStmt	$t_{10,0}$
7	fnCall :: = IDENT '(' const ')'	$\rho_{S(\tau_{1,7,0}) \circ D}\pi_{S(\tau_{1,7,0})} \circ 0(\tau_{1,7,0}))$
8	const :: = INTCONST | STRCONST	
9	expr :: = expr '*' expr | expr '/' expr	
	| IDENT | fnCall| const	
10	loopStmt :: = 'loop' stmtLst 'end'	$\rho_{S(t_{5,0}) \circ D}$ $(\pi_{S(t_{5,0})} - D \circ \mathfrak{F}_{MAX(t_{5,0}.D)} + 1(t_{5,0}))$

Conceptually, we formulate this query in the following way. We access all the statements in the code. The non-loop statements get the depth attribute assigned to zero. The loop statements compute their depth in the following way. Let s_l be a loop statement, and S_e be the set of statements s_l encloses. Let m be the maximum depth of s_l statements. Then s_l gets $m + 1$ depth.

The annotations of the query are shown in Table 3. Rules 4 and 7 define the depth of non-loop statements. Rule 10 defines the depth of the loop statements. Rules 2, 5, 6, and 7 combine the result. The following is the answer to the query for our running example:

$$\frac{\overline{\overline{t_{2,0}}}}{\frac{\overline{V}}{\overline{3}}}$$

5   Implementation of QVESTOR

5.1   Architecture

We propose and implemented a modular system to answer such declarative queries. Our system consists of the following key modules: declarative query, query rewriter, SQL database, and versions of some programming language source code. The *declarative query* module inputs the query from the user. This is expressed in terms of the rules and attributes of the grammar. Then the *query rewriter* inputs the declarative query and transforms it into an SQL query over the database of versions with the help of *annotated grammar*. The detailed architecture of the system is depicted in Figure 4.

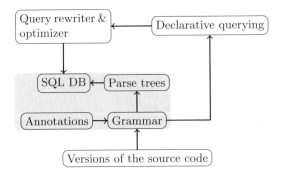

Fig. 4. Architecture of the system

We were reusing expertise and components from the state-of-the-art and standard systems as much as possible. For example, the parse trees were stored in the PostgreSQL database; user defined grammar annotations were translated into database views. This allowed to both achieve efficient storage of the data and obtain efficient query execution plans from query optimizers.

5.2   DB Schema

We store all our data in an SQL database. This allows us to reuse most of the database functionality including ease of expression of queries, query optimization, and effective and efficient storage of the data.

On the database level we keep all the data in the PARSETABLE table. The schema of the table is the same as the schema of the relations of the alternatives of the rules (see Section 3.4), while the table integrates all the node data (see Definition 1).

The tuples for Version 1 (Figure 2(a)) of our running example are given in Table 4.

Table 4. PARSETABLE

ID	V	L	C	N	S	P
19	1			start		
18	1			fnDefLst		19
17	1			fnDef		18
2	1			type	3	17
1	1	1	'int'	INT		2
3	1	2	'a'	IDENT	7	17
7	1			varDecl	13	17
5	1			type	6	7
4	1	4	'int'	INT		5
6	1	5	'v'	IDENT		7
13	1			stmtLst	16	17
12	1			stmt		13
11	1			fnCall		12
8	1	8	'printf'	IDENT	10	11
10	1			const		11
9	1	10	'3'	STRCONST		10
16	1			expr		17
15	1			const		16
14	1	15	'0'	INTCONST		15

5.3 Query Formulation, Translation, and Execution

The implemented system allows for the user to formulate the queries in the SQL-like language. This allows all declarative constructs of the SQL including the SELECT, FROM, WHERE, GROUP BY clauses and all supported predicates of PostgreSQL. In addition we allow the use of symbols $\tau_{k,i,j}$ and $t_{i,j}$ as relation names in the queries (see Section 4). Due to the space constraints we do not define the extended language and hope that the reader gets the spirit of the language.

Given a formulated query in our SQL-like language, we transform it into SQL and send it to PostgreSQL for optimization and execution. The translation of the formulated query is basically achieved in two steps. First, we scan all queries and replace annotations using $\tau_{k,i,j}$ and $t_{i,j}$ with select queries over the PARSETABLE(s) with WHERE clauses. Second, we scan all queries for the second time and create views in the query statements.

As an example consider the query that retrieves the names of the functions that return an integer type. The following is the query expressed with the help of annotations:

- $t_{3,0} = \tau_{1,3,0}$
- $t_{2,0} = \Pi_{\tau_{2,2,0}.C}\big(t_{3,0} \bowtie_{t_{3,0}.P=\tau_{2,2,0}.P} \tau_{2,2,0}\big)$

This query should be formulated in the following SQL-like way:

- $t_{3,0} =$ SELECT * FROM $\tau_{1,3,0}$;
- $t_{2,0} =$ SELECT $\tau_{2,2,0}$.C FROM $t_{3,0}$, $\tau_{2,2,0}$ WHERE $t_{3,0}$.P $= \tau_{2,2,0}$.P;

After the 1st stage the queries become:

```
SELECT p.ID AS ID,
   k.V AS V, k.L AS L, k.C AS C,
   K.N AS N, K.S AS S, p.P AS P
FROM PARSETABLE k, PARSETABLE p
WHERE k.P = p.ID AND
   k.N = 'INT' AND
   p.N = 'type';
```

```
SELECT k.C AS C
FROM t_{3,0} t, PARSETABLE k,
   PARSETABLE p
WHERE t_{3,0}.P = k.P AND
   t_{3,0}.P = p.ID AND
   k.P = p.ID AND
   p.N = 'fnDef' AND
   k.N = 'IDENT';
```

After the 2nd stage the queries become:

```
WITH t_{3,0} AS (
SELECT p.ID AS ID,
   k.V AS V, k.L AS L, k.C AS C,
   K.N AS N, K.S AS S, p.P AS P
FROM PARSETABLE k, PARSETABLE p
WHERE k.P = p.ID AND
   k.N = 'INT' AND
   p.N = 'type'),
```

```
t_{2,0} AS (
SELECT k.C AS C
FROM t_{3,0} t, PARSETABLE k,
   PARSETABLE p
WHERE t_{3,0}.P = k.P AND
   t_{3,0}.P = p.ID AND
   k.P = p.ID AND
   p.N = 'fnDef' AND
   k.N = 'IDENT');
```

6   Conclusions and Future Work

In this paper we presented a system to formulate and answer declarative queries over the versioned source code. Annotations of the grammar rules are the key that allows the declarative querying and connection of the components in the system: first, the annotations allow to naturally formulate queries over the source code (compared, for example, to regexp), and second, allow to translate the queries into SQL and answer them efficiently. Our system consists of declarative query, query rewriter, SQL database, and versions of source code. Tight coupling with the components from state-of-the-art database systems allowed for effectively and efficiently both store and query the data.

Future work will concentrate on the introduction of a sequence model for source code versions. This model will account for the fact that versions are not necessarily available in fully materialized but compressed form. We will introduce a model that describes code evolution with the help of differences between the pairs of versions. It is not necessary to generate a parse table for every version of the code depending on the user's query formulated through the annotations. By defining rules to rewrite the user's query and annotations we see potential for query optimization (e.g. save parsing of code versions that can not participate in the result).

As we have already implemented a (sub-) operator for the generation of the PARSETABLE we will further proceed with integrating the automatic view generation from grammar annotations by implementing an operator tightly into the DBMS. Currently this rather done with an external program than being an extension to SQL.

References

1. Abiteboul, S., Cluet, S., Milo, T.: Querying and updating the file. In: VLDB 1993, pp. 73–84 (1993)
2. Aguilera, V., Cluet, S., Veltri, P., Vodislav, D., Wattez, F.: Querying xml documents in xyleme (2000)
3. Chen, Y.-F., Nishimoto, M., Ramamoorthy, C.: The c information abstraction system. IEEE Transactions on Software Engineering 16, 325–334 (1990)
4. Fischer, M., Pinzger, M., Gall, H.: Populating a release history database from version control and bug tracking systems. In: ICSM 2003 (2003)
5. Kanne, C.-C., Moerkotte, G.: Efficient storage of xml data. In: ICDE 2000 (December 1999)
6. Kemerer, C., Slaughter, S.: An empirical approach to studying software evolution. IEEE Transactions on Software Engineering 25, 493–509 (1999)
7. Krishnamurthy, R., Li, Y., Raghavan, S., Reiss, F., Vaithyanathan, S., Zhu, H.: Systemt: a system for declarative information extraction. ACM SIGMOMOD Record 37, 33–44 (2008)
8. Müller, H.A., Orgun, M.A., Tilley, S.R., Uhl, J.S.: A reverse-engineering approach to subsystem structure identification. Software Maintenance: Research and Practice 5, 181–204 (1993)
9. Paul, S., Prakash, A.: Querying source code using an algebraic query language. In: Proceedings International Conference on Software Maintenance ICSM 1994, pp. 127–136 (1994)
10. Storey, M.-A., Müller, H.: Manipulating and documenting software structures using shrimp views. In: Proceedings in Software Maintenance, pp. 275–284 (1995)

Subsuming Multiple Sliding Windows for Shared Stream Computation

Kostas Patroumpas[1] and Timos Sellis[1,2]

[1] School of Electrical and Computer Engineering
National Technical University of Athens, Hellas
[2] Institute for the Management of Information Systems, R.C. "Athena", Hellas
{kpatro,timos}@dbnet.ece.ntua.gr

Abstract. Shared evaluation of multiple user requests is an utmost priority for stream processing engines in order to achieve high throughput and provide timely results. Given that most continuous queries specify windowing constraints, we suggest a multi-level scheme for concurrent evaluation of time-based sliding windows seeking for potential subsumptions among them. As requests may be registered or suspended dynamically, we develop a technique for choosing the most suitable embedding of a given window into a group composed of multi-grained time frames already employed for other queries. Intuitively, the proposed methodology "clusters" windowed operators into common hierarchical constructs, thus drastically reducing the need for their separate evaluation. Our empirical study confirms that such a scheme achieves dramatic memory savings with almost negligible maintenance cost.

1 Introduction

Continuous query execution has emerged over the last decade as a novel paradigm for processing transient, fluctuating and possibly unbounded *data streams* [3]. Such information is surging in many modern applications, like telecom fraud detection, financial tickers or traffic monitoring; what's more, data must be used to offer real-time response to numerous user requests that remain active for long. To cope with such pressing requirements and avoid dealing with the entire stream history, most processing engines opt for evaluation policies that only examine finite *windows* of memory-resident data. Windows are declared along with the submitted queries and get repeatedly refreshed with the most recent stream items, so as to provide timely, incremental results [13]. Typically, users specify *sliding windows*, expressing interest in recent periods (e.g., items received during past hour) or a fixed tuple count each time (e.g., 1000 fresh items). Through properties inherent in the data, like ordering by arrival time or sequential numbers, windows can actually restrict the amount of inspected information.

But in presence of multiple requests, it is most likely that similarities exist among various windowing constraints. Several window expressions may involve overlapping –if not identical– stream portions: one user may be interested in data over the past hour, whereas another may focus on those of the last quarter only.

J. Eder, M. Bielikova, and A.M. Tjoa (Eds.): ADBIS 2011, LNCS 6909, pp. 56–69, 2011.

Given that most requests are long-running, this key remark opens up a new prospect for multi-query optimization [12]. Discovering commonalities among windows offers a sound basis for their collective handling with significant benefits to sharing system resources [2], particularly memory space for retaining stream tuples and processing overhead for aggregates or joins over similar windows.

In this work, we set out to detect opportunities for shared evaluation among time-based sliding windows. We introduce a subsumption criterion determining whether a pair of such windows have similar properties that enable their common maintenance. To put it simply, a window should not be subsumed under another even if their time horizons overlap, unless their refresh frequencies match as well. We develop a method that embeds windows into jointly evaluated groups, heuristically choosing the "best-fitting" candidate. In order to inspect each stream tuple only once per group, we exploit multi-granular window semantics and adjust a hierarchical structure we proposed in [10], distantly reminiscent of the well-known *"matryoshkas"*, the Russian nesting dolls. Not only can this scheme cope with a fixed window collection, but also with arbitrarily registered or suspended windowed queries, as it actually occurs in a streaming context. Embedded windows can be also used to efficiently evaluate costly operations, like computation of scalar aggregates (SUM, MAX, etc.) or discovery of distinct stream items.

To the best of our knowledge, this is the first approach to subsumption of multiple sliding windows over data streams with the following contributions:

- We identify potential embeddings from a mixture of sliding windows with diverse specifications and dynamically allocate them into similar groups.
- We propose a framework for efficient maintenance and smooth updating of embedded windows, substantially reducing memory consumption.
- We further investigate its advantages on continuous execution of typical operators like aggregation, duplicate elimination and join.
- We empirically demonstrate that this scheme is robust against varying query workloads and scalable with massive volumes of streaming data.

The remainder of this paper is organized as follows. In Section 2, we survey fundamental concepts and related work on window specification and evaluation. In Section 3, we develop a framework for dynamically subsuming windows into groups. Section 4 discusses shared evaluation against multiple nested windows. Experimental results are reported in Section 5. Section 6 concludes the paper.

2 Background and Related Work

2.1 Window-Based Stream Processing

A crucial difference between stream items and relational tuples is that *ordering* must be established among such interminably flowing data. A timestamp value is usually assigned to each streaming item, either at its source (e.g., when sensors issue their readings) or upon admission to the processing engine. Typically, timestamps are drawn from a common *Time Domain* $\mathbb{T}$, i.e., an infinite set of

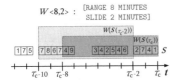

Fig. 1. A time-based sliding window

Fig. 2. Composing sliced windows

discrete instants τ with a total order $\leq$. In effect, a data stream S is an ordered sequence of items $\langle s, \tau \rangle$, where s is a relational tuple with a fixed schema and τ its associated timestamp value [11]. Yet, a problem persists: processing must be carried out in online and incremental fashion, preferably in main memory. Since system resources could hardly sustain the rising volume of accumulating items, it turns out that the amount of data probed each time should be restricted.

Among other suggestions like punctuations [14] or synopses [3], it is *windowing* that prevails as the most suitable means of bounding data streams in query execution. Acting as a *Stream-to-Relation* operator [1], a window W repetitively provides a temporary relation by filtering stream S with specific constraints on a designated windowing attribute [9]. At any time $\tau_i \in \mathbb{T}$, this transient relation is the current *window state* $W(S(\tau_i))$ consisting of a finite portion of stream items that qualify to constraints, most usually based on timestamps (refer to [11] for a complete taxonomy). Indeed, *time-based sliding windows* are indispensable to stream computing [7], since the primary concern is on recently received data. The scope of such a window W is ruled by the following parameters [11] :

- A temporal *range* spanning ω units backwards from current time τ_c. Only data items with timestamps $t \in (\tau_c - \omega, \tau_c]$ qualify for actual state of W.
- A *sliding step* of β time units controls transition to the next state, i.e., how frequently to check for qualifying stream items. Upon sliding, fresh tuples are being accepted into the new state of W at the expense of expiring items discarded from the rear (i.e., remotest) bound of the window.
- *Initiation time* $\tau_0 \in \mathbb{T}$ denotes when W was firstly applied against the stream.

Typically, a time-based sliding window $W\langle \omega, \beta \rangle$ can be declared in CQL [1] with a clause like [RANGE ω SLIDE β]. Figure 1 illustrates two successive states of a window W that ranges over $\omega = 8$ time units and slides every $\beta = 2$ units, assuming that at most $\rho = 2$ items of stream S arrive per minute. Since $\omega > \beta$, overlaps (i.e., common tuples) may occur between successive states.

2.2   Sharing Window State and Computation

Windows are inherent in continuous queries, but they are mostly useful for unblocking operators in query execution plans. Problematic operators [14] that must be coupled with windows include *joins* (since each fresh item cannot be probed against the entire stream history) and *aggregates* (otherwise not a single result could be emitted unless the stream were exhausted).

Strategies for effective scheduling of multiple windowed joins [6] start execution either from the smallest window range or the largest one and accordingly exploit their results for the rest. Still, the focus was on prioritizing tuples that would serve the maximum number of pending queries, ignoring the effect of sliding steps on restructuring window states. Schedule synchronization among multiple aggregation queries with different sliding windows was explored in [5]. To amortize computation cost, they allowed reevaluation of some queries more often (i.e., before their slide occurs) so as to take advantage of aggregates just computed for other windows. However, this approach alters the usual semantics of slide β as a fixed progression step [11], essentially redefining it as an upper bound of the interval between two successive query reevaluations.

Resource sharing among sliding windows has been examined mostly for aggregates. As first shown in [2], indexing of partial aggregates over intervals at multiple resolutions enables the result to be suitably composed from a union of varying-size intervals. In [9], subdivision of a window state into equal-sized slices and performing aggregation on such fundamental *"panes"* was proposed. Given a window $W\langle\omega, \beta\rangle$, the size of each pane is the greatest common divisor (*gcd*) of ω and β, so W is sliced uniformly into $\omega/gcd(\omega, \beta)$ disjoint panes. Partial aggregates are calculated only once per pane, and then "rolled up" to provide the final result. Nonetheless, each window is handled in isolation, determining the size and number of its panes irrespectively of other similar specifications.

Towards sharing state among multiple windows, the *"paired"* approach for streamed aggregation [8] suggests that a window can be split into unequal parts and perform faster than a uniformly sliced one. Aggregates over a time-based $W\langle\omega, \beta\rangle$ comprise partial results obtained from a series of disjoint extents of alternating sizes $s_2 = \omega \mod \beta$ and $s_1 = \beta - s_2$. Intuitively, tuples in any extent s_2 qualify for two consecutive window states, so there is no need to recalculate algebraic or distributive aggregates for interval s_2. Further, such paired windows from multiple queries can be composed into a single window, sliced appropriately as a vector with many edges at a common period. In order to fit all participating scopes, this period is set to the least common multiple (*lcm*) of their sliding steps β. As shown in Fig. 2, to allow for windows with diverse sliding steps, their slice vectors are stretched, simply repeating themselves (twice for W_1, thrice for W_2) until they all obtain the same period (*lcm* = 12). Results are emitted on-the-fly, but this state-of-the-art algorithm is specifically tailored for aggregates only, hence ineffective for generic windowed operations (e.g., joins or distinct items).

2.3   Multi-granular Windowing

In [10] we introduced a multi-level sliding window W that specifies a set of time frames at diverse user-defined *granularities* and then concurrently evaluates a single continuous query over several stream slices of varying size. Subwindow W_k at level k has its own time horizon ω_k and refresh frequency β_k; it ranges in $(t_k, \tau_c]$, yet nested under the widest W_{n-1}, as exemplified in Fig. 3 for $n = 3$ levels. A hierarchy of n subsumed frames is created when $\beta_{k-1} \leq \beta_k$ and $\omega_{k-1} < \omega_k$, for each level $k = 1, \ldots, n - 1$. For smooth transition between successive

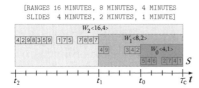
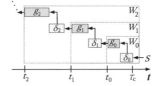

Fig. 3. State of a 3-level sliding window   **Fig. 4.** Stairwise processing scheme

states at any W_k, it holds that $\omega_k = \mu_k \cdot \beta_k$ for $\mu_k \in \mathbb{N}^*$, so a frame at level k consists of a fixed number of primary blocks (*granules*) of size β_k units each.

Beyond semantics, such a hierarchical scheme can be maintained consistently and incrementally. Normally, each subwindow W_k subsumes others in lower levels, so it could suffice to retain in a queue g_k only those *delta tuples* in interval $(t_k, t_{k-1}]$ not covered by subordinating window frames. But since each level prescribes its own slide β_k, nested subwindows may not be always aligned with current time τ_c, hence not concurrently refreshed. Thus, each queue g_k must be combined with an auxiliary one δ_k to buffer items expiring from the preceding subwindow at level $k - 1$. This *"stairwise"* scheme (Fig. 4) of alternating "buffer" δ_k and "core" queues g_k can seamlessly maintain the overall window state without loss of tuples in transit between successive levels. Moreover, it can also answer advanced continuous requests (like multi-grained aggregates, online regression, and recurring stream items) across multiple time horizons [10].

Next, we attempt to generalize this framework and achieve nesting of individual single-level sliding windows into judiciously chosen multi-level constructs with considerable resource savings in memory space and processing overhead.

3   A Multi-subsumption Framework for Sliding Windows

3.1   Problem Specification

Let a *pool* of n sliding windows $\mathcal{W} = \{\langle \omega_k, \beta_k \rangle, k = 1..n\}$ currently specified against a data stream S. Given that various continuous queries are active at any time, their window states may actually overlap; thus, a stream tuple s can qualify for multiple windows, depending on their exact scope. Users may register new requests or revoke existing ones, so pool $\mathcal{W}$ gets dynamically modified. Without loss of generality, we assume that window initiation τ_0 occurs at fixed clock ticks (e.g., minutes), as a means of synchronization among diverse scopes.

We further assume that every window from $\mathcal{W}$ is associated to the same operator. In other words, all windows are used for computing the same aggregate (like SUM or MAX) or duplicate elimination or join, but not a mixture of operations. This may look like a too tight constraint in terms of window state maintenance; diverse operators may specify identical windows, so why not handle a single state for all of them? Similarly to [2,5,8], we aim to use windows' contents to perform calculations on them and not just retain their qualifying tuples, so the adjoined operator definitely matters as it dictates completely different computation.

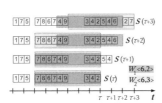

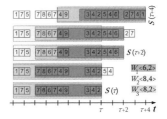

Fig. 5. Windows with different progression   **Fig. 6.** Subsumable sliding windows

Our objective is to find a grouping $\mathcal{G} = \{g_1, g_2, \dots\}$ where each window from $\mathcal{W}$ is assigned into a single group g_i. Each group g_i gathers together specifications with similar parametrization (ω, β), enabling their insertion one after the other into a common structure that could jointly retain their states and also provide incremental output to the associated operator. Inevitably, queries with distinct windowing parameters must be evaluated separately. Our key idea is *subsumption of smaller scopes into broader ones*, provided that such a nesting remains intact with time and always fits all participating windows. Formally:

Definition 1. *A sliding window $W_i\langle\omega_i, \beta_i\rangle$ is subsumable under another one $W_j\langle\omega_j, \beta_j\rangle$ if and only if $\mathsf{mod}(\beta_j, \beta_i) = 0$ and $\omega_j \geq \omega_i$. We denote that $W_i \sqsubseteq W_j$.*

Likewise, windows are subsumable into a group if a pairwise symmetric shift of their frames holds; then, no qualifying tuples can ever be missed for any window state and consistent results are issued by the respective operator. Subsuming multiple windows with arbitrary specifications is an intricate issue, primarily because each window is allowed to slide forward at its own pace. As exemplified in Fig. 5, even if windows W_1, W_2 share state at time τ, it is difficult to coordinate their common tuples in the future as frames get refreshed at varying frequencies. Next, we introduce a heuristic, advocating that windows with similar sliding steps should be grouped together as a sequence ordered by range values ω.

3.2 Identifying Suitable Embeddings among Sliding Windows

A simple approach to subsumption would attempt to classify windows by their sliding steps. Windows with identical slide β_k and overlapping ranges ω could be combined into a single queue q_k with a common head (i.e., most recent stream item) but multiple tails, one for each participating frame. Due to their common step β_k, all queue demarcators (head and tails) slide always in tandem. This scheme creates as many queues as the distinct sliding steps found in the current window pool. Yet, each window still maintains a detached state of qualifying items, which are not reused by other frames in the same group. As an amendment, the unified queue could keep account of *delta substates* only [10], so that narrower ranges readily provide their tuples to successively wider scopes.

Even so, such a policy misses certain nesting opportunities that can lead to much less and more compact groups. As implied from Definition 1, not

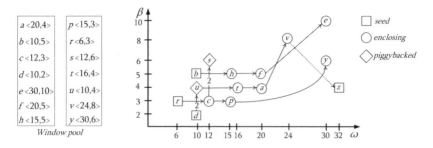

Fig. 7. Graph partitioning for sequences of subsumable windows

necessarily all frames in a group g_i obey the same slide β_i. Therefore, we may even allow a window with $\beta_k = \lambda_k \cdot \beta_i, \lambda_k \in \mathbb{N}^*$, i.e., progressing at a multiple of the base slide β_i, provided that its range ω_k can be embedded among existing ones without breaking the sequence. Referring to the example in Fig. 6, suppose that we want to compute the SUM over each window. Apparently, we can first respond to the smallest W_1 and then readily exploit this partial result for W_3, since both windows slide together every two time units. As for W_2, there is no need for recalculation, as its aggregate is exactly the one given for W_3; the only difference is that it is emitted less frequently (every fourth timestamp).

To identify subsumable windows and then coordinate their evaluation, we introduce a Window Manager that accepts submitted windows and allocates them into suitable groups. We presently examine a fixed pool $\mathcal{W}$ of windows, deferring discussion about dynamic changes at window specifications to Section 3.3. Intuitively, if we plot window scopes in a (ω, β) axis system (Fig. 7), we can simulate recognition of commonalities with a vertical sweepline that moves rightwards and stops at each ω position. Each scope $\langle \omega_k, \beta_k \rangle$ is compared to those already assigned into groups with equal or aliquot slides. By virtue of the sweep, only the last window in each candidate group g needs checking; if its range exceeds ω_k, then subsumption to group g is ruled out. Otherwise, $\langle \omega_k, \beta_k \rangle$ can be inserted as the last window in that group. If all candidate groups have been probed without success, then window $\langle \omega_k, \beta_k \rangle$ starts out a new group. This task resembles to *topological sorting*, as it creates a partitioned graph with distinct sequences of nodes that represent subsumable scopes. Each resulting subgraph enumerates windows that can be evaluated together in the same multi-level construct. Nodes in each group g can be either characterized as:

(i) *Seeds* represent rudimentary window scopes at every β-level that cannot fit in any other existing group. Apparently, windows W_i with a prime slide β_i can become seeds of a group, such as nodes d, r and b illustrated in Fig. 7.

(ii) *Enclosing* is a window that fully subsumes an already visited scope in group g, because it has a greater range and the same or a multiple sliding step. Hence, a series of enclosing windows can share partial states with no missing tuples when sliding takes place. As shown in Fig. 7, window e can be partially served by the state of f, as it moves forward by a double step; upon sliding, e owes its most fresh items to all subordinating frames, whereas also accepting expired tuples in transit from f (as explained in Section 2.3).

(iii) *Piggybacked* windows have identical temporal range with their preceding node, but slide forward less frequently. As depicted in Fig. 7, window u has the same range as d, but moves at a double pace. Since they share states every other step, we may entirely collapse u and instruct d to periodically provide its state to u. This kind of *piggybacking* comes at no extra cost; we simply annotate the edge connecting such equal-range windows with the quotient $q \in \mathbb{N}^*$ of their respective sliding steps. So, the edge from d to u is marked with $q = \beta_u/\beta_d = 2$, to indicate how frequently (at q^{-1}) the result computed for d should be propagated to its piggybacked u as well.

The core functionality of the proposed Window Manager is outlined in Algorithm 1. More concretely, process *SubsumeWindows* partitions the given pool $\mathcal{W}$ into groups following a policy of progressive "expansion": it first attempts to allocate smaller scopes before continuing with wider ones. So, it starts by collecting non-assigned windows in a priority queue $\mathcal{H}$ ordered by ascending range ω and, in case of ties, by slide β as well (Line 4). Then, it iterates through each remaining scope from $\mathcal{H}$, first attempting to append it into an existing group (Lines 9-13). The visiting order of candidate groups is guided by descending β of their seed nodes (Line 7), while only the lastly inserted window should be probed for each group (Lines 10-11) thanks to the implicit ordering of window inspection. If subsumption fails, then a new sequence is created, having the examined scope as a seed (Line 14). When called, function *canEmbed* checks whether window $\langle \omega_i, \beta_i \rangle$ can either be piggybacked on another $\langle \omega_j, \beta_j \rangle$ already assigned into group g (Lines 19-22) or fully subsume the latter (Lines 23-26). In both cases, the examined sliding step β_i should be equal or integer multiple of the compared one β_j (Line 18), otherwise no embedding is possible.

Concerning query evaluation, as soon as each group gets finalized, a corresponding stairwise scheme (Fig. 4) is dedicated to keep account of states for all its members. Seed scopes form the bottommost stairs of the multi-granular scheme built for each new group, while only delta substates are maintained for upper levels that represent enclosing windows. Piggybacked windows are not materialized, but are represented only implicitly according to annotations specified for other embedded scopes. Note that we relax construction features from [10], allowing many successive stairs with the same slide step. For the setting in Fig. 7, windows b, h and f will be placed in consecutive stairs. But since all three proceed in tandem, there is no need for intermediate 'buffer' queues, so items expiring from b are immediately transferred to the substate of h and so on.

Overall, a given window is actually assigned to a suitable group depending on already identified seeds and the current arrangement of inspected windows. Multiple seeds could appear with the same β, though. Indeed, if the seed node of a sequence got nested within scopes of a wider slide step, then a new group should be created for the examined window, which becomes a seed itself. As exemplified in Fig. 7, when window $z\langle 32, 4 \rangle$ is submitted, it is turned into a seed, since it cannot get along jointly with a preceding scope like $v\langle 24, 8 \rangle$.

Algorithm 1.   Window Manager

1: **Procedure** *SubsumeWindows*
2: **Input:** A collection of sliding windows $\mathcal{W} = \{\langle \omega_k, \beta_k \rangle, k = 1..n\}$.
3: **Output:** A grouping $\mathcal{G} = \{g_1, g_2, ...\}$; each window assigned into a single group g_i.
4: $\mathcal{H} \leftarrow$ a priority queue of windows from $\mathcal{W}$, arranged by increasing scope $\langle \omega, \beta \rangle$.
5: **while** $\mathcal{H}$ is not empty **do**
6:     $\langle \omega_k, \beta_k \rangle \leftarrow \mathcal{H}.\text{pop}()$;   *//Currently examined window*
7:     $\mathcal{C} \leftarrow \{g \in \mathcal{G}: \text{mod}(\beta_k, g.\beta) = 0\}$;   *//Candidate groups ordered by descending β*
8:     $nested \leftarrow$ **false**;
9:     **repeat**
10:         $g \leftarrow$ group of $\mathcal{C}$ with the next greater base slide β;
11:         $\langle \omega_m, \beta_m \rangle \leftarrow$ last window W_m (i.e., currently the broader) inserted into g;
12:         $nested \leftarrow canEmbed(g, \langle \omega_k, \beta_k \rangle, \langle \omega_m, \beta_m \rangle)$;   *//Subsumable into group g?*
13:     **until** *nested* **or** $\mathcal{C}$ is exhausted;
14:     **if not** *nested* **then** Append into $\mathcal{G}$ a new $g \leftarrow \{\langle \omega_k, \beta_k \rangle\}$; *//Seed of new group*
15: **end while**
16: **End Procedure**

17: **Function** *canEmbed* (group g, window $\langle \omega_i, \beta_i \rangle$, window $\langle \omega_j, \beta_j \rangle$)
18: **if** $\text{mod}(\beta_i, \beta_j) = 0$ **then**
19:     **if** $\omega_i = \omega_j$ **then**
20:         Collapse $\langle \omega_i, \beta_i \rangle$;   *//Piggyback window W_i on existing one W_j*
21:         Annotate $\langle \omega_j, \beta_j \rangle$ with $q = \beta_i/\beta_j$;
22:         return **true**;
23:     **else if** $\omega_i > \omega_j$ **then**
24:         $g \leftarrow g \cup \{\langle \omega_i, \beta_i \rangle\}$;   *//Subsume window W_j under newly inserted W_i*
25:         return **true**;
26:     **end if**
27: **end if**
28: return **false**;
29: **End Function**

30: **Function** *embedWindow* (window $\langle \omega_k, \beta_k \rangle$)
31: $nested \leftarrow$ **false**;
32: **for each** group $g \in \mathcal{G} \wedge (\text{mod}(\beta_k, g.\beta) = 0 \vee \text{mod}(g.\beta, \beta_k) = 0)$ by descending β **do**
33:     $\langle \omega_m, \beta_m \rangle \leftarrow$ first window W_m having $\omega_m > \omega_k$ when traversing members of g;
34:     **if** $\langle \omega_m, \beta_m \rangle = $ nil **then**
35:         $\langle \omega_m, \beta_m \rangle \leftarrow$ broader window W_m inserted into g;
36:         $nested \leftarrow canEmbed(g, \langle \omega_k, \beta_k \rangle, \langle \omega_m, \beta_m \rangle)$; *//Insert W_k as last in group g*
37:     **else if** $\text{mod}(\beta_m, \beta_k) = 0$ **then**
38:         $nested \leftarrow canEmbed(g, \langle \omega_m, \beta_m \rangle, \langle \omega_k, \beta_k \rangle)$; *//Subsume W_k under existing W_m*
39:     **else if** $\text{mod}(\beta_k, \beta_m) = 0$ **then**
40:         $nested \leftarrow canEmbed(g, \langle \omega_k, \beta_k \rangle, \langle \omega_m, \beta_m \rangle)$; *//Subsume existing W_m under W_k*
41:     **end if**
42:     **if** *nested* **then** break;
43: **end for**
44: **if not** *nested* **then** Append into $\mathcal{G}$ a new $g \leftarrow \{\langle \omega_k, \beta_k \rangle\}$;   *//Seed of new group*
45: return g;
46: **End Function**

3.3   Embedding Windows at Runtime

In practice, as user requests dynamically join in or leave the system, so do their window specifications with respect to pool $\mathcal{W}$. So, the initially derived graph partitioning $\mathcal{G}$ is subject to changes and existing groups could be rearranged. We employ a runtime policy that allows "on-the-fly" embedding and suspension of window specifications, thus avoiding recalculation of $\mathcal{G}$ from scratch.

Function *embedWindow* (Algorithm 1) outlines treatment of a newly submitted window W_k. Not all existing groups should be checked for possible subsumption, but only those with seeds of either multiple or aliquot sliding step (Line 32). Candidate groups are deliberately visited in a descending order of their base slides as dictated by the "expansion" nature of the heuristic, such that the substate of W_k would get the most out of already retained ones. When examining a candidate group, its member window W_m with the least, but still greater range than ω_k is identified. This W_m denotes the ideal place where W_k could potentially fit with the minimal distortion at the existing sequence. If such window W_m is not found, then it is worth trying to append W_k as the last in that candidate group and check if it can successfully subsume all its current members (Lines 34-36). In case that window W_m is found, the question whether to attempt embedding W_k before (Lines 37-38) or after W_m (Lines 39-40) depends on a simple comparison between their respective sliding steps. Of course, if all candidate groups have been inspected without success, then W_k becomes the seed of a new group (Line 44). For example, if a new window $k\langle 21, 3 \rangle$ appears in the setting depicted in Fig. 7, it can be easily placed between existing nodes p and y in full conformance with the subsumption criteria.

Revoking an active window is much simpler, as it stipulates its withdrawal from the group where it was initially allocated. Suspending piggybacked nodes incurs nothing but erasing annotations. Removal of an enclosing node is also easy to handle, since its delta substate is delegated to the next node in the sequence. In case its removal leaves orphan piggybacked nodes, the one with the smallest slide becomes its replacement in the sequence. Finally, new seed nodes may be appointed. In Fig. 7, if d gets suspended, node u takes over as the new seed at a base slide of $\beta = 4$ units with no further repercussions on that group.

4   Shared Evaluation over Subsumed Windows

Admittedly, this subsumption framework is limited to requests specifying exactly the same operation over windows, as pointed out in Section 3.1. But with hundreds or thousands of continuous queries, chances are mounting that both operators and windows actually match. So, we can first distinguish windowed queries into collections according to their prescribed operation, and then initiate a separate window manager for each collection. Next, we outline how the proposed stairwise scheme facilitates shared execution of costly operations inextricably liaised to sliding windows. In all cases, the underlying principle is to avoid duplication in computations and reduce update cost for greater scopes, benefiting from intermediate results available from subsumed ones.

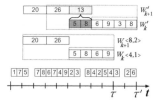  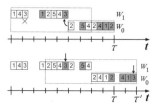

Fig. 8. Shifting partial SUM aggregates   **Fig. 9.** "Phantom" duplicate items

Aggregation. As we suggested in [10], the multi-level stairwise structure supports efficient calculation of typical *scalar aggregates* (MIN, MAX, SUM, COUNT and AVG) by taking advantage of incrementally computed, fine-grained partial results. In brief, each level k retains a fixed number of sub-aggregates, each one derived over β_k time units (i.e., the granularity of the respective level). No wonder that such a scheme seems ideal to handle shared aggregates among subsumed windows, practically without any modification. When subwindow W_k slides forward (Fig. 8), it accepts a new sub-aggregate over items spanning β_k units from its lower level $k-1$ or the raw stream. Then, a similar sub-aggregate expires and is temporarily buffered waiting for W_{k+1} to slide; when this occurs, a corresponding sub-aggregate shifts to level $k+1$ discarding buffered values. Favorably to total execution cost, the number of partial aggregates is expected to shrink toward higher levels, as greater sliding steps translate to less and coarser granules.

Duplicate Elimination. This operator [11] reports distinct items within the current state of a sliding window W. When it comes for multiple windows nested in each other, it suits better to keep the most recent appearance of a distinct item, replacing the previous one with the same attribute value(s). As illustrated in Fig. 9, such an "eager" policy possibly causes frequent short-term replacements and thus incurs additional overhead for the bottommost window W_0. Had a distinct item been parsimoniously retained until its expiration from W_0, this extra cost could have been avoided. Yet, we prefer retaining the most recent distinct items for W_0, as it pays off in upper levels of the hierarchy. Indeed, only distinct items expiring from W_0 with no substitute (as such an item has not appeared ever since) should be propagated to the upper level representing W_1. So, partial results for W_1 include older distinct items that sometime ceased to qualify for W_0; likewise, the same pattern repeats in all stairs.

Overall, a great deal of computation can be spared at every but the first level. This seems affordable enough, because W_0 is definitely the smallest in range and slides forward more frequently. Besides, each subsequent level is only responsible for distinct elements in the time interval not covered by its subordinate windows. However, sets of distinct items in any two successive levels may not be always disjoint, as depicted in Fig. 9. Value 3 was distinct in level $k = 0$ but expired at time τ with no substitute; so, this 3 migrated to the subset of distinct items kept for $k = 1$ and replaced its elder appearance there. At time τ', a fresh value 3 arrives and qualifies as distinct for W_0. Item 3 has then a duplicate for W_1 and all its enclosing windows $W_2, W_3, \ldots$ We opt for lazy elimination of such "phantom" items later on, as it is worth tolerating them and avoid reinspection of enclosing frames upon every tuple arrival. When reporting, enumeration of distinct items starts always from the bottommost level, so any remaining duplicates encountered upwards can be removed with negligible cost.

Join. A join operation between two sliding windows emits a result as soon as a newly arrived item in either stream matches a tuple contained in the opposite window state. Join is a stateful operator in streams [14], so each window must retain all its qualifying tuples. But joined results must be canceled as soon as one of their constituent tuples gets evicted from the window it belongs to. Since sliding window joins are *weak non-monotonic* [4,11], it suffices to attach suitable expiration timestamps to all emitted results, hence delimiting their validity.

This policy can be also adapted for subsumable windows, provided that we deal with joins of common signatures, i.e., involving the same pair of input streams and specifying the same join predicates [6]. This condition guarantees that joined items produced from common execution are meaningful and were actually requested from users. In terms of actual evaluation, each incoming tuple s must be probed only once against the group of n windows specified over the opposite stream. Search for potential matches starts from the bottommost window (i.e. the seed of the group) and continues upwards to wider scopes inspecting state items in descending chronological order. As soon as a matching tuple s' is found, say at level k, then a total of $n - k$ joined results $\langle s, s' \rangle$ are issued; remember that a tuple s' retained at the substate of W_k also belongs to $W_{k+1}, \ldots, W_{n-1}$ (Fig. 3). Therefore, each result is emitted with a suitable expiration timestamp that depends on the range of the corresponding windows and the actual timestamps of its constituent tuples.

5   Experimental Evaluation

In this section, we report indicative results from an empirical validation of the proposed multi-subsumption framework for a mixture of query workloads applied against varying stream arrival rates.

Experimental Setup. We distinguish three classes of window specifications on the basis of their maximum sliding step. In detail, *smooth* windows stipulate small slides of $\beta \leq 10$ timestamps, *medium* windows have $\beta \leq 20$ units, whereas *abrupt* windows may move forward by $\beta \leq 30$ time units. For all scopes, range ω fluctuates up to a limit of 60 timestamps, but always $\beta < \omega$ to imitate genuine sliding behavior. According to these rules, we artificially generate four query workloads for each class; these workloads respectively contain 1000, 2000, 5000, and 10 000 window specifications. Simulations at stream rates up to $\rho = 100\,000$ tuples/sec were performed against a real dataset[1] tracing wide-area TCP connections with the Lawrence Berkeley Laboratory. Algorithms were implemented

---

[1] Available at `http://ita.ee.lbl.gov/html/contrib/LBL-CONN-7.html`

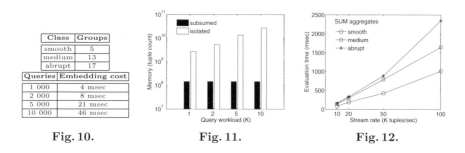

Class	Groups
smooth	5
medium	13
abrupt	17

Queries	Embedding cost
1 000	4 msec
2 000	8 msec
5 000	21 msec
10 000	46 msec

Fig. 10.   Fig. 11.   Fig. 12.

in C++ and simulated on an Intel Xeon 2.13GHz CPU running GNU/Linux with 12GB of main memory. Results are averages of actual measurements per timestamp over complete (i.e., not "half-filled") windows.

Experimental Results. The first set of experiments demonstrates the strong subsumption capabilities of the proposed framework. As shown in Fig. 10, the number of resulting groups depends on the actual window specifications, and on the variance of their sliding steps β in particular. Just a handful of groups are maintained in any setting, whereas the total overhead for their effective allocation is practically negligible, even for increasing number of windowed queries.

But the real benefit of window subsumption lies with the tremendous resource savings it offers. Figure 11 illustrates the accumulated memory footprint of all window states for the medium class. Subsumption reduces memory utilization by orders of magnitude (mind the log scale!); space cost remains practically stable no matter the number of windows, because only delta substates are retained within the multi-level structure. In practice, the space consumption of a stairwise scheme is equivalent to that of its widest scope at the top level (if considered in isolation), plus the cost for buffer queues that hold transitional tuples expiring from each level. In contrast, a conventional approach is much more costly, since each window state is maintained in isolation. As Fig. 11 verifies, memory consumption grows linearly with increasing query workloads, clearly demonstrating the inefficiency of non-shared maintenance for sliding windows.

In the interest of space, we discuss results concerning evaluation cost only for the case of stream aggregation, varying the actual stream arrival rate (in tuples/sec). Specifically, we calculate SUM values for all three classes of sliding windows, each with 10 000 varying specifications. As plotted in Fig. 12, computation time per timestamp is small for aggregates over numerous subsumed windows and escalates almost linearly with the stream rate. Not surprisingly, the cost reflects the variance of slides; the larger the progression steps, the more expensive the aggregation, as more sub-aggregates must be retained and propagated amongst levels. It appears that β is perhaps the most influential factor in smooth reevaluation. Anyway, significant time savings must be attributed to sharing computation across hierarchically organized substates. In most cases, it takes less than 2 sec to provide all responses, a fact that clearly fulfills expectations for real-time answering to multiple long-running queries.

6   Conclusions and Future Work

In this paper, we introduce a framework that identifies opportunities for shared processing among overlapping window states in data streaming applications. Our proposed heuristic subsumes similar sliding window specifications and appropriately coordinates their joint reevaluation through multi-level groups, achieving substantial memory savings and minimal maintenance overhead. We also delineate how to orchestrate fast execution of costly operations over windows, by examining stream tuples only once per group.

In the future, we plan a thorough investigation of operations against nested windows to get more insight on their properties and discover further optimizations. It would be interesting to analyze feasibility of near-optimal partitioning of windows that could lead to a minimal number of groups. Resolving system degradation when many queries get arbitrarily registered or suspended is also a challenging task. To avert potential miscasts to groups, perhaps a scoring approach could be employed to gauge suitability among alternative subsumptions.

References

1. Arasu, A., Babu, S., Widom, J.: The CQL Continuous Query Language: Semantic Foundations and Query Execution. VLDB Journal 15(2), 121–142 (2006)
2. Arasu, A., Widom, J.: Resource Sharing in Continuous Sliding-Window Aggregates. In: VLDB, pp. 336–347 (2004)
3. Babcock, B., Babu, S., Datar, M., Motwani, R., Widom, J.: Models and Issues in Data Stream Systems. In: PODS, pp. 1-16 (2002)
4. Golab, L., Tamer Özsu, M.: Update-Pattern-Aware Modeling and Processing of Continuous Queries. In: SIGMOD, pp. 658-669 (2005)
5. Golab, L., Bijay, K., Tamer Özsu, M.: Multi-Query Optimization of Sliding Window Aggregates by Schedule Synchronization. In: CIKM, pp. 844-845 (2006)
6. Hammad, M., Franklin, M., Aref, W., Elmagarmid, A.: Scheduling for Shared Window Joins over Data Streams. In: VLDB, pp. 297-308 (2003)
7. Jain, N., Mishra, S., Srinivasan, A., Gehrke, J., Widom, J., Balakrishnan, H., Çetintemel, U., Cherniack, M., Tibbetts, R., Zdonik, S.: Towards a Streaming SQL Standard. In: VLDB, pp. 1379-1390 (2008)
8. Krishnamurthy, S., Wu, C., Franklin, M.J.: On-the-Fly Sharing for Streamed Aggregation. In: SIGMOD, pp. 623-634 (2006)
9. Li, J., Maier, D., Tufte, K., Papadimos, V., Tucker, P.: No Pane, No Gain: Efficient Evaluation of Sliding-Window Aggregates over Data Streams. ACM SIGMOD Record 34(1), 39–44 (2005)
10. Patroumpas, K., Sellis, T.: Multi-granular Time-based Sliding Windows over Data Streams. In: TIME, pp. 146-153 (2010)
11. Patroumpas, K., Sellis, T.: Maintaining consistent Results of Continuous Queries under Diverse Window Specifications. Information Systems 36(1), 42–61 (2011)
12. Sellis, T.K.: Multiple-Query Optimization. ACM TODS 13(1), 23–52 (1988)
13. Stonebraker, M., Çetintemel, U., Zdonik, S.: The 8 Requirements of Real-Time Stream Processing. ACM SIGMOD Record 34(4), 42–47 (2005)
14. Tucker, P., Maier, D., Sheard, T., Fegaras, L.: Exploiting Punctuation Semantics in Continuous Data Streams. IEEE TKDE 15(3), 555–568 (2003)

Revisiting the Partial Data Cube Materialization

Nicolas Hanusse, Sofian Maabout, and Radu Tofan

University of Bordeaux
LaBRI, CNRS-UMR 5800, Inria
France
{hanusse,maabout,tofan}@labri.fr

Abstract. The problem of selecting views and/or indexes to materialize has been extensively studied in the context of query optimization.
Traditionally, the problem is formalized as follows: given a set of queries
and a budget e.g., an available memory space, find the objects to materialize (views and/or indexes) that (1) satisfy the given budget and (2)
minimize the query cost. In this paper, we depart from this setting by
adopting a user-centric point of view: given a constraint on query evaluation, namely a maximal query cost the user does accept, find the objects
(1) whose materialization needs the minimal storage space and (2) that
guarantee the query evaluation constraint. We study this problem in the
data cube setting and provide exact and approximate solutions.

1 Introduction

Materialized views have been recognized as an effective query optimization technique for a long time [3]. Since data cubes [7] are sets of special views, their
full and partial materialization have been studied as soon as this concept was
proposed. As it is described in [15], the selection of the data cube part to be
materialized is a multi-criteria task. We recall some of the most studied ones:
the materialization granularity (full vs fragments of views), the constraints taken
into account (available storage space and/or time window allowed to incremental update), presence or not of target queries (workload), complexity of the view
selection algorithm, dynamic workload and the possibility or not to use indexes.
Most of the preceding proposals formalize the view selection problem so that
the returned solution should minimize the *average query cost* while satisfying
the imposed budget space and/or update time constraints. Minimizing average
query cost, or equivalently total query cost, may lead to solutions where some
queries are well optimized while others are not.

In this paper, we propose a new formalization of the view selection problem.
Given a fixed required query evaluation time, find the minimal solution that
satisfies it. As advocated by the position paper [6], data base optimization should
be revisited: users are ready to pay if the system is able to satisfy their required
performance. Of course, they want to pay the minimum. In this new trend, the
goal turns out to minimize the amount of memory achieving the wanted quality
of service.

J. Eder, M. Bielikova, and A.M. Tjoa (Eds.): ADBIS 2011, LNCS 6909, pp. 70–83, 2011.
© Springer-Verlag Berlin Heidelberg 2011

Contributions. In this paper, we make the following contributions:

1. We introduce a new formalization of the materialized view selection in the context of data cubes.
2. we provide exact and approximate solutions when data are static.
3. We investigate the *robustness* of our solutions against data updates and give some sufficient conditions under which the solutions are not perturbed by these data updates.
4. We conduct some experiments on a real data set to confirm our proposal.

Paper organization and contribution. In the next section we introduce some definitions and recall some of the related work. Then, we provide algorithms for solving the view selection problem. We use integer linear programming (ILP) to formalize the exact solution and a polynomial time approximate algorithm. Due to the size of the problem, even the approximate algorithm may be inefficient. So we propose a technique to reduce the search space while we still guarantee an approximation factor of the solution memory size. Then, we show how indexes can be added to the solutions and we provide an exact resolution procedure with the help of ILP. Furthermore, we introduce the *stability* property of the solutions: given a solution $\mathcal{S}$ with the desired properties obtained at time t, how much the underlying data should change due to updates before $\mathcal{S}$ stops being competitive. The paper ends with some experiments showing that the approximation factor upper bounds are far from being reached in practice giving evidence that approximate algorithms are competitive. We also show that the solutions turn to be quite stable w.r.t. updates (assuming target queries are fixed).

2   Preliminaries

2.1   Notations

We assume the reader is aware of the definition of data cubes (please see [7] for a detailed description). We recall here some definitions and notations useful for the rest of the paper. We consider a table $T(D_1, \ldots, D_n, M_1, \ldots, M_p)$ where $D'_i s$ are the dimensions attributes and $M'_i s$ are the measures attributes. Table 1 shows an instance of such table and will constitute our running example throughout the paper.

Example 1. Table *Sales* contains 3 dimensions P, D and S and it contains one measure U. Each row represents the number of units sold per product, date and store.

$Dim(T)$ denotes the dimensions of T, e.g., $Dim(Sales) = \{P, D, S\}$. The data cube associated to T, denoted $\mathcal{C}(T)$, is the set of queries of the form SELECT Dimensions, measures FROM T GROUP BY Dimensions where Dimensions is a subset of $Dim(T)$, e.g., the view defined by SELECT P, D, SUM(U) FROM Sales GROUP BY P, D is a view of $\mathcal{C}(Sales)$. Clearly, if $|Dim(T)| = n$ then $|\mathcal{C}(T)| = 2^n$

Table 1. The fact table *Sales*

Sales	P(roduct)	D(ate)	S(tore)	U(nits sold)
	P1	D1	S1	10
	P1	D2	S1	15
	P2	D2	S2	10
	P1	D1	S2	11
	P1	D3	S1	10
	P1	D3	S2	15

e.g., $C(Sales)$ contains 8 views. The views of a data cube are called *cuboids*. In this paper we consider the queries of the form SELECT Dimensions, measures FROM T WHERE Conditions GROUP BY Dimensions where the conditions are conjunctions of equalities of the form $Dim_i = Constant$. Hence, cuboids are special a case of queries (just remove the WHERE clause). If q is a query then $Att(q)$ is the set of dimensions attributes appearing either in the SELECT or the WHERE clause, e.g., let q be the query SELECT P, SUM(U) FROM Sales WHERE D=D1 GROUP BY P, then $Att(q) = \{P, D\}$. The size of a cuboid c corresponds to its number of rows and is denoted $size(c)$. If S is a set of cuboids then $size(S) = \sum_{c_i \in S} size(c_i)$. If c is a cuboid then $Dim(c)$ denotes the dimensions of c. The data cube lattice is induced by a partial order $\preceq$ defined as follows: $c_1 \preceq c_2$ iff $Dim(c_1) \subseteq Dim(c_2)$. In this case, c_2 is an *ancestor* of c_1. Furthermore if $|Dim(c_2)| = |Dim(c_1)| + 1$ then c_2 is a *parent* of c_1. In this paper we consider just algebraic measures [7] such as if $c_1 \preceq c_2$ then c_1 can be computed from c_2 (SUM, COUNT, AVG are algebraic functions). Let q be a query and c be a cuboid. By notation abuse we extend the relation $\preceq$ to queries as follows: $q \preceq c$ iff $Att(q) \subseteq Att(c)$.

2.2   Cost Model and Performance Measures

Let $S \subseteq C$ be the set of materialized cuboids and let q be a query. Then, $S_q = \{c' \in S | q \preceq c'\}$. The cost of evaluating q wrt. S is defined as follows: if $S_q = \emptyset$ then $cost(q, S) = \infty$ otherwise $cost(q, S) = \min_{c' \in S_q} size(c')$, i.e q is computed from its smallest materialized ancestor. This is the usual cost model (e.g [11, 16, 17]). Since we do not consider for the moment indexes[1], the minimal cost of a query q is proportional to the size of the smallest cuboid c from which it can be computed i.e., $Att(c) = Att(q)$. In this paper the materialized views are only those corresponding to cuboids.

Definition 1 (Performance factor). *Let Q be a set of queries, S a set of materialized cuboids and $q \in Q$. The performance factor of S wrt. q is $pf(q, S) = \frac{cost(q, S)}{size(c_q)}$.*

Intuitively, the performance factor measures the ratio between the response time for evaluating a query q using S over its minimal evaluation time. Beside the

---

[1] These are discussed in Section 4.

performance factor, we take into account other parameters to assess our proposal. More precisely, we consider the minimization of the storage memory space needed by our solutions, the robustness of our solutions against updates i.e., should we compute a new solution each time materialized cuboids are refreshed due to source data updates and finally the time complexity of the view selection algorithm itself. Hence, we aim at getting a trade off between the *quality* of the solutions and the efficiency by which they are computed.

2.3   Related Work

To the best of our knowledge, no previous work has tackled the materialized view selection under the same perspective as us. Most of previous works consider a *budget constraint*, typically the storage memory space available and try to find the best views that fit in the available memory space while minimizing the *total* query evaluation cost of the queries belonging to the workload. In [10] we showed via some experiments that reducing the total query cost may leave some individual queries poorly optimized. More precisely, we showed the deviation from the minimal cost of queries can be quite large. [11] is a seminal paper on partial materialization of data cubes. It considers the following formalization: Given a set of queries Q and an available storage space M for materializing the data cube, find a subset S of cuboids such that (i) $size(S) \leq M$ and (2) $\sum_{q \in Q} cost(q, S)$ is minimal. The authors show that the problem is NP-Hard and turned it to a *gain maximization*. The gain of a cuboids set S is simply the difference between the worst cost and the cost provided by S. They proposed a greedy algorithm which guarantees at least 63% the gain of the optimal solution. As noted by [12], a guarantee on the cost gain does not imply a guarantee about the cost. Nevertheless, the approach has influenced many subsequent papers by extending it to different contexts for example index selection [8], materialized cuboids maintenance [9] or exact solution by using integer linear programming (ILP) techniques [14,17]. Besides the fact that some queries are poorly optimized, another problem with this formalization is that minimizing the total cost tends to come up with solutions that use the maximal available storage space (the more we use memory, the less is the total cost). Hence, as soon as new data is to be inserted and propagated to the materialized views, the former solution may become too large and a new one must be calculated. On another hand, using as much memory space as possible has a negative impact on materialized views maintenance since this last task has a cost proportional to the materialized views sizes. Dynamat [13] dealt with both data and workload dynamic. Its principle can be described as follows: each time a query is submitted, first find the best plan to evaluate it then decide whether its result could be kept among the already materialized views. In a sense, the system computes a new workload each time a new query is evaluated. In another side, when *batch* updates arrive, the system may have a time constraint in order to perform the propagation to the views. Hence, it choses the *most beneficial* that it can update within the allowed window time interval. This is constrained by the available memory and may trigger the removal of old views.

3   View Selection Problem

The problem we want to solve is formalized as follows: given (i) a real number $f \geq 1$ and (ii) a set of queries $\mathcal{Q}$, find a set of cuboids $\mathcal{S}$ such that (1) for each $q \in \mathcal{Q}$, $pf(q, \mathcal{S}) \leq f$ and (2) $size(\mathcal{S})$ is minimal. We denote this problem VSPF (View Selection under Performance Factor constraint). Intuitively, we want to find the smallest (in terms of storage space) set $\mathcal{S}$ that guarantees a performance factor for each target query $q \in \mathcal{Q}$.

3.1   View Selection as Minimal Weighted Vertex Cover

Theorem 1. *The VSPF problem is NP-Hard.*

Proof (Sketch). Otherwise, there exists a polynomial time algorithm solving the view selection problem under space constraint, a problem which has been shown NP-Hard [11]. Indeed, it suffices to use a binary search with a f ranging from 1 to size of the base cuboid. The algorithm stops when the smallest f providing a solution not exceeding the space budget is found.

Hence, an approximate solution is more viable. For this purpose, we show that our solution is actually the solution of a Minimal Weighted Vertex Cover (MWVC) instance. We first give some definitions. For each $q \in \mathcal{Q}$ we denote by $\mathcal{A}_f(q)$ the set of cuboids c such that (1) $q \preceq c$ and (2) $size(c) \leq f * size(c_q)$. We call this set the f-ancestors of q. $\mathcal{A}_f(\mathcal{Q}) = \bigcup_{q \in \mathcal{Q}} \mathcal{A}_f(q)$. Clearly, the solution of our problem belongs to $\mathcal{A}_f(\mathcal{Q})$.

Definition 2 (Search Graph). *Let $G(f, \mathcal{Q}) = (V, E, \mathbf{w})$ be the graph defined as follows: $V = \mathcal{A}_f(\mathcal{Q})$, $(v_1, v_2) \in E$ iff (i) $v_2 \in \mathcal{Q}$ and (ii) $v_1 \in \mathcal{A}_f(v_2)$. $\mathbf{w}$ is a weight function defined as $\mathbf{w}(v) = size(v)$. We denote by $V_{\mathcal{Q}}$ the nodes of G that correspond to an element of $\mathcal{Q}$.*

The set of out-neighbors of a vertex v in G is noted $\Gamma(v) = \{v' \in \mathcal{Q} | (v, v') \in G\}$ and the weight of a set of nodes $\mathcal{S} \subseteq V$, denoted $\mathbf{w}(\mathcal{S})$, is equal to $\sum_{v \in \mathcal{S}} \mathbf{w}(v)$. A set $\mathcal{S}$ **covers** $V_{\mathcal{Q}}$ iff $\bigcup_{v \in \mathcal{S}} \Gamma(v) \supseteq V_{\mathcal{Q}}$. So the solution to our problem consists in finding a subset $\mathcal{S} \subseteq V$ such that $\mathcal{S}$ covers $V_{\mathcal{Q}}$ and $\mathcal{S}$ is of minimal weight. This is an instance of the MWVC which is known to be NP-Hard.

3.2   Exact Solution

In this section, we propose an ILP program to solve our problem. Let us first give some notations: For each $c_i \in \mathcal{A}_f(\mathcal{Q})$, the constant s_i designates the size of c_i, for each $c_i \in \mathcal{A}_f(\mathcal{Q})$, the variable $x_i \in \{1, 0\}$ means respectively that c_i belongs to the solution or not, $y_{ij} \in \{1, 0\}$ means that the query $q_i \in \mathcal{Q}$ uses cuboid $c_j \in \mathcal{A}_f(q_i)$ or not. The linear program is:

$$\min \sum_{j:c_j \in \mathcal{A}_f(\mathcal{Q})} x_j * s_j \tag{1}$$

$$\forall i : q_i \in \mathcal{Q} \sum_{j : c_j \in \mathcal{A}_f(q_i)} y_{ij} = 1 \tag{2}$$

$$\forall i : q_i \in \mathcal{Q}, \forall j : c_j \in \mathcal{A}_f(q_i) \; y_{ij} \leq x_j \tag{3}$$

$$\forall j : c_j \in \mathcal{A}_f(\mathcal{Q}) \; x_j \in \{0,1\} \tag{4}$$

$$\forall i : c_i \in \mathcal{Q}, \; \forall j : c_j \in \mathcal{A}_f(q_i) \; y_{ij} \in \{0,1\} \tag{5}$$

The program above is denoted $ILP(G(f, \mathcal{Q}))$. The objective function is the minimization of the solution's size. Constraint (2) imposes that q_i uses exactly one materialized cuboid and constraint (3) means that the query q_i cannot use c_j whenever c_j is not materialized. Constraints (4) and (5) say that the variables are binary. The following is a straightforward result characterizing the exact solution of our problem.

Proposition 1. *Let $G = G(f, \mathcal{Q})$ be a search graph. Let Sol be a solution of $ILP(G)$, i.e. Sol is assignment function of x_i's and y_{ij}'s variables of $ILP(G)$. Let $\mathcal{S}^* = \{c_i \in \mathcal{A}_f(\mathcal{Q}) | Sol(x_i) = 1\}$. Then $\mathcal{S}^*$ is an optimal solution.*

For notation convenience, we consider $\mathcal{S}^* = ILP(G(f, \mathcal{Q}))$. It is an optimal solution. The solution to our problem is the set of $c_j \in \mathcal{A}_f(\mathcal{Q})$ such that $x_j = 1$. Current solvers cannot handle these linear programs when the number of variables is too large (this number grows exponentially w.r.t. f and $|\mathcal{Q}|$) rapidly reach thousands). This motivates a different approach that is more efficient in terms of execution time but returns an approximate solution.

3.3 Approximate Solution

In this section, we borrow the greedy algorithm of [4] to solve our problem. We first define the **load** of a vertex $v \in V$, noted $\ell(v)$, as $\frac{\mathbf{w}(v)}{|\Gamma(v)|}$.

Function PickFrom$f$Ancestors$(G(f, \mathcal{Q}))$

V = nodes of $G(f, \mathcal{Q})$
$S = \emptyset$
While $V_{\mathcal{Q}} \neq \emptyset$
    $c^* = \arg \min_{c \in V} \ell(c)$
    $S = S \cup \{c^*\}$
    $V = V \setminus \left(\Gamma(c^*) \cup \{c^*\} \right)$
End While
Return S
End.

The algorithm choses at each iteration the vertex with minimal load and adds it to the solution. The following theorem is a direct consequence of [4] result.

Theorem 2. *Let $f \geq 1$, $\mathcal{S} = PickFromf Ancestors(G(f, \mathcal{Q})$ and $\mathcal{S}^* = ILP(G)$. Then:*

- *For each $q \in \mathcal{Q}$, $pf(q, \mathcal{S}) \le f$;*
- *$size(\mathcal{S}) \le (1 + \ln \Delta) * size(\mathcal{S}^*)$ where Δ is the maximal out-degree of G;*

Even if the complexity of this algorithm is polynomial in the size of $G(f, \mathcal{Q})$, the size itself may be exponential depending on f and the cardinality of $\mathcal{Q}$. Thus, reducing the search space is important in both cases (ILP and PickFrom$f$Ancestors). The first obvious simplification consists in removing all candidates c_j such that $size(c_j) > size(\Gamma(c_j))$. This simplification, which has also been suggested in [17], does not change the solutions of both techniques. Still, there may be too many remaining candidates. In the next section, we propose an additional reduction of the search space.

3.4   Reducing the Search Graph

Intuitively, the simplification we consider here consists in keeping for each query q among its f_ancestors, only those that are maximal. More precisely, $\mathcal{B}_f(q) \subseteq \mathcal{A}_f(q)$ denotes the maximal elements of $\mathcal{A}_f(q)$, i.e if $c' \in \mathcal{B}_f(q)$ then for each $c \in \mathcal{A}_f(q)$, either $size(c) > size(c')$ or $c' \not\preceq c$. $\mathcal{B}_f(q)$ is the f_border of q and $\mathcal{B}_f(\mathcal{Q})$ is the union of the f_borders. The intuition behind this heuristic is that keeping maximal f_ancestors tends to keep the ancestors that cover the maximal number of queries (this of course is not always true). The partial search graph is now defined as follows.

Definition 3 (Partial Search Graph). *Let $G_p(f, \mathcal{Q})$ be the graph $(V_p, E_p, \mathbf{w})$ where $V_p = \mathcal{B}_f(\mathcal{Q}) \cup \mathcal{Q}, (v_1, v_2) \in E_p$ iff $v_2 \in \mathcal{Q}$ and $v_1 \in \mathcal{B}_f(v_2)$ and $\mathbf{w} : V_p \to \mathbb{N}$ defined as $\mathbf{w}(c) = size(c)$.*

Example 2. Figure 1 shows the search graph $G(f, \mathcal{Q})$ where $f = 10$ and $\mathcal{Q} = \{\texttt{select} * \texttt{from B}, \texttt{select} * \texttt{from C}, \texttt{select} * \texttt{from D}\}$. The dimensions of the underlying datacube are A, B, C and D. All the nodes do not belong to $G(f, \mathcal{Q})$. They are present just for a sake of clarity. Dashed arrows are not present in the partial graph.

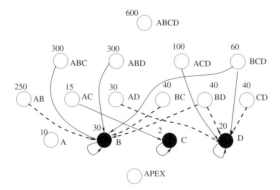

Fig. 1. Global and partial search graphs

If we consider G_p as the search space then both solutions of $ILP(G_p)$ and of PickFrom$f$Ancestors(G_p) guarantee (1) a performance factor less than f and (2) an approximation factor of the solution's size. Indeed,

Theorem 3. *Let $\mathcal{S}^* = ILP(G)$ be the optimal solution. Let $\mathcal{S}_1 = ILP(G_p)$ and $\mathcal{S}_2 = PickFromfAncestors(G_p)$. Then (1) $size(\mathcal{S}_1) \leq f * size(\mathcal{S}^*)$ and (2) $size(\mathcal{S}_2) \leq f * (1 + \ln \Delta) * size(\mathcal{S}^*)$ where Δ is the maximal out degree of G. Recall that $|\Delta| \leq |\mathcal{Q}|$.*

4   Introducing Indexes

Whenever indexes are allowed, the search space includes not only cuboids but also indexed ones. Virtually, to each cuboid with d dimensions we can associate $\sum_{i=1}^{d} \frac{d!}{(d-i)!}$ i.e., each permutation of each subset of its attributes may define an index. Moreover, the cost model should be changed. Indeed, now the minimal cost for a query q does not correspond to the size of its associated cuboid c_q. Therefore, we should modify the cost factor accordingly. We borrow the cost model of [8]. The queries are of the form $\gamma_{Dim_1} \sigma_{Dim_2}$ where Dim_1 are the attributes in the SELECT clause and Dim_2 are those appearing in the WHERE clause. Conditions are conjunctions of equalities $Att_i = value$. Let $q = \gamma_{Dim_1} \sigma_{Dim_2}$ and I be an index over c on the sequence of attributes $\overrightarrow{Dim}$. Let E be the largest subset of Dim_2 such that $\overrightarrow{Dim}$ form a prefix (not necessarily proper) of Dim_2. Assume that $(Dim_1 \cup Dim_2) \subseteq Att(c)$ that's q can be answered from c. Then the cost of answering q from c using the index I is $Cost(q, c, I) = \frac{size(c)}{size(E)}$. The minimal cost for a query is obtained when q is evaluated from a cuboid c such that $Dim_1 \cup Dim_2 = Att(c)$ and there exists an index I over c on Dim_2. Therefore, $MinCost(q) = \frac{size(Dim_1 \cup Dim_2)}{size(Dim_2)}$. Note that if $Dim_2 = \emptyset$ then $size(Dim_2) = 1$ since it corresponds to the Apex cuboid. Now we are ready to define the cost factor of evaluating q from cuboid c using index I. $pf(q, c, I) = \frac{Cost(q,c,I)}{MinCost(q)}$. Finally, we assume the memory size of each index is equal to the cuboid upon which it is defined. Now that we have formalized the performance factor of evaluating a query from a materialized view using an index, we can extend the search space accordingly: The pair (c, I) is an f_ancestor of q iff $pf(q, c, I) \leq f$. The results presented so far can now easily be generalized in this new setting.

4.1   Exact Solution

In this section, we propose an ILP program to solve exactly the problem of selecting views and indexes to materialize. We use the following notations: $y_{ij} \in \{0, 1\}$ represents the fact that index j on cuboid i is selected to be materialized or not. A special case is when $j = 0$ which represents actually just the cuboid i. The variables s_{ij} represent the size of index j of table i. Again, s_{i0} represents the size of cuboid i. Finally, $u_{kij} \in \{0, 1\}$ means that target query k uses index j of cuboid i. The ILP program is as follows:

$$\min \sum_{i \in V} \sum_{j \in Index(i)} y_{ij} * s_{ij} \tag{6}$$

$$\forall i, j : \; y_{ij} \le y_{i0} \tag{7}$$

$$\forall k \in Q, \; \sum_{i \in V(k)} \sum_{j \in Index(i)} u_{kij} = 1 \tag{8}$$

$$y_{ij} \in \{0, 1\} \tag{9}$$

$$u_{kij} \in \{0, 1\} \tag{10}$$

The objective function (6) tends to minimize the total storage space. Constraint (7) means that an index may be selected only if its underlying cuboid is selected too. Constraint (8) guarantees that each target query may be computed efficiently i.e., with a performance factor less than f.

Proposition 2. *Let P be the above ILP program. Let $\overline{S} = \{indexes\ j\ and\ views\ i\ such\ that\ y_{ij} = 1\ in\ the\ solution\ of\ P$. Then $\overline{S}$ occupies the minimal storage memory space such that for each $q \in Q$, $pf(S, q) \le f$.*

We leave approximation for solving this problem to future work.

5   Dynamic Maintenance

In our solutions, we assume that the running time of the view selection algorithms is not a problem. However, whenever some updates on the fact table or the dimensions are performed, not only we should propagate them but may be we have to compute a new set of views in order to guarantee a performance factor below f for target request. We first analyze the *stability* of our solution. Intuitively, this property tells that the query performance factor of a solution computed at time t remains almost unchanged at time $t + 1$ after some updates. Thus, the set of materialized views S can remain unchanged (one only have to maintain it). At the end of this section, we show how to handle this property in order to refresh and to maintain dynamically the views selection with a light cost of computation and materialization.

5.1   Stability

Let us consider a query q belonging to Q and its smallest ancestor $c \in S$ in term of size. We know that $pf(q, \{c\}) \le f$. We aim at computing the number of tuples to insert into (or to delete from) c so that $pf(q, \{c\})$ becomes greater than $f + 1$. We first start with easy lemmas in order to prove in Theorem 4 some sufficient conditions ensuring the stability of our solutions.

Definition 4 (Stability). *Let T and T' be two successive instances of a fact table. T' is obtained from T by performing some insertions and/or deletions. Let Q and S such that $\forall q \in Q$, $pf(q, S) \le f$. S is stable between T and T' if and only if $\forall q \in Q$, $pf(q, S) \le f + 1$ wrt T'.*

The insertion of n tuples in c implies the insertion of m tuples in c_q with $0 \leq m \leq n$. The worst case, from query performance perspective, is when $m = 0$. Indeed, in this case, the size of c increases and that of c_q remains unchanged. The following lemma gives the minimal number of tuples to insert into c to break the stability of c.

Lemma 1. *Let c be a cuboid and q be a query such that $pf(q, \{s\}) \leq f$. The insertion of at least $size(c_q)$ tuples into c is required in order to get $pf(q, \{c\}) \geq f + 1$.*

Using the same argument as before, we can easily see that the deletion of n tuples from c may trigger the deletion of m tuples from c_q with $0 \leq m \leq n$. From the performance factor point of view, the worst situation corresponds to the case $m = n$.

Lemma 2. *Let c be a cuboid and q be a query such that $pf(q, \{c\}) \leq f$. The deletion of at least $\frac{size(c_q)}{f}$ tuples from c is required in order to get $pf(q, \{c\}) \geq f + 1$.*

In our experiments, it turns out that, for the majority of the target queries, the solution is stable after a very large number of updates. This phenomenon can also be explained theoretically by the following result. Let us first give some definitions. Let $Dom(c)$ denotes the domain of a cuboid c and $m(c) = |Dom(c)|$ denotes its cardinality. Clearly, $size(c) \leq m(c)$. c is *saturated* iff $size(c) = m(c)$. c is a *small cuboid* wrt a parameter f iff $size(c) < \frac{|T|}{2f \ln 4f}$. Under some assumption of data distribution and given T:

- There exists a threshold β_1 such that for any *small* cuboid $q \in \mathcal{Q}$ of size larger than β_1, any insertion of tuples does not break the stability property of the solution $\mathcal{S}$;
- There exists a threshold β_2 such that if $|T| \geq \beta_2$, then for any $q \in \mathcal{Q}$ if c_q is a *small* cuboid then the solution $\mathcal{S}$ is stable whatever the number of tuples we insert.

More precisely, without any attempt of optimization of the constant factor β, we have:

Theorem 4. *Let $\{Dom_i\}_{i \in [1,D]}$ be a multi-set. Set $\beta = 64f^2$. Let T be a fact table in which tuples are chosen uniformly at random within the Cartesian product $\Pi_{i=1}^{D} om_i$. Let $\mathcal{Q}$ be a set of queries and $\mathcal{S}$ a set of cuboids such that $\forall q \in \mathcal{Q}$, $pf(q, \mathcal{S}) \leq f$ and c_q is a small cuboid. After any sequence of insertions into T and with probability $1 - 2/|\mathcal{Q}|$:*

- *If $\forall q \in \mathcal{Q}$, $size(c_q) > \beta \ln |\mathcal{Q}|$, then $\mathcal{S}$ is stable.*
- *If $|T| > \beta f \ln |\mathcal{Q}|(\ln \beta + \ln \ln |\mathcal{Q}|)$, then $\mathcal{S}$ is stable.*
- *If we have less than $\frac{|T|}{2f \ln 4f}$ insertions of tuples in T, then $\mathcal{S}$ is stable.*

6   Experiments

We used the *US Census 1990 data*. Here after, the time parameter represents the cumulated time for constructing the search graph and the resolution time for obtaining a solution S. The different performance measures of a solution S depend heavily on the target queries Q. We studied three random generation methods of Q. Each of which corresponds to special properties: (i) **Uniform generation (UNIF)**: All possible queries $q \in C$ have the same probability to belong to Q; (ii) **Queries generated between level 1 and level d_{max} (DMAX)**: We fix the maximal number of dimensions then we iterate over the levels 1 and $dmax$ each time we pick uniformly a query from level i and (iii) **DESC**: We use the same principle as DMAX but here, each time we pick a query q from level i, we add to Q the 2^i queries $\{q_j\}_{j \in [1, 2^i]}$ where q is an ancestor of q_j. In order to check the effectiveness of the approximate algorithms, we compared their results to the exact solutions in terms of computation time and storage space. We generated Q using UNIF and compared the space memory required for storing Q with the memory needed by the exact and that of PickFrom$f$Ancestors($G(f, Q)$) with $f = 10$. Figure 2 shows that PickFrom$f$Ancestors (solution S) behaves very well w.r.t the optimal solution (S^*) in terms of memory gain while being much faster: it took 2 seconds to find S and more than an hour for S^* using the CPlex software. We should mention that this experiment was performed with only 10 dimensions. With more than 10, the computation time of S^* prohibitive. The base cuboid has 500K rows and 20 dimensions. For each query generation method, we varied $|Q|$ and f. For each combination of the three preceding parameters, we computed a solution with both G and G_p.

The execution times (expressed in seconds) are illustrated in Figure 3(a) to 3(c). The partial search graph offers an interesting compromise in terms of computation time and the memory gain. Indeed, the execution time is about 4 times less than that of G while keeping the memory gains comparable. We encountered however one exception (cf. Figures 3(f) and 3(c)). In that case, the partial search graph does not summarize well the set Q because it finds a solution with too

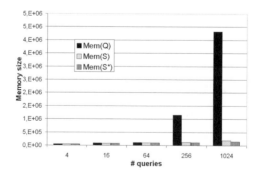

Fig. 2. Storage space of the query targets (Q), the approximate (S) and the exact ($S*$) solutions

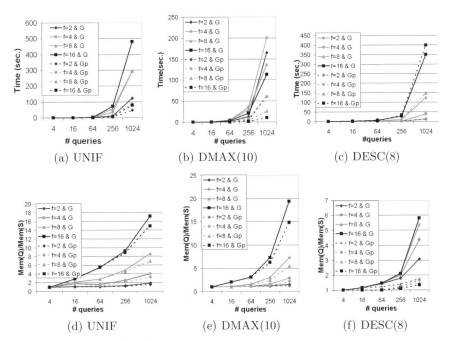

Fig. 3. Execution time and memory gain

much cuboids. Since the computation time is in $O(\Delta|V(G)| \cdot |\mathcal{S}|)$, it depends on the number of returned cuboids. This explains why, in this case, the computation time of G_p is larger than that of G. It also gives a hint about the importance of the way the workload is built. The results depicted in Figures 3(d) to 3(f) concern the same experiments as previously. First, it is clear that with G, the storage space is always more reduced. We also note that in most cases (the first three), the reduction ratios obtained with G or G_p are comparable. The last experience exhibits a different behavior (Figure 3(f)) since we have a drastic difference. Recall that this case is also the one where the execution time with G is better than that with G_p (see 3(c)).

6.1   Stability Analysis

In order to analyze the stability of our solutions, we conducted some experiments whose principle can be described as follows: We fix the factor f, the number of target queries and the generation method. We execute the approximation algorithm PickFrom$f$Ancestors$(G(f, \mathcal{Q}))$ on different data sets $file_1, \ldots, file_n$ such that $file_1 \subset \ldots \subset file_n$. For each $1 < i \leq n$, we obtain a solution $\mathcal{S}_i$ and for each $q \in \mathcal{Q}$, we compute $pf(q, \mathcal{S}_{i-1})$. This allows us to verify in what extent the performance of our solutions worsen from $file_i$ to $file_{i+1}$. The retained criteria of comparison are (1) the number of target queries of $\mathcal{Q}$ whose performance factor become beyond f (2) and for these queries, we measure the difference between the new pf and the fixed f. This represents the amount of overtaking.

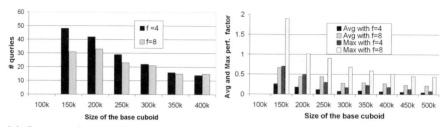

(a) Queries whose performance factor is beyond the threshold

(b) Deviations from the fixed performance factor threshold

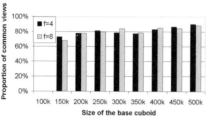

(c) Percentage of common views

Fig. 4. Stability analysis

We present the obtained results when $|\mathcal{Q}| = 512$. The queries are generated using UNIF and the performance threshold takes two values $f = 4$ and $f = 8$. Figure 4(a) shows the number of queries whose performance factor becomes larger than the threshold f. We note that this number is decreasing while the size of the base cuboid increases. Figure 4(b) illustrates the deviations from the fixed threshold, it shows the maximal and the average deviations, i.e., average and maximal values of $pf(q, \mathcal{S})$ for those target queries q whose pf is greater than the fixed f. Again, we note that these parameters (max and average) decrease while the size of $file_i$ increases. Figure 4(c) shows the proportion of views that are kept in the next solution. We measured $|\mathcal{S}_i \cap \mathcal{S}_{i+1}|/|\mathcal{S}_{i+1}|$. Around 80% of the views selected in the previous solution belong to the next one. This shows that even when we have to recompute a new solution, only few views will be calculated from scratch; the majority will need at worst to be refreshed.

7   Conclusion and Future Work

We presented a new formalization of the materialized view selection problem in the context of data cubes. Some extensions of the present work are straightforward, e.g. integrating dimensions hierarchies is made easy because hierarchies are themselves lattices. Furthermore, it is not required to have a unique f for all queries. It suffices to consider $\mathcal{A}_{f_i}(q_i)$ so that, without changing the algorithms, the obtained solutions guarantee $pf(q_i, \mathcal{S}) \leq f_i$ where f_i reflects the importance of the query (lower is f_i more important is q_i). As future research, we plan to analyze the stability property more in depth depending on data distributions.

A prior knowledge of data distribution and/or dimension dependencies could be helpful in this case [5]. [1, 2] provided a solution to the selection of binary join indexes to optimize star join queries under storage space constraint. We believe our solution can easily be adapted in order to select instead of cuboids, the join indexes to materialize.

References

1. Aouiche, K., Darmont, J., Boussaid, O., Bentayeb, F.: Automatic selection of bitmap join indexes in data warehouses. In: Tjoa, A.M., Trujillo, J. (eds.) DaWaK 2005. LNCS, vol. 3589, pp. 64–73. Springer, Heidelberg (2005)
2. Bellatreche, L., Boukhalfa, K.: Yet another algorithms for selecting bitmap join indexes. In: Bach Pedersen, T., Mohania, M.K., Tjoa, A.M. (eds.) DAWAK 2010. LNCS, vol. 6263, pp. 105–116. Springer, Heidelberg (2010)
3. Blakeley, J.A., Larson, P., Tompa, F.: Efficiently updating materialized views. In: Proc. SIGMOD (1986)
4. Chvàtal, V.: A greedy heuristic for the set covering problem. Mathematics of operation research 4(3) (1979)
5. Ciaccia, P., Golfarelli, M., Rizzi, S.: On estimating the cardinality of aggregate views. In: DMDW Workshop (2001)
6. Florescu, D., Kossmann, D.: Rethinking cost and performance of database systems. SIGMOD Record 38(1) (2009)
7. Gray, J., Bosworth, A., Layman, A., Pirahesh, H.: Data cube: A relational aggregation operator generalizing group-by, cross-tab, and sub-total. In: Proceedings of ICDE Conference (1996)
8. Gupta, H., Harinarayan, V., Anand, A., Ullman, J.: Index selection for olap. In: Proc. of ICDE (1997)
9. Gupta, H., Mumick, I.S.: Selection of views to materialize in a data warehouse. IEEE TKDE 17(1), 24–43 (2005)
10. Hanusse, N., Maabout, S., Tofan, R.: A view selection algorithm with performance guarantee. In: Proc. of EDBT (2009)
11. Harinarayan, V., Rajaraman, A., Ullman, J.D.: Implementing data cubes efficiently. In: SIGMOD (1996)
12. Karloff, H., Mihail, M.: On the complexity of the view-selection problem. In: Proc. of PODS (1999)
13. Kotidis, Y., Roussopoulos, N.: A case of dynamic view management. ACM TODS 26(4), 388–423 (2001)
14. Li, J., Talebi, Z.A., Chirkova, R., Fathi, Y.: A formal model for the problem of view selection for aggregate queries. In: Eder, J., Haav, H.-M., Kalja, A., Penjam, J. (eds.) ADBIS 2005. LNCS, vol. 3631, pp. 125–138. Springer, Heidelberg (2005)
15. Morfonios, K., Konakas, S., Ioannidis, Y., Kotsis, N.: Rolap implementations of the data cube. ACM Computing Surveys (2007)
16. Shukla, A., Deshpande, P., Naughton, J.F.: Materialized view selection for multidimensional datasets. In: Proc. of VLDB (1998)
17. Talebi, Z.A., Chirkova, R., Fathi, Y., Stallmann, M.: Exact and inexact methods for selecting views and indexes for olap performance improvement. In: Proc. of EDBT (2008)

Mining Preferences from OLAP Query Logs for Proactive Personalization

Julien Aligon[1], Matteo Golfarelli[2],
Patrick Marcel[1], Stefano Rizzi[2], and Elisa Turricchia[2]

[1] Laboratoire d'Informatique – Université François Rabelais Tours, France
{julien.aligon,patrick.marcel}@univ-tours.fr
[2] DEIS – University of Bologna, Italy
{matteo.golfarelli,stefano.rizzi,elisa.turricchia2}@unibo.it

Abstract. The goal of personalization is to deliver information that is relevant to an individual or a group of individuals in the most appropriate format and layout. In the OLAP context personalization is quite beneficial, because queries can be very complex and they may return huge amounts of data. Aimed at making the user's experience with OLAP as plain as possible, in this paper we propose a proactive approach that couples an MDX-based language for expressing OLAP preferences to a mining technique for automatically deriving preferences. First, the log of past MDX queries issued by that user is mined to extract a set of association rules that relate sets of frequent query fragments; then, given a specific query, a subset of pertinent and effective rules is selected; finally, the selected rules are translated into a preference that is used to annotate the user's query. A set of experimental results proves the effectiveness and efficiency of our approach.

1 Introduction and Motivation

Personalization has attracted a lot of attention in the database community during the last few years, and also raised plenty of interest in the OLAP area. The goal of personalization is to deliver information that is relevant to an individual or a group of individuals in the most appropriate format and layout, and in the OLAP area it has been pursued using different approaches:

- *Query recommendation*: Based on the current query and on the past sessions, the system suggests further queries to help users navigating the cube [1].
- *Personalized visualization*: Users specify a set of constraints that are used to determine a preferred visualization [2].
- *Result ranking*: Query results are organized in a total or partial order so that the user visualizes the most relevant data first [3].
- *Query contextualization*: The query is enhanced by adding preference predicates that depend on the query context [4].

These approaches differ from different points of view, in particular:

J. Eder, M. Bielikova, and A.M. Tjoa (Eds.): ADBIS 2011, LNCS 6909, pp. 84–97, 2011.
© Springer-Verlag Berlin Heidelberg 2011

- Formulation effort: personalization criteria for queries may be either manually specified by users, or transparently inferred from the context and from the user profile.
- Prescriptiveness: personalization criteria may either be used as "hard" constraints that are added to queries, or be meant as "soft" constraints, i.e., preferences.
- Proactiveness: some approaches propose new queries to the user based on the query log and on the context, while others change the current query or post-process its results before returning them to the user.

With reference to the above, the user's experience with OLAP can be made as plain as possible by decreasing the formulation effort (i.e., having query personalization criteria inferred), providing low prescriptiveness (i.e., annotating queries with preferences rather than constraints), and enhancing proactiveness (i.e., transparently changing the current query). The result ranking approach we propose in this paper goes in this direction by coupling an MDX-based language for expressing OLAP preferences to a mining technique for automatically deriving a set of preferences for a user's query from the log of past MDX queries issued by that user. This is done in four steps:

1. The user's query log is mined off-line to extract a set of association rules that relate sets of frequent query fragments (such as group-by attributes, returned measures, selection predicates).
2. When the user formulates a query q, among the rules whose antecedent matches with q, a subset of rules is selected whose cardinality depends on a parameter set by the user to express the desired personalization degree, i.e., the complexity of the preference that will be formulated.
3. The selected rules are translated into an OLAP preference p concerning the group-by set for aggregating data, the measures to be returned, and the values of levels or measures.
4. Query q is annotated with p and executed. The results returned are ranked according to p, so that the user can more effectively explore them by focusing on the most relevant data first.

Remarkably, like in the other result ranking approaches, the overall set of tuples returned by q annotated with p is the same set of tuples that would be returned by q without annotation, because p expresses a soft constraint. This guarantees that the user's intentions are preserved, and makes our approach non-invasive.

The paper outline is as follows. After summarizing the related work in Section 2, we introduce a formal setting to manipulate multidimensional data in Section 3. In Section 4 we describe the main features of the MYMDX language we adopt to express OLAP preferences, while Section 5 describes in detail our approach. Section 6 shows an implementation and reports the results of some experimental tests we performed to test our approach for effectiveness and efficiency.

2   Related Work

Several approaches to personalization were devised in the OLAP context.

In the field of *profile-based personalization*, we mention [2], that presents a framework for providing personalized visualization of OLAP results based on user profiles in form of constraints, and [4], that achieves OLAP personalization by dynamically enhancing queries with context-aware user preferences. Both approaches are proactive and demand low formulation effort, but in both cases the user profile is given, nothing being said on its construction. A recommendation framework for OLAP systems is presented in [5]; new queries are suggested to users based on the current analysis context and on the user's profile. Though the authors mention that the profile could be mined from the user's previous behavior, no specific suggestion is given to this end. A non-prescriptive approach is presented in [3,6], where the MYOLAP algebra for formulating and evaluating OLAP preferences is introduced; the proposed algebra is very expressive, but at the cost of a substantial formulation effort.

The term *history-based personalization* is borrowed from [7], and refers to approaches that suggest a new database query based on the past actions recorded in a log file. The following approaches fall into this category and do not rely on a user profile; they are proactive and demand no formulation effort —like our approach—, but they are prescriptive. The approaches in [1,8] are aimed at suggesting OLAP queries based on a comparison between the current session and former sessions stored in a query log. Also [9] has a similar goal in the context of SPJ queries; here, recommendations are computed based on the presence of tuples in sessions. This approach is further improved in [10] by relying on query fragments instead of tuples. A query log is exploited in [11] to support users in writing new SQL queries; the log is transformed into a graph of query fragments, where edges are labelled with the conditional probability of having one fragment given another fragment. Noticeably, all these work generally assume that history is taken from a query log shared by all users.

To the best of our knowledge, our work is the first that proposes to extract preferences from database query logs. However, the same idea has been used in other contexts. In the context of information retrieval, [12] presents algorithms to extract association rules at query time from a set of documents. These rules are used to associate the documents retrieved by a query to a relevance class and eventually to rank them. In the context of the web, [13] introduces algorithms for preference extraction from web logs, with a targeted preference language. Extraction is based on the frequency of the terms appearing in the log, and clustering is used for identifying preference constructs. A comprehensive overview of the techniques using data mining for personalization can be found in [14].

3   Preliminaries

3.1   Schemata and Instances

Our datacube formalization involves hierarchies; however, to keep the formalism simpler, and without actually restricting the validity of our approach, we will consider hierarchies without branches, i.e., consisting of chains of levels.

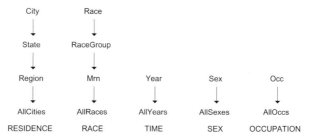

Fig. 1. Roll-up orders for the five hierarchies in the **CENSUS** schema (Mrn stands for MajorRacesNumber)

Definition 1 (Multidimensional Schema). *A multidimensional schema (or, briefly, a* schema) *is a triple* $\mathcal{M} = \langle A, H, M \rangle$ *where:*

- *A is a finite set of* levels, *each defined on a categorical domain $Dom(a)$;*
- *$H = \{h_1, \ldots, h_n\}$ is a finite set of* hierarchies, *each characterized by (1) a subset $Lev(h_i) \subseteq A$ of levels (such that the $Lev(h_i)$'s for $i = 1, \ldots, n$ define a partition of A); (2) a roll-up total order $\succeq_{h_i}$ of $Lev(h_i)$;*
- *a finite set of* measures *M, each defined on a numerical domain $Dom(m)$.*

For each hierarchy h_i, the top level of the order determines the finest aggregation level for the hierarchy. Conversely, the bottom level has a single possible value and determines the coarsest aggregation level.

A group-by set includes one level for each hierarchy, and defines a possible way to aggregate data. A coordinate of a group-by set is a point in the n-dimensional space defined by the levels in that group-by set.

Definition 2 (Group-by Set). *Given schema $\mathcal{M} = \langle A, H, M \rangle$, let $Dom(H) = Lev(h_1) \times \ldots \times Lev(h_n)$; each $G \in Dom(H)$ is called a* group-by set *of $\mathcal{M}$. Let $G = \langle a_{k_1}, \ldots, a_{k_n} \rangle$ and $Dom(G) = Dom(a_{k_1}) \times \ldots \times Dom(a_{k_n})$; each $g \in Dom(G)$ is called a* coordinate *of G.*

Example 1. The **CENSUS** schema includes the five hierarchies whose roll-up orders are shown in Figure 1, and measures AvgIncome, AvgCostGas, and AvgCost-Elect. It is City $\succeq_{\text{RESIDENCE}}$ State; examples of group-by sets are:

$$G_0 = \langle \text{City, Race, Year, Occ, Sex} \rangle$$
$$G_1 = \langle \text{Region, Mrn, Year, Occ, Sex} \rangle$$
$$G_2 = \langle \text{AllCities, AllRaces, AllYears, AllOccs, AllSexes} \rangle$$

A schema is populated with facts, each recording a useful information for the decision-making process. A fact is characterized by a group-by set G that defines its aggregation level, by a coordinate of G, and by a value for one measure.

Definition 3 (Fact). *Given schema* $\mathcal{M} = \langle A, H, M \rangle$, *a group-by set* $G \in Dom(H)$, *and a measure* $m \in M$, *a fact is a couple* $f_{G,m} = \langle g, v \rangle$, *where* $g \in Dom(G)$ *and* $v \in Dom(m)$. *The space of all facts for* $\mathcal{M}$ *is*

$$\mathcal{F}_{\mathcal{M}} = \bigcup_{G \in Dom(H), m \in M} (Dom(G) \times Dom(m))$$

Example 2. An example of fact is $f_{G_1, \mathsf{AvgIncome}} = \langle \langle \text{'Pacific'}, \text{'White'}, \text{'2008'}, \text{'Dentist'}, \text{'Male'} \rangle, 600 \rangle$.

Finally, an instance of a schema (*datacube*) is a set of facts $D \subseteq \mathcal{F}_{\mathcal{M}}$ such that no two facts characterized by the same coordinate and measure exist in D.

3.2 Queries

The MDX (*MultiDimensional eXpressions*) language is a de-facto standard for querying multidimensional databases [15]. Some of its distinguishing features are the possibility of returning query results that contain data with different aggregation levels and the possibility of specifying how the results should be visually arranged into a multidimensional representation. In this paper we consider MDX queries that aggregate data at one or more group-by sets, optionally select them using a predicate in CNF, and return one or more measures. The semantics of such an MDX query is that of a union of GPSJ queries[1] whose group-by sets are the cross product of n sets of levels, one for each hierarchy. This semantics corresponds to the following subset of MDX:

- Clauses **SELECT**, **FROM**, **WHERE** are supported.
- All functions for navigating hierarchies are supported: **AllMembers**, **Ancestor**, **Ascendants**, **Children**, etc.
- All functions for manipulating sets of members or tuples are supported (**Crossjoin**, **Except**, **Exists**, **Extract**, **Filter**, **Intersect**, etc.) except the union.
- All functions for manipulating members/tuples are supported.

To effectively use association rules for modeling frequent portions of queries, we formally split MDX queries into fragments as explained below.

Definition 4 (Query Fragment, Query, Log). *Given schema* $\mathcal{M} = \langle A, H, M \rangle$, *a query fragment is either a level in* A, *a measure in* M, *or a simple Boolean predicate involving a level and/or a measure. A qf-set is a set of query fragments. A multidimensional query (briefly, query) is represented by a qf-set that includes at least one level for each hierarchy in* H *and at least one measure in* M. *A log is a set of multidimensional queries.*

---

[1] A GPSJ query takes form $\pi_{a_{k_1}, \dots, a_{k_n}, Aggr} \sigma_p(\chi)$ where, in our context: χ is the star join between the fact table and the n dimension tables; p is a selection formula in CNF; $\{a_{k_1}, \dots, a_{k_n}\}$ is a group-by set; and $Aggr$ is a list of aggregations of the form $\alpha_j(m_j)$, where m_j is a measure and α_j is an aggregation operator.

Representing an MDX query as a qf-set q means:

1. Including a fragment m in q for each measure m returned by the MDX query.
2. Including a fragment a in q for each level a used in the MDX query to aggregate data.
3. Including a fragment $(a \in V)$ in q for each simple predicate on a level/measure a used in the MDX query to filter data.

Example 3. The MDX query on the CENSUS schema

SELECT AvgIncome **ON COLUMNS,**
          Crossjoin(OCCUPATION.members,
                Crossjoin(**Descendants**(RACE.AllRaces,RACE.Mrn),
                Descendants(RESIDENCE.AllCities,RESIDENCE.Region))) **ON ROWS**
FROM CENSUS **WHERE** TIME.Year.[2009]

is the union of four GPSJ queries:

$$\pi_{\text{AllCities,AllRaces,Occ,Year,AllSexes,}AVG(\text{AvgIncome})} \sigma_{\text{Year}=2009}(\chi_{\text{CENSUS}})$$

$$\pi_{\text{AllCities,Mrn,Occ,Year,AllSexes,}AVG(\text{AvgIncome})} \sigma_{\text{Year}=2009}(\chi_{\text{CENSUS}})$$

$$\pi_{\text{Region,AllRaces,Occ,Year,AllSexes,}AVG(\text{AvgIncome})} \sigma_{\text{Year}=2009}(\chi_{\text{CENSUS}})$$

$$\pi_{\text{Region,Mrn,Occ,Year,AllSexes,}AVG(\text{AvgIncome})} \sigma_{\text{Year}=2009}(\chi_{\text{CENSUS}})$$

and is represented by the qf-set $q = \{$Region, AllCities, Mrn, AllRaces, Occ, Year, AllSexes, AvgIncome, (Year $\in$ 2009)$\}$.

4 The MYMDX Preference Language

The language we adopt in this paper to express OLAP preferences is MYMDX [6], an extension of the MDX language based on the MYOLAP algebra [3]. In this section we summarize its features of interest for this work.

A (qualitative) preference on a datacube is a *strict partial order* (i.e., an irreflexive and transitive binary relation) on the space $\mathcal{F}_{\mathcal{M}}$ of all facts. In the MYOLAP algebra, preferences are inductively engineered by writing a *preference expression* that can be either a *base constructor* or a *composition operator* applied to two preference expressions. The constructors used in this paper are[2]:

- POS(a, V), where $V \subset Dom(a)$, that operates on level values; facts for which a takes a value in V are preferred to the others.
- BETWEEN(m, v_{low}, v_{high}), where m is a measure and $v_{low}, v_{high} \in Dom(m)$, that operates on measure values. Facts whose value of m is between v_{low} and v_{high} are preferred; the other facts are ranked according to their distance from the $[v_{low}, v_{high}]$ interval.

---

[2] The constructors we adopt are actually a generalization of those presented in [3] from two points of view. Firstly, the CONTAIN constructor is extended to work also on a fake hierarchy including all measures. Secondly, all constructors except BETWEEN are extended to operate on sets of values rather than on single values.

- CONTAIN(h, L), where h is a hierarchy and $L \subset Lev(h)$, that operates on levels. Facts whose group-by set includes a level in L are preferred to the others.
- CONTAIN(measures, $Meas$), where $Meas \subset M$, that operates on measures. Facts whose measure is in $Meas$ are preferred to the others.

Preference composition relies on the Pareto operator ($\otimes$), that gives the same importance to both the composed preferences. Remarkably, the Pareto operator is closed on the set of preferences.

The MYMDX language allows an MDX query to be annotated with a preference expression through a PREFERRING clause.

Example 4. The MDX query in Example 3 can be annotated with preference expression BETWEEN(AvgIncome,500,1000) $\otimes$ POS(Occ,'Engineer') $\otimes$ CONTAIN (RESIDENCE, Region) to state that facts aggregated by region and related to engineers with average income between 500 and 1000 kiloeuros are equally preferred. The corresponding MYMDX query is:

SELECT AvgIncome **ON COLUMNS,**
          Crossjoin(OCCUPATION.members,
                    Crossjoin(**Descendants**(RACE.AllRaces,RACE.Mrn),
                    Descendants(RESIDENCE.AllCities,RESIDENCE.Region))) **ON ROWS**
FROM CENSUS **WHERE** TIME.Year.[2009]
PREFERRING AvgIncome **BETWEEN** 500 **AND** 1000
AND Occ **POS** 'Engineer' **AND** RESIDENCE **CONTAIN** Region

5   A Proactive Approach to OLAP

As sketched in the Introduction, our approach relies on four steps:

1. *Log mining.* For efficiency reasons this step is executed off-line, before the current query session starts. It consists in running a data mining algorithm on the user's query log to extract the set R of association rules whose support and confidence are above a given threshold.
2. *Rule selection.* When that user formulates an MDX query q, a subset $R_q \subseteq R$ of rules is selected. Each rule in R_q is *pertinent*, meaning that its antecedent matches with q, and *effective*, meaning that the preference it would be translated into can actually induce an ordering on the facts returned by q. Then, let a positive integer *personalization degree* α be chosen by the user to express the desired preference complexity. A qf-set F_α is generated from R_q in such a way that α base constructors are included in the overall preference expression the fragments of F_α will be translated into.
3. *Fragment translation.* Each fragment in F_α is translated into a base constructor; the resulting base constructors are then coalesced and composed using the Pareto operator into a preference expression p.

Algorithm 1. Extract rules with support and confidence adjustment

---

Input: *Log*: A set of queries; *minSup, minConf*: Floats
Output: *R*: A set of association rules
Uses: *mine(set, float, float)*: An association rule extractor
Variables: *stop*: A Boolean; *confidence, support*: Floats; *Covered*: A set of qf-sets
1: *stop* =false
2: *confidence* = 1
3: *support* = 1
4: **while** !*stop* **do**
5:    *R = mine(Log, support, confidence)*   ▷ Mine rules above *support* and *confidence*
6:    *R = R \ {r ∈ R* s.t. *|r.cons| > 1}*   ▷ Only keep rules with singleton consequent
7:    *Covered =* ∅
8:    **for** each rule *r ∈ R* **do**
9:       *Covered = Covered ∪ {q ∈ Log|r.ant ∪ r.cons ⊆ q}*
10:    **if** *Covered = Log* **then**   ▷ If all queries in the log are covered in *R* stop...
11:       *stop* =true
12:    **else**   ▷ ...else mine again with lower thresholds
13:       *confidence = confidence − 0.1*
14:       **if** *confidence < minConf* **then**
15:          *support = support − 0.1*
16:          *confidence = 1*
17:          **if** *support < minSupp* **then**
18:             *stop = true*
19: **return** *R*

---

4. *Querying.* Query *q* is annotated with *p*, translated into MYMDX, and executed. As shown in [6], the user can effectively explore query results by visually interacting with a graph-like structure that emphasizes the better-than relationships induced by *p* between different sets of facts. Preferred facts are then displayed in a multidimensional table.

The following subsections explain in detail how steps 1, 2, and 3 are carried out. For details about step 4, see [3,6].

5.1 Log Mining

We now briefly describe the mining step. The input of this step is a set of qf-sets that represents the user's query log, while the output is a set *R* of association rules.

Interestingly, the problem of associating a query with a set of fragments representing user preferences bears resemblance to the problem of associating objects with a set of most relevant labels. This problem, named *label ranking*, is a form of classification. Both label ranking and classification have been proved to be effectively handled by association rules (see for instance [16,17]). In this context, rules have a set of features that should match the object to be classified as antecedent, and one label as consequent. We adopt a similar approach here, and we search for rules having exactly one item as consequent, so each rule *r ∈ R* takes the form *ant → cons*, where *ant* is a qf-set and *cons* is a single query fragment. In the following, *r.cons* (resp., *r.ant*) denotes the consequent (resp., antecedent) of rule *r*, and *conf(r)* its confidence.

The mining step is done off-line, and uses any classical association rule extractor that is parametrized by support and confidence thresholds (e.g., Apriori [18]). The only issue in this step is to extract rules that faithfully represent the user's query log. Since the user is not involved at this step, support and confidence have to be adjusted automatically [16]. Algorithm 1 is used for this purpose, and it extracts rules until the whole log is covered by the set of rules extracted. More precisely, the algorithm starts extracting rules with confidence and support equal to 1 (lines 2,3). If the set of rules covers the entire log, then the algorithm stops (line 11,12). Otherwise, extraction starts again with a lower confidence (line 13), and confidence is decreased until the log is entirely covered or the confidence is considered too low (line 14). In this case, confidence goes back to 1 and support is decreased (line 16,17), and extraction is launched again. If both support and confidence are considered too low, then the algorithm stops.

Algorithm 1 needs two thresholds, $minConf$ and $minSupp$. Realistic values for these thresholds can be learned by training the algorithm on query logs, or be derived from log properties like size and sparseness.

5.2   Rule Selection

The output of the mining step, R can be a large set. In this section we present the algorithm that first selects, among the rules in R, the subset R_q of pertinent and effective rules for query q, and then returns a qf-set F_α including a subset of the query fragments that appear as consequents of the rules in R_q. These fragments will be used for annotating q with a preference.

Following the approach presented in [12], the selection of query fragments is made by associating a score to each group of rules in R_q having the same fragment φ as consequent. This score is the average confidence of the rules in the group, i.e., $score(\varphi) = avg_{r \in R_\varphi} conf(r)$ where $R_\varphi \subseteq R_q$ is the subset of rules having φ as a consequent. The selected query fragments are those with highest scores, and are limited by the number α of base preference constructors that the user wants to annotate her queries with.

Given schema $\mathcal{M} = \langle A, H, M \rangle$ and a qf-set F, we adopt the following notation:

- $F.hier(h) = F \cap Lev(h)$ is the set of levels of hierarchy $h \in H$ in F;
- $F.meas = F \cap M$ is the set of measures in F;
- $F.val(a) = \bigcup_{(a \in V_k) \in F} V_k$ denotes the set of selected values for level/measure $a \in A \cup M$ in F.

Algorithm 2 selects, among the set R of association rules mined from the log, the consequents of rules that will be used to annotate the current query with preferences. It starts by removing from R all non-pertinent rules (i.e., those whose antecedent does not match q — line 1), and some non-effective rules (those whose consequent, if it is an attribute or a measure, does not appear in the list of group-by attributes or returned measures of q — line 2). The remaining rules are grouped by their consequent and the score of each group is computed (line 3). Then the top consequents corresponding to α base constructors are returned

---

Algorithm 2. Select Consequents

Input: R: A set of rules; q: A query represented as a qf-set; α: A user-defined *personalization degree*
Output: F_α: A qf-set that will be used to annotate q with a preference
Variables: $numBC$: The current number of base constructs; R_q: The set of pertinent and effective rules; F, F_{sim}: Two qf-sets

1:   $R = R \setminus \{r \in R | r.ant \not\subseteq q\}$ ▷ Drop non-pertinent rules
2:   $R_q = R \setminus \{r \in R | r.cons \in A \cup M, r.cons \notin q\}$ ▷ Drop non-effective rules
3:   $F = \{r.cons | r \in R_q\}$ ▷ Consequents of the rules in R_q
4:   $F_\alpha = \emptyset$
5:   $numBC = 0$
6:   **while** $numBC \leq \alpha$ and $F \neq \emptyset$ **do**   ▷ Iteratively construct F_α...
7:     let $\varphi = ArgMax_F \; score(\varphi)$ ▷ ...starting with the fragment having highest score
8:     $F = F \setminus \{\varphi\}$
9:     **if** $makesIneffective(\varphi, F_\alpha, q)$ **then**   ▷ If φ drives the preference ineffective...
10:       $F_{sim} = \{\varphi' \in F_\alpha | similar(\varphi, \varphi')\}$ ▷ ...find the similar fragments, if any...
11:       $F_\alpha = F_\alpha \setminus F_{sim}$ ▷ ...and drop them
12:       **if** $F_{sim} \neq \emptyset$ **then**
13:         $numBC - -$
14:     **else**
15:       **if** $\exists \varphi' \in F_\alpha | similar(\varphi, \varphi')$ **then** ▷ Other similar fragments were already added to F_α...
16:         $F_\alpha = F_\alpha \cup \{\varphi\}$ ▷ ...so $numBC$ must not be increased
17:       **else**
18:         **if** $numBC < \alpha$ **then**   ▷ Add φ only if this does not violate the α constraint
19:           $F_\alpha = F_\alpha \cup \{\varphi\}$
20:           $numBC + +$
21: **return** F_α

Function 3. makesIneffective

Input: φ: A fragment; F_α: A qf-set; q: a query represented as a qf-set
Output: A Boolean

1:   **if** $\exists h \in H | \varphi \in Lev(h)$ **then**   ▷ φ is a level
2:     **if** $(F_\alpha.hier(h) \cup \{\varphi\}) = q.hier(h)$ **then**   ▷ All query hierarchies are preferred
3:       **return** true
4:   **if** $\varphi \in M$ **then**   ▷ φ is a measure
5:     **if** $(F_\alpha.meas \cup \{\varphi\}) = q.meas$ **then**   ▷ All query measures are preferred
6:       **return** true
7:   **if** $\varphi = (a \in V)$ **then**   ▷ φ is a predicate
8:     **if** $q.val(a) \neq \emptyset$ and $!((F_\alpha.val(a) \cup V) \subset q.val(a))$ **then**   ▷ All values for a are preferred
9:       **return** true
10: **return** false

Function 4. similar

Input: φ_1: A fragment; φ_2: A fragment
Output: A Boolean

1:   **if** $\exists h \in H | \varphi_1 \in Lev(h)$ and $\varphi_2 \in Lev(h)$ **then**   ▷ Two levels of the same hierarchy
2:     **return** true
3:   **if** $\varphi_1 \in M$ and $\varphi_2 \in M$ **then**   ▷ Two measures
4:     **return** true
5:   **if** $\varphi_1 = (a \in V_1)$ and $\varphi_2 = (a \in V_2)$ **then**   ▷ Two predicates on the same attribute
6:     **return** true
7: **return** false

(lines 4-21). If a fragment φ that is about to be selected drives the preferences ineffective because it states that *all* the query results are preferred (Function 3), it is removed together with the other similar fragments (lines 10-13).

Example 5. Consider the qf-set of Example 3, $q = \{$Region, AllCities, Mrn, AllRaces, Occ, Year, AllSexes, AvgIncome, (Year $\in 2009)\}$. Let the set R of rules extracted from the log be as follows:

$$r_1: \quad (\text{Region} \in \{\text{'Pacific','Atlantic'}\}) \rightarrow \text{Year} \quad (0.8)$$
$$r_2: \quad \text{Year} \rightarrow \text{Region} \quad\quad\quad\quad\quad\quad\quad\quad (0.80)$$
$$r_3: \quad \text{Year} \rightarrow \text{AllCities} \quad\quad\quad\quad\quad\quad\quad (0.60)$$
$$r_4: \quad \text{AvgIncome} \rightarrow \text{Region} \quad\quad\quad\quad\quad (0.60)$$
$$r_5: \quad \text{Year} \rightarrow \text{Sex} \quad\quad\quad\quad\quad\quad\quad\quad\quad (0.90)$$
$$r_6: \quad (\text{Year} \in 2009) \rightarrow \text{Region} \quad\quad\quad\quad (0.70)$$
$$r_7: \quad \text{Year} \rightarrow (\text{Year} \in 2009) \quad\quad\quad\quad\quad (0.50)$$
$$r_8: \quad \text{Year} \rightarrow (\text{AvgIncome} \in [500, 1000]) \quad (0.55)$$
$$r_9: \quad \text{AvgIncome} \rightarrow \text{Mrn} \quad\quad\quad\quad\quad\quad (0.45)$$
$$r_{10}: \text{Occ} \rightarrow \text{Region} \quad\quad\quad\quad\quad\quad\quad (0.70)$$
$$r_{11}: \text{Occ} \rightarrow \text{Year} \quad\quad\quad\quad\quad\quad\quad\quad (0.10)$$
$$r_{12}: \text{AvgIncome} \rightarrow \text{Year} \quad\quad\quad\quad\quad (0.70)$$

and let Algorithm 2 be called with $\alpha = 2$. First, the algorithm removes r_1 (non pertinent) and r_5 (non effective). Then the remaining rules are grouped by their consequents, resulting in the set of fragments $F = \{\text{Region}, \text{AllCities}, (\text{AvgIncome} \in [500, 1000]), (\text{Year} \in 2009), \text{Mrn}, \text{Year}\}$ (listed by decreasing order of score). The fragments in F are now orderly explored. The first two fragments are not selected since, together, they drive the preference ineffective (they are exactly the fragments of hierarchy RESIDENCE included in q). Fragment (AvgIncome $\in$ [500, 1000]) is selected. Fragment (Year $\in$ 2009) is not selected since it corresponds precisely to the selection on Year of q. Then fragment Mrn is selected and, finally, Algorithm 2 outputs $F_\alpha = \{(\text{AvgIncome} \in [500, 1000]), \text{Mrn}\}$.

5.3   Fragment Translation

The output F_α of Algorithm 2 is a qf-set used to annotate the current query q with a preference. To this end, each query fragment $\varphi \in F_\alpha$ is translated into a base constructor (see Section 4); the resulting base constructors are then coalesced and composed using the Pareto operator.

The rules for translating fragment φ are explained below:

- if φ is a level $a \in A$, it is translated into a constructor $\mathsf{CONTAIN}(h, a)$, where h is the hierarchy a belongs to.
- If φ is a measure $m \in M$, it is translated into a constructor $\mathsf{CONTAIN}$ (measures,m).
- If φ is a Boolean predicate on a level, $(a \in V)$, it is translated into a constructor $\mathsf{POS}(a, V)$.
- If φ is a Boolean predicate on a measure, $(m \in [v_{low}, v_{high}])$, it is translated into a constructor $\mathsf{BETWEEN}(m, v_{low}, v_{high})$.

The resulting base constructors are coalesced by merging all $\mathsf{CONTAIN}$'s on the same hierarchy, all POS's on the same level, and all $\mathsf{BETWEEN}$'s on the same measure.

Example 6. The preference expression that translates the qf-set F_α in Example 5 is $p = \mathsf{BETWEEN}(\text{AvgIncome},500,1000) \otimes \mathsf{CONTAIN}(\text{RACE, Mrn})$. The MYMDX formulation for q annotated with p is:

SELECT AvgIncome **ON COLUMNS,**
        Crossjoin(OCCUPATION.members,
                Crossjoin(Descendants(RACE.AllRaces,RACE.Mrn**),**
                Descendants(RESIDENCE.AllCities,RESIDENCE.Region**))) ON ROWS**
FROM CENSUS **WHERE** TIME.Year.[2009]
PREFERRING AvgIncome **BETWEEN** 500 **AND** 1000 **AND** RACE **CONTAIN** Mrn

6   Experimental Results and Conclusions

In this paper we proposed a proactive approach to personalization, where mining techniques are applied to transparently annotate OLAP queries with preferences. This section briefly describes the implementation of our approach and reports the results of tests assessing its efficiency and effectiveness.

The approach was implemented in Java, using the Mondrian API for handling MDX queries, the Weka implementation of Apriori for rule extraction, and the MYOLAP tool for evaluating preferences [6]. The tests were conducted starting from synthetic MDX logs generated through Algorithm 5, that uses the *Diff* operator proposed in [19]. This operator explores the reasons why an aggregate is significantly lower in one fact compared to another. It takes as parameters two facts f and f' and an integer N, and looks into the two isomorphic sub-cubes C and C' that detail the two facts (i.e., that are aggregated to form f and f'). As a result, it summarizes the differences in these two sub-cubes by providing the top-N informative pairs of cells. Our generator simulates OLAP sessions on a datacube by starting from a random query q and then deriving the subsequent queries in the session using the result of the *Diff* operator applied to q. The Java implementation of *Diff* was obtained from [20]; N is set to 20 to simulate OLAP sessions including no more than 20 queries.

Algorithm 5. Generate a log

Input: $minSize$: Minimum log size
Output: Log: A set of queries
Uses: $Diff(cell, cell)$: The *Diff* operator defined in [19]
Variables: q: A query ; $nbGenerated$: Integer
 1: $nbGenerated = 0$
 2: **while** $nbGenerated < minSize$ **do**
 3:     randomly generate a query q on a sub-cube
 4:     $Log = Log \cup \{q\}$
 5:     $nbGenerated + +$
 6:     let f_1, f_2 be facts that show the maximum difference in the result of q
 7:     **for** each pair $\langle f_1', f_2' \rangle \in Diff(f_1, f_2)$ **do**
 8:         let q' be the drill-down of q to the group-by set of f_1' and f_2'
 9:         $Log = Log \cup \{q'\}$
10:         $nbGenerated + +$
11: **return** Log

The architecture used for testing is an Intel Core 2 Duo 3 GHz, with 4GB RAM. All tests were made on the CENSUS schema, using real data extracted from the IPUMS database [21], corresponding to about 10^7 facts stored on Oracle 11g. For our tests, we generated a log of about 1000 queries; the initial query of

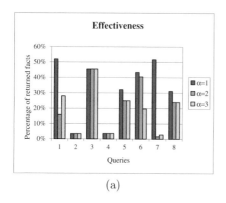 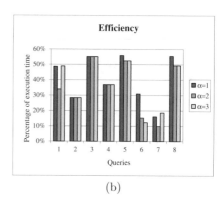

(a)   (b)

Fig. 2. Effectiveness and efficiency of our approach

each session was generated randomly by selecting group-by sets, measures and selections from a small pool. A small selection pool (3 selections on different dimensions) is used to simulate the log of a single user querying a sub-cube. Then, 8 queries to be personalized were extracted randomly from the log and removed from it. Minimum support and confidence were adjusted with Algorithm 1 to 0.6 and 0.7, respectively, resulting in 20 rules that cover the log and have an average support and confidence of 0.63 and 0.85, respectively. The confidence ranges from 0.76 to 1, with a standard deviation of 0.063.

As to effectiveness, Figure 2.a reports, for each query in the benchmark, the ratio between the number of preferred facts returned by the annotated query (i.e., those included in the *best-match only* result of the query [3]) and the one returned by the original query, when the personalization degree ranges between 1 and 3. Our approach is always effective in reducing the number of facts returned to the user. Though in general the reduction gets stronger as the personalization degree is increased, two different trends are apparent. In some cases (queries 2, 3, and 4) the reduction is independent on the personalization degree since only one pertinent and effective fragment was found. In other cases (queries 1 and 7), as the complexity of the preference increases, there are no facts that fully satisfy it so a larger set of facts that partially satisfy the preference are returned.

As to efficiency, we point out that the log mining step was executed in less than 4 secs, while the time for rule selection and fragment translation never exceeded 5 msecs. Figure 2.b reports the ratio between the time taken to execute each annotated query and the time to execute the original query. The reduction is always above 40%, and it is not relevantly affected by the personalization degree. Overall, we can conclude that our approach to personalization not only puts no overhead on the querying process, but it significantly reduces query response times.

While in this paper we used preference mining for result ranking, in our future work we will attempt to generalize it to address query recommendation as well. Besides, we will investigate the feasibility of extending our approach to incrementally manage OLAP sessions, i.e., to take delta queries into account at runtime without having to mine the log from scratch.

References

1. Giacometti, A., Marcel, P., Negre, E.: Recommending multidimensional queries. In: Pedersen, T.B., Mohania, M.K., Tjoa, A.M. (eds.) DaWaK 2009. LNCS, vol. 5691, pp. 453–466. Springer, Heidelberg (2009)
2. Bellatreche, L., Giacometti, A., Marcel, P., Mouloudi, H., Laurent, D.: A personalization framework for OLAP queries. In: Proc. DOLAP, Bremen, Germany, pp. 9–18 (2005)
3. Golfarelli, M., Rizzi, S., Biondi, P.: myOLAP: An approach to express and evaluate OLAP preferences. In: IEEE TKDE (to appear 2011)
4. Jerbi, H., Ravat, F., Teste, O., Zurfluh, G.: Management of context-aware preferences in multidimensional databases. In: Proc. ICDIM, London, UK, pp. 669–675 (2008)
5. Jerbi, H., Ravat, F., Teste, O., Zurfluh, G.: Applying recommendation technology in OLAP systems. In: Filipe, J., Cordeiro, J. (eds.) ICEIS. LNBIP, vol. 24, pp. 220–233. Springer, Heidelberg (2009)
6. Biondi, P., Golfarelli, M., Rizzi, S.: Preference-based datacube analysis with myOLAP. In: Proc. ICDE (to appear 2011)
7. Stefanidis, K., Drosou, M., Pitoura, E.: "You May Also Like" results in relational databases. In: Proc. PersDB, Lyon, France (2009)
8. Giacometti, A., Marcel, P., Negre, E., Soulet, A.: Query recommendations for OLAP discovery driven analysis. In: IJDWM (to appear 2011)
9. Chatzopoulou, G., Eirinaki, M., Polyzotis, N.: Query recommendations for interactive database exploration. In: Winslett, M. (ed.) SSDBM 2009. LNCS, vol. 5566, pp. 3–18. Springer, Heidelberg (2009)
10. Akbarnejad, J., Chatzopoulou, G., Eirinaki, M., Koshy, S., Mittal, S., On, D., Polyzotis, N., Varman, J.S.V.: SQL QueRIE recommendations. PVLDB 3(2), 1597–1600 (2010)
11. Khoussainova, N., Kwon, Y., Balazinska, M., Suciu, D.: Snipsuggest: Context-aware autocompletion for SQL. PVLDB 4(1), 22–33 (2010)
12. Veloso, A., de Almeida, H.M., Gonçalves, M.A., Meira Jr, W.: Learning to rank at query-time using association rules. In: Proc. SIGIR, Singapore, pp. 267–274 (2008).
13. Holland, S., Ester, M., Kießling, W.: Preference mining: A novel approach on mining user preferences for personalized applications. In: Lavrač, N., Gamberger, D., Todorovski, L., Blockeel, H. (eds.) PKDD 2003. LNCS (LNAI), vol. 2838, pp. 204–216. Springer, Heidelberg (2003)
14. Mobasher, B.: Data mining for web personalization. In: Brusilovsky, P., Kobsa, A., Nejdl, W. (eds.) The Adaptive Web, pp. 90–135. Springer, Heidelberg (2007)
15. Microsoft: MDX reference (2009), http://msdn.microsoft.com/
16. Sá, C., Soares, C., Jorge, A., Azevedo, P., Costa, J.: Mining association rules for label ranking. In: Huang, J.Z., Cao, L., Srivastava, J. (eds.) PAKDD 2011, Part II. LNCS, vol. 6635, pp. 432–443. Springer, Heidelberg (2011)
17. Li, W., Han, J., Pei, J.: CMAR: Accurate and efficient classification based on multiple class-association rules. In: Proc. ICDM, pp. 369–376 (2001)
18. Agrawal, R., Srikant, R.: Fast algorithms for mining association rules in large databases. In: Proc. VLDB, pp. 487–499. Santiago de Chile, Chile (1994)
19. Sarawagi, S.: Explaining differences in multidimensional aggregates. In: Proc. VLDB, Edinburgh, Scotland, pp. 42–53 (1999)
20. Sarawagi, S.: I3: Intelligent, interactive inspection of cubes (2009), http://www.cse.iitb.ac.in/sunita/icube/
21. Minnesota Population Center: Integrated public use microdata series (2008), http://www.ipums.org

A Novel Integrated Classifier for Handling Data Warehouse Anomalies

Peter Darcy, Bela Stantic, and Abdul Sattar

Institute for Integrated and Intelligent Information Systems
Griffith University
{P.Darcy,B.Stantic,A.Sattar}@griffith.edu.au

Abstract. Within databases employed in various commercial sectors, anomalies continue to persist and hinder the overall integrity of data. Typically, *Duplicate*, *Wrong* and *Missed* observations of spatial-temporal data causes the user to be not able to accurately utilise recorded information. In literature, different methods have been mentioned to clean data which fall into the category of either deterministic and probabilistic approaches. However, we believe that to ensure the maximum integrity, a data cleaning methodology must have properties of both of these categories to effectively eliminate the anomalies. To realise this, we have proposed a method which relies both on integrated deterministic and probabilistic classifiers using fusion techniques. We have empirically evaluated the proposed concept with state-of-the-art techniques and found that our approach improves the integrity of the resulting data set.

1 Introduction

Duplicate, *Wrong* and *Missing* data anomalies have continually hindered commercial sectors in the past resulting in error-prone data collection which seriously influence business processes. When the information is entered manually, based on additional information already recorded, the erroneous data can quite easily be discarded or approximated within a data set. Unfortunately, this is not the case within automatically recorded Spatial-Temporal data sets as the anomalies that persist are harder to rectify. These anomalies may include False-Positive readings such as *Duplicate* and *Wrong* observations or False-Negative readings such as *Missed* observations. This is especially found within automated data collection technology, such as Radio Frequency Identification (RFID), in which the anomalies can represent faults in sensors, intrusions, or missing objects.

Previous approaches have attempted to correct these anomalies at a deferred stage using deterministic or probabilistic approaches to identify and remove anomalies. However, in cases in which the persistent anomalies are particularly ambiguous, there is a need for a more intelligent methodology to clean the data. Noticing that there was a need for a fusion of both deterministic and probabilistic approaches to handle the most ambiguous anomalies, we have presented an integrated classifier that combines Non-Monotonic Reasoning, Bayesian Networks and Neural Networks to intelligently clean the error-prone observations. Through experimental evaluation, we have found that the Non-Monotonic Reasoning and Bayesian Network fusion methods resulted in

J. Eder, M. Bielikova, and A.M. Tjoa (Eds.): ADBIS 2011, LNCS 6909, pp. 98–110, 2011.

the highest achieving integrated classifier for both the False-Positive and False-Negative anomalies respectively, and provided a higher cleaning rate when compared to other state-of-the-art techniques.

The remainder of this paper is organised as follows: Section 2 will contain background information including RFID, Non-Monotonic Reasoning, Bayesian Networks and Neural Networks. We will introduce our methodology within Section 3 and highlight the motivation, architecture, intended scenario and assumptions. The Experimental Evaluation we performed are presented in Section 4, followed by our Conclusions found in Section 5.

2   Background

To conduct our experimentation and to determine if our approach of integrating classifiers together would significantly improve spatial-temporal databases, we have decided to test it on RFID data. We have chosen to integrate the Non-Monotonic Reasoning, Bayesian Network and Neural Network classifiers to create a novel and intelligent means of cleaning the anomalies. In the following section, we will provide a brief introduction to RFID, Non-Monotonic Reasoning, Bayesian Networks and Neural Networks.

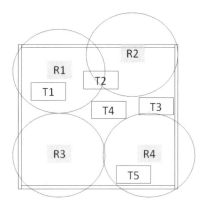

Fig. 1. An example of how in an enclosed environment, there is the possibility of Duplicate (T2), Wrong (T3) or Missing (T4) readings

2.1   RFID

Radio Frequency Identification (RFID) is a convenient technology employed in a wide array of commercial sectors which uses radio waves to allow communication between tagged items and readers [1]. RFID technology has already been employed to be used in various commercial sectors such as air package tracking, airport luggage monitoring and automatic pet identification. There are three different types of tags that may be utilised, the active, semi-active and passive. However, of the three, the passive tag is the easiest and most cost effective to implement due to its reasonable price and no battery being required [2].

Table 1. The recordings that took place from the example in Figure 1 and the observations that should have been recorded

What is Recorded		
Tag EPC	**Timestamp**	**Reader ID**
T1	22/12/2010 10:32:43	R1
T2	22/12/2010 10:32:43	R1
T2	22/12/2010 10:32:43	R2
T3	22/12/2010 10:32:43	R4
T5	22/12/2010 10:32:43	R4
What is supposed to be Recorded		
Tag EPC	**Timestamp**	**Reader ID**
T1	22/12/2010 10:32:43	R1
T2	22/12/2010 10:32:43	R1
T3	22/12/2010 10:32:43	R2
T4	22/12/2010 10:32:43	R2
T5	22/12/2010 10:32:43	R4

Unfortunately, due to various anomalies found predominantly within the passive architecture, the cost effective passive RFID systems are only applied to a fraction of its potential utilisation. These anomalies include: Duplicate Readings, which occur when a tag is scanned twice where it should have only been recorded once; Wrong Readings, where data is found where it should not have been; and Missed Readings in which data is not recorded where it should be [3], [4]. If the problems were able to be eliminated, applications such as automatic scanning of items in a trolley in a supermarket may be implemented allowing saving in both cost and effort. Within Figure 1, R1-R4 represents the four readers of a small enclosed area such as a livestock area with mounted scanners and T1-T5 represent tagged objects within the area such as cattle. Both T1 and T5 are read correctly as they are intended, however T2's reading is duplicated, T3 has a wrong reading and T4 is missed completely. The observational data found from this example, along with the readings that were supposed to be read may be found in Table 1.

In this example, the various anomalies which occur can produce hazardous conclusions for the users of the RFID system. The duplicate anomaly in which T2 is discovered in both R1 and R2 would cause the owner to not know the area which the cattle would be present in as it could be in either. This would be most problematic if the livestock area has designated zones for various animals. If the exact location is not known, the user would have to manually check the locations of the cattle. Similarly, if T3 appears in the quadrant where R2 is situated but is accidently read by R2, the owner will assume the animal is in R4's zone and not R2 which could result in unnecessary work needed to check if the animal with the T3 tag attached is actually in the area where R4 is located. Finally, the most problematic anomaly would be T4 in which the observation is missed completely in the database, resulting in the owner believing that the animal may have escaped.

2.2 Non-Monotonic Reasoning

Non-Monotonic Reasoning (NMR) is a type of logic specifically designed to commence with many conclusions, and, as new information is presented, derive the correct solution. Clausal Defeasible Logic (CDL) is a type of NMR which was designed to be specifically run on a computer. It allows the option to use one of five proof algorithms, each with various amounts of ambiguity permitted. The different formulae allowed include the μ formula which will only allow factual information; the π formula which allows ambiguity to propagate; the β which blocks ambiguity; the α which allows allow the conjunction of π and β; and the δ which only allows the disjunction of π and β [5], [6].

2.3 Bayesian Networks

A Bayesian Network is a means of probabilistically finding the most correct solution when given several pieces of information. Each of the probabilities of certain conclusion will be the product of all the information given up until that point. After each of the probabilities have been found, the conclusion achieving the highest probability is found to be the most accurate solution [7].

2.4 Neural Networks

An Artificial Neural Network refers to an intelligent classification technique which has been designed to emulate the processes of the human brain. Information is processed into a feature set which is then fed into the input nodes, passed through various amounts of hidden nodes and hidden layers which each have different weight calculations and is then passed into the output layer. It uses various training techniques such Genetic Algorithms to train the weights of the neurons to accurately classifier the information as the correct output [8].

3 Proposed Methodology

To correct the missed, wrong and duplicate readings found in various spatial-temporal databases including automatic capture technology such as RFID, we have chosen to employ the use of an integrated classifier architecture. Within this system, we take conclusions drawn from the three classifiers and use fusion techniques, such as a Non-Monotonic Reasoning algorithm, a Bayesian Network or taking the Majority answer, to derive a highly intelligent output. In this section, we identify the motivation behind developing our methodology, the architecture we created, the intended scenario of our approach and the assumptions needed for our concept run as intended.

3.1 Motivation

Radio Frequency Identification has been found to have limited functionality due to problems in the system such as data anomalies [9]. If these anomalies were eliminated, the applications that may benefit from RFID would be increased to various other commercial sectors thereby saving cost and effort. Previous approaches have been utilised to eliminate easily found anomalies, such as middleware algorithm used to determine a duplicate observation recorded in the same location in under a second, however these methodologies lack the intelligence needed to properly correct the stored observations to its maximum integrity. Additional past literature has individually stated that there is a need to use both deterministic and probabilistic methodologies to adequately clean the data [10], [11], [12]. With this in mind, we propose an approach that took advantage of both probabilistic and deterministic approaches to bring RFID data cleaning to a higher level of integrity. We did this because we fundamentally believed that missing data require a level of probability to find the absent information. In contrast, we believe that both wrong and duplicated data will need to have a deterministic approach due to having the information already present and there is less need to rely on probability. We specifically chose two probabilistic approaches (the Bayesian Network and Neural Network) and one deterministic approach (Non-Monotonic Reasoning) to give a probabilistic advantage to the former methods. This is also the reason as to why we chose the global fusion of the classifiers as opposed to pairwise combinations. To counter this, we chose a the novel deterministic Non-Monotonic Reasoning as a fusion technique which permits additional bias to the the Non-Monotonic Reasoning conclusion in its logical rules.

3.2 Architecture

We have divided our methodology into four core components, the *Feature Set Definition*, *Classification*, *Classifier Integration* and *Loader*. Due to the vast differences between the false-positive and false-negative anomalies, we have different classifier integrations for both the duplicate/wrong data, and the missing data. As seen in Figure 2, the Original Data containing the RFID observations, along with the Geographical Data, is passed into the Feature Set Definition where crucial analytical features of the data are identified. This analytical information is then passed into the Classification component where the Non-Monotonic Reasoning, Bayesian Network and Neural Networks are used to determine if a reading is valid or not. The results of the Classifiers are then

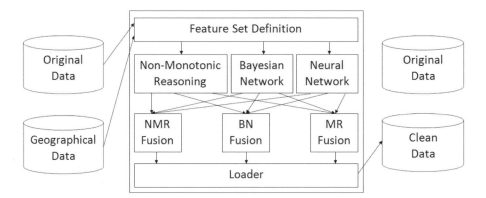

Fig. 2. A high level diagram describing the information flow in our methodology and the steps that takes place from when the data extracted to when it is loaded back into the database

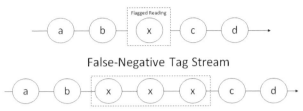

Fig. 3. The visual representation of how the tag's data is broken up into streams and analysed for both the false-positive and false-negative anomalies with the crucial readings (A-D) around the suspicious data found

passed into the Classifier Integration which uses three fusion methods to intelligently determine the validity of each suspicious reading. Finally, after all the information is gathered, the methodology will finally either delete, keep or insert the correct values into the data set within the loader component.

Feature Set Definition. The first action that the Feature Set Definition takes is to divide up the data into streams that follow the geographical path of each tag using the geographical data passed into the system. Once this is done, suspicious readings are found based on the geographical data supplied at the beginning. Only suspicious readings will be flagged by the system and all other observations will be ignored by our system. For example, if a reading occurs in two locations not within proximity concurrently, it will be flagged as suspicious as it may be a duplicate anomaly. Figure 3 describes the data found for both the false-positive and false-negative anomalies, and the data recorded for each. A major difference between both is that, due to the possibility of a duplicate observation, the spatial and temporal locations of A-D are needed for false-positive anomalies whereas only the spatial locations are needed for the false-negative analysis. After these values have been calculated, various binary (true or false) analyses are performed on the values obtained from the Tag Streams. These mathematical operations may be found in Table 2.

Table 2. Each of the operations performed on the Tag Stream data to pass to the classifiers

False-Positive Tag Stream	False-Negative Tag Stream
$b.loc \leftrightarrow x.loc$	$a == b$
$c.loc \leftrightarrow x.loc$	$b \leftrightarrow c$
$b.time == x.time$	$b == c$
$c.time == x.time$	$d == c$
$b.loc == x.loc$	$n == (s-2)$
$c.loc == x.loc$	$n > (s-2)$
$a.loc \leftrightarrow x.loc$	$n > (s-2)$
$d.loc \leftrightarrow x.loc$	
$b.time \leftrightarrow x.time$	
$c.time \leftrightarrow x.time$	

Within Table 2, there are three main comparisons. The first is the equivalence comparison $==$ which will check if the left value is equal to the right. The second comparison is if the left value is within proximity $\leftrightarrow$ to the right value in both spatial and temporal natures. The third comparison we make is to determine if the left value is greater than $>$ the right value. When determining if a value is within proximity, spatially this will mean if the geographical location of the two readers are physically close to each other whereas in the temporal sense, it will check if the time values are within a user-defined threshold. As a default, we have set it to 30 seconds. With regards to the False-Negative Tag Stream operations, n is the number of missing anomalies and s is the shortest path between both b and c. The reason we compare n to $s - 2$ is that the shortest path also contains b and c which may not be needed within the flagged values.

Classification. After the crucial analytical data has been found, it will be passed on as a feature set to the classification component of the methodology to be determined if the suspicious observations should be deleted, kept, or inserted into the database. For the false-positive anomalies which contain either a suspected duplicate or wrong reading, there are two conclusions that may be drawn from the classifiers: either delete or keep the values. With regards to the false-negative anomalies, there are five possible conclusions that may be made each with different reader values combinations. When handling missing data in various data sets, there is a need to impute the data back into the database (i.e. generate possible answers to be used when lacking factual information). We have named permutations to be imputed back into the database as seen in Figure 4. The first two permutations consist of substituting the values of readers b and c for all missing data. The third includes finding the shortest path between readers b and c which will be s, and inserting it into the middle of the missing values with b and c values added around it if the shortest path does not cover all missing recorders. Finally, the fourth and fifth permutation places the shortest path s to either the earliest or latest missing observations and substituting values c or b respectively.

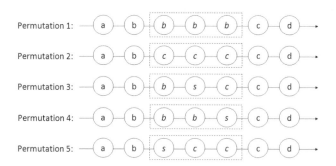

Fig. 4. The five possible Permutations that may be chosen to be imputed for the missing RFID reader values. Please note that s in this figure refers to the shortest path between reader values b and c, and may occupy more than one observational record.

With regard to the classifiers used throughout our experimentation, we have followed the configurations mentioned in literature for correcting anomalies in databases [12]:

- As it has been shown to provide the highest cleaning rate, we have utilised only the α Non-Monotonic Reasoning formula rules for both the false-positive and false-negative data anomalies.
- We trained the Neural Network used for false-positive anomalies with a genetic algorithm that had a large amount of both chromosomes and generations.
- For the Bayesian Network with the false-positive anomalies, as well as both the Bayesian and Neural Networks used for the false-negative anomalies, we implemented a Genetic Algorithm which had a low amount of chromosomes which trained for a large amount of generations [12].
- All training consisted of the information described in the feature set definition with its respective correct output being processed by the various classifiers and modifying the networks to enhance the conclusions being drawn.

Classifier Integration. In this work, we have proposed various methods of combining the classifiers together to develop a new intelligent means increasing the integrity of the conclusions being made. To this end, we have introduced three main fusion techniques to integrate the classifiers, the Non-Monotonic Reasoning Fusion (NMR Fusion), Bayesian Network Fusion (BN Fusion) and Majority Rules Fusion (MR Fusion). For the False-Positive anomalies, since the returned determination is either to keep or delete a value, we only need to integrate the classifiers once. However, due to the False-Negative anomalies finding five varied conclusions, we need to run the fusion algorithms five different times for each permutation.

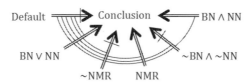

Fig. 5. A visual representation of the rules we implemented when we created the logic engine to deterministically integrate the classifiers in the Non-Monotonic Reasoning Fusion

With regard to the Non-Monotonic Reasoning Fusion, we took all the conclusions made in the classification component and put them into the logic engine depicted in Figure 5. Each of the values represent either the Bayesian Network (BN), Neural Network (NN) or Non-Monotonic Reasoning (NMR), and the arc between the antecedents gives higher weighting to the value anti-clockwise (i.e. $\sim NMR$ will be have its conclusion overwritten if $BN \wedge NN$ is also proven). We also have made sure that the Non-Monotonic Reasoning will have a slightly higher bias than the probabilistic techniques as this fusion method is deterministic in nature. Table 3 contains the values and weighting we have given to each of the classifiers to be used in the Bayesian Network Fusion. Similarly to the NMR Fusion, we wish to give a slightly higher bias to the

Table 3. A Table depicting the configuration of the values found within the Bayesian Network Fusion technique

Conclusion	BN		NN		NMR	
	T	F	T	F	T	F
Positive	60%	40%	60%	40%	40%	60%
Negative	40%	60%	40%	60%	60%	40%

Bayesian and Neural Networks as the Bayesian Network Fusion is itself a probabilistic technique. The final technique we have employed, the Majority Rules Fusion, is an unbiased approach which will use the conclusion voted most by the three classifiers as its determination. We hope that by having deterministic, probaiblistic, and non biased fusion methods, we will observe varying results among the different anomalies. In the unlikely event that none of the permutations have been found to be chosen more than once for the false-negative anomalies, a weighting system is employed based on the scale of most unbiased to biassed (3>1>2>4>5).

Loader. After the decision has been made by the integrated classifier, our methodology will then proceed to either delete, keep, or insert the correct values in the data set. We have made the option to either modify the Original Data set if the user is comfortable with the enhanced data sets or to create a new data set keeping the original data set separate for added integrity. Being that this entire process is at a deferred stage of the capture cycle where all the data has been stored, in this work we did not consider the cost of cleaning. However, in the future, we would like to implement a version of this concept that will run in real-time at the stage of data capture.

3.3 Intended Scenario

We have intended to create our methodology for a scenario in which many readers are mounted around a known environment and tags are passing through the area to be scanned. Applications in which this is already conducted include a hospital in which surgical patients are monitored, airports which track luggage and the transportation of various items in a supply chain. It is crucial that a known environment is used in the scenario as the geographical locations of each of the readers and their proximity to one another must be recorded in the system.

3.4 Assumptions

There are two assumptions we have identified for this scenario relating both to the identification of the false positive and false negative anomalies. The first is that, as stated in the intended scenario, the geographical locations of the readers must be known to the readers so that a tag which is recorded at abnormal locations may be flagged as a suspicious reading. The second assumption we make is that the time used to flag a missed reading is less than the time it takes for the tagged object to move from one readers scan range to another. Both of these assumptions are crucial as they provide the rules that our system follows to identify a suspicious set of observations to be corrected.

4 Experimental Evaluation

To properly test our integrated classifier, we have devised four experiments to determine the overall effectiveness and advantages it has over existing techniques. The first two experiments will be carried out to test the effectiveness of each of the fusion techniques for solving both false positive anomalies (duplicate and wrong data), and false negative anomalies (missing data). The second two experiments will take the best performing integrated classifier to compare it to state-of-the-art techniques currently used to enhance the integrity of RFID data. These experiments will be performed on multiple test beds to determine its effectiveness on varying amounts of anomalies.

4.1 Environment

To properly evaluate the effectiveness of our methodology, we have used simulated test cases of the information obtained from readers. We have created five test beds with 500, 1,000, 1,500, 2,000 and 2,500 test cases each to observe the performance of the approaches where there are various amounts of anomalies. Each of the test cases present within the test bed represent a found anomaly within the data sets. All code used in our methodology was written in the C++ language and executed in Microsoft Visual C++ 6.0. The computer used for this experimentation was a Microsoft Windows XP machine with Service Pack 3 Intel (R) Core 2 Duo CPU E8400 @ 3 GHz 2.99 GHz with 4 GB of RAM.

4.2 Results

The first experiment we ran included testing the percentage of clean data for the Bayesian Network, Neural Network, Non-Montonic Reasoning, Fused Non-Monotonic Reasoning, Fused Bayesian Network and Fused Majority Rules classifiers when attempting to clean 500, 1,000, 1,500, 2,000 and 2,500 False-Positive test cases. From the False-Positive results found in Figure 6, the highest performing classifier average has been found to be the Fused Non-Monotonic Reasoning classifier. The absolute highest performing classifier was also the Fused Non-Monotonic Reasoning classifier when attempting to clean 500 test cases. The least performing classifier for the False-Positive experiment was the Bayesian Network when attempting to clean 500 test cases. We believe that the advantage the Fused Non-Monotonic Reasoning classifier had was due to its deterministic architecture and the nature of False-Positive anomalies.

In the second experimentation we conducted, we took the same classifiers and test case amounts, however we used False-Negative anomalies rather than False-Positive. The results, which may be viewed in Figure 7, has shown that the highest performing classifier average has been found to be the Fused Majority Rules classifier. The highest performing clean on the data sets has been found to be the Fused Majority Rules classifier as well when attempting to clean 1,500 test cases. The least achieving classifier for the False-Negative anomalies has been found to be the Bayesian Network classifier when attempting to clean 500 test cases. The results have shown that the unbiased nature of the Fused Majority Rules classifier has given it a clear advantage when cleaning False-Negative anomalies. It is also important to observe that as highlighted in the

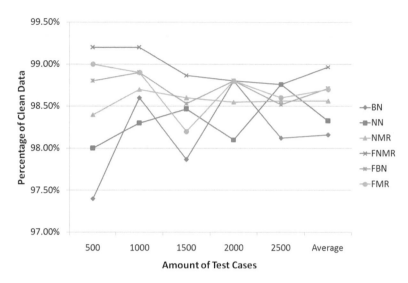

Fig. 6. The False-Positive results of Bayesian Network (BN), Neural Network (NN), Non-Monotonic Reasoning (NMR), Fused Non-Monotonic Reasoning (FNMR), Fused Bayesian Network (FBN) and Fused Majority Rules (FMR) when tested against, 500, 1,000, 1,500, 2,000, 2,500 test cases and the average

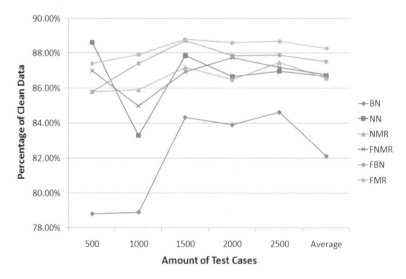

Fig. 7. The False-Negative results of Bayesian Network (BN), Neural Network (NN), Non-Monotonic Reasoning (NMR), Fused Non-Monotonic Reasoning (FNMR), Fused Bayesian Network (FBN) and Fused Majority Rules (FMR) when tested against, 500, 1,000, 1,500, 2,000, 2,500 test cases and the average

above results, of the two types of anomalies, it is harder to correct the False-Negative anomalies.

The Non-Monotonic Reasoning fused classifier was able to achieve the highest false-positive clean as its deterministic nature makes it ideal to clean wrong and duplicate data whereas probabilistic techniques would introduce an additional level of ambiguity. With regards to the false-negative anomalies, the Majority Rules fused classifier gained the highest cleaning rate due to it being able to accept all three classifiers without any bias. Additionally, we believe we may have obtained a higher result if we introduced a dynamically trained fused Bayesian Network or created a Fused Neural Network classifier. As this methodology is designed to be applied at a deferred stage of the RFID capture cycle, our experimentation was not concerned with the runtime performance. However, we would like to extend and modify our approach in the future to allow real-time processing in which case we will be taking the processing time into consideration for each classifier and focusing on the amount of time needed to generate rules or train the networks.

5   Conclusion

In this paper, we presented a methodology to clean anomalous Spatial-Temporal data using an integrated classifier. For this study, we investigated RFID technology as our case study as it continues to generated anomalies within the recorded data sets and has a need to be rectified before it can be employed in various other commercial sectors. Through experimental evaluation, we have found that the highest performing fusion type for wrong and duplicate data was the Fused Non-Monotonic Reasoning classifier, while the Fused Majority Rules approach was the most effective for cleaning missing readings. We then compared each of the highest achieving integrated classifiers against state-of-the-art and currently utilised approaches and found that our technique provides superior integrity. With regard to future work, we would like to investigate other fusion approaches of additional classifiers such as the Support Vector Machine and other classifier training techniques. We would also like to apply our technique to various databases as we believe that our methodology is not limited to merely cleaning RFID data and may be applied to other spatial-temporal data collections as well. Also, as mentioned earlier, we would like to employ a real time implementation of this concept.

References

1. Yang, Q.: Activity recognition: linking low-level sensors to high-level intelligence. In: Proceedings of the 21st International Joint conference on Artifical intelligence (IJCAI), pp. 20–25 (2009)
2. Chawathe, S.S., Krishnamurthy, V., Ramachandran, S., Sarma, S.E.: Managing RFID Data. In: VLDB, pp. 1189–1195 (2004)
3. Jeffery, S.R., Garofalakis, M.N., Franklin, M.J.: Adaptive Cleaning for RFID Data Streams. In: VLDB, pp. 163–174 (2006)
4. Darcy, P., Stantic, B., Mitrokotsa, A., Sattar, A.: Detecting Intrusions within RFID Systems through Non-Monotonic Reasoning Cleaning. In: Intelligent Sensors, Sensor Networks and Information Processing (ISSNIP 2010), pp. 257–262 (2010)

5. Billington, D.: Propositional Clausal Defeasible Logic. In: Hölldobler, S., Lutz, C., Wansing, H. (eds.) JELIA 2008. LNCS (LNAI), vol. 5293, pp. 34–47. Springer, Heidelberg (2008)
6. Billington, D.: An Introduction to Clausal Defeasible Logic. David Billington's Home Page (August 2007), http://www.cit.gu.edu.au/~db/research.pdf
7. Zio, M.D., Scanu, M., Coppola, L., Luzi, O., Ponti, A.: Bayesian Networks for Imputation. Journal Of The Royal Statistical Society Series A 167(2), 309–322 (2004)
8. Blumenstein, M., Verma, B.: A Neural Based Segmentation and Recognition Technique for Handwritten Words. In: The 1998 IEEE International Joint Conference on Neural Networks Proceedings, vol. 3, pp. 1738–1742 (May 1998)
9. Darcy, P., Stantic, B., Derakhshan, R.: Correcting Stored RFID Data with Non-Monotonic Reasoning. Principles and Applications in Information Systems and Technology (PAIST) 1(1), 65–77 (2007)
10. Rao, J., Doraiswamy, S., Thakkar, H., Colby, L.S.: A Deferred Cleansing Method for RFID Data Analytics. In: VLDB, pp. 175–186 (2006)
11. Khoussainova, N., Balazinska, M., Suciu, D.: Probabilistic Event Extraction from RFID Data. In: International Conference on Data Engineering, pp. 1480–1482 (2008)
12. Darcy, P., Stantic, B., Sattar, A.: Correcting Missing Data Anomalies with Clausal Defeasible Logic. In: Catania, B., Ivanović, M., Thalheim, B. (eds.) ADBIS 2010. LNCS, vol. 6295, pp. 149–163. Springer, Heidelberg (2010)

Variable Granularity Space Filling Curve for Indexing Multidimensional Data

Justin Terry[1], Bela Stantic[1], Paolo Terenziani[2], and Abdul Sattar[1]

[1] Institute for Integrated and Intelligent Systems
Griffith University, Brisbane, Australia
[2] Universita' del Piemonte Orientale, Alessandria, Italy

Abstract. Efficiently accessing multidimensional data is a challenge for building modern database applications that involve many folds of data such as temporal, spatial, data warehousing, bio-informatics, etc. This problem stems from the fact that multidimensional data have no given order that preserves proximity. The majority of the existing solutions to this problem cannot be easily integrated into the current relational database systems since they require modifications to the kernel. A prominent class of methods that can use existing access structures are 'space filling curves'. In this study, we describe a method that is also based on the space filling curve approach, but in contrast to earlier methods, it connects regions of various sizes rather than points in multidimensional space. Our approach allows an efficient transformation of interval queries into regions of data that results in significant improvements when accessing the data. A detailed empirical study demonstrates that the proposed method outperforms the best available off-the-shelf methods for accessing multidimensional data.

1 Introduction

In current database applications there is an increasing need to efficiently handle multi-dimensional data such as temporal, spatial, spatio-temporal, multimedia, scientific, and medical data [1]. Multidimensional relational data can be represented as points/vectors in a multidimensional space, where each attribute corresponds to a dimension.

Multidimensional databases are usually very large in size. Such a large and increasing volume of data needs efficient access methods to support it, otherwise the improvements of more complex data representation and reasoning may be lost due to inefficient access. It is well known that with traditional multidimensional access methods [2] performance deteriorates rapidly as the dimensions increase [3], thus they typically do not scale well to higher dimensions.

The difficulties associated with multidimensional data grow with the number of dimensions. Once data have more than three or four dimensions, additional problems begin to arise, loosely termed the *'curse of dimensionality'*, which can severely deteriorate an access method's performance. At higher dimensionality (10 dimensions or higher) the existing methods do not work well, in the sense that a sequential scan of the table becomes faster (less time and/or less block accesses) than using the index to answer most queries [4]. At higher dimensionality space and data become very sparse

J. Eder, M. Bielikova, and A.M. Tjoa (Eds.): ADBIS 2011, LNCS 6909, pp. 111–124, 2011.

and distance metrics lose their meaning. For above 10-15 dimensions the number of dimensions that are not partitioned can become large as there are simply not enough data to require all dimensions to be split. This causes nodes to waste space on redundant information on these unpartitioned dimensions. Selectivity in unpartitioned dimensions is then not supported and the interior nodes can contribute little to the selectivity of the index tree. To cope with high number of dimensions dimensionality reduction techniques have been applied, which reduce the original space to a much lower dimensional subspace [5]. However, the transformation of data or queries requires additional resources and typically only approximate the original data. Therefore dimensions reductions is not a solution in many application domains, and a need for an efficient access method to manage medium to high dimensional vector data remains.

Several types of approaches have been developed in order to cope efficiently with multidimensional data. In particular, Space Filling Curve (SFC henceforth) methods play a prominent role in the area. SFC methods, e.g. Z-order curve [6], Hilbert Curve [7], and Gray Codes [8] partition data into multidimensional pixels according to the bottom granularity, and employ a curve that passes through all pixels in the multidimensional space. This curve produces a total order of pixels in space. This ordering enables the use of existing efficient one dimensional access structures, such as B^+-trees. Leaf pages of the access structures then represent data on a segment of the curve, producing a primary index where nearby data are clustered with a high probability. The main disadvantages of SFC's methods are that they are CPU intensive and that they suffer from high overlap between pages (curve segments) and the query interval. The UB-Tree [9] integrates a space filling curve and a B^+-Tree creating a primary index for multidimensional data. It is a paginated index where each leaf node represents a block of data on a segment of the curve. It divides the space into linear segments of a Z-curve (or any SFC). Disadvantages of the UB-Tree are that it requires modification to the DBMS kernel for integration and like other SFC's the segments are typically not hyper-cubic and may even represent disjoint space. One of the most prominent d dimensional point data structures is the K-D-Tree and its variants: the *hB-Tree* [10], the *BD-Tree* [11], the *hybrid tree* [12] and the quad-Tree. The K-D-Tree is a binary search tree that uses a recursive subdivision of the data space into partitions by means of $(d - 1)$-dimensional hyperplanes. A disadvantage common to all K-D-Tree methods is that for certain distributions, no hyperplane can be found that divides the data objects evenly. Like the K-D-Tree, the quad-tree [13] decomposes the universe by means of iso-oriented hyperplanes. An important difference however, is the fact that quad-trees are not binary trees anymore. The subspaces are decomposed until the number of objects in each partition is below a given threshold. Quad-trees are therefore not balanced, and the subtrees of densely populated regions need to be deeper than sparsely populated regions, giving a bad worst case behavior.

In this paper, we are interested in multidimensional access structures that efficiently support basic vector data operations, in particular interval (window) queries as such queries play a prominent role in many contexts. It has been shown that space partitioning and employing the virtual structure is beneficial for efficient management of temporal data [14], [15]. In this work our focus is on methods that scales well at medium dimensionality (from 4 to 18 dimensions). Also, a fundamental requirement is that our approach should be easily intergraded into current Relational Database Management Systems (RDBMS) to take advantage of the in built industrial strength concurrency and recovery. Specifically, we aim at developing an approach that can be implemented without any modification of the kernel.

In this work, we propose an SFC based method, termed "VG-Curve" method, where "VG" stands for "Variable Granularity", overcoming some of the limitations of existing methods. In our approach, the multidimensional space is partitioned into regions of different dimensions, depending on the distribution of the population in the multidimensional space. Thus, while standard SFC methods chose one granularity to partition space, the VG-Curve method works with variable-granularity regions, so that many pixels can be grouped in the same region. In particular, scarcely populated parts of the space can be enclosed into larger regions, and empty regions do not even need to be stored. Then, the curve (VG-Curve) connects such regions thus achieving an ordering of multidimensional data similar to a SFC so that nearby objects are physically clustered together with a high probability. As a consequence, the advantages of SFC methods are preserved by our approach, which, on the other hand, is more efficient, since less entities (regions) are connected by the Curve.

2   The Variable Granularity Space Filling Curve (VG-Curve)

We assume that the universe of discourse (the data space) is a d-dimensional hyper rectangle with a side length of h_i and volume $v = \prod_{i=1}^{d} h_i$. The data space is assumed to have a non uniform (real world) distribution of data with some empty and some heavily populated areas. Entities in the data space are called *objects*.

Definition 1. *An* object *is a d-dimensional tuple with d indexed attributes, a unique object key, and any number of other non indexed attributes.*

In our approach, the multi-dimensional space is partitioned into hyper-rectangular parts called *regions*. We cope with regions of different sizes. Specifically, a given order is assumed for the dimensions two child regions can be obtained by orthogonally splitting the parent region in two along the current dimension, considering the order of dimensions and this is done in a cyclical way. As a consequence, a *region* is defined as follows in our approach.

Definition 2. *A* region *is an area representing a d-dimensional interval with the first j dimensions (in order) having a side length of x and the next k dimensions, where $k = d - j$, having a side length of 2x. The length of the ith dimension of a region will be $\frac{h_i}{2^n}$ with $1 \leq n \leq \frac{max_{split}}{d}$, where max_{split} is the maximum number of splits allowed.*

A minimum granularity is fixed for regions.

Definition 3. *A* pixel *is the finest granularity of regions, dictated by the choice of max_{split}.*

In our approach, each region can be uniquely identified by an *address*, which is, roughly speaking, a compact binary representation of the sequence of splits that have generated it. Region addresses are obtained by bit interleaving of a N-order curve decomposition e.g. for $d = 2$ the order for quadrants is SW, NW, SE, NE, though any other SFC partitioning strategies may be used. Regions are open on the high side and closed on the low side, i.e., [min, max). A region address is the key for all objects in that region. The volume r_v of a region decreases exponentially ($r_v = v * (2^{1-L})$) with its address length L and volume v. We therefore obtain a fine partitioning of the multidimensional space with relatively short addresses.

Definition 4. *Region addresses form a complete order called* VG-Curve.

In the following, we discuss how such abstract notions can be implemented in our approach, in order to enhance efficiency in the treatment of multidimensional data. Being a complete order, the VG-curve is suitable for indexing with one dimensional index. In short, the VG-curve is implemented by a *base relation* that is managed by a directory relation combined with control processes. The base relation contains the unique object key, the region address where object belongs, and one column for each dimension. It may also contain other (not indexed) columns. Additionally, for the sake of efficiency, we also adopt a *directory*, which is a compressed representation of the base relation containing the addresses of non-empty regions and their population.

2.1 Partitioning Method

The starting point of our approach is a multidimensional space, populated by a set of objects. The task of the partitioning algorithm is to partition such a space into variable-dimension regions, depending on the distribution of the objects in the space, in order to achieve efficient data management.

Partitioning needs to take into account different parameters. First of all, the dimension of *pixels* need to be fixed. Such a parameter is usually chosen by considering the value which cannot be any more further subdivided, since they represent a bottom granularity. The maximum number of splits max_{split} is thus defined accordingly by $max_{split} = log_2 (v/p_v)$ where p_v is the volume of a pixel.

In our approach, an important point is to decide when a region is populated enough in order to be split. Let $bf = \frac{bd}{od}$ be the *blocking factor*, i.e., the maximum number of objects that can be contained into a physical block (where bd and od denote the dimensions of blocks and objects respectively). We choose to split regions whenever their population exceeds the blocking factor. In such a way, we partially enforce the correspondence between physical blocks and regions, to enhance efficiency. However it is worth stressing that in our approach we do not strictly enforce a one-to-one correspondence between regions and blocks, not to suffer the low block utilization due to possible sparse data.

In Partition algorithm, DV is a vector in which dimensions are ordered, $DV[cur]$ indicates the current splitting dimension, and the *next* function is used in order to move from one dimension to the following one, looking at the vector in a circular way. Partitioning operates in a recursive way, by splitting each region in two along the current dimension, until either pixel regions or regions with population smaller than the blocking factor are obtained. At each stage, the region is split in two along the current dimension, considering the following split position:

$$SplitPosition = \frac{r_{high}(s) - r_{low}(s)}{2} \tag{1}$$

where s is the current dimension, $r_{high}(s)$ - the region's s dimension high boundary, $r_{low}(s)$ - the region's s dimension low boundary. The first child region gets all the parents objects that lay below or on the new partition and the high child gets the data that lies above it. At each partition, the address of the first (second) child region is obtained by concatenating '0' ('1') to the address of the current region. Additionally, the directory is updated in order to consider the new regions (while the parent region is removed). In such a way, a tree of addresses and split conditions is virtually generated by the partition process, as shown in Figure 1 and 2.

Algorithm 1. Algorithm 1 Partition

Input: region R, address of region A, directory D, blocking factor BF, current depth CD, max number of splits max_{split}, dimension vector DV, current dimension i
begin
if population of R $>$ BF **then**
  if CD $< max_{split}$ **then**
    partition R along the dimension DV[i]
    Let LeftRec and RightRec the first and second regions obtained;
    remove from D the entry for R;
    if population of LeftRec > 0 **then**
      add into D the entry for LeftRec (address: A.'0');
    end if
    if population of RightRec > 0 **then**
      add into D the entry for RightRec (address: A.'1');
    end if
    Partition(LeftRec, A.'0', D, BF, CD+1, max_{split}, DV, next(i,DV));
    Partition(RightRec, A.'1', D, BF, CD+1, max_{split}, DV, next(i,DV));
  else
    Allow population to grow beyond the blocking factor
  end if
end if
end

Notice that when a CD is equal to max_{split} the partition has reached its maximum allowed depth, i.e., we have reached the pixel level. When a pixel becomes overfull it will not split and it's population is allowed to grow beyond the blocking factor similar to the concept of super-nodes for X-tree high-dimensional indexing [16]. This is possible since the physical storage of a region is not limited to a block but is clustered in order of its address.

As a simple running example, we use a two dimensional domain (with dimensions x and y) where the blocking factor is 3, each dimension has a range from 0 to 100, and the dimensions are ordered x first then y. There are seventeen data objects labelled 'a' to

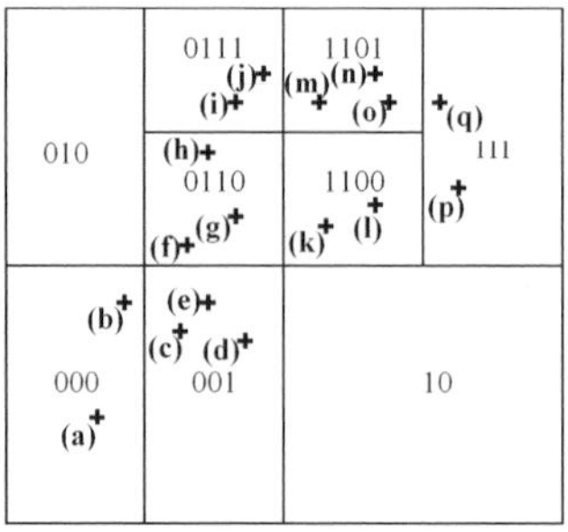

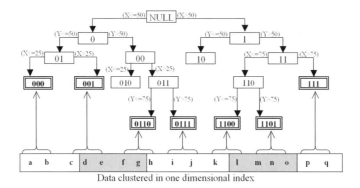

Fig. 1. The data space is recursively divided based on data population density. Example after 17 objects are inserted causing 6 splits (BF = 3).

Fig. 2. Running example, virtual tree nodes are in single border boxes, directory regions are in double border boxes. The objects reference the regions of the directory and are stored in order of the region they reference.

'q' distributed unevenly over the space to show how different distributions are handled. Figure 1 shows the results of partitioning on such data, assuming max_{split} equal to 4.

The upper part of Figure 2 shows the virtual tree produced by partitioning. For the two dimension example the whole space region (represented by '1') is first split with a vertical partition, splitting the space along the x-dimension. The first split which is at $x = \frac{100-0}{2} = 50$ replaces region '1' with two regions. The first child has address '10', representing all objects with an x value $\geq$ 0 and $<$ 50 and a y value $\geq$ 0 and $<$ 100. The second child (address equal to '11') represents all objects with an x value $\geq$ 50 and $<$ 100 and a y value $\geq$ 0 and $<$ 100. If the region '11' still contains more than three objects, then it will be split on the next dimension i.e., the y-dimension at the partition value of 50. Dividing into two regions '111' and '110' that replace region '11', and so on. As shown by the upper part of Figure 2, the partitioned space can be represented as an unbalanced binary tree where data are totally ordered. The leaf nodes of the tree contain the addresses of the regions, whose objects are stored by the DBMS in contiguous blocks of the base relation. These blocks are denoted by alternative shading in the lower part of Figure 2.

It is worth stressing that the binary partitioning tree is only virtual. As a matter of fact, the partitioning algorithm we propose has predictable split positions and split dimensions. Therefore, the partitioning tree needs not to be stored, since region bounds can be easily evaluated when needed. Actually, in our approach, only the leaf nodes of the partitioning tree need to be stored. They are stored in the directory which, besides the addresses of regions, also contains their population.

The directory resulting from the example is shown in Table 1. The directory performs the functions of an index i.e., it is a compressed representation of the data used to efficiently access the data itself, but it does not store pointers to the block(s)where data are stored. It is worth noticing that the directory does not contain the entries concerning empty regions (which are only implicitly represented).

Table 1. Directory containing the binary regions with the population for the running example

address	1000	1001	10110	10111	11100	11101	1111
population	2	3	3	2	2	3	2

2.2   Insertion Method

The identification of the region where a new spatial object has to be inserted (called insert_region henceforth) is conceptually easy when standard SFC methods are used, since they partition space into regions having a fixed known dimension (i.e., pixels). In our approach, on the other hand, regions have different dimensions. Nonetheless, through the addresses stored in the directory, and exploiting the virtual partitioning three, the insert_region, where the data will be clustered in, can be efficiently determined.

Given the coordinates of an object in the multidimensional space, the region containing it can be determined as described by the *Insertion* algorithm.

In the first part of the algorithm, the address of the region where the object should belong (called target_region) is computed. The address of the target_region is computed by first evaluating, for each dimension, a normalized binary value. We obtain such a value following three steps. First, we apply the equation 2, to get a natural number b_i.

The ith dimension's normalized natural value b_i is defined as:

$$\textbf{if } (v_i - min_i) \,=\, 0 \textbf{ then } \quad b_i = 0$$

$$\textbf{else} \quad b_i = \lceil \frac{v_i - min_i}{max_i - min_i} * 2^{(\lceil (curr_{depth}-1)/d \rceil)} - 1 \rceil \qquad (2)$$

where v_i is the object coordinate in the ith dimension, max_i is the maximum value in the ith dimension, min_i is the minimum value in the ith dimension, $curr_{depth}$ is the current depth of the virtual partition tree, and d is the number of dimensions.

Second, the normalized natural value is converted into the corresponding binary number $binary_i$. Since at most $\lceil (curr_{depth} - 1)/d \rceil$ splits have been done along each dimension, in the third step only the leftmost $\lceil (curr_{depth} - 1)/d \rceil$ bits of $binary_i$ are retained (in case $binary_i = 0$ the result is a string with $\lceil (curr_{depth} - 1)/d \rceil$ of '0').

Once these normalized binary strings are obtained for each dimension, the address of the target_region is obtained by bit interleaving them (e.g., the bit interleaving of '100' and '011' is '100101') and by prefixing the result with '1' (to represent the root of the tree). The bit interleaving is similar to the Z-curve bit interleaving (see also [6]), except the value in each dimension is normalized via $\frac{v_i - min_i}{max_i - min_i}$ to a fraction of that dimensions domain range.

The final result is the address of the target_region. Since no region exists below the current depth of the tree, the target_region represents the lowest possible region of the tree where the object should be inserted. Given a target_region of address b, two cases are possible: (1) the directory already contains a region whose address a is equal to b. Such a region is thus the insert_region, (2) the directory already contains a region whose address a is a longest prefix of b. This means that such a region properly contains the target_region, and the new object must be inserted into it (i.e., region a is the insert_region). Once the insert_region has been determined, the new object is inserted into it. In case the resulting population of the insert region exceeds the blocking factor, the insert_region is split.

3 Query Answering: Interval Queries

In this work we focused on the efficient processing of interval queries IQ on medium to high dimensional point data (d = 2-18) as well as the exact match query, as it is a specific type of interval query. Multidimensional range searching, such as interval queries, plays an important role in the way modern applications query their data.

In our approach, interval queries are processed following the primary index two stage query process . In the approximate filter the curve is preprocessed to remove some regions that cannot contain answers, then the remaining regions from the directory are hierarchically searched. The result of such a search are two sets of regions: O, consisting of all the overlapping regions (i.e., regions in the directory that intersect the interval query, but are not completely contained into it), and C, consisting of the regions entirely contained into the interval query. Contained regions only have objects that must be part of the result, whereas overlapping regions will need to have their objects checked for false hits by the exact filter.

Preprocessing trims the curve of regions that the search will examine. It removes from consideration all regions before the first and after the last pixel that can contribute to the answer. We calculate the first and last pixel of interest by bit interleaving (see Equation 2) the minimum and maximum corners of the query interval. The minimum corner will be the point representing the minimum of the interval restriction in all dimensions and similarly for the maximum corner. We prune the directory by retrieving only the regions that cover the curve between and including these pixels, and search this reduced set of regions. This is a fast and simple technique to reduce the load on the approximate filter . Preprocessing removes regions without consulting the directory.

The algorithm to search the directory is shown below. For each region in the directory, the algorithm visits the virtual partition tree level by level, starting from the root. This visit is implemented in the algorithm using the variable L (representing the length of addresses, and, thus, the depth in the virtual tree). Given a region F in the directory,

---

Algorithm 2. Search Directory Algorithm

---

begin
  Input: Preprocessed directory D , Interval Query I_Q
  Output: Containing Regions C, Overlapping Regions O
  Add all regions in D to $LIST$, ordered by address;
  Initialize C and O to the empty set
  Let length L be 1
  while $LIST$ is not empty **do**
    Let F be the first region in $LIST$
    Let R be the region in the virtual partition tree such that $R = prefix(F,L)$
    if R is contained within I_Q **then**
      Move from $LIST$ to C all regions a such that $R = prefix(a,L)$
      Set L to 1
    else if R is disjoint from I_Q **then**
      Remove from $LIST$ all regions a such that $R = prefix(a,L)$
      Set L to 1
    else if R equals F **then**
      Add R to O
      Remove R from $LIST$
    else
      Increment L
    end if
  end while
end

---

and given a level L, the algorithm searches for the L-level ancestor of F. Let R be such a region of the partition tree. R is compared with the binary addresses of the extreme points of the interval query, to check whether it is disjoint, contained or overlapping the interval query I_Q. If R is disjoint from I_Q, the search discards all the directory regions beginning with the address of R (i.e., such that $R = prefix(a,L)$). If R is contained, F and all the other regions in the directory starting with the address of R are put into the set C of contained regions. Otherwise, R overlaps the interval query. If R is equal to F, then F is an overlapping region, and is inserted into O. Otherwise the search must be further refined, by going deeper in the virtual tree (i.e., by incrementing L). The process is repeated until a subtree of disjoint regions is excluded or a subtree of contained regions is included or the full region is tested and classified as disjoint, contained or overlapped. The treatment of exact match queries is a special and easy case of the above. The result of preprocessing of exact match query gives as result a pixel. The region that contains such a pixel, if it exists, is then read to find the objects it contains. If such a region does not exist, the pixel does not contain any object, and the result of the query is empty.

4 Experiment

In order to evaluate the performance of the VG-Curve method, in this section we experimentally compare it (as suggested in the UB-Tree experiment [17]) with two of the best available methods in off-the-shelf commercial RDBMS for medium to high dimensional data, i.e. compound indexes and table scans. We could not directly compare our results with UB-Tree because it requires modification to the kernel. While R-tree methods are commonly available in commercial RDBMS their performance is well known to deteriorate above 5 dimensions so we could not use them as we are interested in medium to high dimensional data (up to 18 dimensions). On the other hand, the performance of basic SFC methods (e.g. Z-curve) deteriorate rapidly when the number of dimensions increases or the query interval grows, due to a blow out in CPU operations, as we confirmed in initial testing, so we found the Z-curve unsuitable for this experiment.

All experimental results presented in this Section are computed on a Sun Fire V880 server with 8 x UltraSPARC-III 900MHZ CPU using 8GB RAM, running Oracle 10g RDBMS. Database block size was 8K and SGA size was 500MB. At the time of testing database server had no other significant load. We used built-in methods for statistics collection, analytic SQL functions, and the PL/SQL procedural runtime environment. All queries had the buffers flushed before running.

We derived a data set of 5.8 million records from the the UCI KDD Archive US forest cover type for 30 x 30 meter cells obtained from US Forest Service (USFS) Region 2 Resource Information System (RIS) data. All relations had a unique identifier and a column for the derived key added.

Queries were randomly generated hypercubes with edge lengths from 20% to 80% of the respective dimensions range. We generated 100 random queries per 10% increment for each (2-18d) data set. The two parameters used in the VG-Curve are the *blocking factor* and max_{split} which was 100 for all experiments. The blocking factor was varied widely to test the sensitivity of the VG-Curve to this parameters setting.

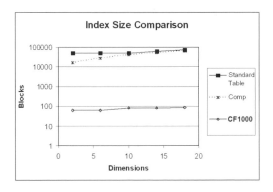

Fig. 3. Total blocks used for the standard table is shown as a reference, a compound index on indexed dimensions and the VG-Curve directory (BF=1000) for 2 to 18 Dimensions on real data

4.1   Results and Analysis

Experiments consider our VG-Curve method, table scan and the compound index method. The measured behavior of queries became less stable as the dimensions grew, as can be seen in Figures 6 and 7. This was due to the reduction in non empty result sets for queries at higher dimensions.

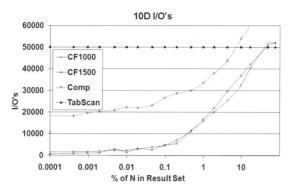

Fig. 4. VG-curve average I/O's for all methods on 10 dimensions of real data

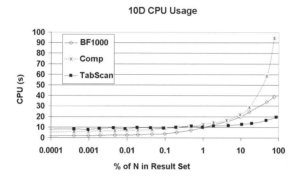

Fig. 5. VG-curve average CPU's for all methods on 10 dimensions of real data

As expected, the the VG-curve approach, has clear advantages over both table scan and compound index methods as regards space complexity. The size of the VG-curve directory is a small fraction of the space required by the compound index. This can be seen in Figure 3 where the size of the VG-Curve directory managing 5.8 million objects is up to 733 times smaller than its corresponding compound index and was always less than 100 blocks.

Due to the space limitations in this paper we only show results for I/O and CPU time of the VG-Curve, table scan, and compound index methods considering real data of 10 dimensions (Figures 4 and 5).

The I/O costs of Figure 4 clearly show that the VG-curve, for blocking factors of 1000 and 1500, outperforms both the compound and the table scan methods by up to a factor of 12. The CPU costs in Figure 5 indicate the VG-curve outperforms both compound and table scan methods for queries with result sets of less than 1% of total number of rows, and is still competitive for queries with result sets of up to 10% of total number of rows.

To avoid the effects of the distribution of data, as mentioned before, we have run multiple tests; results are presented grouped together based on their average result set size so that query performance can be compared as the dimensions of the data increase.

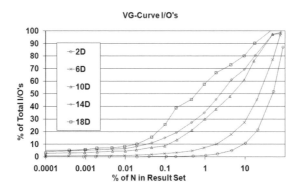

Fig. 6. Comparison of average disk I/O's, as a % of table blocks, for VG-curve BF=1000 from 2 to 18 dimensions on real data

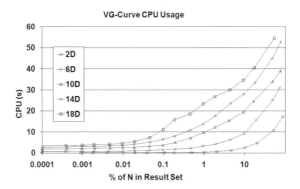

Fig. 7. Comparison of average CPU's for VG-curve BF=1000 from 2 to 18 dimensions on real data

Typically, as for other high-dimensional indexes, index structure performs better for result sets of up to 20% of N_o. However, small result set queries are more important and more common in the management of high-dimensional data. In case of answer sets larger than 20% of all objects, due to the overheads of using an index, the full table scan will usually perform better. Similarly, the VG-Curve becomes worse than full table scan for larger result set, due to the overhead costs of VGC directory I/Os repeated fetching of same blocks (due to page aging).

Also, CPU time for the VG-Curve, with a large result set, is worse than for the full table scan approach. However, it is worth stressing that I/O is a better measure of efficiency than CPU, since I/O is typically the bottleneck for query performance [18].

We have compared our approach with the compound index approach also considering scalability, when the number of dimensions grows from 2 to 18. The performance of VG-curve was not heavily affected by increasing dimensions as can be seen for I/Os in Figure 6 and CPU's in Figure 7. This is particularly the case for queries returning less than 0.1% of N_o. This is the case because the efficient representation of regions in the directory is barely affected by the increase in dimensions.

5 Conclusion and Future Work

In current database applications there is an increasing need to efficiently handle multidimensional data. The difficulties associated with multidimensional data grow with the number of dimensions. In this paper, we have proposed the VG-Curve, a new approach to the treatment of multidimensional data that can be easily integrated into the RDBMS since it does not require modifications to the kernel. The VG-Curve approach is a SFC method since it partitions the multidimensional space into regions and exploits the linear order induced on the regions to take advantage of index structures such as the B^+-tree. However, while SFC methods 'blindly' partition the space into regions of the minimum granularity (pixels), the VG-curve approach adopts a partitioning algorithm which is sensitive to the density of population. It accomplishes this by splitting the multidimensional space in a limited number of hyper-rectangular regions of different sizes. Only non-empty regions are explicitly maintained and considered in the VG-Curve, which has positive effects on the space, CPU and I/O complexity.

More specifically this study makes the following contributions to the field:

- We have presented a method to efficiently index *multidimensional point data*;
- We have shown that multidimensional data can be organised in way suitable for employing a primary index structure, which guarantees better performance;
- We have drawn a set of experiments, empirically demonstrating that our VG-curve is superior to the best available off the shelf RDBMS index for handling points in high dimensional space;
- We demonstrated that the VG-curve is resilient to increasing dimensions;
- Our approach is immediately suitable for full integration as it can be constructed from off-the-shelf RDBMS without modification to the kernel.

References

1. Hamelryck, T.: Efficient identification of side-chain patterns using a multidimensional index tree. Proteins: Structure, Function, and Genetics 51(1), 96–108 (2003)
2. Gaede, V., Gunther, O.: Multidimensional access methods. ACM Computing Surveys 30(2), 170–231 (1998)
3. Orlandic, R., Yu, B.: A retrieval technique for high-dimensional data and partially specified queries. Data Knowl. Eng. 42(1), 1–21 (2002)
4. Weber, R., Schek, H.J., Blott, S.: A quantitative analysis and performance study for similarity-search methods in high dimensional spaces. Proc. of VLDB (1998)
5. Fodor, I.K.: A Survey of Dimension Reduction Techniques. Technical Report Lawrence Livermore National Laboratory (LLNL),UCRL. ID-148494 (2002)
6. Orenstein, J.A., Merrett, T.H.: A class of data structures for associative searching. In: PODS 1984: Proceedings of the 3rd ACM SIGACT-SIGMOD Symposium on Principles of Database Systems, pp. 181–190. ACM Press, New York (1984)
7. Faloutsos, C., Roseman, S.: Fractals for secondary key retrieval. In: PODS 1989: Proceedings of the eighth ACM SIGACT-SIGMOD-SIGART Symposium on Principles of Database Systems, pp. 247–252. ACM Press, New York (1989)
8. Faloutsos, C.: Multiattribute hashing using gray codes. In: SIGMOD 1986: Proceedings of the 1986 ACM SIGMOD International Conference on Management of Data, pp. 227–238. ACM Press, New York (1986)

9. Berchtold, S., Böhm, C., Kriegel, H.-P., Michel, U.: Implementation of multidimensional index structures for knowledge discovery in relational databases. In: Mohania, M., Tjoa, A.M. (eds.) DaWaK 1999. LNCS, vol. 1676, pp. 261–270. Springer, Heidelberg (1999)

10. Lomet, D., Salzberg, B.: The hb-tree: A robust multiattribute search structure. In: Proc. IEEE International Conference on Data Engineering, vol. 5, pp. 296–304 (1989)

11. Ohsawa, Y., Sakauchi, M.: Bd-tree: A new n-dimensional data structure with efficient dynamic characteristics. In: Proceedings of the Ninth World Computer Congress, IFIP, pp. 539–544 (1983)

12. Chakrabarti, K., Mehrotra, M.: The hybrid tree: An index structure for high dimensional feature spaces. In: Proceedings of the 15th International Conference on Data Engineering (ICDE 1999), pp. 440–447 (1999)

13. Samet, H.: The quadtree and related hierarchical data structures. ACM Computing Surveys 16(2), 187–260 (1984)

14. Stantic, B., Topor, R.W., Terry, J., Sattar, A.: Advanced indexing technique for temporal data. Comput. Sci. Inf. Syst. 7(4), 679–703 (2010)

15. Stantic, B., Terry, J., Topor, R.W., Sattar, A.: Indexing Temporal Data with Virtual Structure. In: Catania, B., Ivanović, M., Thalheim, B. (eds.) ADBIS 2010. LNCS, vol. 6295, pp. 591–594. Springer, Heidelberg (2010)

16. Berchtold, S., Keim, D.A., Kriegel, H.-P.: The x-tree: An index structure for high-dimensional data. In: Proceedings of the 22nd International Conference on Very Large Data Bases, pp. 28–39 (1996)

17. Bayer, R., Markl, V.: The UB-tree: Performance of multidimensional range queries. Technical report (1998)

18. Hellerstein, J., Koutsupias, E., Papadimitriou, C.: On the Analysis of Indexing Schemes. In: 16th ACM SIGACT-SIGMOD-SIGART Symposium on Principles of Database Systems (1997)

MOLAP Cube Based on Parallel Scan Algorithm

Krzysztof Kaczmarski and Tomasz Rudny

Faculty of Mathematics and Information Science,
Warsaw University of Technology,
pl. Politechniki 1, 00-661 Warsaw, Poland
{k.kaczmarski,t.rudny}@mini.pw.edu.pl

Abstract. This paper describes a new approach to multidimensional OLAP cubes implementation by employing a massively parallel scan operation. This task requires dedicated data structures, setting up and querying algorithms. A prototype implementation is evaluated in aspects of robustness and scalability for both time and storage.

1 Introduction

Sequential scan algorithm originally proposed for APL in 1962 by Iverson [1] and then reborn in nineties with modern parallel machines is currently one of the most important operations among parallel primitives. Its applications vary from data transformation, sorting, string comparison to tree operations and solving recurrence equations [2]. Especially segmented scan opened a lot of new possibilities in computational programs. The algorithm is known for its very good scalability and SIMD (Single Instruction Multiple Data) processors utilization.

GPU processors as massively parallel machines also benefit from this scalable and robust general algorithms building block. There were already several variants of its parallel implementation leading to a time and space efficient solution O(n) [3]. One of the challenges was to achieve effectiveness - measured in number of primitive operations done by all processors - to be asymptotically not higher then number of operations in a sequential implementation. Another problem was poor organization of segmented scan up-sweep and down-sweep operations when divided between threads in GPU. This has also been significantly improved in works by Sengupta et al. [4].

Currently the most efficient implementation of scan algorithms for GPU is published in CUDPP library [5]. It is known for being able to run scan operation on millions of values in just a few milliseconds. In this paper we investigate its application in MOLAP cube creation and querying, and analyse its effectiveness.

1.1 Real-Time OLAP Cube

On-line Analytical Processing (OLAP) cubes, since their initial definition given by Codd [6] in year 1993, have become an important element of Business Intelligence (BI) systems. These multidimensional data structures contain aggregated data at different levels of aggregation. For example, if sales data are analyzed

J. Eder, M. Bielikova, and A.M. Tjoa (Eds.): ADBIS 2011, LNCS 6909, pp. 125–138, 2011.
© Springer-Verlag Berlin Heidelberg 2011

by years, months or days then an OLAP cube would store them as calculated aggregations for all levels in the hierarchy from days, through months to years. Thanks to preprocessing, reports can be created almost immediately and ad-hoc analyses can be very quick ("on-line"). All the calculations are performed upon the creation of the OLAP cube. This process usually is quite time-consuming and therefore OLAP cubes have to be restricted or limited in their size. Also cubes are usually build when the system is less loaded or even offline at nights. This may lead to situations when business reports can in fact only be created on the 'next day'. If the cube creation process takes longer, the time gap between real database and BI reports may be unacceptably long.

Therefore, companies seeking for improvement in making strategic decisions tend towards Real-Time OLAP (RTOLAP) systems. However, different products offer different understanding of a real time OLAP. It may be:

- Storing only the lowest level ('leaves') of aggregations and calculating an answer to a query in real time upon this cashed data in RAM.
- Ability to perform near real-time updates to the cube data from changing fact tables.
- Ability to perform near real-time changes in cube dimensions upon changes in dimensions tables.

Near real-time in this case means that a server tries to perform the update but it may take unpredicted time. If updates happen too often a server may respond in errors, significant slowdown or "cache thrash". Also changing dimensions may be very costly in time and space and should be done "sparingly" [7].

Solutions of this kind use two basic improvements to standard OLAP products. First, they are RAM-based. All the cube data, highly compressed is kept only in a computer's memory. Second, they calculate only basic data which is used to calculate the final answer to the query given by users. One of the examples of such efficient systems is IBM Cognos TM1 [8] .

In the RAM based system there are obvious storage limitations. The question is, how big cubes can be then created. One of the vendors suggests that the RAM memory should be two times larger than imported CSV file, which is a significant limitation concerning that current GPU devices have up to 6GB of memory [9].

This device's memory capacity problem is a huge issue for developers. We propose a solution which offers the biggest data volume, without any additional helper structures and indexes, containing almost only pure measures, but possibly with more expensive querying. However, the timing results even for complicated queries are still satisfactory.

1.2 OLAP Cubes and Graphical Devices

The data in the OLAP cubes is organized into dimensions and measures. Dimensions are categorical variables like year, month, product, region etc. The measures are numerical values e.g. sales amount, transactions count, customers count etc. Within dimensions there are hierarchies that serve the purpose of navigating through the data. For example the user may want to first have a look at the yearly sums, then to drill-down to monthly values, then to drill-down to daily ones, thus navigating along the YEAR-MONTH-DAY hierarchy. The measures which the aggregates are calculated for may be any additive functions. Typically these are sum, average, maximum, minimum, median, count. It happens that, apart from the other more database-like possibilities, input data for an OLAP cube is given by a single flat, denormalized data table, a CSV file or even RAW values send via a network connection. It is often also the simplest, yet most efficient way to create a cube.

The time of OLAP cube creation can be an issue on a heavily used configuration or when the cube has to be frequently updated (Multidimensional OLAP requires rebuilding to update). Even if this is not the case, the possibility of fast OLAP cubes generation can be useful for ad-hoc analyses and reports performed on local machines (even on laptops). On the other hand, modern GPU devices offer tremendous computational power which is already widely used in many applications. Also their very fast development promises even better performance in near future. Our experience proved that a modern GPU device could be well suited to handle OLAP cubes under several circumstances:

1. The cube is stored on the GPU device side or occasionally copied to CPU RAM memory. This kind of copying over PCI-Express bus consumes too much time and could kill effectiveness of the solution.
2. Queries to the cube are send to the GPU and processed on the GPU side. This is sensible for two reasons. First is similar to the previous one, we need to avoid costly transfer of large volumes of data. Second, GPU can process queries much faster than CPU because a cube is well shaped and contain uniform array of values ready for massively parallel computations.
3. A plain or compressed cube may fit in the GPU memory. Currently the most powerful devices offer up to 6GB of memory, which may give us for example a cube of more than 14 millions intersections containing 100 measures of 4 bytes each with each intersection uniquely described by an 8 byte hash code.
4. The cube must be often updated due to constantly changing situation and the system must immediately be able to process queries across all the cube.
5. Graphical results could be directly drawn in the screen by GPU without any CPU processing time.
6. Knowledge on the input data is very limited and the system has to analyse data first. Here we can also benefit from enormous instruction and memory throughput. A device may load a portion of data and within milliseconds calculate all statistics necessary to create a cube later on.

In our previous studies [10] we created a solution based on low level CUDA programming which was a first approach to OLAP cubes build on GPUs. Although it was very fast it suffered from being not general and limited in available data types. This paper improves this by propose a new more general algorithm for MOLAP (Multidimensional On Line Analytical Processing) cube creation and querying which is based on high level computational primitives. The main advantage is the parallel scan operation as a fundamental building block of most of the algorithms assures that parallelism is pushed to the limits working on all available processors achieving excellent scalability. The very same program when run on a device with more streaming processors will automatically use all available cores. This ability places our solution in the centre of interest of industrial business intelligence.

The rest of the article is organized as follows. Section 2 describes our MOLAP cube structure, set up and query algorithms. In section 3 we present important implementation details and run-time experiments including various queries executed against cubes in memory. Finally, section 4 concludes.

2 Multidemensional OLAP Cube Structure and Algorithms

An efficient OLAP cube implementation for a graphical device requires adjusting memory structures and algorithms in order to maximize benefit from GPU architecture and type of parallelism which can be employed. Already mentioned memory space limitation is an obvious problem. It is therefore needed to save the memory even for the price of longer computation times. In many cases it may be better to perform more instructions than to read from memory a value which may be computed.

2.1 Compressed In-Memory OLAP Cube

Typically, a multidimensional cube is stored as a n-dimensional array with k measure values in each element. Storing a cube requires normalization of dimensional attributes by mapping them to integer values. A tuple of n integers may be then used to select a single intersection of a cube. Accessing content of an intersection requires then $O(k)$ time (devoted to measure values memory reads) if only address in the array may be calculated in constant time. In most cases due to limitations of memory addressing calculating an intersection placement may require additionally $O(n)$ operations. Since both n and k are rather small values we can say that accessing a cube represented in multidimensional array can be done in approximately constant time. In industrial databases number of dimensions and possible ranges of dimension values may lead very sparse cubes in which most of places in the cube are containing no measure values. There are many techniques to construct an efficient sparse cube representation [11] by changing normalization of data, using various heuristics or sorting indices.

One of the simplest techniques is to store only measure values together with dimensional attributes omitting empty intersections. An advantage here is being independent of data density characteristics but for giving up direct access to any intersection measures as described above. Because of memory limitation we decide to go one step further and to compress cube memory representation to one dimensional array containing hash keys and measure values. All dimensional attributes for a given intersection are exchanged with a single hash value.

If a single intersection was described by: $d_0^j, \ldots d_{n-1}^j, m_0^j, \ldots m_{k-1}^j$, where d_i^j denotes i-th dimension and m_i^j denotes i-th measure values the compressed cube contains structures $\{h^j, m_0^j, \ldots, m_{k-1}^j\}$ (see fig. 1). For the details on the hash function see section 2.3.

Since we do not know which intersections are filled with data and which are empty, we cannot predict the place in the compressed cube for given dimensional values. Therefore, accessing a cube intersection requires at least a binary search over the sorted set of hash codes. This process has logarithmic complexity and may be performed by only one thread at a time. For a SIMD processor we propose an alternative solution (see section 2.4).

	d_0	d_1	m_0			2008	2009		h	c
r_0	2008	10	23		10	23	0		0	23
r_1	2008	12	5		11	0	0		2	48
r_2	2008	12	43		12	48	0		5	8
r_3	2008	15	8		13	0	0		15	90
r_4	2009	15	90		14	0	0		17	21
r_5	2009	17	21		15	8	90		19	6
r_6	2009	19	3		16	0	0			
r_7	2009	19	3		17	0	21			
					18	0	0			
					19	0	6			

Fig. 1. From the left: 1. Sample data with two dimensions and one measure. **2.** Fragment of a direct cube representation in n dimensional array (in this case $n = 2$). **3.** Compressed cube with hashing in one dimensional array.

Our compressed representation of a multidimensional cube makes it very well suited for graphical devices and their coalesced memory access technique. It works if a set of threads (typically 16 - a so called half-warp of threads) access a limited area in memory called a segment, usually of 32, 64 or 128 bytes. Coalescing means that all threads will get their data in the very same memory access instruction. On the contrary, if threads within a half-warp access memory randomly exceeding the same memory segment, their reading (or writing) instructions are serialized and lead to degradation of memory bandwidth. Obviously, our flat array of structures $[\{h_0, m_0^0, \ldots, m_0^{k-1}\}, \ldots, \{h_p, m_p^0, \ldots, m_p^{k-1}\}]$ may be converted to a structure of arrays $\{[h_0, \ldots, h_p], [m_0^0, \ldots, m_p^0], \ldots, [m_0^{k-1}, \ldots, m_p^{k-1}]\}$ for full coalescing. No other structure could stand coalescing requirements.

2.2   Scan-Based Parallel Primitives

For the purpose of cube creation and querying we shall use the following scan-based primitive operations.

Reduce operation takes a binary associative operator $\oplus$, and an array of n elements $[x_0, x_1, \ldots, x_{n-1}]$, and returns a single element $x_0 \oplus x_1 \cdots \oplus x_{n-1}$.

Scan operation takes a binary associative operator $\oplus$, and an array of n elements $[x_0, x_1, \ldots, x_{n-1}]$, and returns the array $[x_0, (x_0 \oplus x_1), \ldots, (x_0 \oplus x_1 \cdots \oplus x_{n-1})]$, while **prescan** operation taking the same array would return array $[I, x_0, (x_0 \oplus x_1), \ldots, (x_0 \oplus x_1 \cdots \oplus x_{n-2})]$.

Segmented scan operation takes an array of input values $[a_0, \ldots, a_{n-1}]$ and array of boolean flags $[f_0, \ldots, f_{n-1}]$ and returns an array $[x_0, \ldots, x_{n-1}]$ satisfying the equation:

$$x_i = \begin{cases} a_0 & i = 0 \\ \begin{cases} a_i & f_i = 1 \\ (x_{i-1} \oplus a_i) & f_i = 0 \end{cases} & 0 < i < n \end{cases}$$

Segmented scan works like scan but it is limited to segments denoted by true f_i flags placed at beginnings of segments. *Inverted segmented scan* works like segmented scan but $\oplus$ operation is calculated in opposite direction, from the end of the segment to the beginning.

Pack or compact operation (pack) having an input array $[x_0, x_1, \ldots, x_{n-1}]$ and boolean flags array $[f_0, f_1, \ldots, f_{n-1}]$ returns an output array containing all elements x_i for which f_i is true.

All the above primitive operations if implemented efficiently [2,3] have time complexity $O(n)$. Memory requirements will be discussed later since they are connected to details of implementation.

2.3 Massively Parallel Creation Algorithm

We assume that the input data set is organized in p rows $r^0, \ldots r^{p-1}$ containing plain $n + k$ integer values each: $r^j = d_0^j, \ldots d_{n-1}^j, m_0^j, \ldots m_{k-1}^j$, where d_i^j denotes i-th dimension and m_i^j denotes i-th measure values.

Considering only integer values in the input data set is not a limitation, since most of dimension data usually falls to a small range of possible values and therefore is stored and encoded in dictionary tables. E.g. 16 districts of Poland can be easily enumerated, so there is no obstacle in having their indices in a dictionary table rather then full names (string data type).

In many cases this input set is already sorted by dimensions or may be easily sorted on the GPU side before processing. Cube creation of unsorted data sets without sorting cannot be done with scan operation, needs different approach, and will be addressed by another paper.

First, we analyse data to find number of values in each dimension which are necessary to describe the cube. This is based on a simple min and max value search for each dimension d_i. As a result we get a collection of n minimums $(d_i{}^{min})$ and n maximums $(d_i{}^{max})$. We assume that each dimension d_i may have up to $d_i{}^{max} - d_i{}^{min} = s_i$ different values. s_i is a size of i-th dimension. This step can be done with $2n$ reduce operations over p values each:

$$d_i{}^{min} = \text{reduce}_{min}(d_i^0, \ldots d_i^{p-1}), \quad d_i{}^{max} = \text{reduce}_{max}(d_i^0, \ldots d_i^{p-1})$$

The second step of the algorithm finds data ranges, a set of tuples which should be aggregated for each cube intersection. A number of v rows $r^j, \ldots, r^{(j+v-1)}$ belong to the same data range (to the same scan segment) if no dimension attribute changes in all given rows. To find segments we run a parallel computation comparing a set of dimensions values within subsequent rows. Beginning of new segments are defined by flags $f^j \in \{0, 1\}$ where

$$f^0 = 1$$
$$f^{j>0} = \begin{cases} 0 & \text{if} \quad \forall_{i=0\ldots n-1} d_i^j = d_i^{j-1} \\ 1 & \text{otherwise} \end{cases}$$

The segments flags are used to calculate the number of existing intersections, and in the same time a place of each intersection in the compressed cube, by calling prescan operation on the flags vector. An example result of this operation is illustrated in fig 2, see column f for flags and ps_f for prescan effects.

All the process of calculating k measures in cube intersections can be done by k executions of inverted segmented scan operation on subsequent measures $m_0, \ldots m_{k-1}$ using flags f. This step ends calculations of the final content of a cube. Now, to finalize cube creation, we must calculate hash codes for all intersections and reorganize the data to create the final cube storage.

Having all necessary information about the input we calculate a hash code for each row using a perfect hash given by formula:

$$h^j = \sum_{i=0}^{n-1} \left((d_i^j - d_i^{min}) \prod_{r=i+1}^{n-1} s_r \right)$$

This hash function is actually an index of an intersection in a multidimensional array cube (see fig. 1) and has two very important properties. Firstly, it is conflict-free and secondly, it is fully reversible. Having a hash code h^j for given intersection in a cube and characteristics of data in d_i^{min} and d_i^{max}, we will be able to reverse the hash in time $O(n)$ and calculate values of all $d_0^j, \ldots, d_{n-1}^j$ dimension attributes later.

Finally, we are ready to pack results into a compressed cube array by calling pack operation for all k columns produced by scans on measures and adding hash codes for all intersections. Again this process is shown in fig. 2.

It is obvious that all the above steps have time complexity $O(n)$ since in each step we perform only two kinds of operations: we concurrently access all input data rows (one or at most two times each) or call scan operation for all the elements. In the same way regarding number of dimensions or number of measures the creation algorithm is always only linear.

2.4 Querying Multidimensional OLAP by Linear Searching

Efficient implementation of OLAP cube querying is a hard task for disk-based systems. It is important to optimize number of disk accesses and page swapping. Many structures like statistics tree [12] were proposed to solve this problem

	d_0	d_1	d_2	d_3	d_4	m_0		f		ps_f	iss_fm_0		h	c	
r^0	2008	10	04	190	6	23		1		0	79		132	79	c^0
r^1	2008	10	04	190	6	5		0		1	56		134	111	c^1
r^2	2008	10	04	190	6	43		0		1	51		135	6	c^2
r^3	2008	10	04	190	6	8		0		1	8				
r^4	2008	10	04	190	8	90		1		1	111				
r^5	2008	10	04	190	8	21		0		2	21				
r^6	2008	10	05	164	4	3		1		2	6				
r^7	2008	10	05	164	4	3		0		3	3				

Fig. 2. A sample cube creation process. $r^0, \ldots, r^7$–rows of input data. $d_0, \ldots, d_4$–dimensions. m_0–single measure column. f–flags indicating segments(ranges) in input data. ps_f–result of prescan operation on f. iss_fm_0–result of inverted segmented scan done on m_0 using flags f. This column is calculated for each measure m_i. h–hash codes in compressed cube. c–measure values in the resulting cube.

by minimizing a search path along dimensions. Also a large number of dimensions may lead to disk storage explosion and serious degradation of performance. Condensed [13,14] or minimal cubing [15] addresses this problem. Especially the later one introduces many additional structures which are not acceptable due to limited device's memory space. Surprisingly, just a flat array used as a cube container may be sufficient if only the search algorithm is efficiently implemented allowing to use all available SIMD processors.

Although compressed cube hashes in the memory array are sorted, binary search cannot be used due to parallelism degradation. A classical binary search of an array assumes that there is only one thread performing nested divisions of searched array. In each step the remaining array is divided into two equal parts and the process continues for one of them. Since we have only one thread working in one query it is not well suited for SIMD processor like a GPU device. Even running a number of queries in the same time would not improve efficiency because each thread would focus on different part of the cube array and coalesced memory reads could not be achieved.

Slicing and dicing is easy for a n dimensional cube array. In the first case we just set given number of dimension indices iterating over the rest of dimensions. In the second, again it is enough to set boundaries for dimension values and read intersections inside. Compressed cube representation make this process more complicated. Except for very special situations it is impossible to predict where intersections for given dimension values are located. This leads us to a conclusion that a compressed cube query probably has to access all elements of the cube in order to create a slice or dice.

Hopefully, graphical device memory bandwidth is much higher (achieving around 192GB/s) than in case of a classical PC CPU memory. Therefore, we can just try to access all the available intersections at once in parallel threads and select the ones which satisfy a query sent to all the threads. This leads us to the following brute force linear search algorithm:

1. Convert a query to a form ready for processing by single SIMD thread.
2. Run as many threads as distinct entries in the compressed cube array.
3. In each thread: read single cube intersection, reverse hash code and decide whether it satisfies the condition, mark selection flag true or false.
4. If all threads finished then calculate result by calling pack operation on the cube array using the recently created selection flags.
5. If necessary perform a reduction operation on the result.

All the procedure is illustrated in fig. 3. The most important part of this algorithm is to decide if a given intersection should be marked for the final result set. A single thread calculates all dimensions values by reversing hash code in its intersection in time $O(n)$. Unfortunately brute force query has bad memory accessing complexity since all the cube must be read. However, memory representation of the cube allows for coalesced reads and for the newest devices we can achieve 32 threads reading subsequent memory locations in the same time.

There are evidences that brute force algorithms can be successful in other applications like for example password recovery, MD5 cracking or nearest neighbour search [16].

	h	c	f	p	r
c_0	132	79	1	79	98
c_1	134	111	0	6	
c_2	135	6	1	98	
c_3	137	15	0		
c_4	190	98	1		
c_5	196	4	0		

Fig. 3. A sample cube querying process. $c_0, \ldots, c_5$–intersections of the cube. h–hash codes for intersections. These codes are used to select intersections satisfying a query. c–single measure in cube. f–flags indicating query results. p–result of pack operation called on c using f. r–eventually created aggregated result if a query defines this kind of output. This is done by executing a reduce operation on p. In this case, reduction was done using max operator.

3   Implementation and Run-Time Experiments

The algorithm's implementation is based on CUDPP library in version 1.1 [5] currently shipped with NVIDIA CUDA SDK 3.2. Among other functions, this library contains efficient procedures for scan and segmented scans with variants allowing them to work as all elements scans or prescans. Additionally a multiple row scan may perform parallel scans of multiple row arrays allowing for better instruction and memory throughput. However, we found this method not significantly faster then a set of single column scans run one after another.

3.1   Cube Creation with Scan Primitives

In this section we analyse how parallel scan primitives were used to implement cube creation algorithm. Below we enumerate all the steps of the algorithms with additional remarks on memory complexity and notes on CUDPP utilization (n – number of dimensions, k – number of measures, p – number of records in the input database, typically much larger than $n + k$, l – number of existing intersections in the cube).

1. *Find minimum and maximum for each dimension in the input data.* This task could be done by $2n$ reduction operations. However, currently reduction is not implemented in CUDPP. A normal scan operation must be used instead. Both operations have linear time complexity but scan perform two times more steps. Time complexity: $O(np)$. Memory complexity: $O(p)$, since we perform single scan at a time which requires memory storage for partial scan sums. It is reused in all scans in this step and may be freed afterwards.

2. *Marking segments in the input data by flags.* This task has to be coded manually, since there is no appropriate operation in CUDPP library. It requires $2pn$ memory reads (each of p threads – one thread for each input record – must read n dimension attributes of two subsequent input records) and p memory writes. Time complexity: $O(p)$. Memory complexity: $O(p)$ for p flags marking segments. Flags could be stored as single bits, however this is not supported by CUDPP and could not be used in next steps of the algorithm. Unfortunately, flags have to be kept in memory up to the very end of the procedure (see fig. 2 array f).

3. *Enumerate ranges in the input data.* Done with CUDPP prescan operation over flags of segments. Time complexity $O(p)$. Memory complexity: $O(p)$ because we need additional storage for scan results (see fig. 2 array ps_f). It is possible to reuse flags storage in this step but for the price of more memory reads in the next steps. Time complexity: $O(p)$. Memory complexity: $O(p)$. After this step we know how many intersection constitute resulting cube and the storage for the compressed cube array may be allocated.

4. *Reduce measure values inside segments in the input data to get content of the output cube intersections* This step requires a segmented reduction repeated for each measure. Again, this kind of operation is not implemented in CUDPP. A segmented scan must be used instead. To simplify the next step we perform k times a variant called inverted segmented scan. The advantage is that for each segment a result of aggregation in placed in the position of the flag indicating the segment beginning (see fig. 2 array iss_fm0). Time complexity: $O(kp)$. Memory complexity: $O(p)$ for scan results in each step. Memory may be freed after results are copied to the output cube array.

5. *Copy output of step 4 into the resulting cube.* This step requires $2p$ memory reads for flags and segment enumerations plus kl memory reads and kl memory writes. Time complexity: $O(p + kl)$. Memory complexity: $O(1)$ No additional storage is required except for the final compressed cube array of size $l(k + 1)$ (k measures and 1 for hash code in each intersection).

6. *Calculate hashes column for the resulting cube and write it into the output cube.* In the final step we calculate hash code for each segment (n reads for each segment marked by flag) and write it into the cube array. Time complexity: $O(p + nl)$. Memory complexity: $O(1)$

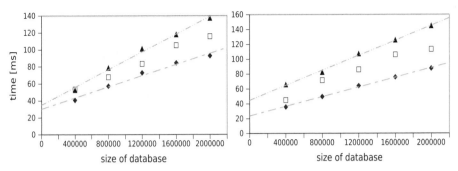

Fig. 4. Cube creation times. X axis presents number of records in input database. Y axis presents creation time in milliseconds. Experiment performed for various dimensions and measures. **Left:** Number of dimensions: 5, 10 and 15. **Right:** Number of measures: 5, 10 and 15.

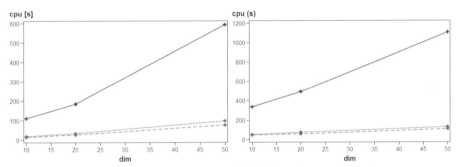

Fig. 5. Cube creation times in a commercial CPU-based system. X axis presents number of dimensions, while Y axis presents creation time in seconds. Experiment performed for various dimensions and measures. **Left:** Number of measures: 5, number of intersections (aggregations): 130k, 400k, 4M. **Right:** Number of measures: 50, number of intersections (aggregations): 130k, 400k, 4M.

The overall time complexity of this algorithm is linear regarding the number of input records, dimensions and measures but with excellent factors which may be easily observed in figure 4. Increasing size of the database input by two times increases time by 1.5. Similarly increasing dimensions or measures number by three changes the time by only 1.75. As far as we know there is no commercial system having similar performance. E.g. we have measured the cube creation times for a highly-esteemed commercial system SAS OLAP Server 9.1.3 and obtained the results presented on figure 5. Although the test environment was limited (IBM PC 2 GHz, 1 GB RAM) which leaves some room for improving the performance of the commercial system, yet we may observe an increasing influence of number of dimensions and measures on the cube creation time.

Another important observation is that this approach is not slower then our previous low-level and much less flexible OLAP cube creation implemented without the scan primitive [10]. Time measurements indicated very similar results but now we can add more dimensions and measures (limited only by memory capacity) witch is an important improvement.

3.2   Massively Parallel Cube and Database Queries

As it was explained earlier, abilities of graphical devices allows for brute force querying. This experiment was performed on three different cubes created from different databases. Therefore size of a queried cube varied from about 21k to about 174k of intersections (see fig. 6 left).

Here we present implementation for each step of the querying algorithm explaining usage of CUDPP library. (n – dimensions, k – measures, l – intersections in the cube, v – query results)

1. *Marking intersections in the cube as partial query results.* The first step of brute force querying requires opening each cube intersection, reversing its hash code and deciding if it satisfies the query. For each thread: reading the hash code is single operation, reversing the hash is done in n steps, writing the decision in one step. Time complexity: $O(nl)$. Memory complexity: $O(l)$, we need to store a boolean flag for each intersection indicating whether it satisfies query or not (see. fig. 3 array f).

2. *Enumerate results.* This task uses prescan operation run on flags array (see. fig. 3 array ps_f). Time complexity: $O(l)$. Memory complexity: $O(l)$ for the results. Prescan can be also run on flags array directly without any additional memory consumption but with more time needed to copy results of the query.

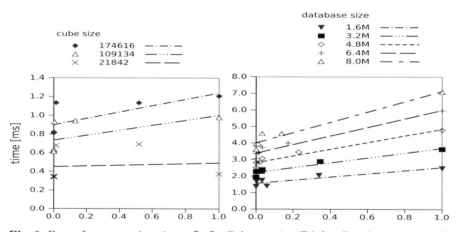

Fig. 6. Brute force querying times. **Left:** Cube queries. **Right:** Database queries. Although, cube size is much smaller than a database size query has to additionally perform hash function inversion for each intersection. X axis: query selection ratio – almost 0 means that a single record satisfied query criteria while 1 means that all records were selected as a result. Y axis: query time in milliseconds. Each query was selecting records and then reducing (summing) to a single value.

3. *Copying measure values from cube to query results.* This step requires $2l$ reads of flags and results enumeration and kv writes. Time complexity: $O(l + kv)$. Memory complexity $O(1)$.

The overall complexity is obviously linear in the size of the cube and the number of measures and dimensions. Very good results of cube querying encouraged us to test the same method for querying the whole database stored in a graphical device. We managed to run experiment with five different databases of sizes from 1.6M to 8.0M of records (see fig. 6 right).

4   Conclusions and Future Works

We have presented how OLAP cubes can be created and queried efficiently using the massively parallel scan-algorithm. The results prove that GPU-based data structures perform way better than classical CPU implementations. Moreover, we have achieved great scalability of the presented algorithms. A compressed cube may be easily divided between several devices in the same machine or even between a number of computers in a cluster.

Our novel algorithm based on parallel scans not only allows any number of input dimensions and measures but also behaves very well resulting in linear dependency between time and number of dimensions or measures in cube creation. Surprisingly compressed cube querying with a brute force algorithm responses in a few milliseconds.

The only drawback of the parallel scan is its memory consumption. Due to its internal construction it has to store temporary computations data aside scanned table. This requires $O(n)$ additional storage. However, when well organized the processed data may fill half of the available memory allowing processing storage in the rest of the space.

Another problem is CUDPP library which is still under development and does not implement several important operations like for example a simple reduction or its segmented version. This is expected to be improved in the future.

Among future open topics we could count: measure values compression for even higher memory savings; compressing the whole cube in memory for extremely demanding systems; improvement in cube storage by exchanging a flat array with other structures and the most important ability of accepting unsorted input when creating a cube. The last requirement may be easily achieved by creating a dynamic cube which is reorganized for each new database row appearing in the input. In a typical system reorganization of a MOLAP cube by changing number of dimensions or number of different values in a dimension is very expensive. In case of GPU this process could be much faster an possibly done in the real time. This means that the cube could be recreated between the subsequent records are read from a hard drive or network connection. This needs further investigations and probably changes in memory structures.

References

1. Iverson, K.E.: APL: A Programming Language. John Wiley & Sons, Chichester (1962)
2. Blelloch, G.E.: Prefix sums and their applications. In: Sythesis of parallel algorithms, pp. 35–60. Morgan Kaufmann Publishers Inc., San Francisco (1990)
3. Sengupta, S., Harris, M., Zhang, Y., Owens, J.D.: Scan primitives for GPU computing. In: Segal, M., Aila, T. (eds.) Graphics Hardware, pp. 97–106. Eurographics Association, San Diego (2007)
4. Sengupta, S., Harris, M., Garland, M.: Efficient parallel scan algorithms for GPUs. Tech. Rep. NVR-2008-003, NVIDIA Corporation (December 2008)
5. CUDA Data Parallel Primitives Library (2011), http://code.google.com/p/cudpp/
6. Codd, E.F., Codd, S.B., Salley, C.T.: Providing OLAP (On-Line Analytical Processing) to User-Analysts: An IT Mandate. E. F. Codd and Associates (1993)
7. Kennedy, D.: The reality of real-time OLAP. tech. rep., Microsoft Corporation, Microsoft SQL Server 2000 Analysis Service (February 2003)
8. IBM Cognos TM1, high-performance enterprise planning software for budgeting, forecasting and analysis. IBM Corporation (2011), http://www-01.ibm.com/software/data/cognos/products/tm1/
9. Palo GPU Accelerator: whitepaper, tech. rep., Jedox (2011), http://www.jedox.com
10. Kaczmarski, K.: Comparing GPU and CPU in OLAP cubes creation. In: Černá, I., Gyimóthy, T., Hromkovič, J., Jefferey, K., Králović, R., Vukolić, M., Wolf, S. (eds.) SOFSEM 2011. LNCS, vol. 6543, pp. 308–319. Springer, Heidelberg (2011)
11. Kaser, O.: Compressing molap arrays by attribute-value reordering: An experimental analysis. tech. rep., Dept. (2002)
12. Fu, L., Hammer, J.: Cubist: A new algorithm for improving the performance of ad-hoc OLAP queries. In: DOLAP, pp. 72–79 (2000)
13. Wang, W., Lu, H., Feng, J., Yu, J.X.: Condensed cube: An efficient approach to reducing data cube size. In: ICDE, pp. 155–165. IEEE Computer Society, Los Alamitos (2002)
14. Feng, J., Fang, Q., Ding, H.: Prefixcube: prefix-sharing condensed data cube. In: Song, I.-Y., Davis, K.C. (eds.) DOLAP, pp. 38–47. ACM, New York (2004)
15. Li, X., Han, J., Gonzalez, H.: High-dimensional OLAP: A minimal cubing approach. In: Nascimento, M.A., Zsu, M.T., Kossmann, D., Miller, R.J., Blakeley, J.A., Schiefer, K.B. (eds.) VLDB, pp. 528–539. Morgan Kaufmann, San Francisco (2004)
16. Garcia, V., Nielsen, F.: Searching high-dimensional neighbours: Cpu-based tailored data-structures versus gpu-based brute-force method. In: Gagalowicz, A., Philips, W. (eds.) MIRAGE 2009. LNCS, vol. 5496, pp. 425–436. Springer, Heidelberg (2009)

Real-Time Computation of Advanced Rules in OLAP Databases

Steffen Wittmer[1], Tobias Lauer[2], and Amitava Datta[3]

[1] Jedox AG, Freiburg, Germany
steffen.wittmer@jedox.com
[2] University of Freiburg, Germany
lauer@informatik.uni-freiburg.de
[3] University of Western Australia, Perth, Australia
datta@csse.uwa.edu.au

Abstract. In Online Analytical Processing (OLAP) users view data through a multidimensional model known as the *data cube*, allowing the aggregation of information along different attributes and operations such as slicing and dicing. In-memory OLAP systems keep all relevant data in main memory and also support efficient updates of cube data, enabling interactive planning, forecasting, and what-if analysis. Since usually only the base data is stored and all aggregations and other calculations are computed on the fly, complex computations may seriously downgrade performance. We present an approach that uses graphics processing units (GPUs) as parallel coprocessors for high performance in-memory OLAP operations. In particular, our method accelerates the calculation of compute-intensive *rules*, which represent business dependencies that are more complex than mere aggregates. In addition to the data structures and algorithms, we describe how to extend the approach to multi-GPU systems in order to scale it to larger data sets.

1 Introduction

In Online Analytical Processing (OLAP) users can view data from a data warehouse through a multidimensional model known as the *data cube* [8], allowing them to aggregate information along different attributes and to perform operations such as slicing and dicing as well as roll-up and drill-down to different levels of detail along dimensional hierarchies. One key requirement for OLAP systems is performance – users ideally want responses to even complex queries within seconds.

In addition to standard aggregations, more advanced business dependencies may be part of the user's data model. Common examples from the business context include computation of sales figures from prices and quantities, currency conversions, and many more. We refer to these dependencies as *rules*. Rules may be computed on demand or pre-calculated and stored in the database before loading the data into the OLAP system. The major disadvantage of the latter approach is that any update to the database requires re-computing all dependencies, which is problematic when it comes to users' real-time changes to the OLAP data in planning, forecasting, or what-if scenarios. In order to facilitate data consistency after updates, it makes sense to only

J. Eder, M. Bielikova, and A.M. Tjoa (Eds.): ADBIS 2011, LNCS 6909, pp. 139–152, 2011.

keep base data in the database and compute all aggregates and other dependencies on the fly, i.e. when requested by the user. Another advantage of online aggregation is a drastically reduced memory consumption, since aggregates do not have to be stored (except for caching). Usually, OLAP cubes are very sparse and the number of possible aggregates usually exceeds the number of existing (i.e. non-zero) base cells by many times.

Online aggregation is the preferred domain of *in-memory* OLAP databases such as [12-14], which keep all relevant data in main memory, hence allowing more efficient updates compared to secondary storage-based systems. On the other hand, very efficient algorithms are required to compute the requested aggregates and rule results.

In this paper, we describe the utilization of graphics processing units (GPUs) in order to accelerate OLAP rule computations. The following section provides related work and important preliminaries on OLAP rules. Section 3 briefly recapitulates our approach to parallel OLAP aggregation using GPUs, before we turn to the main part, GPU-based calculations of advanced cube rules, in section 4. In the fifth section, we extend our approach to multiple GPUs and outline how the involved algorithmic steps have to be augmented. Section 6 provides results of first performance tests, before we conclude this work and give some directions for future research.

2   Background and Related Work

A solid introduction of OLAP and the data cube would be beyond the scope of this paper, and we refer interested readers to literature such as [3]. However, for the purposes of this work, it is useful to situate our approach within the diverse landscape of OLAP systems. A rough division can be made between relational (ROLAP) and multidimensional (MOLAP) approaches with our focus clearly being on the MOLAP side.

2.1   Parallel Database and OLAP Computation

Parallelization of data cube computation in the ROLAP world has been studied extensively by Dehne et al. [5]. In their work, the focus is on optimization of cube construction methods (i.e. pre-computation of aggregates) using load-balancing on computer clusters. Although our approach could also be applied to clusters in order to scale for larger amounts of data, the parallel method as well as its application scenario is different. We apply the massive data-parallelism of GPUs on the level of individual computation steps in order to better support real-time ad-hoc OLAP analysis and planning.

Uses of GPUs in databases include the works by Govindaraju and colleagues [7][10]. These authors have worked on a number of database primitives such as relational joins, sorting, predicates, range queries and achieved some remarkable speedups compared to efficient CPU counterparts of their algorithms.

As a practical example, Bakkum and Skadrum [2] have designed a GPU extension for SQLite in-memory tables. Their approach for aggregation is quite similar to our own previous work published in [16] and briefly described below in section 3.2.

However, the data types usable in their method are very limited. In addition, storage restrictions of GPUs pose a severe limitation to the use of their method as they do not support multiple GPUs.

Recently, Kaczmarski [15] compared the use of GPUs and CPUs for parallel OLAP cube pre-aggregation, i.e. offline computation of all possible aggregates. His method works on fully denormalized fact tables and uses data parallelism on three levels of the cube computation: finding dimension intersections, calculation of different aggregates, and within each individual aggregation. The latter two are similar to the pure aggregation part of our approach [16]. However our data structure does not require full denormalization of tables and is hence more memory-efficient, so cubes can reside in GPU memory, avoiding costly data transfer through the PCIe bottleneck.

Other uses of GPUs in fields related to OLAP and data analysis include the approach by Andrzejewski and Wrembel [1] who apply GPUs to the WAH compression technique for bitmap indexes used in data warehousing.

2.2   Rules in OLAP Cubes

A rule can be understood as an advanced function beyond simple aggregations. It basically gives users the option to apply formula on cells of a data cube (which resembles the common spreadsheet functionality of applying formulas on cells in a spreadsheet). Rules are commonly supported by in-memory OLAP servers and can be found in many major systems of this class, such as [12-14].

We explain the concept by the following example: Consider a data cube containing the dimensions "Products" and "Zipcode" with cardinalities 500 and 10,000, respectively, and the rule-dimension "Measures", with elements *turnover*, *quantity*, and *price*. However, since turnover is a function of quantity and price, data cells for *turnover* are not stored but computed by the following rule:

$$\text{Measures}[turnover] = \text{Measures}[price] \text{ x Measures } [quantity].$$

The computation of such a rule can be imagined as illustrated in Figure 1. For each "Measures [*turnover*]" cell in the data cube the values of the corresponding *price* and *quantity* cells are multiplied. In the example, all cells in the lowest horizontal slice of the data cube are [*turnover*] cells, such that 500 x 10,000 = 5,000,000 multiplications will be done in this rule computation on a fully filled data cube. Theoretically, this number is determined by the product of the cardinalities of all dimensions but the rule dimension (in our case, "Measures"). For real-world data cubes, the computation of a single rule can therefore involve millions of cells.

In terms of performance, this implies high workloads for the MOLAP Server. Rules are typically computed in the context of aggregate queries, for example when the overall turnover for all products or all zip codes is requested. In general, the rule operation and the aggregation operator (SUM, AVERAGE, COUNT, etc.) cannot be applied in arbitrary order. In our example, the aggregate turnover of all products is not the same as the aggregated price multiplied by the aggregated quantities. Hence, the rule has to be applied on the base level before aggregation.

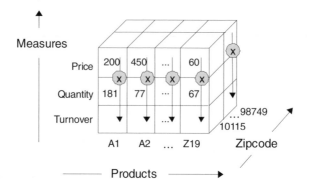

Fig. 1. Application of a multiplication rule on the data cube

3 OLAP Computation on GPUs

In the context of Online Analytical Processing (OLAP) the utilization of GPUs for enhancing performance is quite promising. The main focus of our work is the implementation of hardware efficient GPU algorithms to provide an accelerated computation of OLAP aggregations and rules.

3.1 GPU Cube Data Structure

The data structure we employ to handle OLAP cube data on the GPU device(s) is based on the one described in [16]. While information about the dimensions (such as hierarchies etc.) reside in main memory, the cube base cells (facts) are stored on the GPU in a fact table fashion using a combination of arrays. The cell coordinates (the *path* of a fact) are transcoded and compressed and then inserted column-first in one array. Measure *values* are stored in a separate array. Corresponding paths and values are implicitly connected via their absolute positions in the respective array. Figure 2 shows a simplified schematic of the basic structure. On the left-hand side, input facts are given as tuples containing the coordinates and the value of a cell.

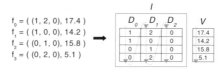

Fig. 2. Data structure for storing a fact table on a GPU

3.2 OLAP Aggregations on GPUs

Since only base facts are stored in our cube model, all aggregate values are computed on demand (unless they are found in a query cache). We have previously designed a GPU aggregation method utilizing the massively parallel architecture of graphics processors, which is very efficient compared to sequential algorithms [16].

The basic procedure is a data-parallel version of a full table scan identifying all relevant facts (i.e. the ones contributing to the requested aggregate) and a parallel reduction of the respective values.

In performance tests, we measured speedup factors of up to 40x compared to an optimized sequential algorithm (involving a sophisticated hybrid data structure) running on the same system. One nice property of the described approach is that it scales seamlessly to multiple GPUs in one system, which is achieved by splitting the fact table evenly and distributing it over the available GPUs (cf. [16]). The aggregation results of all GPUs are collected and again aggregated to the final result. This method comes with a double advantage: First, by distributing the data over several GPUs, the approach can accommodate larger data sets. Second, if the amount of data remains the same, the performance scales up with the number of GPUs, as each graphics processor's work is reduced to only its part of the fact table.

We will see that multi-GPU support is more challenging when calculating rules rather than just simple aggregations.

4   GPU Rule Computation

As mentioned in section 2.2, the calculation of rule results can be extremely compute-intensive. As an example, we use a prototypical rule of the form $D_R[c] = D_R[a] \circ D_R[b]$ which semantically defines that each cell value of a cell that is defined by the element ID c in the rule-dimension D_R (or, shorter, a $D_R[c]$ cell), is assigned with the result of a given binary arithmetic operation applied on corresponding $D_R[a]$ and $D_R[b]$ cell values (the approach can be easily transferred to operators with only one operand or more than two operands). Corresponding cells, or *matching facts,* are those cells that share with the given cell the same element IDs in all dimensions but D_R. We say a cell is *affected by a rule* if it is a $D_R[c]$ cell.

Although our above sample rule is simple in its appearance, its sequential computation can be cumbersome, especially when many cells are affected. When an aggregated cell is requested by the client, the OLAP server determines all base facts that contribute to the aggregation and checks sequentially for each fact whether it is affected by a rule. If so, the server applies the rule for that fact and returns the result as an input for the aggregation. The computation of rules on base facts is costly, as for each affected cell (the left-hand side of a rule) the corresponding operand facts (on the right-hand side) need to be looked up in the fact table as well. As data cubes are typically very sparse, a large amount of operand facts will not even be existent, such that only a small fraction of the required lookup operations are relevant in that they return non-zero values, which actually contribute to a rule result. Note that the hierarchical structure of dimensions allows for nested aggregations, i.e. an aggregated cell can be part of another aggregated cell at a higher level in a hierarchy. The algorithm for finding base facts is therefore recursive.

If such a query is processed in a top-down fashion, i.e. driven by the aggregate, in a sparse cube millions of non-existent values will be individually looked up, making online rule computation practically infeasible. An alternative ("bottom-up") approach would be an index which points, for each rule, to only the existent facts needed for the computation of that rule. In practice, this method is usually much faster, but it requires extra memory for the index and introduces some problems if non-existent (i.e. zero-value) cells also contribute to the rule result.

4.1 Finding and Matching Input Facts

We have designed a data-parallel algorithm for computing such rules efficiently on the GPU data structure discussed above. As a simple example, we use the following rule:

$$D_1[0] = D_1[1] \circ D_1[2].$$

Recall that the element with ID 0 in dimension 1 should be calculated as a function of elements 1 and 2 in the same dimension, where the function is specified by the operator $\circ$. In order to calculate an aggregate query over many cells whose D_1 coordinate is 0, we have to find the corresponding facts with 1 and 2 in that coordinate, match them, and calculate the result. On the GPU, this can be done in parallel for many facts at a time.

As illustrated by the vertical arrows in Figure 3 (left), matching $D_1[1]$ and $D_1[2]$ cells are those that share the same dimension indexes for all dimensions but D_1.

The task of finding matching $D_1[1]$ and $D_1[2]$ facts that need to be processed together is non-trivial, as the fact table is assumed to be unordered. Figure 3 (right) displays the corresponding fact table of the data cube on the left.

It is obvious that no information can be gained on where to find a matching fact for an arbitrarily chosen fact. Moreover, it is even uncertain if such a matching fact exists in the fact table at all, due to cube sparsity. In general, if either $D_R[a]$ or $D_R[b]$ does not have a matching fact we label them *single facts*. When a single fact is found, it depends on the arithmetic operation how this fact is processed. Since a non-existing fact is regarded to have the value zero, in the case of addition a single fact already provides the result of the operation. In many other cases, like multiplication, the result of an operation involving only a single fact is defined to be zero.

Searching matching facts in the fact table implies that for each $D_R[a]$ and $D_R[b]$ fact the whole fact table needs to be scanned for finding the matching fact or for validating the non-existence of such a fact. Clearly, an unsorted fact table is an inappropriate data structure regarding the search of facts.

However, we cannot ensure that the fact table is always sorted, especially when frequent updates to the cube occur, as is common in interactive planning scenarios. Our approach therefore utilizes a series of filters and only (re-)sorts small parts of the fact table when it is required.

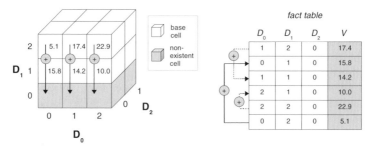

Fig. 3. Matching facts in data cube and in a fact table

4.2 A Massively Parallel Solution

The basic algorithm for computing a binary rule operation on the GPU consists of four parallel computation steps graphically represented in Figure 4.

1. Filter all involved facts (as given by the right-hand side of the rule) and copy them to separate arrays A and B.
2. Sort the B array in lexicographic order (yielding B').
3. Do a binary search for all A facts (in parallel) in B' to find matching and single facts. Perform the arithmetic operation as defined in the rule and prepare the following copy step.
4. Copy all result facts to array C.

In the following these steps are explained in more detail.

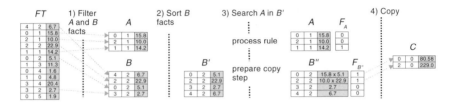

Fig. 4. Basic rule computation steps

Finding Relevant Facts (Filter Step). As a first step, all facts required for computing the rule result must be found and collected. In order to achieve this, we use a variant of stream compaction, a standard building block in parallel computing, which, from a given set of data, returns all data points satisfying a specified condition. In the OLAP case, the condition is specified by a query range and the rule definition. In our example, it would consist of the given query range in all dimensions except D_1. In D_1, the original coordinate (the left-hand side of the rule; in our case, 0) is replaced by each of the coordinates on the right-hand side (here, 1 or 2).

The basic parallel algorithm for extracting the $D_1[1]$ facts works as follows. For each entry of the fact table a CUDA thread is started that checks for its assigned fact whether the element ID of dimension D_1 is 1. If such a fact is found, a flag is set at the corresponding position in the flag array F as depicted in Figure 5.

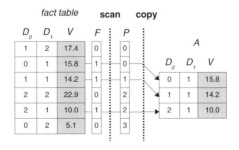

Fig. 5. Finding facts with $D_1 = 1$

In order to copy the flagged facts to array A, each flagged fact must be assigned a position in A such that the copy operation can be performed in parallel. Determining the positions is done by computing the parallel prefix sum of F, which is stored in P. Our CUDA implementation is based on the parallel algorithms as proposed in [9] and [11].

The same method is applied for finding the $D_1[2]$ facts and copying them into the array B. Hence, the result of the first step consists of two subsets of the fact table (one for each of the operands of the rule function). Together, those subsets consist of all relevant facts required for the rule calculation.

Sorting Step. If the fact table has not been sorted already, sort B in lexicographic order (i.e. within each dimension in a predefined nested order of dimensions), such that a binary search can be performed on the sorted table B'. We use a fast CUDA implementation of radix sort [17] as the basis for our sorting algorithm.

Search, Process and Flag. In order to find matches we replace for each A and B' fact in their rule dimension the element ID with the element ID as given in the left-hand side of the rule. All A facts then search (in parallel) their own coordinate index in B'. We base our binary search kernel on the parallel GPU algorithm presented in [6]. If the B fact is found the rule operation can be performed on the values and we store the result of the operation in the value of the B' fact, yielding a modified table B''. Moreover, a flag is set at the corresponding position in a temporary array $F_{B''}$ to keep track of all computed facts such that after this step we can distinguish between B facts that had a matching A fact (flagged) and single B facts without an A counterpart (unflagged). Conversely, single A facts are flagged in a temporary array F_A after an unsuccessful search in B'.

Copy. As we have replaced the element IDs of the rule dimension for all A and B facts in the previous step, we get the result array C by simply copying subsets of A and B'' to C. The decision which elements to copy solely depends on the rule operation. In an operation like addition in which single facts contribute to the result, we copy all *single* (flagged) A facts and *all* B'' facts to C. (For subtraction we basically do the same, but additionally have to multiply each single B'' fact by -1, as it stands on the right side of the operator). For operations like multiplication we can neglect all single facts, such that we only need to copy all flagged B'' facts.

As a result, array C contains all rule results relevant for the given query based on the existing facts in the database. These can now be used for any aggregations required in the view.

Special Cases. The above approach works only if the result value of a rule will always be 0 when all input values are 0. However, rules like $D_R[0] = D_R[1] + c$ (where $c \neq 0$) will not give a correct result for the 0 values of $D_R[1]$, as these facts are not represented in the table and thus will never be "seen" by the rule.

This problem can be solved by assigning a *default* value to each such rule, which will be selected for each non-existing result. Since the overall number of values in a cube query is known (by the size of the query region) and the number of existing values is given by the size of array C, the number of records with the default value can easily be deduced and incorporated in an aggregation.

5 Utilizing Multiple GPUs

In order to handle very large data volumes we have extended our algorithm such that it runs on multiple GPUs in parallel utilizing the GPU RAM of all devices. The basic idea is to distribute a fact table equally over all N devices (graphics cards) and execute, in parallel, an augmented version of the presented algorithm on each card. As a consequence the memory capacity can be increased by simply adding more devices. The distribution of the fact table is depicted in Figure 6.

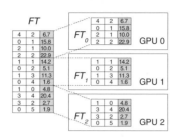

Fig. 6. Distributing the fact table

There are two options of getting our approach to work with multiple GPUs. The most desirable option would be to avoid device-to-device data transfer altogether by a suitable distribution of the data across devices. We will show in the next subsection that this is indeed possible within certain limitations. For all other cases, the basic algorithm of the last section needs to be adapted to the new data organization, which is described in section 5.2.

5.1 Data Distribution

In order to avoid data transfer between GPUs for rule calculation, it is necessary to distribute the data across the devices in such a way that the matching cells for each rule calculation will be located on the same device. Note that in the general case, especially in the presence of multiple rules, it may be impossible to achieve such a distribution for all rules simultaneously. Nevertheless, any reduction of unnecessary data transfer is desirable.

Since data distribution is carried out on server startup (and later on during all updates of the cube), a preprocessing step of the data is required. As mentioned before, the GPU data storage we use supports sorting of the cube data. The default sorting order is a nested sorting determined by a given order of the dimensions, $D_1, D_2, \ldots, D_d$. This dimension ordering is completely arbitrary and can be changed without affecting the correctness or performance of the algorithms.

Since each rule is usually defined only on a (small) subset of dimensions and works across all other dimensions equally, we propose a simple reordering of the dimensions to ensure matching cells will be on the same device.

Consider the following simple example of a cube with 3 dimensions, 2 rules and a hardware system with k GPUs:

D_1 ("Years"):   '2009', '2010', '2011', '2012'
D_2 ("Products"):   'Product 1', 'Product 2', ..., 'Product N'
D_3 ("Datatypes"):   'Actual', 'Budget', 'Overhead%'

R_1: ['2012']   = MAX(['2011'],['2010'],['2009']})
R_2: ['Overhead%']   = 100 * ['Actual'] / ['Budget']

The main idea is to split the cube data (for distributing them to the k devices) solely according to the "Products" dimension (which is not manipulated by any of the rules), meaning that the parts we create all consist of "slices" (sub-cubes) containing all data for one or more products. If we assume that for each product there are (roughly) the same number of filled base cells in the cube, we will simply divide the "Products" dimension into k equal portions each containing $x = N/k$ elements, and put all data for the first x products on GPU_0, the next x on GPU_1, and so on. If the data are skewed along the "Product" dimension, we will have to create slices of possibly unequal width to make their actual sizes (i.e. the number of existing base cells for the respective range of product elements) as equal as possible.

Since for any single base rule computation the source base cells must come from the same product, we can do each of these computations using data residing on the same GPU and will never have to move base data between devices. This is because the data are split along a dimension that is not targeted in any of the rules. Whenever a '2011' or 'Overhead%' cell is calculated, the matching pair of base cells for one computation must belong to the same product (because some element in that dimension has to be chosen in the query) and hence must also be on the same GPU.

The necessary condition for this approach to work is that there is at least one dimension in the cube not targeted by any rule. Furthermore, it must be possible to split the cube along element boundaries in this dimension such that k roughly equally sized parts are created. While this may be true for the majority of real-life OLAP cubes, it cannot be always guaranteed. Therefore, we also need to be able to calculate rules across multiple GPUs.

5.2 Augmented Algorithm

Unfortunately, the extension of the basic rule computation in the general case of distributed facts over multiple GPUs is not as straightforward as for the basic aggregation approach. In particular, the task of finding matching facts, and determining single facts becomes more complicated on distributed data. For a better understanding let us consider the constellation as shown in Figure 7.

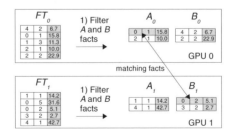

Fig. 7. Distributed matching facts

If we execute the basic algorithm on each of the devices we have to expect to get, as a result of the first step, some facts in A_0 that have their matching fact on another GPU, e.g. in B_1. As each algorithm can only see the data stored on its own assigned device, no matching fact would be found in this case and the algorithm would treat the fact as a single fact, thereby producing incorrect results.

The first two steps of the basic algorithm remain unchanged, such that as a result of the filter step we get on device i the arrays A_i and B_i holding facts that have been filtered out of the partial fact table $FT_{i,}$. After the sort step we have the sorted B_i-facts in B_i'. This is shown in the center column of Figure 8.

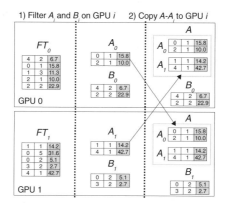

Fig. 8. Preparation of A on each device

Before the binary search is carried out, we insert a new step in which we distribute the complete A array to each device. We do this by copying in parallel the A_i part of each GPU to all other devices. The resulting configuration can be seen on the left hand side of Figure 9.

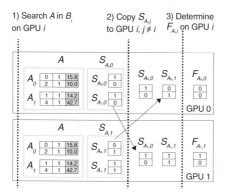

Fig. 9. Preparation of F on each device i

We continue with the parallel binary search of all A facts in B_i' on device i (done on all devices at the same time). If for an A fact we find a matching fact in B_i' we process the facts normally. As on each device all A facts search on the B_i' facts it is guaranteed that all matching facts are found (each match on exactly one of the devices).

However, if no match is found on device i we cannot directly flag the involved A fact as single because it can possibly have a match on another device. Although we still flag it, we understand the flagged facts not as single facts, but as single fact *candidates*. Only if an A fact is a single fact candidate on *all* devices it really is a single fact. As shown in Figure 9 for each part A_j of A on device i we store the single-candidate flags in the array $S_{Aj,i}$. We then copy to device i the respective flag array $S_{Ai,j}$ for A_i from all other devices j ($j \neq i$) such that device i can determine the single A_i facts by ANDing all single-candidate flags for each A_i fact. Hence, in the end each device has determined the single facts of its original part of the A facts.

It remains to copy, on each device i, the A_i and B_i values to a result array C_i just like in the basic algorithm. All required aggregations can then be performed in parallel with the multi-GPU aggregation algorithm (cf. section 3.2).

6   Performance Evaluation

In order to assess the performance of our GPU rule evaluation method, we have conducted a first set of tests, which confirms the feasibility and usefulness of the approach, though they are certainly not conclusive (only the single-GPU algorithm has been tested so far). The tests were carried out on a realistic data set (a 7-dimensional OLAP cube with roughly 3 million filled cells), but with "artificial" rules in order to better control the setting. The rule was a simple division, computing the deviation of actual sales figures from budgeted ones. The GPU used was an Nvidia Tesla C1060 computing processor with 4 GB of memory (although the cube also fits easily in the memory of a standard consumer graphics card).

Table 1 shows the timings for complete rule computation (i.e. rule operation on base facts followed by aggregation) on CPU and GPU, depending on the number of facts participating in the computation, i.e. used as input for the rule operation.

Table 1. Performance of rule processing

#Facts in Rule	CPU Time	CPU Time (Index)	GPU Time	Speedup
0.51 million	2655 ms	271 ms	17 ms	156x / 16x
0.77 million	7922 ms	829 ms	21 ms	377x / 39x
1.5 million	15890 ms	1527 ms	34 ms	467x / 45x
3.0 million	31610 ms	2954 ms	54 ms	585x / 55x

Some explanation is required for the CPU times. We have integrated our GPU algorithm in the open-source MOLAP server *Palo* [8]. This software offers two (CPU-based) ways for calculating rules: The default algorithm is driven by the aggregation to be computed and recursively searches all base cells involved in the rule. Since sparsity is not taken into account, those searches include empty (non-existing) cells, which explains the long CPU overall times, taking several seconds. In addition, *Palo* allows the use of so-called "markers" to handle cube sparsity by adding a special rule index which boosts the performance by roughly factor 10, but comes at the cost of increased memory usage. Our GPU approach computes the same rules in 17-54 milliseconds, without any additional index. Though the results are preliminary, they show that the GPU approach is very promising, as the whole computation remains well within a time suitable for real-time interactive OLAP analysis.

7 Conclusions and Future Work

We have presented a massively parallel method for online computing of advanced OLAP rules, using GPU hardware as coprocessing units. Our algorithms outperform comparable sequential computations by several orders of magnitude and do not require any extra index.

We have extended the approach to multi-GPU systems, which can be used in order to accommodate larger data volumes. The problem of inter-device communication is solved with relatively moderate transfer costs.

Next steps include the extension of the approach to more complex rules, which can be broken down into smaller parts and handled by the simpler cases presented here. In addition, thorough performance testing is required, in particular for the multi-GPU approach, and a comparison with other OLAP systems supporting the rules concept.

Future research could address the extension to clusters of GPU servers, which have recently been proposed as inexpensive high-performance computers.

Acknowledgments. Parts of the research presented here are funded by the German Research Foundation (DFG) within its technology transfer program and by Jedox AG, Freiburg. The authors would like to thank Christoffer Anselm, Zurab Khadikov, and Jiří Junek for valuable input and criticism.

References

1. Andrzejewski, W., Wrembel, R.: GPU-WAH: Applying GPUs to compressing bitmap Indexes with word aligned hybrid. In: Bringas, P.G., Hameurlain, A., Quirchmayr, G. (eds.) DEXA 2010. LNCS, vol. 6262, pp. 315–329. Springer, Heidelberg (2010)
2. Bakkum, P., Skadron, K.: Accelerating SQL database operations on a GPU with CUDA. In: Proceedings of GPGPU 2010, pp. 94–103. ACM Press, New York (2010)
3. Chaudhuri, S., Dayal, U.: Data warehousing and OLAP for decision support. In: Proceedings of SIGMOD 1997, Tucson, AZ. ACM Press, New York (1997)
4. CUDA website, http://www.nvidia.com/object/cuda_home_new.html
5. Dehne, F., Eavis, T., Rau-Chaplin, A.: The cgmCUBE project: optimizing parallel data cube generation for ROLAP. Distributed and Parallel Databases 19(1), 29–62 (2006)
6. Fernando, R.: GPU Gems: Programming Techniques, Tips and Tricks for Real-Time Graphics. Pearson Higher Education, London (2004)
7. Govindaraju, N.K., Lloyd, B., Wang, W., Lin, M.C., Manocha, D.: Fast computation of database operations using graphics processors. In: Proceedings of SIGMOD 2004, Paris, France, pp. 206–217. ACM Press, New York (June 2004)

8. Gray, J., Chaudhuri, S., Bosworth, A., Layman, A., Reichart, D., Venkatrao, M., Pellow, F., Pirahesh, H.: Data cube: a relational aggregation operator generalizing group-by, cross-tab, and sub-totals. Data Mining and Knowledge Discovery 1(1), 29–53 (1997)
9. Harris, M., Sengupta, S., Owens, J.D.: Parallel Prefix Sum (Scan) with CUDA. In: Nguyen, H. (ed.) GPU Gems 3, pp. 851–876. Addison Wesley, Reading (August 2007)
10. He, B., Lu, M., Yang, K., Fang, R., Govindaraju, N.K., Luo, Q., Sander, P.V.: Relational query coprocessing on graphics processors. In: Transactions on Database Systems, vol. 34, ACM Press, New York (April 2009)
11. Horn, D.: Stream reduction operations for GPGPU applications. In: Pharr, M. (ed.) GPU Gems 2, pp. 573–589. Addison Wesley, Reading (March 2005)
12. IBM Cognos TM1,
    `http://www.ibm.com/software/data/cognos/products/tm1/`
13. Infor PM10, `http://www.infor.com/solutions/pm/pm10/`
14. Jedox Palo Suite, `http://www.palo.net`
15. Kaczmarski, K.: Comparing CPU and GPU in OLAP cube creation. In: Černá, I., Gyimóthy, T., Hromkovič, J., Jefferey, K., Králović, R., Vukolić, M., Wolf, S. (eds.) SOFSEM 2011. LNCS, vol. 6543, pp. 308–319. Springer, Heidelberg (2011)
16. Lauer, T., Datta, A., Khadikov, Z., Anselm, C.: Exploring graphics processing units as parallel coprocessors for online aggregation. In: Proceedings of DOLAP 2010, Toronto, Canada, ACM Press, New York (October 2010)
17. Satish, N., Harris, M., Garland, M.: Designing efficient sorting algorithms for manycore GPUs. In: Proceedings of the IEEE International Symposium on Parallel & Distributed Processing, pp. 1–10 (May 2009)

Designing a Flash-Aware Two-Level Cache

Ioannis Koltsidas[1] and Stratis D. Viglas[2]

[1] IBM Research, Zurich, Switzerland
iko@zurich.ibm.com
[2] School of Informatics, University of Edinburgh, UK
sviglas@inf.ed.ac.uk

Abstract. The random read efficiency of flash memory, combined with its growing density and dropping price, make it well-suited for use as a read cache. We explore how a system can use flash memory as a cache layer between the main memory buffer pool and the magnetic disk. We study the problem of deciding which data pages to cache on flash and propose alternatives that serve different purposes. We give an analytical model to decide the optimal caching scheme for any workload, taking into account the physical properties of the flash disk used. We discuss implementation issues such as the effect of the flash cache block size on performance. Our experimental evaluation shows that questions on systems with flash-resident caches cannot be given universal answers that hold across all flash disks and workloads. Rather, our cost model should be applied per case to provide an optimal setup with confidence.

1 Introduction

With growing capacities, improved I/O performance, and constantly dropping prices, flash disks, or solid-state drives (SSDs), are now a viable storage option not only in personal computing, but also in the server market. In some cases, SSDs have completely replaced magnetic hard-disk drives (HDDs) in the enterprise [13]; elsewhere, SSDs have been used along with HDDs to boost database performance [11]. Our work stems from the low latency and high random read efficiency of SSDs. By comparing the price and performance characteristics of SSDs to those of DRAM and HDDs, it follows that an SSD is ideal as a cache layer between the main memory and the HDD; this implies a 3-tier memory hierarchy. We study various aspects of such a system and provide analytical tools that aid the designer to decide with high confidence the optimal system configuration.

SSDs are arrays of flash memory chips packaged with a controller in a single enclosure that provides a common interface (*e.g.*, SATA). Applications targeting SSDs should account for the I/O characteristics of flash memory; prominently, no mechanical moving parts and, thus, no mechanical latency. Access latency is irrespective of the access pattern and orders of magnitude less than that of HDDs. The electrical properties of flash memory make reading the value of a bit faster than changing it. To aggravate matters for writes, to update an already written sector one needs to first erase it. Erasures are carried out in *erase units*,

J. Eder, M. Bielikova, and A.M. Tjoa (Eds.): ADBIS 2011, LNCS 6909, pp. 153–169, 2011.
© Springer-Verlag Berlin Heidelberg 2011

i.e., blocks typically consisting of 256 sectors. Each erasure is two orders of magnitude more expensive than a read or a write, so updating a sector is costly. The erase-before-write limitation of SSDs means that on-disk caches can do little to help. Thus, writes perform poorly. The random read efficiency of flash is its greatest advantage, while its random write inefficiency is its greatest bottleneck.

When designing a system with a 3-tier memory hierarchy like the one we discuss, a salient decision is determining the sizes of the main memory and the flash disk caches. As of February 2011, the cost of DRAM is about $16/GB; the cost of SSDs varies from about $1.6/GB for the low-performance ones [17], to about $8/GB for the high-performance consumer SSDs [8], and to about $30/GB for enterprise-level solutions [6]. The performance of SSDs in this price range varies by two orders of magnitude for reads and four orders of magnitude for random writes. Considering the price/performance trade-off for the two types of cache, and given a specific budget, minimizing the ratio for main memory and SSD capacities is not straightforward. Should an SSD be used as a cache, or is it better to invest in DRAM memory? Should one buy a small but fast SSD or a large but slow one? Such questions are crucial for performance and cannot be given universal answers. If one can buy enough DRAM to fit the working set of the workload, then this is the way to go. Similarly, for write-intensive workloads one should invest in a high-performance SSD to use as a cache, instead of a cheaper one. However, it is not safe to decide based on intuition: with the characteristics of SSDs constantly changing, decisions should constantly be re-evaluated.

Next, the designer should decide which data will be cached on the flash disk. Contrary to buffering in main memory, pages do not need to be brought into the flash cache before being processed. That is, a page may go directly from the magnetic disk to the memory and may well never be written to flash. Thus, deciding how data flows from one level of the memory hierarchy to the others is not straightforward. A set of rules dictates the flow of data pages across levels: we term this a *page flow scheme*. A related issue is how the workload of a page affects the decision about caching the page or not. For instance, in the ZFS filesystem [18] dirty pages are never cached on flash. From an implementation perspective, questions arise about the directory of pages cached on the flash disk and the optimal page size to use on flash. We show why these questions are crucial and provide the tools to address them. Our proposals and results are independent of the page replacement algorithm used by either cache.

Contributions and Organization. We present how an SSD can efficiently act as a page cache between the main memory and the HDD. Our contributions are:

– We study the problem of deciding which data should be placed in the flash cache of a system. We identify three invariants for the sets of pages cached either in main memory or on flash. For each invariant, the flow of pages between levels of the memory hierarchy is different. We present the page flow scheme of each invariant and an analytical model of the I/O cost it incurs (Section 3).

– We discuss several implementation issues that arise when using a flash disk as a cache: (*a*) the page directory for the cache, (*b*) the size of flash pages, and (*c*) the caching only of pages that satisfy specific predicates; we show the correlation between each alternative and the properties of the flash disk (Section 4).

– We have implemented and evaluated our techniques. Our results show that questions on flash-resident caches cannot be given answers with confidence, unless one uses our cost model on a per-case basis (Section 5).

2   Related Work

The main problem of flash memory is its random-write inefficiency due to its erase-before-write limitation. In [4] the authors study different write patterns on SSDs. They argue that (*a*) latency greatly affects performance, (*b*) I/O in larger blocks can substantially improve random writes, (*c*) I/O blocks should be aligned to flash pages, and (*d*) if random writes exhibit spatial locality they can be performed almost as efficiently as sequential ones. The authors also point out that SSDs are more complex than "bare" flash chips. This is due to the on-disk DRAM and controllers of SSDs, which mainly aim to improve random writes. To that end, they employ parallelism when accessing flash chips along with elaborate *Flash Translation Layer* (FTL) algorithms (see also [1,3,5,10]).

To improve write efficiency in flash-based databases, [13] proposes *in-page logging* (IPL): data changes are separately logged and each data page and its log records are in the same erase unit. When out of room, log records and data pages are merged into a new erase unit. Simulation shows that IPL substantially improves performance. In [16] the authors argue that for flash writes one should avoid in-place updates and sub-block deletions, while random writes should be replaced with semi-random ones; blocks can be written to in any order, but sectors belonging to the same block are written sequentially from the start of the block. The results show these techniques boost random write performance. In [11] the authors study systems equipped with both an SSD and an HDD and propose placing read-intensive data pages on flash and update-intensive pages on the magnetic disk, thus alleviating the high cost of random flash writes.

When the SSD is used for persistent storage BPLRU has been proposed as a replacement policy for the on-disk cache [9]. The buffer is treated as a write cache and RAM buffers are grouped in blocks equal in size to the flash erase unit; page replacement is performed at erase unit granularity (using LRU). If not all sectors of a dirty victim page are present in memory, the missing ones are read from disk so that the whole block can be written to a new flash location without the need for an in-place update. Additionally, a block that was written sequentially is moved to the tail of the LRU list and becomes the next victim. Evaluation shows this technique to be very promising. The authors of [19] propose that the buffer cache choose for replacement a clean page over a dirty one thus trading writes for reads. This is generalized in [11] when the buffer pool holds pages from both the SSD and the HDD. Not only the dirtyness of the page, but also its access history and the read/write costs are considered when choosing a victim.

In [15] the authors propose an offline-only tool that uses multiple metrics, *e.g.*, performance or energy efficiency, to decide the optimal storage configuration for a workload. The tool does not address the relationship between the data cached in flash and in RAM. Our proposals go beyond the offline choice of the optimal hardware, to the online decision of which data should be cached in flash with respect to what is cached in RAM. Not strictly database-related, is the ZFS filesystem [14,18]. The SSD acts as a cache for the HDD to improve the performance of random read workloads. There is no eviction from main memory to the SSD. The flash cache is asynchronously filled with clean pages only, thereby avoiding write latencies on main memory evictions. Our work is an analytical study of the behavior of flash caches; the techniques of ZFS are complementary.

3   Page Flow Schemes

We describe page flow schemes for systems employing an SSD as a cache between the main memory and the HDD. The schemes are independent of the replacement policies used by the main memory buffer pool and by the page cache on flash.

3.1   Problem Statement

Consider a database, or any other data processing system, with three data storage and staging components: (*a*) RAM memory (*e.g.*, DRAM chips), (*b*) one or more SSDs, and (*c*) persistent storage, *e.g.*, a single HDD, or a disk array. Data processing requires demand paging: on referencing, pages are brought into main memory before being processed. Such a system is shown in Fig. 1. We refer to main memory as RAM and to the main memory buffer pool as RAM *cache*. We use FLASH to refer to the system's SSD(s) used as a page cache (the on-flash cache is termed FLASH *cache*); HDD is the underlying long-term storage.

The key decision for a page cache is which pages will be cached; how long pages are cached for is decided by the replacement policy, which we do not consider. For a system of only a RAM cache and an HDD, the former decision is easy: demand paging requires *all* referenced pages be written to the RAM cache. If there is an additional FLASH cache there is no such requirement. To reduce I/O to/from the HDD, the sensible choice is to store in the FLASH cache the "hot" portion of the dataset that cannot fit in RAM. Let $P_{\mathrm{RAM}}(t)$ be the set of pages stored in the RAM

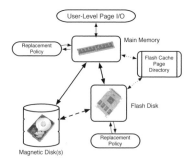

Fig. 1. An overview of our system

cache at time t, and $P_{\mathrm{FLASH}}(t)$ be the set of pages in the FLASH cache (for all practical cases, $|P_{\mathrm{RAM}}(t)| < |P_{\mathrm{FLASH}}(t)|$). We identify three invariants:

1. $\forall t\ P_{\mathrm{RAM}}(t) \bigcap P_{\mathrm{FLASH}}(t) = P_{\mathrm{RAM}}(t)$: Whenever a page is in RAM it is also cached in FLASH, in analogy to the *inclusive* memory hierarchies of CPUs.
2. $\forall t\ P_{\mathrm{RAM}}(t) \bigcap P_{\mathrm{FLASH}}(t) = \emptyset$: No page is stored in *both* RAM and FLASH at any time. A page brought from FLASH to RAM is removed from FLASH (and vice

versa). Specifically, a RAM victim is stored in the frame of the page hit on FLASH, *i.e.*, a RAM page is *swapped* with a FLASH page.

3. $\forall t \ P_{\text{RAM}}(t) \bigcap P_{\text{FLASH}}(t) \subseteq P_{\text{RAM}}(t)$: A page in RAM may or may not be cached in FLASH, depending on user-set criteria or the current workload.

Enforcing any one of the invariants results in a different *page flow scheme* across the levels of the memory hierarchy. Each scheme incurs a different I/O cost for a given workload. We detail the schemes and model their I/O costs. The RAM or FLASH cache page replacement policies are orthogonal to deciding which pages should be cached where; our schemes can be used with any policy.

3.2   The Inclusive Scheme

Under the inclusive scheme, any page cached in RAM is also cached in FLASH. To fetch a page *pg* under inclusive we use Alg. 1. We look up *pg* in the RAM cache directory; if *pg* is found it is served in-memory. Else, we bring it in RAM and evict a page v_r if memory is full. Given the invariant, v_r is also cached in FLASH; it is written back only if dirty. We look up *pg* in the FLASH cache directory and, if *pg* is there, we read it and put it in the RAM cache; else, the page is read from HDD, and written to the FLASH and RAM caches. If the FLASH cache is full, a page v_f is evicted; if dirty, it is written to HDD. Since $|P_{\text{RAM}}(t)| < |P_{\text{FLASH}}(t)|$, v_f will not exist in RAM if both caches use the same replacement policy; otherwise, the FLASH replacement policy must ensure that a page in RAM is never evicted.

Let h_r, m_r, h_f and m_f respectively be the total number of RAM hits, RAM misses, FLASH hits and FLASH misses incurred by the workload. Let F_R, F_W, D_R, D_W be the average cost of a flash read or write, and an HDD read or write, respectively. These include the cost of writing the page to or reading the page from RAM. Consider the probability that a page in RAM is dirty before its eviction and let this probability be p_d. Let R_{RAM} be the cost of running the replacement algorithm for the RAM cache and R_{FLASH} be the corresponding cost for the FLASH cache. We assume constant time replacement algorithms, *i.e.*, R_{RAM} and R_{FLASH} are negligible; still, we include them in the cost formulas for completeness.

Algorithm 1. inclusive
1   **if** *pg in* RAM *cache* **then return** *pg*;
2   **else if** *pg in* FLASH *cache* **then**
3   Evict victim page v_r from RAM;
4   Write v_r to FLASH, iff it is dirty;
5   Read *pg* from FLASH;
6   **return** *pg*;
7   **else**
8   Evict victim page v_r from RAM;
9   Write v_r to FLASH, iff it is dirty;
10   Evict victim page v_f from FLASH;
11   Write v_f to HDD, iff it is dirty;
12   Read *pg* from HDD;
13   Write *pg* to FLASH;
14   **return** *pg*;

Algorithm 2. exclusive
1   **if** *pg in* RAM *cache* **then return** *pg*;
2   **else if** *pg in* FLASH *cache* **then**
3   Read *pg* from FLASH;
4   Pick a victim page v_r from RAM;
5   Replace *pg* with v_r on FLASH;
6   **return** *pg*;
7   **else**
8   Evict victim page v_f from FLASH;
9   Write v_f to HDD, iff it is dirty;
10   Evict victim page v_r from RAM;
11   Write v_r to FLASH;
12   Read *pg* from HDD;
13   **return** *pg*;

A RAM hit incurs no I/O. On a RAM miss either a FLASH hit or a FLASH miss occurs (*i.e.*, $m_r = h_f + m_f$). For a FLASH hit, a page is evicted from RAM with cost $R_{\text{RAM}} + p_d F_W$ and a page is read from FLASH with cost F_R. On a FLASH miss, a RAM page is evicted with cost $R_{\text{RAM}} + p_d F_W$; a FLASH page is evicted with cost $R_{\text{FLASH}} + p_d D_W$ and the referenced page is read from disk and written to FLASH, with cost $D_R + F_W$. The cost C_1 of inclusive is:

$$C_1 = h_f(F_R + R_{\text{RAM}} + p_d F_W) + m_f(R_{\text{RAM}} + p_d F_W + R_{\text{FLASH}} + p_d D_W + D_R + F_W)$$
$$\Rightarrow C_1 = h_f F_R + m_r(R_{\text{RAM}} + p_d F_W)) + m_f(R_{\text{FLASH}} + p_d D_W + D_R + F_W)$$

If the RAM and FLASH page directories are stored in-memory for both caches, the cost of a lookup or an update is $O(1)$ for computationally cheap replacement policies like LRU. However, the page directory of the FLASH cache may require substantial memory. When memory is limited it may be better to store the FLASH directory on FLASH itself. We can similarly account for the directory costs; due to lack of space we omit the details here, but present them thoroughly in [12].

3.3   The Exclusive Scheme

The exclusive scheme enforces Invariant 2: the set of pages cached in RAM and the set of pages cached in FLASH are disjoint. The exclusive algorithm for fetching a page is given in Alg. 2. RAM hits are treated the same as for inclusive. On a RAM miss, we look up *pg* in the FLASH cache directory; if found, the page is read from FLASH. If the RAM cache is full, a page is evicted from RAM. The victim is selected by the replacement policy and written to FLASH (whether it is dirty or not); the referenced page is deleted from FLASH and inserted in RAM. Effectively, we swap the on-flash referenced page with the RAM victim. For a FLASH miss, the RAM victim is written to FLASH and the referenced page is read from the HDD into main memory. If the FLASH cache is full we evict a page from FLASH.

There is no I/O for a RAM hit. A RAM miss results in either a FLASH hit or a FLASH miss. On a FLASH hit the cost of evicting from RAM is $R_{\text{RAM}} + F_W$. The referenced page is read from FLASH with cost F_R and the victim page is written to FLASH with cost F_W. On a FLASH miss, FLASH eviction costs $R_{\text{FLASH}} + p_d D_W$ on top of the $R_{RAM} + F_W$ cost of evicting from the RAM cache; reading the referenced page from HDD adds a cost of D_R. The cost C_2 of exclusive is:

$$C_2 = h_f(R_{\text{RAM}} + F_R + F_W) + m_f(R_{\text{RAM}} + F_W + R_{\text{FLASH}} + p_d D_W + D_R)$$

3.4   The Lazy Scheme

The lazy scheme enforces Invariant 3 by caching an arbitrary set of pages in FLASH. The system decides if a page will be cached in FLASH when it evicts it from RAM, *i.e.*, after there is an indication for the workload of the page. The algorithm is shown in Alg. 3 where we assume that a RAM victim is always written to FLASH and stays there until evicted by the FLASH replacement policy.

A page is served in-memory if found in RAM. Otherwise, we look it up in the FLASH directory. On a FLASH hit, the page is read from FLASH (and the directory's bookkeeping is updated). If the RAM cache is full, a victim is evicted by the RAM replacement policy. If the victim is also in FLASH, it is written back only if it is dirty. If not, a page is evicted from FLASH (and written back to HDD) to make room in FLASH for the RAM victim. On a FLASH miss, a page is evicted from RAM and written to the FLASH cache, as for a FLASH hit. The referenced page is read from HDD and brought in main memory. Note that one can apply any predicate to decide if the page should be cached in FLASH or not. We discuss alternatives later on.

Algorithm 3. lazy

```
1  if pg in RAM cache then return pg;
2  else if pg in FLASH cache then
3  │    Read pg from FLASH;
4  │    Evict victim page v_r from RAM;
5  │    if v_r in FLASH cache then
6  │    │    Write v_r to FLASH, iff it is dirty;
7  │    else
8  │    │    Evict victim page v_f from FLASH;
9  │    │    Write v_f to HDD, iff it is dirty;
10 │    │    Write v_r to FLASH;
11 │    └    return pg;
12 else
13 │    Evict victim page v_r from RAM;
14 │    if v_r in FLASH cache then
15 │    │    Write v_r to FLASH, iff it is dirty;
16 │    else
17 │    │    Evict victim page v_f from FLASH;
18 │    │    Write v_f to HDD, iff it is dirty;
19 │    │    Write v_r to FLASH;
20 │    │    Read pg from HDD into RAM;
21 │    └    return pg;
```

Consider now the cost of the lazy scheme. A main memory victim may or may not exist in the FLASH cache. Let the probability of a RAM victim being on FLASH be q. The cost C_3^V for a RAM victim (*i.e.*, Lines 5-11, 14-20) equals $R_{\text{RAM}} + p_d F_W$ if the page is in FLASH and $R_{\text{RAM}} + R_{\text{FLASH}} + p_d D_W + F_W$ otherwise. Therefore: $C_3^V = R_{\text{RAM}} + q p_d F_W + (1-q)(R_{\text{FLASH}} + p_d D_W + F_W)$. For a FLASH hit the cost is $F_R + C_3^V$ and for a a FLASH miss the cost is $C_3^V + D_R$. The cost of this scheme is therefore: $C_3 = h_f(C_3^V + F_R) + m_f(C_3^V + D_R) = (h_f + m_f)C_3^V + h_f F_R + m_f D_R$.

Recall (Section 3.2) that h_f and m_f are the total hits and misses for pages of the workload. Thus, lookups in the FLASH index for the RAM victim are not accounted for by h_f and m_f. The probability of the RAM victim being in FLASH is expected to be equal to the probability of any referenced page being in FLASH: it does not depend on whether the looked up page was in RAM at the time of the lookup. Therefore, $q = {}^{h_f}\!/_{h_f + m_f}$ and $1 - q = {}^{m_f}\!/_{h_f + m_f}$, giving: $C_3^V = R_{\text{RAM}} + ({}^{h_f}\!/_{h_f + m_f}) p_d F_W + ({}^{m_f}\!/_{h_f + m_f})(R_{\text{FLASH}} + p_d D_W + F_W)$. Then:

$$C_3 = (h_f + m_f)R_{\text{RAM}} + h_f p_d F_W + h_f F_R + m_f(R_{\text{FLASH}} + p_d D_W + F_W) + m_f D_R$$

$$\Rightarrow C_3 = h_f(R_{\text{RAM}} + p_d F_W + F_R) + m_f(R_{\text{RAM}} + R_{\text{FLASH}} + p_d D_W + D_R + F_W)$$

Various criteria can be applied to decide whether a RAM victim page should be cached in FLASH. For instance, flash disks that are poor in random writes can benefit from caching only clean pages (see also Section 5.6). Similarly, the access history for a FLASH page can be maintained by tracking its hits, or the number of times it has been dirtied. The system could then maintain a set of the f hottest pages, where f is the capacity of FLASH in pages. Only these f pages will be cached on flash, thus implementing a frequency-based replacement policy. One may also cache the f pages that have the most read-intensive workload as in [11].

These options can even be combined; however, we will not study them further here as they assume or define some aspects of the cache replacement policy.

3.5   Comparison

We compare the three schemes based on their I/O costs. We assume (for now) that the FLASH cache directory is stored in main memory and do not consider directory maintenance costs. Given the formulas for C_1, C_2, and C_3, one might factor out $h_f F_R + m_f (R_{\text{FLASH}} + F_W + D_R) + m_r R_{\text{RAM}}$. However, this falsely assumes that, for a fixed workload, h_f, m_f remain fixed for all schemes.

Assume that a workload is executed three times, once with each scheme. Throughout, the RAM and FLASH replacement policies are the same. Also, assume a stack replacement algorithm (not a FIFO one), *i.e.*, one that does not exhibit Belady's anomaly [2]. The hit ratio of the cache grows with cache size (that is, with the number of available frames). Let r, f be the maximum capacity, in pages, of the RAM and FLASH caches, respectively. We define the *effective capacity* of a cache at level i as the number of pages cached at level i that are guaranteed *not* to be cached at any level higher than i at the same time. The effective size of the RAM cache is $e_r = r$. For the FLASH cache, its effective size e_f is equal to the number of pages cached in FLASH that are not cached in RAM at the same time. For inclusive the effective size of the FLASH cache is $e_f^1 = f - r$, while for exclusive it is $e_f^2 = f$. For lazy, the subset of FLASH pages also cached in RAM varies with the workload; however, the following always holds: $f - r \leq e_f^3 \leq f$.

Observe that the FLASH cache hit ratio depends on the effective size of the cache, not its capacity. Consider, *e.g.*, inclusive: when it looks a page up in FLASH, it is only likely to find the requested page in $f - r$ pages; if the requested page was any of the r pages cached in RAM, no lookup in FLASH would be needed. Thus, the hit ratio is a function of the replacement policy, the effective size of the cache, and the workload. For a replacement policy Y and a workload W, let the hit ratio be $H = H(Y, W, e_f)$. Thus: $H(Y, W, e_f^1) \leq H(Y, W, e_f^3) \leq H(Y, W, e_f^2)$. Taking into account that $h_f = H \cdot |W|$, we have that: $h_f^1 \leq h_f^3 \leq h_f^2 \Rightarrow m_f^1 \geq m_f^3 \geq m_f^2$ for the three algorithms, since $m_r = h_f + m_f$. The effective size of the RAM cache is the same for all three schemes; the same applies for the RAM hit ratio.

One can only model the hit ratio for a page replacement policy if the characteristics of the workload are priorly known. Our evaluation shows that for a given policy the hit ratio varies widely across workloads. In a real deployment, where the characteristics of the reference pattern are not known *a priori*, one cannot statically determine the optimal page flow scheme. Thus, we continuously monitor the hit ratio with respect to the effective size of the FLASH cache and accordingly adapt the page flow scheme. We keep track of FLASH hits and misses and the rate at which pages are dirtied (p_d). Based on the normalized read and write costs for the SSD and the HDD, which are known or can be measured [11], we periodically evaluate the cost formula for each scheme and adopt the one that minimizes the total cost. In Section 5.7, we discuss workload characteristics by which one can decide the optimal scheme statically and with confidence.

4   Implementation Issues

One important decision is the location of the FLASH cache page directory. Let b bytes be the size of a directory entry, B be the size of a page, S_r be the size of RAM, and S_f be the size of FLASH; the number of FLASH directory entries is $f = S_f/B$. For an in-memory FLASH directory, $S_r - bf$ bytes are left in main memory for caching. So, $S_r - bf/B + b$ pages are cached in RAM. If all memory is used for the RAM cache it fits $S_r/B + b$ pages. Given a replacement policy and a workload, $H\,(Y, W, S_r/B + b) \le H\,(Y, W, S_r - bf/B + b)$ holds. Our experiments show a large difference between these hit ratios as the discrepancy between the RAM and FLASH sizes grows. Larger FLASH pages may alleviate the situation.

Using Larger Pages for Flash. Let B_r and B_f be the RAM and FLASH page sizes respectively; b_r bytes are required for a RAM directory entry and b_f bytes for a FLASH directory one. An entry holds, at least, the HDD offset of the page (acting as its identifier), a pointer to the page in the cache (a main memory pointer for a RAM page, or a disk offset for the FLASH cache) and a dirtyness bit. The replacement policy requires extra bytes for bookkeeping *e.g.*, a pointer to the next LRU page, bits for pinning, mutexes for concurrency control, *etc.*; the same applies for b_r, but we will not detail b_r as it is not our focus.

If $B_f > B_r$, each FLASH page has B_f/B_r RAM pages; we refer to such FLASH pages as *blocks*. All I/O between the SSD and the HDD is in blocks of B_f bytes, while data movement from/to the RAM cache is in pages of B_r bytes. The RAM cache and all in-memory structures use the HDD *offset* δ of a page as its universal identifier. The RAM directory uses δ/B_r as the page identifier; the FLASH directory uses δ/B_f as the identifier of a block stored at δ on HDD. Thus, $\log_2 \delta/B_f$ bits are required to identify a page in the FLASH directory.

By knowing the RAM directory identifier of a page, one can use B_f and B_r to obtain the identifier of the host FLASH block. Let p_r be a RAM page of FLASH block p_f. For each reference to p_r, we look it up in the FLASH directory. If p_f is there, then p_r is located at offset $(p_r B_r \mod B_f)$ in p_f, at location $(p_r B_r \div B_f)$. Else, p_f is read from HDD into FLASH and p_r is computed the

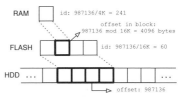

Fig. 2. Using larger FLASH pages

same way. FLASH evictions take place with B_f granularity. The case for $B_r = 4$kB and $B_f = 16$kB is shown in Fig. 2.

If p_r is evicted from RAM to FLASH but p_f is not cached in FLASH at that time (*i.e.*, under exclusive or lazy), writing page p_r of block p_f to FLASH is not straightforward. A solution is to fetch p_f from HDD into FLASH and overwrite its p_r page incurring one extra HDD read; we term this overwriting. When fetching the whole block from the HDD, some pages of the block may already be cached in RAM, thus compromising Invariant 2 of exclusive. Under inclusive this never arises: any page cached in RAM will have its host FLASH block cached in FLASH.

An alternative is to assign a block to p_f on FLASH, invalidate all its pages but p_r, and overwrite p_r. If block p_f is later read from HDD, only the invalid pages will be overwritten on FLASH; if it is written to HDD, only the valid pages will be written. We term this technique **invalidating**. Except for a slight implementation complexity the main drawback of this solution is that a large number of pages in a flash block may become invalid and waste space. This is especially true if the reference pattern exhibits poor spatial locality. A solution is for invalid pages not to be stored on FLASH blocks, but only *marked* as invalid in the FLASH directory.

Directory sizes for various FLASH blocks sizes for a 128GB flash disk are shown in Table 1. Using larger FLASH pages saves considerable memory, which can be used for caching in RAM to increase the RAM hit ratio. Larger flash pages reduce the paging granularity; so the flash hit ratio will drop, especially for workloads with poor spatial locality (see Section 5.5). Writing to flash using a large block size (*e.g.*, 32kB or 64kB) increases bandwidth and random write efficiency [4].

Table 1. FLASH directory size

Flash page size	overwriting	invalidating
4kB	568MB	N/A
8kB	280MB	284MB
16kB	138MB	142MB
32kB	68MB	72MB
64kB	33.5MB	37.5MB
128kB	16.5MB	20.5MB

Thus, large flash blocks not only shrink the FLASH directory and increase RAM hits, but also speed up random writes to flash. An alternative is to store the FLASH directory (or a part of it) in FLASH instead of RAM. Due to lack of space this is not discussed here; more details can be found in [12]. **How much flash? How much RAM?** Without a FLASH cache, assume h_r RAM hits and m_r RAM misses occur for a workload. The total cost C_0 for this case is $C_0 = m_r(D_R + p_d D_W)$. One can simulate the cache behavior of a system with varying RAM and FLASH cache sizes (or even with no FLASH cache). By simulating the workload for various cache sizes, we can collect values for h_r, m_r, h_f, m_f, and p_d. These values, along with the read/write costs of specific flash and magnetic disks, can determine which storage/cache configuration is the most I/O-efficient for workloads of the given type. The price-to-I/O-cost ratio for each case gives the most cost-efficient solution. Alternatively, the 5-minute rule of [7] can determine the optimal memory and SSD capacities required, assuming prior workload knowledge. Our cost formulas determine the type of SSD that gives the best price/performance ratio for a type of workload. The decision for the size of the main memory and the SSD is an offline one and optimized for specific workloads. However, the optimal page flow scheme can be decided online, on a per-workload basis, by periodically evaluating the cost formulas. Our proposals are also applicable in DBMSs that employ per-file/relation buffer management: by monitoring the workload for each file and calculating the cost of each scheme, our model may lead to different files being buffered using different schemes.

5 Experimental Study

We evaluated our algorithms under various workloads. Our system consists of a main memory buffer pool for caching in RAM, a page cache on an SSD, and an HDD for persistency. Each page is identified by its HDD offset. The system was implemented in C++ and ran on an Intel Pentium 4 at 2.26GHz with 1.5GB of main memory running Linux (2.6.26 kernel). We used two HDDs and one SSD. Our system and the OS ran from one of the HDDs, while the other (referred to as HDD hereafter) was used to store the data. The HDD was a 300GB Maxtor 6L300R0 with 16MB of cache. The SSD was a 32GB MLC NAND Samsung MCAQE32G5APP. To eliminate OS caching we used both media as raw devices.

The SSD we used has a poor write performance and is unsuitable as a cache. Thus, we considered other SSDs, better suited for caching, by using their I/O costs in the equations of Section 3. We used published benchmarks ([6,8,17]) about the efficiency of each disk in IOPS. We present the read/write costs of all considered SSDs

Table 2. Flash disks considered

Disk Model	4kB Read IOPS	4kB Write IOPS	$/GB
Samsung	2500	21	1.6
Intel X25-M	12000	592	8.1
Intel X25-E	35000	3300	20
Fusion ioDrive	102000	101000	30

in Table 2. Random read performance varied by up to two orders of magnitude among disks, while random write performance varied by as much as four orders of magnitude.

We used three different workloads. The first, termed IRP, is an independent reference pattern where all pages in the dataset have the same probability of reference. We varied the probability of a page being read or written to and created workloads of varying dirtyness ratios. For the second workload, referred to as TPC-C, we ran the TPC-C benchmark on the PostgreSQL DBMS and collected a trace of page references, which we translated into HDD offsets. We did the same for the TPC-H benchmark to obtain the third workload. We report the results of executing these workloads on our system after varying its parameters. In all cases, the main memory page size was set to 4kB. For all experiments we used LRU as the page replacement policy (for both the RAM and the FLASH caches).

5.1 Impact of Cache Size on Hit Ratio

We measured the effect of the size of a page cache on its hit ratio, *i.e.*, how $H(Y, W, S)$ varies with S, the effective size of the cache, under LRU. We ran the three workloads for different page cache sizes; we report the hit ratio in Fig. 3.

The x-axis is S as a percentage of the size of the whole dataset. Throughout, the hit ratio grows with S (see also Section 3.5). The growth rate varies with the workload: it is linear for IRP and non-linear for TPC-C and TPC-H. This is due to both TPC-C and TPC-H having working sets (of different sizes), while IRP does not. Observe that, apart from

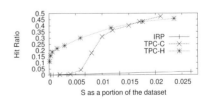

Fig. 3. Varying S in $H(Y, W, S)$

H growing with S, one cannot make assumptions or draw conclusions that hold for all workloads.

5.2   Impact of FLASH Size on RAM Hit Ratio

In our system, the directory for the FLASH cache is stored in main memory. Recall from Section 4 that as the size of the flash cache grows, the available main memory for the RAM cache shrinks and the RAM hit ratio is expected to drop. To show this, we grew the size of the FLASH cache (and thus the FLASH directory) while keeping the size of the RAM cache fixed, and measured the RAM hit ratio H.

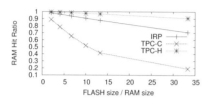

Fig. 4. H/H' for varying FLASH cache

We also ran the same workloads with no FLASH cache (and thus all main memory available to the RAM cache) and measured the RAM hit ratio H'. In Fig. 4 we show H/H' for different sizes of FLASH cache. The hit ratio drops linearly for IRP as it has no working set, and for TPC-H as its working set always fits in RAM. For TPC-C the working set fits in main memory for small FLASH sizes, but not for larger ones; thus, the ratio drops quickly and the curve is the inverse of the TPC-C curve of Fig. 3. In all cases, the main memory given to the FLASH index greatly affects the RAM hit ratio.

5.3   Validation of the Cost Formulas

We now verify the validity of our cost model. We executed a synthetic IRP workload using the Samsung SSD and measured the running time of each scheme. We also used the cost formulas of Section 3 with the I/O costs for the SSD and HDD to estimate the total cost of each scheme. We plot the ratio of the execution time for each physical

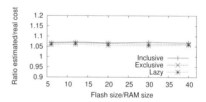

Fig. 5. Validation of the cost model

run over the cost projected by the formulas *for that scheme* in Fig. 5. The ratio remains constant for all FLASH cache sizes. Also, this ratio remains the same across page flow schemes, indicating the consistency of the model. The ratio being $6 - 8\%$ greater than 1 is due to our cost formulas not accounting for the cache warm-up time. Our formulas assume that each RAM miss results in a RAM eviction (and thus either a FLASH hit or a FLASH miss), which does not hold until after the RAM cache is full. The same holds for the warm-up time of the FLASH cache. Although we can estimate after how many references each cache fills up and adapt the formulas to account for this we chose not to do so for simplicity; moreover, this cost is negligible for workloads of interest. Additionally, for very small datasets the on-disk caches of FLASH and HDD affect their read/write costs. For all real-world workloads, however, our formulas were accurate in their cost estimation.

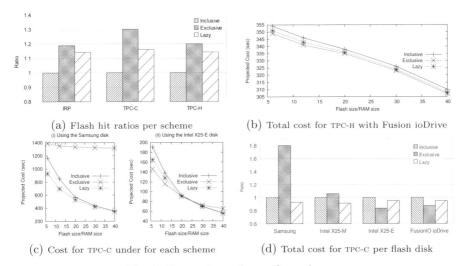

(a) Flash hit ratios per scheme

(b) Total cost for TPC-H with Fusion ioDrive

(c) Cost for TPC-C under for each scheme

(d) Total cost for TPC-C per flash disk

Fig. 6. Comparison of page flow schemes

5.4   Comparison of Page Flow Schemes

Flash Hit Ratio. We measured the FLASH hit ratio for each scheme and workload. We experimented with different RAM and FLASH sizes obtaining similar results; due to space limitations we only report in Fig. 6a the results for a FLASH cache 6 times the size of the RAM cache. All ratios are normalized by the hit ratio of inclusive. As explained in Section 3.5, exclusive has the highest hit ratio for all workloads and inclusive has the lowest. The hit ratio for lazy varies between the two. However, the highest hit ratio for exclusive does entail a lower I/O cost.

Total I/O Cost. We ran TPC-H and TPC-C for a varying FLASH size and a fixed RAM size. We plotted the total I/O cost as calculated using the formulas of Section 3 for different SSDs. We first ran TPC-H with the FLASH cache size varying from 5 to 40 times the size of the RAM cache. The projected I/O cost of the FusionIO ioDrive is shown in Fig. 6b. Here, exclusive outperforms the other two for all FLASH cache sizes; we will see that this is not always the case. Note also that increasing the size of the FLASH cache significantly benefits performance.

We then ran TPC-C for the same FLASH cache sizes and calculated the total cost based on the I/O costs of the Samsung disk; the results are shown in Fig. 6c(i). The exclusive scheme is unsuitable in this case due to the disk's disproportionally high write cost (as for each RAM eviction exclusive pays the cost of a flash write). For inclusive and lazy, while their costs are similar for large FLASH sizes, there is a performance gap for small FLASH sizes (or, big RAM sizes).

We repeated the calculations for TPC-C, but for the Intel X25-E disk; the results are shown in Fig. 6c(ii). When the FLASH cache is less than 15 times the size of the RAM cache, exclusive is the most efficient scheme. Its I/O cost is up to 30% lower than that of inclusive and 14% lower than the I/O cost of lazy. Conversely, for a FLASH cache size more than 35 times that of the RAM cache, lazy is the optimal scheme with an I/O cost that is 16% lower than that of exclusive. Therefore, even for the same FLASH disk and workload, the optimal scheme changes with the ratio of the FLASH cache size over the RAM cache size.

Next, we kept the FLASH and RAM cache sizes fixed and ran TPC-C under each scheme and calculated the total I/O cost for all disks. The results of Fig. 6d show that the optimal algorithm differs for each disk. Lazy is optimal for the MLC disks (Samsung and Intel X25-M), while exclusive is optimal for high-performing SLC devices (Intel X25-E and FusionIO). This confirms our hypothesis: no scheme is optimal across all workloads and disks. Though inclusive appears never to performs best, this is not the case if directory maintenance costs are also added [12].

5.5   Impact of Flash Cache Block Size

We then investigated how the FLASH block size affects performance. The RAM page size was always set to 4kB. We first varied the flash cache block size from 4kB to 128kB. For each block size we ran TPC-H and measured the FLASH hit ratio and the total number of HDD reads, using the overwriting technique of Section 4: upon page eviction from RAM, if the host block is not on FLASH then the whole block is brought from HDD to FLASH. We ran TPC-H under inclusive and lazy.

In Fig. 7 we show each scheme's FLASH hit ratio (top graph) and number of HDD reads (bottom graph). For inclusive, before a page is brought into RAM its flash block is written to the FLASH cache. Subsequent accesses to the block's pages will be served from FLASH. Thus, inclusive's hit ratio grows with the block size and overwriting acts as a prefetching mechanism, greatly affected by locality of reference. For lazy, a block is written to flash when one of its RAM pages is first evicted

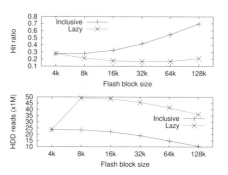

Fig. 7. Impact of block size under overwriting

from RAM. Even for workloads with a high degree of locality, pages of the same flash block will have most likely been read into RAM before one of them is evicted to FLASH. Thus, locality does not affect lazy as much for small block sizes. As the block size grows, so does the granularity at which the replacement policy tracks the reference pattern through access recency (or frequency). Thus, the hit ratio drops (for inclusive this effect is cancelled by the effect of prefetching). As shown in the bottom graph, lazy performs about twice as many HDD reads as inclusive. This is not only due to its lower hit ratio: when a RAM victim is written to FLASH, its host block needs to be read from HDD if it is not cached on FLASH. For inclusive, Invariant 1 guarantees that the host block of the page is on FLASH.

Next, we gauged the performance of overwriting and invalidating under TPC-C and TPC-H as we varied the flash block size from 4kB to 128kB; we used the lazy scheme in all cases. Let h_o and h_i be the hit ratios for overwriting and invalidating. In the top graph of Fig. 8 we report the ratio $\lambda = h_o/h_i$ for the two workloads; in the bottom graph we show the corresponding ratio of HDD reads.

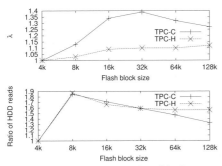

For both workloads, overwriting had a higher hit ratio than invalidating due to the decribed prefetching effect. This was more evident in TPC-H as it has a higher degree of locality than TPC-C. As explained, under the lazy scheme the hit ratio for both overwriting and invalidating shrinks with finer grained replacement (*e.g.*, 4kB blocks). Overwriting performed more HDD reads than invalidating for both workloads (between 1.3 to two times

Fig. 8. overwriting *vs.* invalidating

as many HDD reads). For workloads with locality of reference, overwriting is expected to give a higher hit ratio than invalidating at the cost of extra HDD reads. The optimal choice depends on the read efficiency of the SSD and the HDD.

5.6   Caching Only Clean Pages

Recall from Section 3 that, for the lazy scheme, one may apply any criterion to decide if a RAM victim page will be cached in FLASH or not. Dirty pages cached in FLASH are more likely to result in writes to flash. Thus, if the SSD is inefficient at random writes, it makes sense to restrict FLASH caching to clean pages only. Other criteria can also be used, *e.g.*, the update frequency of a page, but we do not explore these further due to lack of space. We used IRP workloads with different dirtyness ratios, *i.e.*, the probability of a page being dirtied on each next reference. Each workload was executed using the lazy scheme twice: once caching all RAM victims on FLASH and once caching only the clean ones.

We used the inefficient at random writes Samsung disk and measured the hit ratio for the FLASH cache and the total execution time for varying dirtyness. In Fig. 9 we show hit ratios (left graph) and execution times (right graph). The hit ratio drops when caching only clean pages, as some of the hot dirty pages are evicted to HDD. For low dirty-

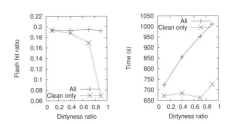

Fig. 9. Effect of caching only clean pages

ness, the hit ratio drops gradually, as the hottest dirty pages fit in RAM. For dirtyness ratios greater than 0.7, the hit ratio drops substantially. However, the execution time is less when caching only clean pages, due to the write inefficiency of the SSD. For dirtyness ratios between 0.1 to 0.7 the running time remains the same when caching only clean pages: the increased miss ratio is counterbalanced by the savings of avoiding flash writes. For higher dirtyness ratios the hit ratio drop results in a 10% increase in execution time.

5.7   Discussion

The I/O cost of a workload depends heavily on: (a) the workload itself, (b) the page flow scheme, and (c) the I/O costs of the flash disk. One cannot confidently decide the optimal scheme *a priori* without evaluating our cost formulas. For instance, exclusive writes to flash once for each RAM miss, regardless of the dirtyness of the victim page; inclusive and lazy do so only for dirty pages. Hence, for write-intensive workloads, exclusive will perform worse, more so if the flash disk is write-inefficient. Then, multiple flash writes can be avoided if only clean pages are cached on FLASH, (see Section 5). For large RAM cache sizes exclusive is likely the best option: no page will be cached on both caches, saving space on FLASH.

The results verify that hit ratios alone do not fully describe the system's I/O efficiency. Recall Figures 6a and 6d: although exclusive has the highest hit ratio for TPC-C, it is not optimal across all SSDs in terms of the total I/O cost. This holds for all workloads we have tested. Even for a specific workload and flash disk, the optimal scheme changes for different FLASH/RAM cache sizes (*e.g.*, Fig. 6c).

6   Concluding Remarks

Low read latencies, growing capacities, and dropping cost, make SSDs ideal for caching data between the main memory and the HDD. We studied the salient aspects of such a setup. We presented: (a) three invariants for the set of pages cached on flash, (b) algorithmic schemes enforcing the invariants, and (c) an I/O-based cost model for the performance of the algorithms. We studied problems such as the optimal page size for a flash cache and how the size of the page directory affects performance. We implemented our proposals and experimented with flash-resident caches. We showed that there is no universally optimal design for a flash cache. To make informed decisions, our analytical tools are necessary.

References

1. Agrawal, N., et al.: Design tradeoffs for ssd performance. In: ATC 2008: USENIX 2008 Annual Technical (2008)
2. Belady, L.A., et al.: An anomaly in space-time characteristics of certain programs running in a paging machine. Commun. ACM 12(6), 349–353 (1969)
3. Birrell, A., et al.: A design for high-performance flash disks. SIGOPS Oper. Syst. Rev. 41(2) (2007)
4. Bouganim, L., Jonsson, B.P., Bonnet, P.: uFLIP: Understanding flash IO patterns. In: CIDR (2009)
5. Chung, T.-S., et al.: System software for flash memory: A survey. In: Sha, E., Han, S.-K., Xu, C.-Z., Kim, M.-H., Yang, L.T., Xiao, B. (eds.) EUC 2006. LNCS, vol. 4096, pp. 394–404. Springer, Heidelberg (2006)

6. FusionIO - the power of 1000 hard drives in the palm of your hand. TGDaily, http://tgdaily.com
7. Graefe, G.: The five-minute rule twenty years later, and how flash memory chenges the rules. In: DAMON (2007)
8. Intel X25-M SSD: Intel Delivers One of the World's Fastest Drives. Anand Tech., http://anandtech.com
9. Kim, H., Ahn, S.: BPLRU: a buffer management scheme for improving random writes in flash storage. In: FAST 2008, Usenix Association (2008)
10. Kim, J., et al.: A space-efficient flash translation layer for compactflash systems. Transactions on Consumer Electronics (2002)
11. Koltsidas, I., Viglas, S.D.: Flashing up the storage layer. Proc. VLDB Endow. 1(1), 514–525 (2008)
12. Koltsidas, I., Viglas, S.D.: The case for flash-aware multi level caching. University of Edinburgh Technical Report EDI-INF-RR-1319 (2009)
13. Lee, S.-W., Moon, B.: Design of flash-based DBMS: An in-page logging approach. In: SIGMOD (2007)
14. Leventhal, A.: Flash storage memory. Commun. ACM 51(7), 47–51 (2008)
15. Narayanan, D., et al.: Migrating enterprise storage to SSDs: analysis of tradeoffs. Microsoft Technical Report MSR-TR-2008-169 (2008)
16. Nath, S., Gibbons, P.B.: Online maintenance of very large random samples on flash storage. Proc. VLDB Endow. 1(1), 970–983 (2008)
17. OCZ Core Series 64GB SATA II 2.5 Solid State Drive. TigerDirect.com., http://tigerdirect.com
18. Sun Microsystems. The Solaris ZFS filesystem
19. Yeong Park, S., et al.: CFLRU: a replacement algorithm for flash memory. In: CASES 2006. ACM, New York (2006)

Declarative Serializable Snapshot Isolation

Christian Tilgner[1], Boris Glavic[2], Michael Böhlen[1], and Carl-Christian Kanne[3]

[1] University of Zurich,
{tilgner,boehlen}@ifi.uzh.ch
[2] University of Toronto,
glavic@cs.toronto.edu
[3] University of Mannheim
kanne@informatik.uni-mannheim.de

Abstract. Snapshot isolation (SI) is a popular concurrency control protocol, but it permits non-serializable schedules that violate database integrity. The Serializable Snapshot Isolation (SSI) protocol ensures (view) serializability by preventing pivot structures in SI schedules. In this paper, we leverage the SSI approach and develop the Declarative Serializable Snapshot Isolation (DSSI) protocol, an SI protocol that guarantees serializable schedules. Our approach requires no analysis of application programs or changes to the underlying DBMS. We present an implementation and prove that it ensures serializability.

1 Introduction

Snapshot Isolation (SI) [3] is a popular multiversion concurrency control (MVCC) protocol, but it permits non-serializable schedules. Fekete et al. [9] showed that every non-serializable SI schedule necessarily contains an access pattern with two consecutive vulnerable edges (see Sec. 2.2), and Cahill et al. [5] presented the *Serializable Snapshot Isolation (SSI)* protocol that ensures serializable schedules by preventing such structures.

We leverage the ideas of SSI, define *pivot structures* and propose the *Declarative Serializable Snapshot Isolation (DSSI)* protocol, a declarative technique that guarantees serializable schedules by preventing pivot structures while maintaining the advantages of SI. We implement DSSI using our declarative scheduling model called *Oshiya*. Oshiya models the scheduler state (including the generated schedule) in so-called *scheduling relations* and formalizes a protocol as a *protocol specification*. A protocol specification is a set of constraints specified as boolean domain relational calculus expressions that have to hold for all scheduling relation states. In Oshiya, a protocol specification is implemented as declarative *scheduling queries*. Request scheduling is performed by applying a generic scheduling algorithm that repeatedly executes the scheduling queries over the scheduling relations. The queries determine the pending requests that can be added to the relation modelling the schedule without violating the protocol specification. We show how to detect and prevent pivot structures using Oshiya and implement the DSSI protocol specification as scheduling queries. Our

J. Eder, M. Bielikova, and A.M. Tjoa (Eds.): ADBIS 2011, LNCS 6909, pp. 170–184, 2011.

implementation is concise and close to the formal protocol specification which enables us to prove its correctness. The main contributions of the paper are:

- We introduce DSSI, a protocol that ensures serializable SI executions, and formalize it as an Oshiya protocol specification.
- Using Oshiya we develop an SQL implementation of DSSI.
- We prove that the implementation ensures serializable schedules.

The paper structure is as follows: Sec. 2 describes SI and reviews the approach applied by the SSI protocol to detect non-serializable schedules. Sec. 3 introduces Oshiya. Sec. 4 shows how we model data snapshots and presents schemata for the scheduling relations. Sec. 5 formalizes the DSSI protocol. Sec. 6 presents the DSSI scheduler implementation. Sec. 7 proves that our implementation ensures serializable executions. We review related work in Sec. 8 and conclude in Sec. 9.

2   Background: Snapshot Isolation and Serializability

We model a transaction t_i as a sequence of read and write requests (denoted as $r_i(x)$ resp. $w_i(x)$ where x stands for the accessed data item). Each transaction finishes with an abort (a_i) or commit (c_i) request. The write-set WS_i of t_i contains all data items written by t_i. A *history* (schedule) is a sequence of interleaved executions of requests from a set of concurrent transactions. The requests in a history are totally ordered. We write $p <^H q$ if request p is executed before request q. Let bot_i denote the begin of t_i (when t_i executed its first request) and eot_i its end (when t_i aborted resp. committed). The execution interval of a committed transaction t_i is $[bot_i, c_i]$, the one of a non-aborted, possibly committed transaction t_i is $[bot_i, l_i]$ (l_i is t_i's latest operation). Two committed transactions t_i and t_j *overlapped* if: $Overlapped_{ij} \Leftrightarrow [bot_i, c_i] \cap [bot_j, c_j] \neq \emptyset$. Two non-aborted (maybe active) transactions t_i and t_j *overlap* if: $Overlap_{ij} \Leftrightarrow [bot_i, l_i] \cap [bot_j, l_j] \neq \emptyset$.

2.1   Snapshot Isolation

SI is a multiversion concurrency protocol that maintains multiple versions of data items (tuples). Each write $w_i(x)$ creates a new version of item x that is visible to other transactions after c_i. Each read $r_i(x)$ accesses the latest version of x written by transactions that committed before bot_i. Moreover, a transaction always sees the versions it created itself. Under SI, reads are never delayed because of write requests of concurrent transactions and vice versa. SI avoids inconsistent read anomalies because transactions never access partial results of other concurrent transactions. SI requires disjoint write-sets of concurrent committed transactions which is, e.g., ensured by the First-Committer-Wins (FCW) rule. FCW specifies that a transaction is aborted if a concurrent transaction with an overlapping write-set already committed. FCW also prevents lost updates. A typical anomaly that leads to non-serializable SI histories is the *Write Skew* [3], detailed in Ex. 1.

Example 1. Consider history H_{ws} in Fig. 1. Initially, data items $x = 50$ and $y = 50$ are consistent and satisfy constraint $C = x + y \geq 0$. Transaction t_1 reads

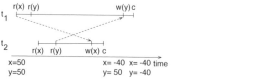

Fig. 1. History H_{ws} **Fig. 2.** MVSG for History H_{ws}

x and y. A concurrent transaction t_2 reads x and y, writes x (after subtracting 90) and commits (after checking C). Finally, t_1 writes y (after subtracting 90) and commits (after checking C). In the final state, C is violated although t_1 and t_2 checked C explicitly before committing. This happens because C is checked on the version of x and y that is visible to t_1 and t_2 and not on the final state resulting from their interleaved execution.

2.2   Detecting Non-serializable Histories

Serializability of SI histories can be checked using a multiversion serialization graph $MVSG = (N, E)$ [5]. The MVSG of a history H is a graph that contains a node for each committed transaction t_i of H: $t_i \in N \Leftrightarrow c_i \in H$. It contains an edge from transaction t_i to transaction t_j with $i \neq j$ if (a) $w_i(x) <^H w_j(x)$, (b) $w_i(x) <^H r_j(x)$, or (c) $r_i(x) <^H w_j(x)$. An edge of type (c) that occurs between two overlapped transactions t_i and t_j is called a *vulnerable edge* [9]. A *pivot structure* is defined as follows: $Overlapped_{ij} \wedge (r_i(x) <^H w_j(x)) \wedge Overlapped_{jk} \wedge (r_j(x) <^H w_k(x))$. Fekete et al. [9] showed that every MVSG of a non-serializable SI history must contain a pivot structure. The existence of a pivot structure is a necessary but not sufficient condition for the non-serializability of an SI history. Thus, an SI history is serializable if its MVSG does not contain pivot structures.

Example 2. Fig. 1 shows history H_{ws}. Vulnerable edges are shown as dotted lines. The MVSG for H_{ws} in Fig. 2 has a node for each committed transaction of H_{ws} (t_1 and t_2) and two edges e: (e_1) from t_1 to t_2 due to $r_1(x) <^H w_2(x)$; (e_2) from t_2 to t_1 due to $r_2(y) <^H w_1(y)$. H_{ws} is not serializable and, thus, the MVSG contains a pivot structure (two consecutive vulnerable edges e_1 and e_2).

2.3   Serializable Snapshot Isolation Protocol

The SSI protocol proposed by Cahill et al. [5] ensures serializability by preventing pivot structures. The main idea is to check SI histories at runtime for structures that can evolve into pivot structures. We call such structures *potential pivot structures*. A potential pivot structure is defined as: $Overlap_{ij} \wedge (r_i(x) <^H w_j(x)) \wedge Overlap_{jk} \wedge (r_j(x) <^H w_k(x)) \wedge \neg(c_i \wedge c_j \wedge c_k)$. I.e., a potential pivot structure is a pivot structure without the requirement that the three (not necessarily distinct) participating transactions have committed. It evolves into a pivot structure once all participating transactions have committed. The set of transactions in potential pivot structures is naturally a superset of the transactions in pivot structures. For each detected potential pivot structure, one of the participating transactions is aborted to prevent it from evolving into a pivot structure. This approach guarantees that the resulting histories are serializable, but it may produce false positives, i.e., not every potential pivot structure finally results in a non-serializable history. Our implementation leverages this idea and aborts transactions that participate in potential pivot structures (see Sec. 6).

3   Declarative Scheduling Model

We propose a *declarative scheduling model* [13] called $Oshiya^1$ to model and implement DSSI. The main ideas of Oshiya are: (1) The state of a scheduler (including the history it produces) is modeled as instances of three *scheduling relations*: *PendingRequests* ($\mathcal{R}$) buffers arriving client requests for scheduling. *RelevantHistory* ($\mathcal{H}$) stores already executed requests in their execution order, and models the schedule generated so far. *Executable* ($\mathcal{E}$) buffers requests that have been scheduled for execution. (2) Oshiya formalizes a protocol as a set of constraints, called *protocol specification*, that have to hold for each generated state of $\mathcal{H}$. (3) The protocol specification constraints are implemented as declarative *scheduling queries*: $Q_{Schedule}, Q_{Revoked}, Q_{Irrelevant}$. Request scheduling is performed by repeatedly executing the *scheduling queries* over the *scheduling relations* to determine which of the pending requests in $\mathcal{R}$ can be added to $\mathcal{H}$ without violating the protocol specification constraints.

Example 3. For presentation purposes, we use simplified schemata for the scheduling relations in this example. Assume the following schema for relations $\mathcal{R}$ and $\mathcal{E}$: (TA, Op, Ob). For each request, TA is the transaction executing the request, Op is the type of operation (e.g., r for a read), and Ob is the data object the operation accesses. Relation $\mathcal{H}$ has an additional attribute ID for recording the request execution order. Using this schema, the scheduler state after scheduling the first request from history H_{ws} (Fig. 1) is as follows:

$\mathcal{R}_1$

TA	Op	Ob	...
1	r	x	

$\mathcal{H}_1$

ID	TA	Op	Ob	...
1	1	r	x	

$\mathcal{E}_1$

TA	Op	Ob	...
1	r	x	

The state of the scheduler is advanced in iterative steps by applying a generic scheduling algorithm (shown on the right) that evaluates scheduling queries over the current instances of the scheduling relations. Each iterative step (one while loop), called *scheduler iteration*, schedules multiple requests at once, resulting in updated instances of the scheduling relations. This is in contrast to

```
1   H = E = R = ∅
2   while true do begin
3       R = R − E ;
4       R = R ∪ N ;
5       R = R − Q_Revoked(H, R) ;
6       E = Q_Schedule(H, R) ;
7       Execute(E) ;
8       H = H ∪ E ;
9       H = H − Q_Irrelevant(H) ;
10  end
```

DBMSs that schedule requests individually. The algorithm is the same for every protocol, but it is parameterized by the protocol specific schema of the scheduling

---

[1] Oshiya refers to the passenger arrangement staff at Japanese train stations who help to fill a train by pushing people onto the train or guiding them to free railway cars.

relations and the scheduling queries. $\mathcal{N}$ is the set of newly arrived client requests. $Q_{Revoked}$ identifies nonexecutable requests (e.g., deadlocked) (line 5). $Q_{Schedule}$, the main scheduling query, identifies the pending requests from $\mathcal{R}$ that should be selected for execution in this iteration (line 6). $Q_{Irrelevant}$ returns requests that are irrelevant for future scheduling decisions. They are removed from $\mathcal{H}$ (line 9). In the remainder of this paper we limit the discussion to $Q_{Schedule}$.

Example 4. Reconsider the scheduler state from Ex. 3. Two new requests got inserted into $\mathcal{R}$ at the beginning of scheduler iteration 2: $r_1(y)$, $r_2(x)$. Assume that running the scheduling queries selected both request from $\mathcal{R}$ for execution. This leads to the following updated scheduler state:

$\mathcal{R}_2$

TA	Op	Ob	...
1	r	y	
2	r	x	

$\mathcal{H}_2$

ID	TA	Op	Ob	...
1	1	r	x	
2	1	r	y	
3	2	r	x	

$\mathcal{E}_2$

TA	Op	Ob	...
1	r	y	
2	r	x	

Applying the scheduling queries to a set of newly arrived requests $\mathcal{N}$, each scheduler iteration produces new instances of the scheduling relations $\mathcal{R}$, $\mathcal{H}$ and $\mathcal{E}$. This yields a sequence of states of $\mathcal{H}$ called history, defined below. We use this definition of history to reason over the properties of a protocol and to prove the correctness of a scheduler implementation.

Definition 1 (History). *Let $\mathbb{I} = < \mathcal{N}_0, \dots >$ be a sequence of sets of input requests. Let q be protocol-specific versions of the scheduling queries. We define the history $\mathbb{H}_q(\mathbb{I})$ generated according to q over input $\mathbb{I}$ as $< \mathcal{H}_0, \dots >$, where $\mathcal{H}_i$, called a history state, is the state of relation $\mathcal{H}$ after the ith scheduler iteration produced using q to parameterize the generic algorithm and $\mathcal{N}_i$ as input $\mathcal{N}$. In the paper, we drop q and $\mathbb{I}$ if it is clear from the context and solely use $\mathbb{H}$.*

In the remainder of this paper, we use $\mathcal{H}$ to denote both the history relation and one history state and drop indices on $\mathcal{H}$ if the scheduler iteration is irrelevant for the discussion (same holds for $\mathcal{R}$, $\mathcal{E}$ and $\mathcal{N}$). According to the algorithm presented above, the history state $\mathcal{H}_i$ is a *cumulative snapshot*, i.e., it includes all previous history states $\mathcal{H}_j$ with $j < i$.

Example 5. For instance, the history states shown below could be the result of scheduling the requests $\mathbb{I} = < \{(1, r, x), (2, r, x)\}, \{(1, r, y)\}, \{(2, r, y)\} >$:

$\mathcal{H}_0$

ID	TA	Op	Ob	...

$\mathcal{H}_1$

ID	TA	Op	Ob	...
1	1	r	x	

$\mathcal{H}_2$

ID	TA	Op	Ob	...
1	1	r	x	
2	1	r	y	
3	2	r	x	

$\mathcal{H}_3$

ID	TA	Op	Ob	...
1	1	r	x	
2	1	r	y	
3	2	r	x	
4	2	r	y	

We model a protocol as a set of constraints called *protocol specification*. A protocol specification constraint is a boolean *domain relational calculus* expression over histories. We allow quantification over scheduler iterations to enable, e.g., constraints that check the order of requests in the history.

Definition 2 (Protocol Specification). *A protocol specification Φ is a set of boolean domain relational calculus expressions over $\mathbb{H}$.*

The formalization of a protocol as logical constraints and its implementation as queries allows us to formally reason about the correctness of an implementation. Given a protocol specification Φ and an implementation of this protocol as a set q of scheduling queries, the definition presented below defines what it means for q to correctly implement Φ. Intuitively, this is the case if for every input $\mathbb{I}$, the history created by our scheduling algorithm using q satisfies Φ. We use this definition in Sec. 7 to prove the correctness of our DSSI implementation.

Definition 3 (Correctness of Scheduling Queries). *Scheduling queries q satisfy a protocol specification Φ, denoted as $q \models \Phi$, if for every input sequence $\mathbb{I}$ the generated history $\mathbb{H}$ produced using q satisfies Φ: $\mathbb{H}_q(\mathbb{I}) \models \Phi$.*

3.1   Assumptions and Notational Remarks

We make the following assumptions: (1) Client requests read resp. manipulate only one tuple. (2) A transaction waits until its current request is executed before issuing new requests. (3) Object identifiers are unique over all relations. (4) Rollbacks of transactions are considered as regular requests issued by clients. Extending Oshiya to schedule complex queries like joins or range queries is an interesting avenue for future work. Assumptions 2-4 simplify the presentation, but can be changed with minor modifications to Oshiya.

Scheduling queries and protocol specifications are given as domain relational calculus expressions. Capital letters denote variables, small letters indicate constants and ϵ denotes *null*. All non-target variables not used in a universal quantification are implicitly existentially quantified. E.g., instead of $\{A \mid \exists B : (I(A, B) \land \neg \exists C : (J(C, A)))\}$ we write $\{A \mid I(A, B) \land \neg J(C, A)\}$. Unrestricted existentially quantified variables are displayed as an underline ("_"), disjunctive use of constants by "|". E.g., for the expression $I(A, B) \land (A = a \lor A = c)$ we use the shortcut $I(a|c, \_)$. We define aggregation as: $\{G, F_1(A_1), \ldots, F_n(A_n) \mid E\}$. E is a domain relational calculus expression, G is a set of attributes on which to group on (can be empty), and each F_i is an aggregate over attribute A_i.

4   Modeling Data Relation Snapshots and Defining the Oshiya Scheduling Relation Schemata for DSSI

In order to implement DSSI with Oshiya, we have to (1) specify the schema of the scheduling relations that model the scheduler state, (2) formalize the protocol specification based on these relations (Sec. 5), and (3) implement the protocol specification as scheduling queries (Sec. 6). In this section, we show how to adapt data relation schemata to support data item versions (Sec. 4.1) and develop protocol-specific schemata for the Oshiya scheduling relations (Sec. 4.2).

4.1   Modeling Snapshots with Data Relations

We model snapshots explicitly by extending the schemata of data relations. This allows us to achieve DB independence and to run DSSI on DBMSs that do not support snapshots. We identify a version of data item x using a tuple (TA, Seq) where TA is the transaction that created the version and Seq is the position of the request within this transaction. Of course, versions can be modeled differently but this is orthogonal to our approach and beyond the scope of this paper. Given a database schema with relations $R_1, \ldots, R_n$, we map each relation R_i to a relation R'_i which has four additional attributes. These attributes store the version identifier for the creator transaction (CTA and $CSeq$) and, if applicable, for the transaction that deleted the data item (DTA and $DSeq$). The primary key of R'_i is the primary key of R_i union the attributes CTA and $CSeq$.

Example 6. Assume a bank stores account data with account numbers and balances in relation $Accounts$(AccNr,Bal). We map this relation to $Accounts'$ by extending its schema with the four additional attributes mentioned above. An example instance shown on the right contains an initial version of object x created by transaction t_1 ($CTA = 1, CSeq = 1$) and two new versions created by t_2 and t_3.

$Accounts'$

AccNr	Bal	CTA	CSeq	DTA	DSeq
x	5	1	1	-	-
x	10	2	2	-	-
x	15	3	1	-	-

4.2   Oshiya Scheduling Relation Schemata

For DSSI, we use the schemata for *scheduling relations* $\mathcal{R}$, $\mathcal{H}$ and $\mathcal{E}$ shown below. For simplicity, we present only attributes needed for scheduling and omit those necessary for request execution (e.g., the value to be written for write requests).

$\mathcal{R}$ (TA,Seq,Op,OID) $\mathcal{H}$ (ID,TA,Seq,Op,OID,OTA,OSeq) $\mathcal{E}$ (ID,TA,Seq,Op,OID,OTA,OSeq)

For each incoming request, we insert a tuple into $\mathcal{R}$ storing an identifier T_i for the transaction t_i that issued the request (TA), the request position within this transaction (Seq), the type of operation (read, write, abort or commit, stored in attribute Op) and the data object the requests is applied to (OID). Transactions identifiers (TA) are ordered, i.e., if $bot_i < bot_j$ then $T_i < T_j$. $\mathcal{H}$ and $\mathcal{E}$ contain additional attributes: ID records the execution order of requests. For read requests, OTA and $OSeq$ store which object version was read by the request. These attributes correspond to the data relations attributes CTA and $CSeq$.

Example 7. Assume the instances of relations $\mathcal{R}$ and $\mathcal{H}$ displayed below. $\mathcal{H}$ contains the requests that produced the state of relation $Accounts'$ from Ex. 6: (1) and (2) Transaction t_1 created the initial version of object x and committed. (3) Transaction t_2 read this version of object x. (4) and (5) t_2 and t_3 wrote new versions of object x. (6) t_2 committed. (7) t_4 read the new version created by t_2. At this iteration, $\mathcal{R}$ contains no pending requests that have to be scheduled.

$\mathcal{R}$

TA	Seq	Op	Ob

$\mathcal{H}$

ID	TA	Seq	Op	Ob	OTA	OSeq
1	1	1	w	x	-	-
2	1	2	c	-	-	-
3	2	1	r	x	1	1
4	2	2	w	x	-	-
5	3	1	w	x	-	-
6	2	3	c	-	-	-
7	4	1	r	x	2	2

5 DSSI Protocol Specification

We now develop the protocol specification for DSSI based on the scheduling relations presented in Sec. 4. Recall from Sec. 3 that a protocol specification models a protocol as a set of domain relational calculus expressions over histories.

To formalize SI with Oshiya, we use views over relation $\mathcal{H}$ to get the relevant information described in Sec. 2. For *bot*, we use view $BOT(TA, ID)$ querying for each transaction (TA) the ID of its first request in $\mathcal{H}$. $EOT(TA, Op, ID)$ selects for each finished transaction t_i (TA) the ID of its final request in $\mathcal{H}$ (corresponds to eot_i) and whether t_i aborted or committed (Op). $Overlap(TA1, TA2)$ contains all pairs of concurrently executed, non-aborted transactions, i.e., they do not have to be committed. $PotPivotStr(TA1, TA2, TA3)$ selects all triples of transactions forming potential pivot structures as described in Sec. 2.3.

C1 (Read Versions). The SI protocol specifies [3,5,14] that a read request $r_i(x)$ of a transaction t_i reads t_i's most recent changes to x. If no such changes exist, then $r_i(x)$ reads the latest version of x created by transactions that committed before t_i started. These conditions are formalized as protocol specification constraint C1 (a) and (b) shown in Fig. 3: **(a)** The first case applies if a transaction T has written object O before reading a version (X, Y) of O:

$$\mathcal{H}(I, T, N, r, O, X, Y) \wedge \mathcal{H}(I_2, T, N_2, w, O, \_, \_) \wedge I_2 < I$$

It follows that T read a version it created itself ($X = T$) and (X, Y) is the latest version produced by T before the read (no newer versions exist):

$$X = T \wedge N_2 = Y \wedge \neg(\mathcal{H}(\_, T, N_2, w, O, \_, \_) \wedge Y < N_2 < N)$$

(b) The second case applies if T has not written O before the read was executed: $\neg(\mathcal{H}(I_2, T, \_, w, O, \_, \_) \wedge I_2 < I)$. It follows that (1) O was written by another transaction X and X committed before T started. (2) (X, Y) has to be the latest version written by X and (3) there may not be another version written by a transaction T_2 that committed after X but before T started:

(**1**) $X \neq T \wedge EOT(X, c, I_3) \wedge BOT(T, I_4) \wedge I_3 < I_4$ (**2**) $\neg(\mathcal{H}(\_, X, N_3, w, O, \_, \_) \wedge N_3 > Y)$
(**3**) $\neg(\mathcal{H}(\_, T_2, \_, w, O, \_, \_) \wedge EOT(T_2, c, I_5) \wedge I_4 < I_5 < I_3)$

C2 (FCW). SI requires disjoint write-sets for all committed concurrent transactions. Protocol specification constraint C2 (see Fig. 3) models this condition as follows. If (1) two overlapping transactions T and T_2 (2) both wrote the same object O and (3) T did already commit, then (4) T_2 did not commit:

(**1**) $Overlap(T, T_2)$ (**2**) $\mathcal{H}(\_, T, \_, w, O, \_, \_) \wedge \mathcal{H}(\_, T_2, \_, w, O, \_, \_)$
(**3**) $EOT(T, c, \_)$ (**4**) $\neg EOT(T_2, c, \_)$

C3 (Serializability). Recall that an SI history is serializable, if it does not contain pivot structures. In constraint C3 (see Fig. 3), we follow the approach outlined in Sec. 2.3: If (1) relation $\mathcal{H}$ contains a potential pivot structure, then we require that (2) at least one of the participating transactions did not commit:

(**1**) $PotPivotStr(T, T_2, T_3)$ (**2**) $\neg(EOT(T, c, \_) \wedge EOT(T_2, c, \_) \wedge EOT(T_3, c, \_))$

(C1) (a) $\forall I, N, O, T, X, Y : \mathcal{H}(I, T, N, r, O, X, Y) \wedge \mathcal{H}(I_2, T, N_2, w, O, \_, \_) \wedge I_2 < I \Rightarrow$
$X = T \wedge N_2 = Y \wedge \neg(\mathcal{H}(\_, T, N_3, w, O, \_, \_) \wedge Y < N_3 < N)$
(b) $\forall I, N, O, T, X, Y : \mathcal{H}(I, T, N, r, O, X, Y) \wedge \neg(\mathcal{H}(I_2, T, \_, w, O, \_, \_) \wedge I_2 < I) \Rightarrow$
$X \neq T \wedge EOT(X, c, I_3) \wedge BOT(T, I_4) \wedge I_3 < I_4 \wedge \neg(\mathcal{H}(\_, X, N_3, w, O, \_, \_) \wedge N_3 > Y) \wedge$
$\neg(\mathcal{H}(\_, T_2, \_, w, O, \_, \_) \wedge EOT(T_2, c, I_5) \wedge I_3 < I_5 < I_4)$

(C2) $\forall O, T, T_2 : Overlap(T, T_2) \wedge \mathcal{H}(\_, T, \_, w, O, \_, \_) \wedge \mathcal{H}(\_, T_2, \_, w, O, \_, \_) \wedge EOT(T, c, \_)$
$\Rightarrow \neg EOT(T_2, c, \_)$

(C3) $\forall T, T_2, T_3 : PotPivotStr(T, T_2, T_3) \Rightarrow \neg(EOT(T, c, \_) \wedge EOT(T_2, c, \_) \wedge EOT(T_3, c, \_))$

Fig. 3. DSSI Protocol Specification

6   DSSI Implementation

Recall that with Oshiya, protocols are implemented as scheduling queries. We implemented all scheduling queries for DSSI, but in this paper we only describe $Q_{Schedule}$. Our prototype implementation of Oshiya requires the scheduling queries to be expressed in SQL. However, for conciseness, domain relational calculus expressions are used throughout this section. $Q_{Schedule}$ is developed in two steps. First we present queries necessary to detect potential pivot structures (Sec. 6.1). Afterwards, we use these queries to implement $Q_{Schedule}$ (Sec. 6.2). Recall that detecting potential pivot structures and aborting one of the participating transactions ensure serializability. However, this approach may detect false positives (see Sec. 2). Studying the trade-off between the number of false positives and the cost of scheduling is an interesting avenue for future work.

6.1   Detecting Potential Pivot Structures

We now discuss how to express BOT, EOT, $Overlap$ and $PotPivotStr$ introduced in Sec. 5 as queries over $\mathcal{H}$. BOT and EOT are defined below. E.g., EOT queries for each finished transaction T its abort resp. commit state (A) and its eot (I) which is equal to the ID of its abort resp. commit request in $\mathcal{H}$.

$$BOT = \{T, I \mid \mathcal{H}(I, T, \_, \_, \_, \_, \_) \wedge \neg(\mathcal{H}(I_2, T, \_, \_, \_, \_, \_) \wedge I_2 < I)\}$$
$$EOT = \{T, A, I \mid \mathcal{H}(I, T, \_, A, \_, \_, \_) \wedge A = a|c\}$$

Overlapping transactions are inferred as specified below. Two (1) non-aborted transactions T_1 and T_2 overlap if (2) $bot_1 <^H bot_2$ and (3) $bot_2 <^H c_1$ (if T_1 has already committed) or (4) the symmetric case holds:

$$Overlap = \{T_1, T_2 \mid T_1 \neq T_2 \wedge \neg EOT(T_1|T_2, a, \_) \wedge \tag{1}$$
$$((BOT(T_1, I) \wedge BOT(T_2, I_2) \wedge I < I_2 \wedge \tag{2}$$
$$(EOT(T_1, c, I_3) \Rightarrow I_2 < I_3)) \vee \tag{3}$$
$$(BOT(T_2, I_2) \wedge BOT(T_1, I) \wedge I_2 < I \wedge (EOT(T_2, c, I_3) => I < I_3)))\} \tag{4}$$

We use $PotVulnEdge$ to query all potential vulnerable edges between concurrent, non-aborted transactions T and T_2 (potential, because T and T_2 might not yet have committed). $PotPivotStr$ detects potential pivot structures by checking for transactions (T_2) that have both an incoming and outgoing $PotVulnEdge$:

$$PotVulnEdge = \{T, T_2 \mid \mathcal{H}(I, T, \_, r, O, \_, \_) \wedge \mathcal{H}(I_2, T_2, \_, w, O, \_, \_) \wedge Overlap(T, T_2) \wedge I < I_2\}$$
$$PotPivotStr = \{T, T_2, T_3 \mid PotVulnEdge(T, T_2) \wedge PotVulnEdge(T_2, T_3)\}$$

Example 8. We show the results of the queries defined above (highlighted) for the history state $\mathcal{H}$ from Ex. 7. For instance, $PotVulnEdge$ contains one potential vulnerable edge from transaction t_2 to t_3, because t_2 and t_3 overlap and t_2 read object x and afterwards t_3 wrote a new version of object x ($r_2(x) <^H w_3(x)$).

$\mathcal{H}$

ID	TA	Seq	Op	Ob	OTA	OSeq
1	1	1	w	x	-	-
2	1	2	c	-	-	-
3	2	1	r	x	1	1
4	2	2	w	x	-	-
5	3	1	w	x	-	-
6	2	3	c	-	-	-
7	4	1	r	x	2	2

BOT

TA	ID
1	1
2	3
3	5
4	7

EOT

TA	Op	ID
1	c	2
2	c	6

Overlap

TA1	TA2
2	3
3	2
3	4
4	3

PotVulnEdge

TAout	TAin
2	3

PotPivotStr

TA1	TA2	TA3

6.2 $Q_{Schedule}$

The DSSI version of $Q_{Schedule}$ implementing the protocol specification constraints C1-C3 is shown in Fig. 4. According to the SI conditions, all write, abort, and read requests from $\mathcal{R}$ may always be selected for execution. $Q_{Schedule}$ selects all of these requests using queries *AbortWrites* and *Reads*. Which commit requests can be selected without violating constraints $C2$ and $C3$ is determined through query *ValidCommits*. In $Q_{Schedule}$, function *GenID()* generates unique values for the *ID* attribute of $\mathcal{H}$ (modelling the execution order of requests).

Read Requests (C1). The *Reads* query uses LVV (last valid version) to select for each read request of transaction T on object O the version (T_2, N_2) that has to be read. Recall that attributes OTA and $OSeq$ of relations $\mathcal{E}$ and $\mathcal{H}$ identify a version of an object O. Version (T_2, N_2) is computed in two steps. $LastOTA$ queries the transaction identifier (T_2) of the transaction that wrote the version of O that has to be read by T. Based on this information LVV determines N_2, the Seq value of the latest write request of T_2 on object O. T_2 is the maximal value from the following union: (a) $T_2 = T$ if T itself created versions of O and (b) transactions that wrote a version of O and committed before T started.

(a) $\mathcal{H}(\_, T_2, \_, w, O, \_, \_) \wedge T = T_2$ (b) $\mathcal{H}(\_, T_2, \_, w, O, \_, \_) \wedge EOT(T_2, c, I_2) \wedge (BOT(T, I) \Rightarrow I_2 < I)$

Example 9. Consider $\mathcal{H}$ from Ex. 8. $r_2(x)$ read the initial version of object x (since $c_1 <^H bot_2$) and $r_4(x)$ read the version written by t_2 (since $c_2 <^H bot_4$).

Commit Requests (C2 and C3). To guarantee that constraints C2 and C3 hold for each history produced by $Q_{Schedule}$, we have to prevent commit requests to be executed if (1) the commit would violate the FCW rule (C2) or (2) the commit would violate serializability (C3). There are two possible ways how the execution of commit requests can violate the FCW rule: (1a) A commit is from a transaction whose write-set overlaps with the one of a concurrent but already committed transaction and (1b) if $\mathcal{R}$ contains commit requests from multiple

$$Q_{Schedule} = \{GenID(), T, N, A, O, T_2, N_2 \mid \mathcal{R}(T, N, A, O) \wedge (ValidCommits(T, N, T_2, N_2)$$
$$\vee AbortsWrites(T, N, T_2, N_2) \vee Reads(T, N, T_2, N_2))\}$$

$$AbortsWrites = \{T, N, \epsilon, \epsilon \mid \mathcal{R}(T, N, a \mid w, \_)\}$$

$$Reads = \{T, N, T_2, N_2 \mid \mathcal{R}(T, N, r, O) \wedge LVV(T, O, T_2, N_2)\}$$
$$LVV = \{T, O, T_2, MAX(N_2) \mid LastOTA(T, O, T_2) \wedge \mathcal{H}(\_, T_2, N_2, w, O, \_, \_)\}$$
$$LastOTA = \{T, O, MAX(T_2) \mid \mathcal{R}(T, \_, r, O) \wedge ((\mathcal{H}(\_, T_2, \_, w, O, \_, \_) \wedge T = T2) \vee$$
$$(\mathcal{H}(\_, T_2, \_, w, O, \_, \_) \wedge EOT(T_2, c, I_2) \wedge (BOT(T, I) \Rightarrow I_2 < I)))\}$$

$$ValidCommits = \{T, N, \epsilon, \epsilon \mid NonForbCs(T, N) \wedge \neg DelayedCs(T, N)\}$$

$$DelayedCs = \{T, N \mid NonForbCs(T, N) \wedge NonForbCs(T_2, \_) \wedge$$
$$\mathcal{H}(\_, T, \_, w, O, \_, \_) \wedge \mathcal{H}(\_, T_2, \_, w, O, \_, \_) \wedge T > T_2\}$$

$$NonForbCs = \{T, N \mid \mathcal{R}(T, N, c, \_) \wedge \neg(ForbCs(T, N) \vee ForbCinPPS(T, N))\}$$

$$ForbCinPPS = \{T, N \mid \mathcal{R}(T, N, c, \_) \wedge PotPivotStr(T_2, T_3, T_4) \wedge (T = T_2 \mid T_3 \mid T_4) \wedge$$
$$\neg(\mathcal{R}(T_5, \_, c, \_, \_) \wedge (T_5 = T_2 \mid T_3 \mid T_4) \wedge T < T_5)\}$$

$$ForbCs = \{T, N \mid \mathcal{R}(T, N, c, \_) \wedge \mathcal{H}(\_, T, \_, w, O, \_, \_) \wedge \mathcal{H}(\_, T_2, \_, w, O, \_, \_) \wedge$$
$$Overlap(T, T_2) \wedge EOT(T_2, c, \_)\}$$

Fig. 4. $Q_{Schedule}$

transactions with overlapping write-sets, then only one of these transaction may commit. Note that in the concrete implementation, commits identified to violate C2 or C3 are selected by $Q_{Revoked}$ and aborted.

We use a two stage approach to select valid commits: In step 1, query *Non-ForbCs* selects commits from $\mathcal{R}$ and filters out commits of case 1a using query *ForbCs* and those of case 2 using query *ForbCinPPS*. *NonForbCs* may still contain sets of commit requests from transactions with overlapping write-sets (case 1b). We only allow the oldest transaction from each set to commit. Therefore, in step 2, query *ValidCommits* selects all requests from *NonForbCs* and uses query *DelayedCs* to keep only the commit request of the oldest transaction for each set of transactions with overlapping write-sets.

Step 1. Query *ForbCs* (case 1a) identifies commits of transactions T that (a) wrote an object also written by an (b) overlapping committed transaction T_2.

    (a) $\mathcal{H}(\_, T, \_, w, O, \_, \_) \wedge \mathcal{H}(\_, T_2, \_, w, O, \_, \_)$ **(b)** $Overlap(T, T_2) \wedge EOT(T_2, c, \_)$

ForbCinPPS (case 2) selects a commit of transaction T from $\mathcal{R}$ if (a) T belongs to potential pivot structure p and (b) $\mathcal{R}$ does not contain a commit request of a younger transaction T_5 (recall that $bot_1 < bot_2 \Rightarrow T_1 < T_2$) also belonging to p. Thus, if $\mathcal{R}$ contains commits of more than one of the transactions belonging to p, we disallow only the youngest one to commit (and abort it using $Q_{Revoked}$).

  (a) $PotPivotStr(T_2, T_3, T_4) \wedge (T = T_2 \mid T_3 \mid T_4)$ **(b)** $\neg(\mathcal{R}(T_5, \_, c, \_, \_) \wedge (T_5 = T_2 \mid T_3 \mid T_4) \wedge T < T_5)$

Example 10. Consider the instances of $\mathcal{R}$ and $\mathcal{H}$ shown below that model history H_{ws} from Fig. 1. To keep the example simple, we do not show the actions of transaction t_0 that created the initial versions of objects x and y. Requests c_1 and c_2 belong to the same potential pivot structure p. Their execution can lead to a write skew violating C3. $Q_{Schedule}$ selects c_1 (smallest TA value). c_2 (commit of youngest transaction) is selected by *ForbCinPPS* and aborted to break p.

R

TA	Seq	Op	Ob	$Q_{Schedule}$	ForbCs	DelayedCs	ForbCinPPS
1	4	c	-	X			
2	4	c	-				X

H

ID	TA	Seq	Op	Ob	OTA	OSeq
1	1	1	r	x	0	1
2	1	2	r	y	0	2
3	2	1	r	x	0	1
4	2	2	r	y	0	2
5	1	3	w	x	-	-
6	2	3	w	y	-	-

Overlap

TA1	TA2
1	2
2	1

PotVulnEdge

TAout	TAin
2	1
1	2

PotPivotStr

TA1	TA2	TA3
1	2	1
2	1	2

Step 2. *DelayedCs* detects case 1b by selecting all transactions T from *NonForbCs* where (a) *NonForbCs* contains another transaction T_2 which (b) wrote an object O that has also been written by T and (c) which is older than T.

(a) $NonForbCs(T_2, \_)$ (b) $\mathcal{H}(\_, T, \_, w, O, \_, \_) \wedge \mathcal{H}(\_, T_2, \_, w, O, \_, \_)$ (c) $T > T_2$

Example 11. Consider the instances of $\mathcal{R}$ and $\mathcal{H}$ displayed below. $Q_{Schedule}$ selects all read ($r_6(x)$) and write ($w_7(y)$) requests. c_3 belongs to *ForbCs* because transaction t_3 wrote the same object as the concurrent but already committed transaction t_2 and is, thus, not allowed to commit. c_4 and c_5 belong to *NonForbCs*, but t_4 and t_5 both wrote the same object x. *ValidCommits* selects only c_4 (oldest transaction from the set $\{t_4, t_5\}$ of transactions with overlapping write-set). c_5 is filtered out by *DelayedCs*.

R

TA	Seq	Op	Ob	$Q_{Schedule}$	ValidCommits	DelayedCs	NonforbCs	ForbCinPPS	ForbCs
3	2	c	-						X
4	3	c	-	X	X		X		
5	2	c	-			X	X		
6	1	r	x	X					
7	1	w	y	X					

H

ID	TA	Seq	Op	Ob	OTA	OSeq
1	1	1	w	x	-	-
2	1	2	c	-	-	-
3	2	1	r	x	1	1
4	2	2	w	x	-	-
5	3	1	w	x	-	-
6	2	3	c	-	-	-
7	4	1	r	x	2	2
8	4	2	w	x	-	-
9	5	1	w	x	-	-

7   Correctness Analysis

We now proof that every history produced under DSSI is serializable. Recall that an SI history is serializable if it does not contain a pivot structure. Thus, we can show this fact by proving that $\mathcal{H}$ cannot contain a potential pivot structure between committed transactions (equivalent after Sec. 2.3). Note that the influence of the other scheduling queries (mentioned in Sec. 3) on the results of $Q_{Schedule}$ and the compliance of C1 and C2 are not in the scope of this paper.

Theorem 1 ($Q_{Schedule}$ Prevents Pivot Structures). $Q_{Schedule} \models C3$

Proof. We omit to prove that the query *PotPivotStr* returns all potential pivot structures contained in $\mathcal{H}$, because the proof is trivial. We proof Theorem 1 by contradiction. Assume the negation of C3 holds:

$$\neg(\forall T, T_2, T_3 : PotPivotStr(T, T_2, T_3) \Rightarrow \neg(EOT(T, c, \_) \wedge EOT(T_2, c, \_) \wedge EOT(T_3, c, \_)))$$
$$\Leftrightarrow \quad \exists T, T_2, T_3 : PotPivotStr(T, T_2, T_3) \wedge EOT(T, c, \_) \wedge EOT(T_2, c, \_) \wedge EOT(T_3, c, \_)$$

Let k be the first scheduler iteration where this equation holds for a fixed T_1, T_2, T_3 and T_4.

$$\Leftrightarrow \exists T, T_2, T_3, k : PotPivotStr_k(T, T_2, T_3) \wedge EOT_k(T, c, \_) \wedge EOT_k(T_2, c, \_) \wedge EOT_k(T_3, c, \_)$$

Without loss of generality, let T_3, the transaction at the third position of the potential pivot structure ($PotPivotStr(T, T_2, T_3)$), be the youngest transaction of the participating transactions. This assumption does not result in a loss of generality, because the position of T_3 is irrelevant for the rest of the proof. There must exist a scheduler iteration $i < k$ where T_3 has not yet committed but already belongs to $PotPivotStr$.

$$\Rightarrow \exists i : PotPivotStr_i(T, T_2, T_3) \wedge T_3 > T \wedge T_3 > T_2 \wedge \neg EOT_i(T_3, c, \_)$$

It follows that the commit request c_3 of T_3 occurs in relation $\mathcal{R}$ at some scheduler iteration j ($i < j < k$). To be executed, c_3 has to belong to the set of non-forbidden commits ($NonForbCs$). We can assume $PotPivotStr_i(T, T_2, T_3) \Rightarrow PotPivotStr_j(T, T_2, T_3)$.

$$\Rightarrow \exists j : PotPivotStr_j(T, T_2, T_3) \wedge T_3 > T \wedge T_3 > T_2 \wedge \neg EOT_j(T_3, c, \_) \wedge NonForbCs_j(T_3, \_)$$

We now replace $NonForbCs$ by its definition and, afterwards, remove terms that are not needed to derive the contradiction:

$$\Leftrightarrow \exists j : PotPivotStr_j(T, T_2, T_3) \wedge T_3 > T \wedge T_3 > T_2 \wedge \neg EOT_j(T_3, c, \_) \wedge$$
$$\mathcal{R}_j(T_3, \_, c, \_) \wedge \neg ForbCs_j(T_3, \_) \wedge \neg ForbCinPPS_j(T_3, \_)$$
$$\Rightarrow \exists j : PotPivotStr_j(T, T_2, T_3) \wedge T_3 > T \wedge T_3 > T_2 \wedge \mathcal{R}_j(T_3, \_, c, \_) \wedge \neg ForbCinPPS_j(T_3, \_)$$

Since c_3 in $\mathcal{R}$ is the commit request of the youngest transaction participating in p, $\mathcal{R}$ cannot contain a commit request of a transaction that is both younger than T_3 and also belongs to p:

$$\Leftrightarrow \exists j : \mathcal{R}_j(T_3, \_, c, \_) \wedge PotPivotStr_j(T, T_2, T_3) \wedge \neg(\mathcal{R}_j(T_4, \_, c, \_) \wedge T_4 = T | T_2 \wedge T_4 < T_3) \wedge$$
$$\neg ForbCinPPS_j(T_3, \_)$$

From the first line of the equation shown above, we can follow $ForbCinPPS_j(T_3, \_)$ which leads to the contradiction and, thus, proves Theorem 1:

$$\Rightarrow \exists j : ForbCinPPS_j(T_3, \_) \wedge \neg ForbCinPPS_j(T_3, \_) \Rightarrow \lightning$$

$\square$

8   Related Work

The *ACTA* framework allows to formalize properties of transaction models using first-order formulas over schedules [6]. Its conciseness and clarity inspired us to implement schedulers based on declarative protocol specifications. The basic ideas of Oshiya have been presented in [13], but this work focused on single-version protocols (2PL) and did not consider correctness. Recent research projects leverage the advantages of declarative languages in various areas [2,4,7,12,15,16]. The *Boom* approach uses Overlog to build distributed systems [2], e.g., a scheduler for MapReduce tasks with policies like First-Come-First-Served. In contrast to our approach, Boom does not focus on DB requests or consistency.

Application analysis techniques have been presented in [10,9] to determine if applications generate serializable executions when running on a system that applies SI. The key idea is that DBAs analyze transaction programs, produce static dependency graphs and manually check for dangerous access patterns leading to non-serializability. Some approaches modify transaction programs to ensure serializable SI schedules: Fekete [9] proposed the techniques *Materialize* and *Promotion* to achieve serializability. Jorwekar et al. [11] tried to automate the check whether non-serializable SI executions can occur. However, this approach still requires manual confirmation and modification. Fekete [8] executes certain transactions of pivot structures under S2PL, others run under SI. This approach requires the underlying platform to support both S2PL and SI. Alomari et al. [1] set exclusive locks in an *External Lock Manager* (ELM) to ensure serializability with SI. In contrast to DSSI, these approaches do not work for ad-hoc transactions and require static analysis or manual program modifications.

Another line of work focused on modifying the SI algorithm of the underlying system to ensure serializability. The closest approach to DSSI is the SSI protocol [5] described in Sec. 2.3. This approach modifies the DB lock manager with an additional type of locks that are used to detect potential pivot structures. DSSI infers all necessary information to detect and prevent these structures from relation $\mathcal{H}$. Our implementation works with DBMSs out of the box. The underlying DBMS does not even need to provide SI since we model data versions in a standard relational schema (see Sec. 4). Using Oshiya, the implementation of DSSI is close to its formal specification, which enabled us to prove its correctness.

9   Conclusions and Future Work

We develop Declarative Serializable Snapshot Isolation (DSSI) using our declarative scheduling model Oshiya. DSSI ensures serializable schedules by avoiding pivot structures and provides DB independence. We formally define DSSI as an Oshiya protocol specification, present a scheduler implementation, and prove that the implementation ensures serializability. In future work, we will experimentally evaluate the performance of DSSI and investigate the trade-offs involved in reducing the amount of false positives.

References

1. Alomari, M., Fekete, A., Röhm, U.: A Robust Technique to Ensure Serializable Executions with Snapshot Isolation DBMS. In: ICDE, pp. 341–352 (2009)
2. Alvaro, P., Condie, T., Conway, N., Elmeleegy, K., Hellerstein, J.M., Sears, R.: Boom Analytics: Exploring Data-Centric, Declarative Programming for the Cloud. In: EuroSys, pp. 223–236 (2010)
3. Berenson, H., Bernstein, P., Gray, J., Melton, J., O'Neil, E., O'Neil, P.: A Critique of ANSI SQL Isolation Levels. In: SIGMOD, pp. 1–10 (1995)
4. Böhm, A., Marth, E., Kanne, C.-C.: The Demaq System: Declarative Development of Distributed Applications. In: SIGMOD, pp. 1311–1314 (2008)
5. Cahill, M.J., Röhm, U., Fekete, A.D.: Serializable Isolation for Snapshot Databases. TODS 34(4), 1–42 (2009)

6. Chrysanthis, P.K., Ramamritham, K.: ACTA: A Framework for Specifying and Reasoning about Transaction Structure and Behavior. In: SIGMOD, pp. 194–203 (1990)
7. Chu, D., Popa, L., Tavakoli, A., Hellerstein, J.M., Levis, P., Shenker, S., Stoica, I.: The Design and Implementation of a Declarative Sensor Network System. In: SenSys, pp. 175–188 (2007)
8. Fekete, A.: Allocating Isolation Levels to Transactions. In: PODS, pp. 206–215 (2005)
9. Fekete, A., Liarokapis, D., O'Neil, E., O'Neil, P., Shasha, D.: Making Snapshot Isolation Serializable. ACM Trans. Database Syst. 30(2), 492–528 (2005)
10. Fekete, A.D.: Serializability and Snapshot Isolation. In: Australasian Database Conference, pp. 201–210 (1999)
11. Jorwekar, S., Fekete, A., Ramamritham, K., Sudarshan, S.: Automating the Detection of Snapshot Isolation Anomalies. In: VLDB, pp. 1263–1274 (2007)
12. Kot, L., Gupta, N., Roy, S., Gehrke, J., Koch, C.: Beyond Isolation: Research Opportunities in Declarative Data-Driven Coordination. SIGMOD Rec. 39, 27–32 (2010)
13. Tilgner, C.: Declarative Scheduling in Highly Scalable Systems. In: EDBT/ICDT Workshops, pp. 41:1–41:6 (2010)
14. Weikum, G., Vossen, G.: Transactional Information Systems. Morgan Kaufmann Publishers, San Francisco (2002)
15. White, W., Demers, A., Koch, C., Gehrke, J., Rajagopalan, R.: Scaling Games to Epic Proportions. In: SIGMOD, pp. 31–42 (2007)
16. Yang, F., Shanmugasundaram, J., Riedewald, M., Gehrke, J.: Hilda: A High-Level Language for Data-Driven Web Applications. In: ICDE (2006)

Resource Scheduling Methods for Query Optimization in Data Grid Systems*

Igor Epimakhov[1], Abdelkader Hameurlain[1], Tharam Dillon[2], and Franck Morvan[1]

[1] Institut de Recherche en Informatique de Toulouse IRIT,
Paul Sabatier University, 118 Route de Narbonne, 31062 Toulouse, France
{Igor.Epimakhov,Abdelkader.Hameurlain,Franck.Morvan}@irit.fr
[2] Curtin University, DEBII Institute, Perth, Australia
Tharam.Dillon@cbs.curtin.edu.au

Abstract. Resource allocation (RA) is one of the most important stages of distributed query processing in Data Grid systems. Recently, a number of papers that propose different methods for RA were published. To deal with specific characteristics of the data grid systems, such as dynamicity, heterogeneity and large-scale, many studies extend classic methods from distributed and parallel databases domains. Others invite fundamentally different methods based on incentives for autonomous nodes. The present study provides a brief description, qualitative comparison and performance evaluation of the most interesting approaches (extended classic and incentive-based) for RA. Both approaches are promising and appropriate for successful data grid systems.

Keywords: Data grid systems, resource allocation, distributed query processing and optimization, incentive-based scheduling, extended classic scheduling.

1   Introduction

Currently, query processing in Data Grid environments is an actual research topic. Resource Allocation is one of the key stages of query processing, which determines the efficiency of the entire system. It constitutes optimal resource allocation for a set of operations of a query execution plan.

To the best of knowledge, in the literature, there is currently no complete comparison of recently proposed resource allocation methods. [13] provided a brief overview of currently discussed issues, such as Resource Allocation algorithms, Fault tolerance, Security. Thus assumptions and metrics, widely used to solve the problem of resource allocation in the Calculation Grid environment, were highlighted. [11] carried out a review and comparison of the basic heuristics, used for selection of the optimal set of resources to solve the resource allocation problem. [1] proposed a classification of Resource Allocation architectures. They were categorized into Centralized, Hierarchical and Decentralized approaches and the first two of these

* This work was supported in part by the French National Research Agency ANR, PAIRSE Project, Grant number -09-SEGI-008.

J. Eder, M. Bielikova, and A.M. Tjoa (Eds.): ADBIS 2011, LNCS 6909, pp. 185–199, 2011.

were discussed in more detail. [14] highlighted three main directions of load balancing improvement in Data Grid: Data replication, Cache techniques and I/O load balancing.

The main limitations of the above-mentioned papers are:

- They take into consideration only particular parts of resource allocation problem, but not the characteristics of the general resource allocation approaches.
- There was lack of attention to specific features of the Data Grid environment (distributed and fragmented relations, complex queries).

In the present article, we will propose a somewhat different point of view to the main Resource Allocation approaches and classification of the methods.

The rest of the paper is organized as follows: in section 2 we will describe the resource allocation problem in the Data Grid environment. In section 3 we will highlight the main criteria that we will use for the comparison of the methods. In section 4 we will propose our classification, briefly describe the main approaches and selected methods and make a comparison using the highlighted criteria. In section 5 we will present simulation experiments for two the main resource allocation approaches. Finally, we will summarize the major findings of the study in section 6.

2 Resource Allocation Problem

For the period of time, Data Grid system receives a set of queries from customers $Q=\{q_1, q_2, ..., q_N\}$, where each query qi consists of a set of operations $O_i=\{o_1, o_2,..., o_M\}$, i = 1,2,...,N. Thus, the set of operations O_i that must be executed by the system is a set O, such that $O_i \subseteq O$, i = 1,2,...,N. The objective of Resource allocation is to assign the set of nodes $U=\{u_1, u_2, ..., u_K\}$ to perform each operation $o_j \in O$, j=1,2,...,K. Note that a part of the operations of a set O accepts as input the results of other operations, and therefore they can be performed only after completion of the latter. That means that we can separate the set O into subsets of independent and dependent operations O_{Ind} and O_{Dep}, where $O = O_{Ind} \cup O_{Dep}$.

The whole set of nodes U in the Data Grid stores a set of database relations $R=\{R_1, R_2,...,R_L\}$. Each relation R_r is distributed among a set of nodes $U_r \subseteq U$, i.e. $R_r \rightarrow U_r$. Each operation $o_j \in O$, j=1,2,...,K require a set of relations $Rop_j \subseteq R$ and uses as a data source a set of nodes Uop_j, i.e. $Rop_j \rightarrow Uop_j$. We consider the problem of resource allocation as a function $F(Q,R,U) = F(O,R_{op},U_{op},U) = \{o_1 \rightarrow u_{x1}, o_2 \rightarrow u_{x2}, ..., o_k \rightarrow u_{xk}\}=M$, such that estimated execution time function $T(M) \rightarrow$ min. It has been proved that the Resource Allocation Problem is a NP-complete task [10].

Several methods consider the objective of Resource Allocation as minimization of execution time of a separate query. Others as the objective consider a minimization of execution time of a set of queries.

Heterogeneity of the Data Grid environment makes the usage of parallel algorithms significantly more complex. Indeed, each node of the system may have different CPU performance, amount of memory, communication throughput and latency. As a result, the algorithm must not only determine an optimal number of nodes for the query execution, but also select exactly which nodes will participate in the execution [5].

3   Background

Here we will define aspects of the Data Grid to facilitate our further discussion. This is a computing infrastructure providing intensive computation and analysis of shared large-scale databases. That system unites multiple servers (nodes of the Grid) and stores large amounts of data within a common theme (for example biomedical). Its main characteristic features are [9]: heterogeneity, dynamicity and large scale.

We will try to highlight some of the requirements that are imposed by these features for the resource allocation mechanism.

3.1   Basic Requirements for Resource Allocation Methods in Data Grid Systems

First of all, it should be noted that a common theme of the data stored in Data Grid is that it involves a limited number of entities of the domain. Although the number of relations is not limited, a large number of similar relations, which involve the same entity (by using ontological transformation schemes), can be regarded as a widely distributed relation. In the general case each relation in the Data Grid is distributed and duplicated among a set of nodes. This means that the resource allocation algorithm must consider distributed relations querying, i.e. allocate resources for queries, all relations of which are distributed and duplicated.

Heterogeneity of nodes imposes upon the Resource Allocation mechanism a new requirement: the necessity to take into account the CPU and I/O performance, amount of local memory and network bandwidth of each node. This greatly complicates the task of parallelization, and not allows use of well-researched algorithms from the domain of Parallel and Distributed databases.

New nodes that appear in the system during the query processing raise the problem of using their data and resources for the processing query. It is important to take into account leaving nodes and to provide an algorithm that allows successful completion of the query execution, even if one or more used nodes disconnect during it.

A large scale system imposes serious restriction on the usage of global centralized Resource Allocation (or global catalogues of meta-data required for Resource Allocation), because when there are large numbers of nodes and queries, the global centralized scheduler can become a bottleneck of the entire system, seriously suppressing performance. In addition, in the event of a central node failure, it may lead to inoperability of the Data Grid. Also, decentralized allocation methods can be promising in the Data Grid environment in terms of scalability.

In a large scale Data Grid environment with dynamicity of nodes, it can be more effective to modify the query execution plan in run-time, taking into account rapidly changing parameters, which leads to using dynamic reallocation techniques. An important criterion for evaluating the Resource Allocation algorithm is also the maximal usage of local data, or data locality principle.

3.2   Performance Requirements

An important general requirement for a Data Grid system is a performance. Naturally, having received a great number of queries, the system must provide results within a reasonable time for each of them. To do this it is necessary to effectively use computational resources and to take advantage of different types of parallelism.

Each query execution plan consists of a set of physical operations, many of which are independent and can be performed simultaneously, i.e. inter-operation independent parallelism (*irop*). For dependent operations one can often be use pipeline parallelism (*pipl*), which, however, requires full completion of resource allocation before execution of the query.The most low-level parallelism is intra-operation (or partitioned) parallelism (*iaop*), which consists in splitting the relation into fragments and fulfillment operations against them on several sites simultaneously. An important factor limiting the usage of this type of parallelism is latency. We believe that the usage of basic forms of parallelism is required for optimal resource allocation in the Data Grid.

While the Data Grid system is able to process a set of queries over certain time, it can have advantage from coordinated resource allocation for all of them at the same time. We will denote that by Multi-query scheduling. Still, this is not the perfect solution in a large scale environment, because it requires a global centralized resource allocation mechanism.

3.3   Methods Comparison Criterions

Those are several criteria that we have identified above; Here we have decided to separate them into two classes namely; Criteria of capability: 1. Distributed relations querying (*Dst*), 2. Using new appeared nodes' data during execution-time (*NN*), 3. Node leaving tolerance (*LT*), 4. Dynamic reallocation (*DR*), 5. Data locality consideration (*DL*), 6. Scalability (*Sca.*).

Criteria of performance: 1. Using inter-operation and intra-operation parallelism (*Prl*), 2. Taking into account heterogeneous nodes' performance (*Het.*), 3. Multi-query scheduling (*Mlt*).

Based on these criteria, we will provide a brief overview and comparison of existing methods of Resource Allocation. As the two main approaches we have identified Extended classic and Incentive-based Resource Allocation. Let us consider both approaches.

4   Analysis of Extended Classic and Incentive-Based Methods for Resource Allocation

We analyze the major activities in the area of resource allocation and identified two main approaches: Extended classic approach and Incentive-based approach. The first is an extension and adaptation of classical methods developed in the area of Parallel and Distributed databases. The second approach is relatively new and represents an attempt to completely change the concept of site interactions in the Data Grid.

4.1   Extended Classic Approach

This is the classic approach to solving the problem, which is widely used in scientific work and practical implementations of Grid systems. Usually, resource allocation is done on the node that initiated a query. Firstly it collects meta-data about relations placement and performance characteristics of nodes etc. Then, using allocation algorithms and cost models, it chooses optimal resource allocation among a set of possible variants and sends the generated query execution plan to the selected nodes. The algorithm is based on the assumption that all nodes of the Data Grid obey the overall discipline and the node-scheduler can directly give the nodes orders to execute operations.

In this approach there are three main resource allocation strategies: static, dynamic and hybrid. The first strategy uses statistical data accumulated in the system during operations execution, i.e. it allows scheduling execution of all query operations with the maximum usage of parallelism. Its disadvantage, however, that in a rapidly changing Data Grid environment statistics may quickly become obsolete and the chosen resource allocation plan may become non optimal. In the second case the scheduler collects the necessary metadata just before the resource allocation. This approach can be considered reasonable in environments with high dynamicity of nodes. However, absence of a preliminary plan does not allow the use of pipeline parallelism, what can be considered as a significant disadvantage. To combine the advantages of both strategies was proposed a combined static resource allocation methods with dynamic reallocation during the execution. It can use all kinds of parallelism and, at same time, promptly adjust the execution plan to changing conditions in the Grid.

Another important feature is the nature of the decision-making module in the process of resource allocation, which can be Centralized, Combined or Decentralized [1]. The first one is not suitable for a large-scale Grid environment due to low reliability and limited scalability. The second one is an extension of a scheduler with a centralized nature, which improves scalability, but does not eliminate its fundamental disadvantages. Decentralized is the most scalable, but also the most complex organization. In this section, we will examine in more detail methods of extended classic resource allocation and we will compare them on the basis of selected criteria.

4.2 Methods Based on Static Strategy

All of the most important static methods implement at least one of forms of the intra-query parallelism. When [15] and [19] consider only inter-operation parallelism, [17] takes advantage also of intra-operation parallelism. While the method of [5] utilize all forms of parallelism, including parallelism pipeline. And it is the only method that considers a Distributed relations querying.

The algorithm [5] sorts operations of an execution plan by cost and successively, starting with the most expensive, it determines their level of intra-operation parallelism. In [17], the authors propose an iterative algorithm, which takes the query bushy tree and parallelizes each operation on the optimal number of nodes. In order to provide Load Balancing, the scheduler selects the first sites that are not currently using the requested relation. This too can be considered as a disadvantage, because the selected node can still be overloaded with queries, that uses others relations.

In [19], the author considers the problem of static resource allocation of a set of independent jobs with intensive usage of large volumes of data. As an objective he takes maximization of throughput. For the calculations the algorithm selects computational resources closest to the data sources. Here a disadvantage is the ignoring of dependent jobs, which is an important factor in query processing in the

Data Grid. The problem was reduced to a Set Covering Problem and used a well-known algorithm for its solution. In [15], the author proposed an algorithm, which allocates all operations only on those nodes, which initially contains the data.

4.3   Methods Based on Dynamic and Hybrid Strategies

In papers [20, 3, 4] the authors proposed and described the system DartGrid. Its scheduler is a classic centralized dynamic approach, which uses only Inter-operation parallelism. They implemented a dynamic iterative scheme of resource allocation, whose principal objective is to minimize the size of intermediate results. The method works as follows: after the logical optimization, the optimizer demands all urgent meta-data (size of relations, the current load of node, etc) from all nodes that participle in query execution. Than it sequentially parallelize and performs all join operations in the query plan. Disadvantages of the system include the impossibility of inter-operation parallelism and pipeline, as well as a centralized scheduler, which collect data from all the sites that participate in query processing after each iteration.

Technique of Adaptive Query Processing (AQP) was proposed in [6, 7]. It consists in monitoring of the current state of resources during query execution and changing the query plan if necessary for maximal performance improvement. It was implemented in system OGSA-DQP [8] with modifications for load balancing and dynamic resource allocation, which allows the usage of new resources that have become available since the beginning of the query execution. Failure recovery, allows the system to react to the shutdown or disconnection of one or more of the nodes during the execution phase, restoring the lost intermediate data.

In [16] the author proposed a greedy resource allocation algorithm, which selects nodes based on their throughput capacity, known from previous query executions. It was implemented using a Dynamic Load Balancing algorithm on the base of the algorithm Eddies, which allows transferring the load between nodes during operation execution without interrupting the operation. Also, it implemented Dynamic Reallocation, which in the process of query execution checks the current throughput of the node, compares it with the preliminary estimated one and, if necessary, makes reallocation.

4.4   Comparison of Classic Resource Allocation Methods

In the Table 1 the above methods fulfillment of previously selected criteria is given. As shown the table, not all methods provide the very important Data Grid function of Distributed Relation Querying. Among those that provide this function, there are no methods that consider the problem of resource allocation for multiple queries (Multi-query scheduling). Also, not all authors have proposed a complete approach, some of them paid attention only to resource allocation algorithms for matching operation on resources, ignoring the problem of Scheduler organization. Considering the Scalability criterion, the most interesting from our point of view is the method [17], which provides a hierarchical model of resource allocation for multiple queries. In other methods the centralized approach dominates, which creates a number of fundamental problems in Large scale Data Grids.

The main conclusion that can be made from the table is that we did not find any method that fully conforms with all our criteria. The most interesting by the whole set of criteria is the method [6, 7, 8] whose main disadvantage is the lack of resource allocation for a set of queries (Multi-query scheduling).

Table 1. Characteristics of Extended classic resource allocation methods

Work	Dst	Mlt	Prl	Het.	NN	LT	Sca.	DR	DL
DartGrid [20, 3, 4]	Yes	No	iaop	Yes (cpu, storage)	No	No	Centralized scheduler reduce scalability	No	Yes
Liu08 [15]	No	No	irop	Yes (cpu, ram)	No	No	Based on centralized metadata services, Not very scalable	No	Yes
Gounaris04 [5]	Yes	No	irop, iaop, pipl	Yes (cpu, ram, i/o, network)	No	No	Low complexity, but architecture is not proposed	No	Yes
Soe05 [17]	No	Yes	irop, iaop	Yes (cpu, ram, i/o, network, storage)	No	No	Hierarchical model for inter-query parallelism is more scalable	No	Yes
Venugopal9 6 [19]	No	Yes	irop	Yes (cpu, storage)	No	No	Algorithms are scalable, but architecture is not proposed	No	Yes
Ogsa-dqp [6, 7, 8]	Yes	No	irop, iaop, pipl	Yes (cpu, ram, i/o, network)	Yes	No	Centralized scheduler reduce scalability	Yes	Yes
DaSilva06 [16]	Yes	No	irop, iaop	Yes (node throughput)	No	No	Centralized approach, not very scalable	Yes	No

4.5 Incentive-Based Approach

This is a relatively new approach, the basic premise of which is the full autonomy of Data Grid nodes. The site administrator can completely determine the policy, and use his own algorithms to estimate the execution time of local operations. An incentive-based system gives nodes positive points for the successful execution of operations, and takes points off of those sites that could not perform the operation in the estimated time. The approach leaves execution time estimation to nodes-candidates so a significant part of computations for resource allocation is distributed among a large set of nodes-candidates.

There are two main methods in this approach: (i) Economic principles based and (ii) Reputation based resource allocation.

The first use such economic tools as money, trade and auctions. It works as follows. The user distributes a query among a set of Data Grid nodes. Each node on the basis of its own policies decides whether it wants to participate in the query processing or not. If yes, the node estimates and informs the user about the time necessary to finish the query, and cost of that work. The user selects the most attractive offer and reports its decision. Selected node executes the query and returns the result to the user. In case of delay in query execution, the node will have to pay to the user specified penalty.

In Reputation based methods, each node has its own reputation, which rises in the case of a successful query execution, and falls in the case of exceeding the defined execution time. The principle is very similar to the previous method, the main difference that the node informs the user only about estimated time of query execution. User, when he selects the best offer, takes into account the reputation of the site and selects the most reliable nodes with good reputation.

4.6 Economic and Reputation Principles Based Resource Allocation

As far as we know, the earliest work in this area is the article [18], in which the authors proposed wide-area distributed database system Mariposa. This was the first time the requirements for large-scale distributed database systems were defined namely: a) scalability; b) data mobility; c) no global synchronization; d) total local autonomy; c) easily configurable policies. Also, in the article the basic principles were described in detail, namely the architecture and algorithm of the economic-based scheduler, including: a virtual bank, auction system, a decentralized distribution system for query announcement and resource allocation, etc. These requirements, the architecture and algorithms without major changes lie at the basis of all the latest proposed economic-based resource allocation methods in the Data Grid domain. We will describe some of the most interesting recent works.

In [21], an Economic-like resource allocation scheme was proposed. The method broadcasts a query among the nodes, which can bid on the query, i.e. make its proposals with the deadline and the virtual "cost" of the query, estimated by itself. The user selects the most profitable offer. This paper does not address directly the Load Balancing problem, but it is supposed that the economic-like system will encourage the sites to trace their load and determine their policies for participating in the query execution.

In [12] was proposed to use the principle of continuous auction, during which candidate nodes can haggle with the customer to achieve the most acceptable price for both sides. This method, however, as the foregoing [21], not provides an implementation within the Data Grid environment.

In [2] a Reputation based resource allocation algorithm was proposed. The method is targeted at a Grid with full autonomy of nodes. Each node on the basis of its own policy proposes itself as a candidate for participating in fulfilling the query, offering its estimated time in which it can fulfill the query. The scheduler selects the best node for query execution taking into consideration its reputation. The reputation of the node depends on how accurate was its previous estimations of the execution time. The method suffers from the lack of distributed relations querying. Also, usage of parallelism for query optimizing is not considered.

4.7 Comparison of Incentive-Based Resource Allocation Methods

In the Table 2 we present the above described incentive-based methods. Two methods [21, 12] do not take into account the principle of Data locality, i.e. strictly speaking, they are not suitable for use in the Data Grid. However, we believe that these methods can be extended, so we include them into consideration.

Table 2. Characteristics of Incentive-based resource allocation methods

Work	Dst	Mlt	Prl	Het.	NN	LT	Sca.	DR	DL
Mariposa [18]	Yes	No	irop	Indirectly, with Price	No	No	Decentralized, good scalability	No	Indirectly, with data migration
Xiao08 [21]	No	No	No	Indirectly, with Price	No	No	Decentralized, good scalability	No	No
Izakian09 [12]	No	No	irop	Indirectly, with Price	No	No	Decentralized, good scalability	No	No
Costa09 [2]	No	No	No	Indirectly, with Rep.	No	No	Decentralized, good scalability	No	Yes (Indirectly)

All the above methods has excellent scalability, however, the criteria set out namely Distributed Relations Querying, Parallelism, and Dynamic reallocation are a weak point practically of all the methods. As shown in the table, only one of the above methods supports Distributed Relations Querying, and this is [18]. We should also note one important feature of the Incentive-based approach namely: during the node selection, once an agreement has been concluded between the user and the nodes, it cannot be altered. This means that using newly appeared nodes' data during execution-time and Dynamic reallocation in this approach is fundamentally impossible.

Comparing the principles of economic and reputation based resource allocation, we have concluded that the latter is much simpler to implement and better reflects the objective of the virtual parameter, introduced in the Incentive-based approach. After all, the task is to represent the quality of the Data Grid nodes functionality. The principle of Reputation can do this directly as it is the most transparent approach for both users and nodes. In the economic model the users do not have any information about the reliability of nodes. Also, the principle of payment generates a steady stream of virtual money from users to the sites, but without the reverse process, all the money of the system eventually will be stored at sites that provide query execution. Without a balanced circulation of money in the economy, such a model cannot operate effectively over a long period of time.

Another problem of the economic model is the support of a Distributed relations querying. Indeed, in the case of the distributed relations, each of the nodes may contain an important part of requested relation, which means that user will not "choose" among the nodes, but "buy up" all parts of the relation of all the nodes.

4.8 Comparison of the Two Approaches

Let's compare the two basic resource allocation approaches by each of the selected criteria (Table 3). As we have seen before, both approaches can in principle have the support of Distributed relations querying, however, its realization in the Incentive-based approach is more complicated. Also, Incentive-based approach loses by the criteria of Parallelism and Taking into account heterogeneous nodes 'performance, Using newly appeared nodes' data during execution-time, Dynamic reallocation.

Table 3. Characteristics of Extended classic and Incentive-based approaches

Work	Dst	Mlt	Prl	Het.	NN	LT	Sca.	DR	DL
Classic	Yes	Yes	Yes	Yes	Yes	Yes	Good	Yes	Yes
Incentive-based	Yes, but more complicated	No	Yes, but much more complicated	Yes (Indirectly)	No	Yes	Very Good	No	Yes (Indir.)

However, by the most important criterion Scalability, this approach is superior to the traditional Classic resource allocation. A key advantage of the Incentive-based approach is node autonomy, which, however, places strong responsibility for the functioning of the Grid system to administrators of nodes and poses many difficulties for them.

5 Performance Evaluation

For more detailed comparison of the highlighted resource allocation approaches, we have decided to perform an experiment. The main objective of the experiment is to study the most general characteristics of both approaches in the Data Grid environment: allocation methods cost and optimality of generated plans. For measuring the two characteristics a detailed implementation of the algorithms studied is required.

We had examined a number of available program Grid simulators and real scientific-purpose Grid systems, and we did not found once that meets all of our requirements which are: very large number of nodes available for the experiment (1000 and more); dependent operations support; relations distribution and fragmentation control support. For example, GridSim does not have dependent operations support and we cannot reserve 1000 or more nodes for our experiment in real Grids. That is why we decided to develop a Data Grid simulator, specialized for our purposes.

5.1 Simulation Model and System Parameters

We created a Data Grid simulator, the main characteristics of which are: large number of nodes; heterogeneity of nodes by CPU, memory, I/O performance, network link connection performance; large number of distributed and duplicated relations.

In our simulator we are interested only in resource allocation and query plan execution phases. It is highly parameterized and we will describe the major parameters and their settings used in this study.

In Table 4 we can see the major parameters of our Data Grid simulator. Values are randomly generated in selected ranges that we found adequate. While the node's parameters define performance of hardware resources of the Data Grid, the relation's parameters define the size of relation, its fragmentation and duplication level and other characteristics. All of them are important for resource allocation decisions.

Table 4. System configuration and database parameters

Parameter	Value
Node CPU performance	10 – 1 000 MIPS
Node I/O performance	10 – 90 Mb/s;
Node memory amount	0,001 – 40 Mb
Node network connection bandwidth	10 – 60 Mbit
Node network connection latency	0.5 s
Relation number of attributes	10
Relation size of attribute	300 Bytes
Relation cardinality of attributes	0.3 – 0.9
Relation size of tuple	3000 Bytes
Relation number of tuples in relation	1000 – 11000
Relation size	3Mb – 33Mb
Relation fragments number	10
Relation duplicates number	10

As a minimum value for a node's memory, we chose a value that is very close to zero for examining algorithms in the presence of highly charged nodes. For the study we used a high level of distribution and duplication: each relation is fragmented to 10 equal fragments, each fragment is duplicated among 10 nodes, which means that each relation uses 100 nodes for storage of its different parts. In general, for a join operation we have 200 node-candidates. This is not applied to intermediate relations, because they have a level of distribution according to the level of join parallelism. Their level of duplication is always equal to 1.

5.2   Performance Analysis

In this study two algorithms were implemented, one for each approach. As a base for the algorithm of Extended classic approach we chose the method [5], implementing it with the combination of the greedy algorithm heuristic. We found it to be the most representative because of using the most common static strategy with all types of intra-operation and inter-operation parallelism.

For the Incentive-based approach we used as a base the method [18], proposed for the Mariposa system. Although it is not the newest algorithm it is one of the better described and implemented for an environment that is very similar to a Data Grid. It is important to mention, that in origin the algorithm of Mariposa intended to optimize two parameters at the same time: response time and query cost. In our study we are not interested in the virtual economic efficiency of the generated plans. That is why we simplified the algorithm for optimizing only the response time, in our study.

One of the principal differences between the capabilities of the above described algorithms is that the Incentive-based algorithm cannot use intra-operation parallelism. We expect that the limitation will bring lesser efficiency and greater response time in most cases. But we can expect also that, because of its lower allocation algorithm complexity and distributed nature, Incentive-based methods will have a smaller resource allocation time.

The query execution plan that is generated in the resource allocation phase has a number of physical operations. Each of them is allocated to the selected node. We developed a query execution system that executes all independent operations simultaneously and uses the pipeline principle for the dependent ones. For the operations that share the same resources, such as CPU, memory, I/O system or network, we provided a concurrency control mechanism. In our experiment we decided to take into account only the two most important logical operations: Join and Read. For join processing, we used the memory-adaptive hybrid hash join algorithm because it is one of the most universal and effective.

We generated a simulated Data Grid with heterogeneous nodes and a set of distributed and fragmented relations. We randomly generated a series of queries of three types of complexity: simple (5 joins), middle (12 joins), complex (20 joins). For each query the two algorithms generated two execution plans. For execution time and resources consumption measurement for each type of query we used the average values of a series of 100 queries.

Allocation Method Cost and Optimality of Generated Plan. In Fig.1 and 2 displayed a dependency between resource allocation time and number of nodes-candidates for Classic and Incentive-based methods. We denote allocation time by Talloc and total allocation time by Talloc total. Talloc allows comparison of the performance of the two algorithms. While Talloc total shows total time consumption for all participating nodes and allows comparision of the summary complexity of the examined algorithms. The following computation formulas were used respectively:

$$T_{alloc} = T_{central} + max(T_{local\ i}) .$$

$$T_{alloc\ total} = T_{central} + \sum T_{local\ i} .$$

Where $T_{central}$ is an allocations time for the central node. $T_{local\ i}$ is the local allocation time for nodes-candidates that participle in the resource allocation process in Incentive-based method. Naturally, since the Classic method is completely centralized and does not share any computations with local nodes, $T_{local\ i}$ is always equal to zero for it and we have the same curve on both graphs. On the other hand for Incentive based method there is a considerable difference that indicates sharing of a large part of computations between a set of nodes.

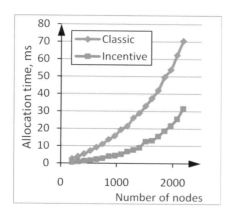
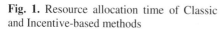
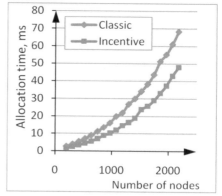

Fig. 1. Resource allocation time of Classic and Incentive-based methods

Fig. 2. Total resource allocation cost of Classic and Incentive-based methods

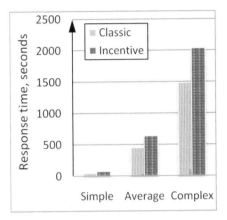

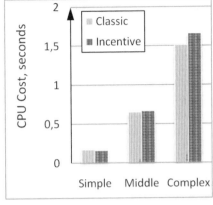

Fig. 3. Query response time of Classic and Incentive-based methods for simple, middle and complex queries

Fig. 4. CPU cost of Classic and Incentive-based methods for simple, middle and complex queries

The diagrams show a significant time consumption superiority of the Classic method over the Incentive-based. Optimality of generated plans.

Because of the computational work division between central and local nodes, the Incentive-based method has a significantly less allocation time. Since the Classic method considers intra-operation parallelism, it requires more computations for creating the query execution plans (Fig.2).

Fig.3 displays the response time for the Classic and Incentive-based methods for simple, middle and complex queries. We used the following computation formula to determine this:

$$T_{response} = T_{alloc} + T_{execution}.$$

Where T_{alloc} is the resource allocation time and $T_{execution}$ is the time of query execution. As we can see in Fig. 3, the Classic method provides a much lower response time for

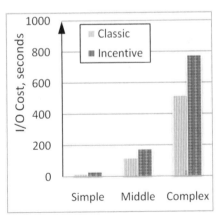

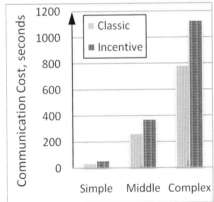

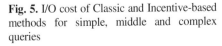

Fig. 5. I/O cost of Classic and Incentive-based methods for simple, middle and complex queries

Fig. 6. Communication cost of Classic and Incentive-based methods for simple, middle and complex queries

all types of queries. From the performance point of view, the Classic method is significantly superior over the Incentive-based method. We found out that the CPU cost is very small for both methods and do not have a significant influence on the response time (Fig. 4). Analyzing Fig. 5 and 6, we can tell that the Classic method has a much lower I/O and communication cost. It can be explained by intra-operation parallelism, which is used by Classic method and raise the efficiency of the generated query execution plans. During our experiment we also identified that the most critical resources in the Data Grid are network and I/O performance.

6 Conclusion

In this paper we have presented our review of recent works in the field of resource allocation in the Data Grid environment. We have highlighted and analyzed two principal approaches: Classic and Incentive-based resource allocation. The first one is well developed and simple in realization, but the second one provides full node autonomy and excellent scalability. In the performance evaluation we found out, that Extended classic method provides more optimal resource allocation, but we believe, that because of calculation sharing, Incentive based resource allocation potentially more scalable and it is a promising avenue of research.

References

1. de Carvalho Costa, R.L., Furtado, P.: Scheduling in Grid Databases. In: 22nd Int. Conference on Advanced Information Networking and Applications – Workshops (2008)
2. de Carvalho Costa, R.L., Furtado, P.: Runtime Estimations, Reputation and Elections for Top Performing Distributed Query Scheduling. In: 9th IEEE/ACM Int. Symposium on Cluster Computing and the Grid (2009)
3. Chen, H., Wu, Z.: DartGrid III: A Semantic Grid Toolkit for Data Integration. In: Proceedings of the First Int. Conference on Semantics, Knowledge, and Grid, SKG 2005 (2005)
4. Chen, H., Wu, Z., Mao, Y., Zheng, G.: DartGrid: a semantic infrastructure for building database Grid applications. Concurrency Computat.: Pract. Exper. 18, 1811–1828 (2006)
5. Gounaris, A., Sakellariou, R., Paton, N.W., Fernandes, A.A.A.: Resource Scheduling for Parallel Query Processing on Computational Grids. In: GRID, pp. 396–401 (2004)
6. Gounaris, A., Paton, N.W., Sakellariou, R., Fernandes, A.A.A.: Adaptive Query Processing and the Grid: Opportunities and Challenges. In: DEXA Workshops, pp. 506–510 (2004)
7. Gounaris, A., Paton, N.W., Sakellariou, R., Fernandes, A.A.A., Smith, J., Watson, P.: Practical Adaptation to Changing Resources in Grid Query Processing. In: ICDE, p. 165 (2006)
8. Gounaris, A., Paton, N.W., Sakellariou, R., Fernandes, A.A.A.: Modular Adaptive Query Processing for Sevice-Based Grids. CoreGRID Tech. Report Number TR-0076 (2007)
9. Hameurlain, A., Morvan, F., Samad, M.E.: Large scale data management in grid systems: a survey. In: IEEE International Conference on Information and Communication Technologies: from Theory to Applications (ICTTA), pp. 1–6 (2008)
10. Ibarra, O.H., Kim, C.E.: Heuristic algorithms for scheduling independent tasks on nonidentical processors. Journal of Association of Comp. Machine 24(2), 280–289 (1977)

11. Izakian, H., Abraham, A., Snásel, V.: Comparison of Heuristics for Scheduling Independent Tasks on Heterogeneous Distributed Environments. CSO 1, 8–12 (2009)
12. Izakian, H., Abraham, A., Tork Ladani, B.: An auction method for resource allocation in computational grids. Future Generation Computer Systems 26, 228–235 (2010)
13. Jiang, C., Wang, C., Liu, X., Zhao, Y.: A Survey of Job Scheduling in Grids. In: Dong, G., Lin, X., Wang, W., Yang, Y., Yu, J.X. (eds.) APWeb/WAIM 2007. LNCS, vol. 4505, pp. 419–427. Springer, Heidelberg (2007)
14. Qin, X.: Design and analysis of a load balancing strategy in Data Grids. Future Generation Computer Systems 23, 132–137 (2007)
15. Liu, S., Karimi, H.A.: Grid query optimizer to improve query processing in grids. Future Generation Computer Systems 24, 342–353 (2008)
16. Da Silva, V.F.V., Dutra, M.L., Porto, F., Schulze, B., Barbosa, A.C., de Oliveira, J.C.: An adaptive parallel query processing middleware for the Grid. Concurrency Computat.: Pract. Exper 18, 621–634 (2006)
17. Soe, K.M., New, A.A., Aung, T.N., Naing, T.T., Thein, N.L.: Efficient Scheduling of Resources for Parallel Query Processing on Grid-based Architecture. In: APSITT (2005)
18. Stonebraker, M., Aoki, P.M., Litwin, W., Pfeffer, A., Sah, A., Sidell, J., Staelin, C., Yu, A.: Mariposa: a wide-area distributed database system. The VLDB Journal 5, 48–63 (1996)
19. Venugopal, S., Buyya, R.: An SCP-based heuristic approach for scheduling distributed data-intensive applications on global grids. J. Parallel Distrib. Comput. 68, 471–487 (2008)
20. Wu, Z., Chen, H., Changhuang, C., Zheng, G., Xu, J.: DartGrid: Semantic-Based Database Grid. In: Bubak, M., van Albada, G.D., Sloot, P.M.A., Dongarra, J. (eds.) ICCS 2004. LNCS, vol. 3036, pp. 59–66. Springer, Heidelberg (2004)
21. Xiao, L., Zhu, Y., Ni, L.M., Xu, Z.: Incentive-Based Scheduling for Market-Like Computational Grids. IEEE Transactions on parallel and distributed systems 19(7) (2008)

A Recommendation Technique for Spatial Data

Barbara Catania, Maria Teresa Pinto, Paola Podestà, and Davide Pomerano

University of Genoa, Italy
{catania,podesta}@disi.unige.it,
{terry.pinto,pome85}@gmail.com

Abstract. Recommendation functionalities have been recently considered in traditional database systems as an approach for guaranteeing a satisfactory interaction with the database also to users with a low or moderate technical skill or in presence of huge volumes of, potentially heterogeneous, data. Recommendation is performed by extending query results with additional and potentially interesting items. Among the proposed techniques, current-state approaches exploit the content and the schema of a query result as well as the database instance in order to recommend new items. While some preliminary current-state approaches have been proposed for relational databases, in this paper, we claim that current-state approaches can also be relevant for providing new ways of interactions in spatial databases. To support our claim, we present a current-state recommendation approach for spatial data and topological queries. The proposed approach exploits the principles of locality and similarity between topological predicates to recommend new spatial objects besides those precisely returned by a query. An index-based query processing algorithm for the proposed recommendation operator is also proposed, to guarantee an efficient computation of recommended items.

1 Introduction

Recommendation techniques have been originally designed for Web search engines with the aim of providing advice to the user with respect to her specific needs. More recently, recommendation functionalities have been considered in traditional database systems as an approach for extending query results with additional and potentially useful items, guaranteeing a more satisfactory interaction with the database in all the cases in which it may be difficult for the user to specify the query in a precise way (e.g., when users have a low or moderate technical skill or in presence of huge volumes of, potentially heterogeneous, data).

Most recommendation approaches rely on the user history [1]. However, in database systems, there may be situations in which no user profile or history is available, for example because the user is occasional, or profile maintenance is considered too expensive. In those cases, the usage of history-based approaches is not viable and other information should be used, e.g., external sources, like the Web, or the database content itself. Techniques based only on database content

J. Eder, M. Bielikova, and A.M. Tjoa (Eds.): ADBIS 2011, LNCS 6909, pp. 200–213, 2011.
© Springer-Verlag Berlin Heidelberg 2011

and query results have been called *current-state* techniques in [15], where a current-state recommendation approach has been proposed for relational data.

The aim of this paper is to present a current-state recommendation approach for spatial queries, In particular, based on their relevance from an application point of view and their support in current spatial query language proposals and standards (e.g., [8,14]), we consider topological selection queries, i.e., selection queries based on a topological predicate (e.g., *disjoint, touch, in, contain, equal, cross, overlap, coveredBy, cover* [17]). We then propose a parametric recommendation operator which provides to the user additional results with respect to those returned by the original query by taking into account the query result and the database content. Additional results are also provided in case the result of the original query is empty, thus providing a solution to the *empty-answer* problem [2].

To motivate the proposed recommendation operator, consider a map, denoted by M_{Italy}, containing information about the provinces (represented as regions), the main rivers and the railways (represented as lines), and the main towns (represented as points) in Italy. Now suppose the user is interested in the rivers that somehow interact with the Bari and Arezzo provinces and suppose that she has obtained with previous computation the polygons representing such provinces. She may come out with the following formulation of the queries: Q_1: *Find the rivers which are contained in the Bari province*; Q_2: *Find the rivers which cross the Arezzo province*. Based on Figure 1 (a), we can see that no river satisfies Q_1, thus the query result is empty. However, objects that satisfy a very similar condition exist, for example rivers exist that *cross* the province of Bari. Such rivers could be recommended to the user since potentially interesting for her and can be detected by executing a *relaxed selection query*, which is able to identify the rivers which best fit the original query. On the other hand, query Q_2 selects those rivers which cross Arezzo, i.e., which intersect the interior of the Arezzo province but are not contained in it. In this case, as we can see from Figure 1 (b), the query result is not empty. Further potentially interesting results may correspond to affluents of the returned rivers, which are close to the boundary of the Arezzo province. Such affluents can be detected by first generating a new query object, starting from the original query results and the original query object, and then executing a (set of) new topological selection queries, possibly in a relaxed way, i.e., by exploiting similarity between topological relations in order to return the results which best fits the given condition.

Starting from the previous examples, we design a flexible recommendation operator which allows the user to specify her preferences in order to compute recommended objects. Preferences can be specified in terms of: (i) a topological similarity function ts, to be used to compare topological relationships; (ii) an object function $\mathcal{F}_O$, to compute the new query object starting from the result of the original query and the original query object; (iii) a predicate function $\mathcal{F}_T$, to determine the new set of topological predicates to be considered in the new topological selection queries; (iv) a semantics, either precise or relaxed, for each new selection query to be executed. Notice that functions $\mathcal{F}_O$ and $\mathcal{F}_T$ usually

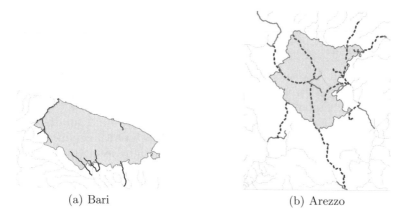

(a) Bari   (b) Arezzo

Fig. 1. (a) The Bari province and the rivers which cross it. (b) The Arezzo province and: (i) the rivers which cross it (dashed lines); (ii) the rivers which may be returned as result of the recommendation process (non-dashed lines).

depend on the considered application domain and are not directly related to spatial object properties. This is why they represent parameters for the proposed recommendation operator.

Based on the specified preferences, the recommendation operator works as follows. If the result of the original query is not empty, correlations between the returned data and data contained in the database are detected. A new query object is then generated from the original query object and the query result through the usage of $\mathcal{F}_O$. The resulting object represents a potentially interesting area for the user to be used for result extension. Then, new selection queries are defined, based on $\mathcal{F}_T$, and executed with respect to the new query object, in order to identify new potentially useful results for the user. In case the query result is empty, a relaxed approach is directly applied on the given condition, without changing the predicate and the query object.

Since in the spatial context query processing algorithms usually rely on the usage of a given index structure (e.g., an R-tree), we also present an index-based query processing algorithm for the proposed relaxed selection, taking into account the way it is executed in the recommendation operator.

The paper is organized as follows. Section 2 surveys related work. Background concepts are provided in Section 3. Section 4 then presents the recommendation operator and discusses all components it relies on. The processing algorithm for executing relaxed selections is presented in Section 5. Finally, Section 6 presents some conclusions and outlines future work.

2   Related Work

Recommendation systems aim at recommending to the users items not in the results of the posed queries but of potential interests. They have been initially proposed for Web services, very recently some approaches have also been proposed for the database context [1]. Motivated by the fact that most recommendation methods are hard-wired into the system, a framework for the declarative specification of the recommendation process over structured data has been proposed in [16]. The techniques reported in [15] and [6] present specific current-state and content-based recommendation processes, respectively. Existing recommendation approaches for spatial data mainly rely on history-based techniques [19].

A current-state recommendation process can be considered a query relaxation approach since the precise result of the query is extended with other potentially interesting items. Preference-based queries like top-k and skyline [5,13] are other examples of query relaxation operations. The recommendation operator presented in this paper relies on the usage of a relaxed selection operator which represents a top-1 query with respect to a specific scoring function. It also corresponds to a variation of the Best-Fit operator provided in [4] to take into account specific recommendation issues.

The aim of a top-k operator is to restrict the number of returned results to a fixed number (k), based on some ranking. Most top-k operators have been proposed for monotone scoring functions, which give the opportunity of optimizing top-k query processing, using some threshold value to prune the visit of non-interesting data. When considering spatial data, spatial relationships, and especially the distance-based ones, are often considered in computing scores. Nearest neighbor operators are examples of top-1 queries, using distance as scoring function [7,12,18]. Other approaches have then been proposed, using different types of scoring functions, based on spatial and non-spatial properties of spatial objects [20,21]. The most efficient algorithms assume that data are indexed and use a Branch and Bound approach to prune index subtrees which cannot provide any answer.

The proposed approach relies on the usage of specific topological similarity functions to determine how close two topological relations are. In the literature, several distance functions for topological relations have been proposed [9,10]. Often, similarity is computed only between pairs of objects with the same dimension and multiple object representations are not considered. More recent proposals extend the distance function proposed in [9] (only for regions) to geometry type-independent set of topological or cardinal relations, computing a value between 0 and 1 based on the matrix representation of the considered relations [4].

3   Background

The spatial data model considered in this paper relies on the concept of *feature type (ft)* [14], that represents a class of real spatial objects (e.g., rivers, roads, towns, etc...). Each feature type has some *descriptive attributes* and a *spatial attribute o*, having a given dimension $d \in DIM = \{0, 1, 2\}$. Values for o are taken from a set of spatial objects $\mathcal{SO}$, which in the paper we assume it contains

Table 1. Definition of the reference set of topological relations (f^o denotes the interior of object f, i.e., the set of points of f which do not belong to the boundary of f)

Name	Definition	Object dimension
disjoint (d)	$f_1 \cap f_2 = \emptyset$	All pairs
touch (t)	$(f_1^o \cap f_2^o) = \emptyset \wedge (f_1 \cap f_2) \neq \emptyset$	All pairs but 0/0
in (i)	$(f_1 \cap f_2^o = f_1) \wedge (f_1^o \cap f_2^o) \neq \emptyset$	2/2, 1/1, 1/2, 0/2, 0/1
contain (c)	$(f_1^o \cap f_2 = f_2) \wedge (f_1^o \cap f_2^o) \neq \emptyset$	2/2, 1/1, 2/1, 2/0, 1/0
equal (e)	$f_1 = f_2$	2/2, 1/1, 0/0
cross (r)	$dim(f_1^o \cap f_2^o) = (max(dim(f_1), dim(f_2)) - 1) \wedge$ $(f_1 \cap f_2) \neq f_1 \wedge (f_1 \cap f_2) \neq f_2$	1/2, 2/1, 1/1
overlap (o)	$dim(f_1) = dim(f_2) = dim(f_1^o \cap f_2^o) \wedge f_1 \cap f_2) \neq f_1 \wedge (f_1 \cap f_2) \neq f_2$	2/2, 1/1
coveredBy (b)	$(f_1 \cap f_2 = f_1) \wedge (f_1^o \cap f_2^o) \neq \emptyset \wedge (f_1 \cap f_2^o \neq f_1)$	2/2, 1/1, 1/2
cover (v)	$(f_1 \cap f_2 = f_2) \wedge (f_1^o \cap f_2^o) \neq \emptyset \wedge (f_1^o \cap f_2 \neq f_2)$	2/2, 1/1, 2/1

point, curve, and surface values (sas described in [14]).[1] We define a *map schema* MS as a set of feature types. An instance M of a map schema MS is a set of features, instances of the feature types in MS. The set of *features* associated with a feature type ft inside a map M is denoted by $M.ft$. The dimension of a feature or feature type f in a map M is unique and is denoted by $dim(f, M)$ ($dim(f)$ when there is no ambiguity).

In this paper, to explain the proposed concepts and techniques, we choose the set of topological relations $\mathcal{T}$ presented in [17] even if the proposed approach works for any set of topological relationships. The semantics of the chosen relations is provided in Table 1. We notice that, independently from the chosen set of topological relations, usually not all of them are defined for any pair of dimensions (see, e.g., column 3 of Table 1). In the following, when we want to remark the dimensions d_1 and d_2 of the pairs of features a topological relation $\theta \in \mathcal{T}$ is applied to, we denote θ with θ_{d_1, d_2} (*typed relation*). Additionally, we denote with $\mathcal{T}_{d_1, d_2}$ the set of typed relations defined for dimensions d_1 and d_2 and with $\mathcal{T}_D$ the set of all typed relations.

Topological relations can be used to define topological selection conditions of the form $c \equiv ft\theta^q O$, where $O \in \mathcal{SO}$, $ft \in MS$, $\theta^q \in \mathcal{T}_{dim(ft), dim(O)}$. Given a map M, instance of MS, the selection operation $\sigma_{ft\theta^q O}(M)$ returns all features $f \in M.ft$ such that $f\theta^q O$ holds.

The proposed recommendation operator relies on the usage of a *topological similarity function*, defined as $ts : \mathcal{T}_D \times \mathcal{T}_D \rightarrow [0, 1]$. The returned value quantifies the similarity between the two input typed topological relations. Of course, if $\theta_{d_1, d_2}^1 = \theta_{d_1, d_2}^2$, then $ts(\theta_{d_1, d_2}^1, \theta_{d_1, d_2}^2) = 1$. In the examples proposed in the remainder of this paper, we consider the topological similarity function presented in [4].

---

[1] In the following, we use the term *object* also to denote features, when the specific meaning is clear from the context.

4 A Current-State Approach to Spatial Recommendation

As pointed out in [15], current-state approaches exploit the result of a user query and the database content in order to recommend to the user further results. In order to define a current-state recommendation operator for topological selection queries, we use a very simple approach: the query object is modified based on the query result and a set of new spatial selections are generated and executed in either a precise or relaxed way. Relaxed selection allows one to determine the features which best fit the given query condition, even if they do not precisely satisfy it. Formally, relaxed selection can be defined as a top-1 query with respect to a scoring function depending on the similarity between the topological relation satisfied by a given feature and the query object and the query predicate.

The previous description points out all the elements required to set-up a recommendation operator. They can be thought as preferences specified when designing the recommendation operation for a given application and can be summarized as follows:

- the function $\mathcal{F}_O : 2^{SO} \times SO \to SO$ used to generate the new query object, based on the result of the original query and the original query object (Subsection 4.2);
- the function $\mathcal{F}_T : T_D \to 2^{T_D}$ used to generate the new query predicates (Subsection 4.3);
- the type of selection semantics used, either precise (traditional) or relaxed, for each new selection query to be executed. To make more evident the type of semantics, we denote by σ^p precise selection and by σ^r relaxed selection (Subsection 4.4).

In the following, we first present the recommendation operator in a parametric way with respect to $\mathcal{F}_O$, $\mathcal{F}_T$, and the chosen selection semantics. Then, we discuss how all such components can be defined.

4.1 Recommendation Operator

The current-state spatial recommendation operator can be defined as follows.

Definition 1 (Recommendation Operator). *Let M be a map. Let $O \in SO$, let $\theta \in T_D$. Let $Q^o(M) = \{f.o | f \in Q(M)\}$. Let $\mathcal{F}_O : 2^{SO} \times SO \to SO$. Let $\mathcal{F}_T : T_D \to 2^{T_D}$. Let $Q() \equiv \sigma_{ft\theta O}()$. The recommendation operator $\gamma_{ft\theta O}$ for Q is defined as follows:*

$$\gamma_{ft\theta O}(M) = \begin{cases} \bigcup_{\theta_i \in \mathcal{F}_T(\theta)} \left[\sigma^{s_i}_{ft\theta_i \mathcal{F}_O(Q^o(M),O)}(M \setminus Q(M)) \right] & \text{if } Q(M) \neq \emptyset \\ \sigma^r_{ft\theta O}(M) & \text{if } Q(M) = \emptyset \end{cases}$$

where $s \in \{p, r\}$. In the following, we denote by $\overline{\sigma}_{ft\theta_i O}(M, Q(M))$ the expression $\sigma^r_{ft\theta_i O}(M \setminus Q(M))$ and we call operator $\overline{\sigma}$ restricted relaxed selection. □

In the previous definition we notice that, when the result of the original query is not empty, the selection operator is executed against a map obtained from the original one by removing the result of the original query. This guarantees that the recommendation operator will always return features which have not been returned by the original query.

The reader may wonder why $Q(M)$ has to be removed from the input map instead of from the query result. The reason is that, while for precise selection the following equivalence holds:

$$\sigma^p_{ft\theta,O'}(M \setminus Q(M)) = \sigma^p_{ft\theta,O'}(M) \setminus Q(M)$$

this is not true for relaxed selection, as we will show in Subsection 4.4, where we also show that the result of a restricted relaxed selection cannot be empty if $M \setminus Q(M)$ is not. We also notice that, when the result of the original query is empty, we assume to execute a relaxed selection based on the original selection condition, in order to provide to the user the most similar results to those she is interested in.

Table 2 presents some examples of application of the recommendation operator, by considering specific preferences in terms of $\mathcal{F}_O$, $\mathcal{F}_T$, and types of executed selection queries.

Table 2. Examples of usage of the recommendation operator. $b(o, d)$ denotes the buffer of object o with radius d.

Original query			
Query expression	$\mathcal{F}_O(Q_i^o(M_{Italy}), O)$	$\mathcal{F}_T(\theta)$	Sel type
Recommendation			
Q_1: Select the rivers that *cross* the province of Arezzo.			
$Q_1 \equiv \sigma_{Rivers\ cross\ Arezzo}(M_{Italy})$	$b(Arezzo, d_1) \cap (\bigcup Q_1^o(M_{Italy}))$	*touch*	r
The affluents of the rivers that *cross* the province of Arezzo, close to the Arezzo province: they touch the new query object (which is a set of portions of rivers) or well approximate this condition.			
Q_2: Select the provinces interested by a certain agricultural area FA.			
$Q_2 \equiv \sigma_{Provinces\ overlap\ FA}(M_{Italy})$	$b(FA, d_1) \cap b(\bigcup Q_2^o(M_{Italy}), d_2)$	*in, overlap*	r, r
The provinces which may be influenced by the agricultural activity performed in FA: they are contained or overlap the new query object (which is a region) or well approximate one of these two conditions.			
Q_3: Select the railway paths containing the railway segment N.			
$Q_3 \equiv \sigma_{Railways\ contain\ N}(M_{Italy})$	$\bigcup Q_3^o(M_{Italy})$	*overlap*	r
The railway paths that have a portion in common with those returned as original query result: they overlap one of the railway path returned by the original query or well approximate this condition.			
Q_4: Select the main towns in a specific polluted area PA.			
$Q_4 \equiv \sigma_{MainTowns\ in\ PA}(M_{Italy})$	$b(PA, d)$	*in*	r
Other cities that may be polluted: cities near the boundary of the polluted area PA.			

4.2   Detection of the New Query Object

In order to generate the new query object, various functions $\mathcal{F}_O$ can be defined. We claim that a good function should rely on two main principles:

- *Principle of the memory:* the result of the original query should be taken into account in creating the new query object.
- *Principle of locality:* the new object must identify a region of space which is close to that containing the original query object.

In order to implement the principle of the memory, we suggest that a given aggregate operator is applied upon the objects returned by the original query. Typical aggregate functions are union and intersection. In order to implement the principle of locality, a buffering operation can be applied to such aggregate object and/or to the original query object.

More generally, function $\mathcal{F}_O$ can be defined by composing the following operators:

- $agg : 2^{\mathcal{SO}} \rightarrow \mathcal{SO}$, where $agg \in \{\bigcup, \bigcap\}$.
- $buffer : \mathcal{SO} \times \mathbb{R} \rightarrow \mathcal{SO}$.
- $\cap, \cup, \setminus$ operators over spatial data values, with the obvious meaning.

The following are some examples of recommendation objects which can be defined:

- *Dataset-based recommendation object.* In this case, each part of the new query object represents a portion of an object returned by the original query. The new query object can be computed from the geometry of the objects contained in the original query result. This is possible, for example, by creating a buffer of the query object and intersecting it with the result of the original query. The new object is composite and its dimension coincides with that of the input features. The following expression implements this behavior (other expressions can however be devised):

$$buffer(O, d) \cap \left(\bigcup Q^o(M) \right), \text{ with } d > 0.$$

  The recommendation operations for queries Q_1 and Q_3 in Table 2 rely on the usage of dataset-based objects.
- *Query-based recommendation object.* In this case, the new query object is obtained by modifying (usually, extending) the original query object. This is possible, for example, by creating a buffer of the query object, a buffer for the union of the result objects and taking the union of the two. In this case, the result will always be a (possibly multi-)region. The following expression implements this behavior (other expressions can however be devised):

$$buffer(O, d_1) \cup buffer\left(\bigcup Q^o(M), d_2 \right) \text{ with } d_1, d_2 > 0.$$

  Another example of query-based recommendation object is provided for query Q_4 in Table 2; in this case, no dataset object is used for the computation.

- *Domain-dependent recommendation object.* It corresponds to any other combination of the *agg* and *buffer* operators, e.g.,

$$buffer(O, d_1) \cap buffer(\bigcup Q^o(M), d_2) \text{ with } d_1, d_2 > 0.$$

The object used in the recommendation operator for query Q_2 in Table 2 is an example of a domain-dependent object.

We notice that in the buffer computation, the radius d, d_1, d_2 may correspond to either a system parameter or to a dynamic value, computed starting from O.

Example 1. Consider query $Q_1 \equiv \sigma_{Rivers\ cross\ Arezzo}(M_{Italy})$.

- A *dataset-based recommendation object* corresponds to the portions of rivers which cross Arezzo, contained in a specific buffer of the object representing the Arezzo province. Such portions could be relevant, in the context of the recommendation, for investigating what happens to rivers that cross Arezzo, before entering in such province.
- A *query-based recommendation object* corresponds to a region enclosing the surroundings of the Arezzo province and of the rivers which cross it. Such new object could be relevant, in the context of the recommendation, for investigating properties of rivers close to the Arezzo province or to one of the rivers previously detected.
- A *domain-dependent recommendation object*, computed as described above, returns the regions of space in the surroundings of Arezzo and of at least one river that crosses Arezzo. Such new object could be relevant, in the context of the recommendation, for investigating properties of rivers close to the Arezzo province and to one of the rivers previously detected.   ◇

4.3   Detection of the New Query Predicates

The new query predicates belong to $\mathcal{T}_{dim(ft),dim(O')}$, where O' is the new query object, computed through function $\mathcal{F}_O$. They can be selected according to various approaches, all depending on a specific topological similarity function ts. We claim that at least the following three approaches should be taken into account:

- *Best-Fit approach*: the topological relations in $\mathcal{T}_{dim(ft),dim(O')}$ which are most similar to the query predicate, based on ts, are chosen.
- *Threshold-based approach*: all the topological relations in $\mathcal{T}_{dim(ft),dim(O')}$ whose similarity with respect to the query predicate is greater than a threshold ρ are selected.
- *Extensional approach*: in this case, function $\mathcal{F}_T$ is defined in a case by case way. This approach can be useful when recommendation is domain dependent. In this way, information about the domain (possibly represented through ontologies) can be taken into account in order to select the new query predicates.

Example 2. Consider query $Q_1 \equiv \sigma_{Rivers \ cross \ Arezzo}(M_{Italy})$. Suppose the new query object is still a region (i.e., its dimension is 2). Different functions for the detection of the new query predicates can be considered:

- *Best-Fit approach*: function $\mathcal{F}_{\mathcal{T}}$ on input $cross_{1,2}$ returns the set of most similar relations to $cross_{1,2}$, defined for pairs composed of one line (the river) and one region (the province). Based on the similarity function proposed in [4], such relations are $\{in, coveredBy\}$.
- *Threshold-based approach*: when considering a threshold, say 0.7, function $\mathcal{F}_{\mathcal{T}}$ on input $cross_{1,2}$ returns the set of topological relations whose similarity with respect to $cross_{1,2}$ is greater than or equal to 0.7, i.e., the set $\{in, coveredBy, touch\}$, based on [4].
- *Extensional approach*: assuming the user is interested in the affluents of the rivers returned by the original query, function $\mathcal{F}_{\mathcal{T}}$ on input $cross_{1,2}$ returns $\{touch\}$. In this case, similarity function is not exploited.   ◇

4.4   Relaxed Selection Operator

The result of a recommendation operator corresponds to the union of a set of either precise or relaxed selection queries. In most contexts in which recommendation can be useful (huge quantity of potentially heterogeneous data for which the user has only a limited knowledge), relaxed selection operators can be quite useful in order to always return a non-empty result which is as closest as possible to the user request, i.e., which best fits the user requests.

The relaxed selection operator σ^r can be defined as a sort of *top-1 operator* with respect to a scoring function which quantifies, for each feature, how well it fits the given query. Since topological relationships are qualitative, in order to define the scoring function, as a first step we need to quantify how similar two features are with respect to a given topological relation.

Definition 2 ($\mathcal{T}$-Based Spatial Similarity). *Let $\theta \in \mathcal{T}_{\mathcal{D}}$. Let ts be a topological similarity function for $\mathcal{T}$. The $\mathcal{T}$-based spatial similarity function s_{ts}^{θ}, based on ts and θ, is defined as $s_{ts}^{\theta} : \mathcal{SO} \times \mathcal{SO} \to [0,1]$, $s_{ts}^{\theta}(o_1, o_2) = ts(\theta, \theta')$ if $o_1 \theta' o_2$.*   □

For example, if θ is the *overlap* relation then $s_{ts}^{overlap}(o_1, o_2)$ is equal to 1 if o_1 and o_2 overlap, otherwise it has a value that measures the similarity between *overlap* and the existing relation θ' between o_1 and o_2.

Based on s_{ts}^{θ} it is now possible to define a scoring function quantifying how well a feature fits a given selection condition and a top-k query based on it.

Definition 3 ($\mathcal{T}$-Based Top-1 Query). *Let $\theta \in \mathcal{T}$. Let $O \in \mathcal{SO}$. Let s_{ts}^{θ} be a $\mathcal{T}$-based spatial similarity function. The $\mathcal{T}$-based scoring function τ_O^{θ}, based on θ, O, and s_{ts}^{θ}, on a feature f is defined as follows: $\tau_O^{\theta}(f) = s_{ts}^{\theta}(f.o, O)$. The relaxed selection operator σ^r can now be defined as a $\mathcal{T}$-based top-1 query as follows: $\sigma_{ft\theta O}^{r}(M) = \{f | f \in M.ft \wedge f$ is a top-1 feature with respect to the $\mathcal{T}$-based scoring function $\tau_O^{\theta}\}$.*   □

It is possible to show that the equivalence provided in Subsection 4.1 for the precise selection operator does not hold for the relaxed one, since the presence in the input map of the objects we do not want to return (i.e, the result of the original query) may alter the result.

Proposition 1. *Let $Q() \equiv \sigma_{ft\theta O}()$. Let $c \equiv ft\theta_i O'$. For relaxed selection, the following equality does not hold:*

$$\sigma_c^r(M \setminus Q(M)) = \sigma_c^r(M) \setminus Q(M).$$

Proof Sketch. Relaxed selection looks for features which best fit c. Suppose that $\sigma_c^r(M) \subseteq Q(M)$, i.e., the features which best fit c in M are contained in $Q(M)$. In this case, the expression on the right will produce an empty result. On the other hand, the expression on the left will never produce an empty result, unless $M \setminus Q(M)$ is empty, i.e., when M and $Q(M)$ coincide.   □

5   Query Processing for the Recommendation Operator

The implementation of the recommendation operator strongly depends on how operator σ^r is implemented, i.e., on how top-1 features are detected. In the following, we therefore propose a specific query processing technique for $\mathcal{T}$-based top-1 queries, as defined in Definition 3. The proposed algorithm is a variation of the one proposed in [4] to take into account specific recommendation issues.

We assume, as usual in the spatial context, that instances of each feature type ft_i in a map M are indexed by an R-Tree R_i [11] or one of its variants, R⁺-trees and R*-trees [3], for guaranteeing a fast access. In the following, for the sake of presentation simplicity, we assume that leaf nodes of any index tree contains objects and not their approximations.

As a first consideration, we notice that the R-tree indexes all instances of a given feature type. However, based on Definition 1 and Proposition 1, not all such instances have to be considered during the processing (instances contained in $Q(M)$ have not to be considered). The proposed recommendation algorithm therefore takes as input, besides the selection condition, also a given set of features $Q(M)$, not to be considered during processing. Thus, it directly implements restricted relaxed selections (see Subsection 4.1).

Similarly to what has been done in [4], the algorithm relies on a Branch and Bound approach. The following structures are therefore maintained during the computation: (i) a priority queue PQ, used to store objects and R-tree nodes visited so far which may potentially produce further results; (ii) a threshold value ρ, representing the lowest score value in PQ. In order to visit first entries which most probably will generate some results, we order entries and objects in PQ based on their key value.

Under the Branch and Bound visit, the R-tree is visited in a breadth-first way (see Algorithm 1). All R-tree nodes and objects to be visited are inserted in PQ, which at the beginning contains the root of the tree. Nodes and objects in PQ are visited starting from the maximum element, based on the considered ordering (see below). Three cases may arise:

1. Given a non-leaf node N, for each entry e in N, we determine whether the visit of the subtree pointed by e has to be performed. The visit of a subtree is avoided if and only if it is possible to establish that objects pointed by its leaves do not belong to the result. To guarantee efficiency, such property has to be checked locally at the corresponding index entry, using some *key value*. A typical approach consists in defining the key value for an entry e, denoted by $key(e)$, as the minimum interval $[smin, smax]$ containing all score values assigned to features indexed by the subtree rooted by e. Ordering in PQ can be defined based on the maximum value of each interval. The visit of the subtree rooted by a given entry e can then be avoided if $key(e).smax$ is lower than ρ, since in this case no object in the subtree can belong to the result (better results have already been detected). In case the visit has to be performed, the entry is inserted into PQ, ρ is updated (only if $key(e).smin > \rho$), and all elements in PQ which cannot produce further results, based on the new ρ, are removed from PQ (i.e., all elements e such that $key(e).smax$ is lower than ρ).

2. Given a leaf node N, for each object o in N, the score for o is computed only if $o \notin Q(M)$. In case $\tau_O^\theta(o)$ is greater than or equal to ρ, this means that o may potentially belong to the result and it is therefore inserted in PQ. Similarly to the entry case, after the insertion ρ is updated, and all elements in PQ which cannot produce further results, based on the new ρ, are removed from PQ.

3. Finally, given an object o, it can be directly returned as result since PQ is ordered and any node or object following the one at hand will generate lower similarity values with respect to the query.

Algorithm 1 presents the approach described above. In order to compute key values for entries, we exploit properties of minimum bounding rectangles and the notion of *compatible topological relation*, presented in [17] for regions and extended in [4] to objects with arbitrary dimension. The basic idea behind compatibility is that, given an object o contained in the subtree pointed by an entry e and a query object O, if o θ O holds, the topological relation between $mbr(e)$ and $mbr(O)$ cannot be arbitrary but must be *compatible* with θ. Compatibility depends on dimension of the objects to which the topological relation is applied. In the following, the set of relations in $\mathcal{T}_D$ which are compatible with relation θ_{d_1,d_2} is denoted by $c(\theta, d_1, d_2)$.

Example 3. In order to illustrate compatibility, we consider its application to precise selection. Let SQ: $\sigma_{Rivers\,overlap\,O}(M)$. Suppose that during the visit of the R-tree indexing rivers in M, we consider an entry e such that $mbr(e)$ *disjoint* $mbr(O)$ holds. Since it is possible to show that the only relation which is compatible with disjoint over a pair of region is *disjoint* (see [4]), objects contained in the subtree rooted by e must be *disjoint* with respect to O, Thus, the subtree rooted by e can be discarded since it cannot produce further query results. $\diamond$

Using the notion of compatibility, key values for R-tree entries, with respect to the query $\sigma_{ft\theta O}^r(M)$, can be defined as follows.

Definition 4. *Let* $Q \equiv \sigma^r_{ft\theta O}(M)$ *be a relaxed selection query. Let* e *be an entry of the R-tree indexing* ft *features in* M. *We define the key value of* e *with respect to* θ *and* O, *denoted by* $key(e, \theta, O)$ *as follows:*

$$key(e, \theta, O) = [\ min\{ts(\theta, \overline{\theta})\ |\ mbr(e)\ \theta'\ mbr(O) \wedge \overline{\theta} \in c(\theta', dim(ft), dim(O))\},$$
$$max\{ts(\theta, \overline{\theta})\ |\ mbr(e)\ \theta'\ mbr(O) \wedge \overline{\theta} \in c(\theta', dim(ft), dim(O))\}\ ]. \quad \square$$

Algorithm 1. Restricted Relaxed Selection

Require: Query predicate θ, query object O, result set $Q(M)$, R-tree R over ft instances in map M

```
 1: PriorityQueue PQ:= empty
 2: PQ.insert(R.root,0)
 3: ρ=0
 4: while PQ is not empty do
 5:     N:=PQ.getMax()
 6:     if N is non-leaf node then
 7:         for all entries e in N do
 8:             compute key := key(e, θ, O);
 9:             let N' be the node pointed by e;
10:             if key.smax >= ρ then PQ.insert(N',key.smax); endif
11:             update PQ and ρ according to e;
12:         end for
13:     else if N is leaf node then
14:         for all objects o in N do
15:             if o ∉ Q°(M) then
16:                 compute score := τ^θ_O(o);
17:                 if score >= ρ then PQ.insert(o,score); endif
18:                 update PQ and ρ according to o;
19:             end if
20:         end for
21:     else
22:         return N;
23:     end if
24: end while
```

---

6   Concluding Remarks

In this paper, we have presented a current-state recommendation approach for spatial data. The recommendation operator is defined in a parametric way with respect to some basic components and relies on the usage of restricted relaxed selection operations. The restricted relaxed selection operator has been formally defined as a sort of top-1 query and an index-based query processing algorithm for its execution has been provided.

The proposed recommendation operator has been implemented in the context of an existing prototype [4]. Based on the realized prototype, future work includes an experimental evaluation of the efficiency and the efficacy of the proposed approach. Concerning the efficacy, we plan to perform experiments with end-users in order to compute quality-based evaluation measures, such as recall and precision, for a set of workload queries. The design of an optimized algorithm for the batch execution of a set of, either precise or relaxed, selection queries is another issue we plan to investigate.

References

1. Adomavicius, G., Tuzhilin, A.: Toward the Next Generation of Recommender Systems: A Survey of the State-of-the-Art and Possible Extensions. IEEE Trans. Knowl. Data Eng. 17(6), 734–749 (2005)
2. Agrawal, S., et al.: Automated Ranking of Database Query Results. In: CIDR (2003)
3. Beckmann, N., et al.: The R*-tree: An Efficient and Robust Access Method for Points and Rectangles. In: SIGMOD Conference, pp. 322–331 (1990)
4. Belussi, A., Catania, B., Podestà, P.: Topological Operators: a Relaxed Query Processing Approach. GeoInformatica (2011),
   http://www.citeulike.org/article/9315987
5. Börzsönyi, S., Kossmann, D., Stocker, K.: The Skyline Operator. In: ICDE, pp. 421–430 (2001)
6. Chatzopoulou, G., Eirinaki, M., Polyzotis, N.: Query Recommendations for Interactive Database Exploration. In: SSDBM, pp. 3–18 (2009)
7. Corral, A., et al.: Closest Pair Queries in Spatial Databases. In: SIGMOD Conference, pp. 189–200 (2000)
8. Egenhofer, M.J.: Spatial SQL: A Query and Presentation Language. IEEE Trans. Knowl. Data Eng. 6(1), 86–95 (1994)
9. Egenhofer, M.J., Al-Taha, K.K.: Reasoning about Gradual Changes of Topological Relationships. In: Frank, A.U., Formentini, U., Campari, I. (eds.) GIS 1992. LNCS, vol. 639, pp. 196–219. Springer, Heidelberg (1992)
10. Egenhofer, M.J., Mark, D.M.: Modeling Conceptual Neighborhoods of Topological Line-Region Relations. International Journal of Geographical Information Systems 9(5), 555–565 (1995)
11. Guttman, A.: R-trees: A Dynamic Index Structure for Spatial Searching. In: SIGMOD Conference, pp. 47–57 (1984)
12. Hjaltason, G.R., Samet, H.: Distance Browsing in Spatial Databases. ACM Trans. Database Syst. 24(2), 265–318 (1999)
13. Ilyas, I.F., Beskales, G., Soliman, M.A.: A Survey of Top-k Query Processing Techniques in Relational Database Systems. ACM Comput. Surv. 40(4) (2008)
14. Open Geospatial Consortium Inc. Opengis Implementation Standard for Geographic Information - Simple Feature Access - Part 1: Common Architecture. OpenGIS Implementation Standard (2010)
15. Pitoura, E., Stefanidis, K., Drosou, M.: You May Also Like Results in Relational Databases. In: PersDB Workshop (2009)
16. Koutrika, G., Bercovitz, B., Garcia-Molina, H.: FlexRecs: Expressing and Combining Flexible Recommendations. In: SIGMOD Conference, pp. 745–758 (2009)
17. Papadias, D., et al.: Topological Relations in the World of Minimum Bounding Rectangles: A Study with R-Trees. In: SIGMOD Conference, pp. 92–103 (1995)
18. Roussopoulos, N., Kelley, S., Vincent, F.: Nearest Neighbor Queries. In: SIGMOD Conference, pp. 71–79 (1995)
19. Ye, M., Yin, P., Lee, W.C.: Location Recommendation for Location-based Social Networks. In: GIS, pp. 458–461 (2010)
20. Yiu, M.L., et al.: Top-k Spatial Preference Queries. In: ICDE, pp. 1076–1085 (2007)
21. Zhu, M., et al.: Top-k Spatial Joins. IEEE Trans. on Knowl. and Data Eng. 17(4), 567–579 (2005)

Processing (Multiple) Spatio-temporal Range Queries in Multicore Settings

Goce Trajcevski*, Anan Yaagoub, and Peter Scheuermann*

Dept. of Electrical Engineering and Computer Science
Northwestern University
Evanston, Il 60208
Tel.: +1-847-491-7069
{goce,anany,peters}@eecs.northwestern.edu.com

Abstract. Research in Moving Objects Databases (MOD) has addressed various aspects of storing and querying trajectories of moving objects: from modelling, through linguistic constructs and formalisms/ algebras, to indexing structures and efficient processing of different query-categories have been subjects to a large body of works. Given the architectural trends of multicore CPUs becoming a commonplace, in this work we focus on efficient processing of spatio-temporal range queries in such settings. We postulate that coupling the semantics of the problem domain into the query processing algorithms in a manner that is aware of the multicore features, can yield performance improvements that surpass the gains obtained by relying solely on the compiler-generated threads parallelization. Towards that end, we present and evaluate heuristics for processing variants spatio-temporal range queries in multicore settings by partitioning the load (i.e., data set) and assigning partial tasks to the individual cores. Our experiments demonstrate that 5-fold speed-ups can be achieved, when compared to the (semi) naive approach which relies on the compiler to generate the multicore-compatible code.

1 Introduction

Moving Objects Databases (MOD) [17] provide a foundation that enables a wide range of applications relying on some form of Location Based Services (LBS) [33]. The typical MOD-tasks evolve around the efficient (storage and) retrieval of the spatio-temporal data representing the motion of a large number of moving objects, along with the efficient processing of various queries of interest (e.g., whereabouts-in-time, range, (k)Nearest-neighbor, similarity, skyline, etc.), towards which various indexing techniques and processing algorithms have been developed [3,23,27,36].

Historically, variants of different problems – both from the perspective of the architectures/environments, as well as the applications' semantics – have prompted researchers to address many issues related to parallel and distributed processing [5] (and the many references therein). MOD-related research has also

---

* Research supported by NSF:CNS-0910952.

J. Eder, M. Bielikova, and A.M. Tjoa (Eds.): ADBIS 2011, LNCS 6909, pp. 214–227, 2011.
© Springer-Verlag Berlin Heidelberg 2011

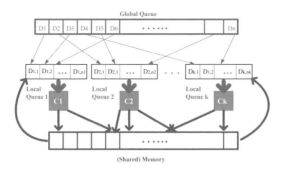

Fig. 1. Typical Multi-core Processing

generated works which, in one way or another, have addressed distributed contexts – from distributed data delivery and delegation of some responsibilities among mobile clients [19, 13, 12], to processing of spatio-temporal queries in sensor networks settings [4, 8].

One of the common trends in recent research is the hardware-software co-design, related to efficient execution of algorithms in multiprocessor/multicore settings [2, 20, 22, 30]. Using multiprocessor and/or multi-core architectures maximizes the benefit of parallelizing various tasks involved in the overall solution [2, 20]. A typical high-level configuration is illustrated in Figure 1: (1) the data that needs to be processed is stored initially in a commonly accessible global structure (e.g., "Global Queue" in Figure 1); (2) the data is then partitioned among the different processing units (cores) which can execute threads in parallel; (3) some of the partial results may need to be stored in a shared memory which, in turn, may be used as additional "feedback" for the subsequent steps/iteration of the respective threads.

While some algorithms are inherently sequential (or, have "sequential bottlenecks" [20]) and some are more amenable to parallelization, one possible avenue of exploiting the parallelism is to rely on the existing compilers for languages which are close in syntax to the existing ones (e.g., "Cilk Plus"[1]). For a given program, they will use a fixed set of translation-rules to maximize the extent of parallelism in the executable code generated for a given hardware platform.

In this work, we postulate that incorporating the semantics of the underlying data may yield considerable benefits in exploiting the multicore parallelization capabilities, when compared to the "vanilla" solutions which rely on the default (compiler-generated) parallelization. We focus on the efficient processing of spatio-temporal range queries in multi-core settings, the basic syntax of which is:
Qr: *Retrieve all the moving objects which are inside the region R between* $[t_1, t_2]$.

Different syntactic variants of **Qr** are possible (e.g., *sometimes* vs. *always* within the time-interval of interest), even incorporating uncertainty [7, 25, 37]). In this work, we specifically investigate the impact of the temporal-validity of interest for a particular query like, for example, the variant:

---

[1] http://software.intel.com/en-us/articles/intel-cilk-plus/

Qr$_\Theta$: *Retrieve all the moving objects which are inside the region R at least Θ % of the time between $[t_1, t_2]$.*

In addition, we consider settings in which multiple such range queries need to be processed, as a Boolean combination:

Qr$_{\Theta,B}$: *Retrieve all the moving objects which are inside the region R_1 and R_2 and ... R_q at least Θ % of the time between $[t_1, t_2]$.*

Our main objective is to devise assignments of partial-processing to the individual cores, similar in spirit to the range-splitting in traditional databases for query processing on multiprocessor machines [15]. In addition, we would like to quantify the benefits of these assignments in comparison with the approaches that rely on the compilers to utilize the threads parallelization. Towards that end our main contributions can be summarized as follows:

- We introduce efficient algorithms for processing variants of spatio-temporal range queries in multicore settings:
- two of the proposed heuristics focus on distributing the MOD data;
- the other heuristics focus on the query-syntax and use Computational Geometry techniques to partition the load among the cores, based on the geographic region(s) specified in the query.
- We conducted extensive experimental evaluations which provide quantitative illustrations of the benefits of the proposed approaches.

The rest of this paper is structured as follows. In Section 2 we review the necessary background. Section 3 presents the details of our proposed approaches. In Section 4 we present the results of the experimental evaluations. Section 5 positions our work with respect to the related literature, summarize the results and outline directions for future work.

2   Preliminaries

We now present the basic concepts and notation used throughout the rest of this paper. In the MOD-literature, the motion of the objects is represented by a *trajectory* [24, 25, 37]:

Definition 1. *A trajectory Tr of a moving object, is a polyline in a 3D space (2D spatial + time), represented as a sequence of points $Tr = (x_1, y_1, t_1), \ldots, (x_n, y_n, t_n)$, where $\forall(i,j)(i < j \Rightarrow t_i < t_j)$. Between two consecutive points (x_i, y_i, t_i) and $(x_{i+1}, y_{i+1}, t_{i+1})$, the object is assumed to move along the straight line-segment $((x_i, y_i)(x_{i+1}, y_{i+1}))$, and with a constant expected speed $v_i = \sqrt{(x_{i+1} - x_i)^2 + (y_{i+1} - y_i)^2}/(t_{i+1} - t_i)$. The expected location of the object at any time-point t ($\in (t_i, t_{i+1})$) is the one obtained via linear interpolation between the endpoints, using the expected speed v_i. The projection of Tr_k in the Euclidian 2D space is called its route.*

According to Definition 1, a trajectory is function from *Time* domain into the 2D Euclidian space (i.e., $f(t) \to \mathbf{R}^2$), and we consider *past* motion [17,25], which is, the entire motion of each objects is known and stored in the MOD. Such data may also correspond to *future motion plan* of a given object, obtained either via some trip-planning tool (e.g., MapQuest or Google Maps), or dictated by some business fleet-planning rules [10]. Other variations of trajectories' representations have been exploited in the literature – e.g., road-network constrained [9, 14, 16]; streaming updates of *(location, time)* data [27]; non-linear interpolation in-between consecutive points [28] – each with their specifics on processing the typical spatio-temporal queries. However, those settings are beyond the scope of this work.

We note that the variants of spatio-temporal range queries considered in earlier works [37] – *sometimes* and *always* within the temporal-interval of interest are special cases of $\Theta \neq 0$ and $\Theta = 100\%$.

We assume multicore settings in a multiple-reader multiple-writer (MRMW) shared memory context [20]. Specifically, each core $\mathbf{C}_i$ can access the different portions of the MOD-data and, when applicable (cf. Section 3) can write in a buffer(s) that can be read by the other cores for the purpose of determining whether a particular thread running on $\mathbf{C}_i$ should be terminated or not, without affecting the correctness of the answer to the query.

Lastly, we note that in this work, the region R of interest for the spatio-temporal range queries are assumed to be convex polygons. Except for the complexity-bounds of the underlying algorithms executed in a particular core, this does not affect the general idea of the proposed approaches.

3   Algorithms in Multicore Context

We now present in detail the algorithms developed for efficient processing of spatio-temporal range queries in multicore settings. In the sequel, we assume that there is a total of m different trajectories ($\{Tr_1, Tr_2, \ldots, Tr_m\}$) stored in the MOD, and that there are k cores ($\{\mathbf{C}_1, \mathbf{C}_2, \ldots, \mathbf{C}_k\}$) available.

3.1   MOD-Level Load Distribution

The first two heuristics partition the MOD data along two complementary dimensions.

The simplest heuristics, called *H1*, is illustrated in Figure 2(a). Essentially, it partitions the MOD into collections of consecutive trajectories – assuming that they are sorted by the unique object-ID – where each collection is essentially a subset consisting of m/k trajectories. In the case that m is not an exact multiple of k, in accordance with the pigeonhole (a.k.a. Dirichlet's) principle, some chunks will consist $(m/k) + 1$ trajectories.

Each of the k cores evaluates the complete query – both in terms of the region R as well as the other parameters (Θ, $[t_1, t_2]$) – on a subset of the trajectories-data. Although, based on Figure 2(a), it may appear that each core $\mathbf{C}_i$ is "in-charge" of consecutive trajectories, we note that in practice, this need not be the case – for as long as each $\mathbf{C}_i$ has m/k trajectories from the entire MOD.

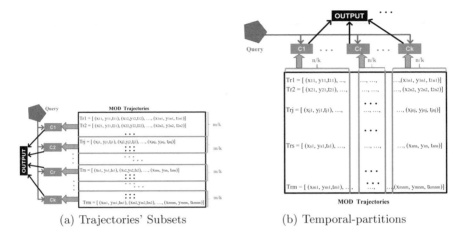

(a) Trajectories' Subsets   (b) Temporal-partitions

Fig. 2. MOD Data Partitions

Formally, each core executes the following algorithm:

Algorithm 1. Execution of H1 in the core $\mathbf{C}_j$

Input: $(\mathbf{Set}_j = \{Tr_{j,1}, \ldots, Tr_{j,(m/k)}\}; R, [t_1, t_2], \Theta)$

1: **for all** $Tr_{j,s} \in \mathbf{Set}_j$ **do**
2:    Calculate the times of the entry/exit intersections of $Tr_{j,s}$ with R.
3:    **if** $((Total\_Time\_Inside)/(t_2 - t_1)) \geq \Theta$ **then**
4:       Add $Tr_{j,s}$ to the Common_Answer
5:    **end if**
6: **end for**

We rely on the algorithms for intersection of a trajectory-segment with a region R of a spatio-temporal range-query in [25] (cf. line 2 of Algorithm 1). Assume that R has l vertices/edges, and that each trajectory has at most n segments. Due to the convexity of R, the time-complexity of obtaining the intersections of a trajectory segment with its boundary ∂R is $O(\log l)$ [31]. Consequently, the worst-case complexity of executing H1 in a given core $\mathbf{C}_i$ is bounded by $O((m/k) \cdot n \cdot \log l)$. Note that there is one shared structure denoted as "Common_Answer", where the different cores are writing each of the trajectories that qualify as an answer to the range query, which was implemented as a file. The contents of the file are subsequently presented to the user as the complete answer-set to the given query.

The basic idea behind the second heuristic – H2, is illustrated in Figure 2(b). In contrast to H1, H2 partitions each *individual trajectory* into k chunks of consecutive segments along its temporal dimension, and assigns $O(n/k)$ such segments from each of the m trajectories to an individual core $\mathbf{C}_i$. Based on this, there is a subtle difference of the algorithm that is executed in the individual cores implementing H2, which is formally specified by:

Algorithm 2. Execution of H2 in the core $\mathbf{C}_j$

Input:

$(\mathbf{S}_{j,1} \qquad = \qquad \{(x_{1,(j-1)\cdot(n/k)}, y_{1,(j-1)\cdot(n/k)}, t_{1,(j-1)\cdot(n/k)}), \cdots \qquad \cdots$
$(x_{1,(j\cdot(n/k))-1}, y_{1,(j\cdot(n/k))-1}, t_{1,(j\cdot(n/k))-1})\};$
$\cdots$
$\mathbf{S}_{j,m} \qquad = \qquad \{(x_{m,(j-1)\cdot(n/k)}, y_{m,(j-1)\cdot(n/k)}, t_{m,(j-1)\cdot(n/k)}), \cdots \qquad \cdots$
$(x_{m,(j\cdot(n/k))-1}, y_{m,(j\cdot(n/k))-1}, t_{m,(j\cdot(n/k))-1})\};$
$R, [t_1, t_2], \Theta, Flag_1, \ldots Flag_m)$

1: **for all Set**$_{j,q}$ **do**
2:      **if** Flag$_q \neq$ true **then**
3:          Calculate the times of the entry/exit intersections of the j-th portion of Tr_q with R.
4:          Add the fraction of the total time inside R to the Buffer$_q$
5:          **if** $(((Time\_Inside\_Set_{j,q})/(t_2 - t_1)) \geq \Theta)$ OR (Total_Time of Buffer$_q \geq \Theta$) **then**
6:              Add Tr_q to the Common_Answer
7:              Flag$_q = true$
8:          **end if**
9:      **end if**
10: **end for**

Observe that Algorithm 2, in addition to the different input-structure due to the partitioning of the MOD data, has an additional collection of shared variables. Namely, for each of the trajectories (e.g., Tr_q), there is a separate *Buffer$_q$* which aggregates the fractions of time that the portions of Tr_q (assigned to various $\mathbf{C}_i$'s) have spent inside R. In addition, there is the collection of *Flag$_q$* variables, initially set to *false*, and changed to *true* only upon detecting that the cumulative time of Tr_q detected by the current core (along with some other cores that have already completed the processing of their portion of Tr_q) has determined that it is part of the answer-set. This, in a sense, adds a "lazy-evaluation" flavor to Algorithm 2, at the expense of having to perform an extra-check of the corresponding flag, before moving to processing the query for the (portion of the) corresponding trajectory. We note that due to the similar reasoning as in Algorithm 1, the worst case time-complexity of the Algorithm 2 is $O((n/k) \cdot m \cdot \log l)$.

We conclude this section with the observation that both H1 and H2 need to repeat the respective algorithms for each separate region in the case of a range query for a Boolean combination of $q(> 1)$ regions $R_1 \ldots, R_q$. This, in turn, increases the upper-bound on the running time complexity to $O(q \cdot (n/k) \cdot m \cdot \log l)$.

3.2 Query-Aware Load Distribution

The last approach that we present, named H3, separates itself from the previous two because it distributes the load among the cores based on the query-parameter R (the region of interest).

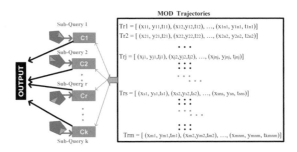

Fig. 3. H3: R-based Load Distribution among Cores

As illustrated in Figure 3, the main intuition behind H3 is to partition the query-region into k non-overlapping regions – $R_1, R_2, \ldots, R_k$, having only a common boundary-edge as intersection between two consecutive sub-regions. The processing of the sub-query pertaining to R_j is assigned to the core $\mathbf{C}_j$. The pseudo-code of the corresponding algorithm is given by:

Algorithm 3. Execution of H3 in the core $\mathbf{C}_j$

Input: (MOD, R_j, $[t_1, t_2]$, Θ, $Flag_1 \ldots, Flag_m$)

1: **for all** $Tr_i \in$ MOD **do**
2:     **if** $Flag_i \neq$ true **then**
3:         Calculate the times of the entry/exit intersections of Tr_i with R_j.
4:         Add the fraction of the total time inside R_j to the Buffer$_i$
5:         **if** $(((Time\_InsideR_j)/(t_2 - t_1)) \geq \Theta)$ OR (Total_Time of Buffer$_i \geq \Theta$) **then**
6:             Add Tr_i to the Common_Answer
7:             $Flag_i = true$
8:         **end if**
9:     **end if**
10: **end for**

In a sense, the general layout of Algorithm 3 is similar to the one of Algorithm 2. However, the main difference is in the structure of the input – while each core $\mathbf{C}_j$ operates over the entire MOD, it is in charge of a sub-region from the original query region R. Assuming that the number of cores in a multicore machine is typically a power-of-2 (e.g., [22], although other multicore architecture exist), we can recursively apply the algorithms for bisecting a given polygon into two (sub)polygons of equal areas [34], so that each core is assigned an equal-area sub-region of R.

Note that, once again, we have the opportunity of a lazy-evaluation, in the sense that whenever a particular trajectory has been determined to satisfy the temporal-threshold required by the query, its processing is no longer needed with respect to the rest of the sub-regions – and by the corresponding cores. As for the complexity, noting that now each sub-region will have $O(\lceil l/k \rceil)$ sides, the worst case time-complexity of Algorithm 3 is bounded by $O(m \cdot n \cdot \log\lceil l/k \rceil)$.

When it comes to processing Boolean conjunction of range queries $\mathbf{Qr}_{\Theta,B}$, one can straightforwardly extend H3 by applying it to each of the regions R_i ($i \in \{1, \ldots, q\}$). However, that may cause a particular core $\mathbf{C}_j$ to be in charge of subsets of regions which are geographically far apart. To minimize that effect, we use a variant of H3 for multiple regions (denoted H3m), which can be specified as follows. Let $A(R_i)$ denote the area of the i-th query region, and let $\mathcal{A}(\mathbf{Qr})$ = $\sum_{j=1}^{j=q} A(R_j)$. Given a reference coordinate system, we traverse the regions $R_1, R_2, \ldots, R_q$ based on the row-major order (X-axis value, followed by Y-axis value) of the coordinates of their the centroids, and we assign the portions of the regions to the cores sequentially, as follows:

(1) **IF** $(A(R_1) > \mathcal{A}(\mathbf{Qr}) / k)$
divide R_1 in two portions, $\mu_1(R_1)$ and $\mu_2(R_1)$, such that $\mu_1(R_1) = \mathcal{A}\mathbf{Qr} / k$. Assign $\mu_1(R_1)$ to $\mathbf{C}_1$.

Recursively, proceed with the assignment, starting with $\mathbf{C}_2$ and considering the boundary of the region $\mu_2(R_1)$ as a first polygon in the new sequence.

(2) **Else_IF** $(A(R_1) = \mathcal{A}(\mathbf{Qr}) / k)$
assign R_1 to $\mathbf{C}_1$.

Recursively proceed with the assignment, starting with $\mathbf{C}_2$ and considering $2R_2$ as a first polygon in the new sequence.

(3) **Else**
Assign R_1 to $\mathbf{C}_1$ and split R_2 into two portions, $\mu_1(R_2)$ and $\mu_2(R_2)$, such that: $\mu_1(R_2) = (\mathcal{A}(\mathbf{Qr}) / k) - A(R_1)$. Assign the region bounding $\mu_1(R_2)$ to $\mathbf{C}_1$.

Recursively, proceed with the assignment, starting with $\mathbf{C}_2$ and considering the boundary of $\mu_2(R_2)$ as a first polygon in the new sequence.

To achieve a weighted-bisection of a given region (e.g., splitting R_i into $\mu_1(R_i)$ and $\mu_2(R_i)$) we need a slight modification of the bisection algorithm in [34] which, once again, can be done in a linear time [1]. In the worst case, assuming that each of the q query-regions is bounded by m-sided convex polygon, this incurs an additional overhead of $O(q \cdot m)$ in the upper-bound on the running time complexity. Once the assignment of geographical regions to cores has been completed, each core proceeds with executing the Algorithm 3 which, essentially yields $O(q \cdot m + m \cdot n \cdot \log\lceil l/k \rceil)$ as an upper bound on the running time complexity of H3m.

4   Experimental Evaluation

We now proceed with presenting the observations from the experimental evaluations of our proposed techniques. Our experiments were conducted on an Intel Core i3 CPU machine with 4GB memory, with 4 dual-core processors at 2.13GHz. We used a task-based parallelization model among the cores, where each task is scheduled among the individual core-queues in a work-stealing manner.

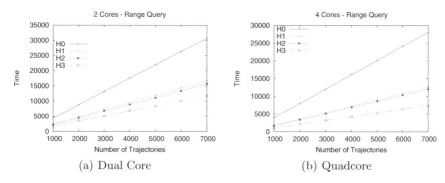

(a) Dual Core   (b) Quadcore

Fig. 4. Existential Variant

Data Sets: The trajectories data-sets were generated using the Random Way-point Model (RWP) [6], with the restriction that the velocity in two consecutive segments would not increase/decrease by more than 50%, and is bounded between *20 mph* and *90 mph*. The change of the angle between two consecutive segments was bounded by $+/-2\pi/3$. The length of the trajectories varied from 10 miles to 100 miles (uniformly distributed), within a 50 x 50 miles^2 area. We generated sets with cardinalities between 1,000 and 7,000 trajectories.

As for the query parameters, we used regular square, hexagon and octagon, with varying sizes (in terms of the percentage of the total area of interest). We ran 10 different simulations for each type of a polygon and each size, and we report the averaged results. The temporal interval of interest for the queries varied between 1h and 2h and, once again, we report the average values of the results.

First, we compared the benefits of each of *H1, H2* and *H3* against each other, as well as against the *baseline*-approach – denoted *H0*. The baseline approach is essentially the sequential algorithm for processing the respective (variants of the) range queries. We report observations in both 2-cores and 4-cores settings and, unless otherwise indicated, the time-values are expressed in *milliseconds*.

The first set of experiments that we report pertains to the *existential* (i.e. $\exists t \in [t_1, t_2]$) variant of the range query. The respective results for each of the heuristics in 2-cores and 4-cores settings are illustrated in Figure 4(a) and 4(b). As shown, the trend is similar in both settings except, as expected, the processing time for each of H1, H2 and H3 is shorter in 4-core settings. In addition, while the 2-core settings provide 3-times faster execution (when comparing H3 to H0), the speed-up factor in 4-core settings is 5.

We note that H3 appears to yield the largest improvements in terms of processing time for the existential queries – a trend which remains the same in all the other experimental settings.

Next, we report the experiments pertaining to different values of the Θ-threshold as a fraction of the temporal interval of interest for the query.

Figures 5(a) and 5(b) present the averaged running times for each of the H1, H2 and H3 (as well as the baseline approach – H0), when the threshold-value for a trajectory to qualify as an answer is 30% of the query time-interval, for

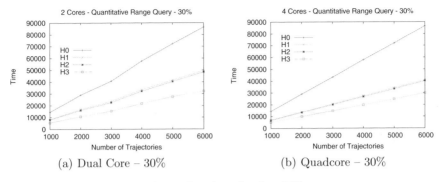

Fig. 5. Speed up for $\Theta = 30\%$

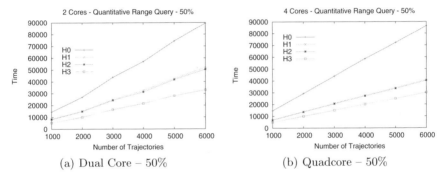

Fig. 6. Speed up for $\Theta = 50\%$

2-cores and 4-cores settings, respectively. Once again, we observe the similar phenomena as in Figures 4(a) and 4(b): the relative speed-up trends are similar, in both 2-cores and 4-cores settings, and H3 yields the largest speed-up.

Figures 6(a) and 6(b) show the outcome of the same experiments when the desired temporal threshold-value for a trajectory to qualify as an answer is 50% of the query time-interval, for 2-cores and 4-cores settings, respectively.

The next set of experiments describes the impact of the size of the query region when the answer-set consists of the trajectories which are inside of it for at least 50% of the time-interval of interest for a given query. As can be seen in Figure 7(a), as the area of the query polygon increases, the speed-up benefits of H3 in 2-core settings are decreasing – equivalently, the processing time taken by H3 is increasing. One of the reasons for this effect is that the benefits of the "lazy-evaluation" of H3 are diminished as larger portion of the individual trajectories need to be examined. In contrast, in 4-core settings, the trend of improvements, although very insignificant, is still present. We leave it for a future work to investigate in greater detail whether there is some correlation between the types of the polygons and the number of cores which could influence this trend.

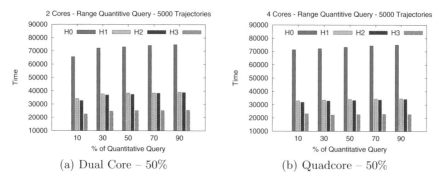

(a) Dual Core – 50%   (b) Quadcore – 50%

Fig. 7. Impact of the Query Region Size

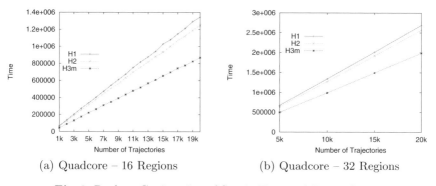

(a) Quadcore – 16 Regions   (b) Quadcore – 32 Regions

Fig. 8. Boolean Conjunction of Spatio-Temporal Range Queries

We conclude this section with the results pertaining to evaluating a Boolean conjunction combination of spatio-temporal range queries (cf. Figure 8). Figure 8(a) presents the running time of H1, H2 and H3m for 16 polygons ($\Theta = 30\%$), as function of the number of trajectories, whereas Figure 8(b) shows the corresponding times for a query consisting of 32 polygons. Note that we do not display the "vanilla" (baseline) approach because its processing time was too long. As can be seen, H3m consistently outperforms H1 and H2 which perform similarly, with a slight advantage of H2.

5   Related Work and Concluding Remarks

MOD research has generated a large body of results pertaining various aspects of the management of spatio-temporal data. Based on the model of the motion (e.g., full-trajectories vs. streams of *(location, time)* updates, possibly with uncertainty); spatial constraints (e.g., road-networks vs. free motion) quite a few indexing methodologies and query processing techniques have been proposed with algorithms addressing various categories of queries of interest (range, (k)Nearest-neighbor, similarity, skyline, etc.) [3, 32, 21, 26, 18, 29, 11, 36, 38] – to mention but a few.

All the above works, in one way or another, have focused on the efficiency of the overall processing of the respective queries, but certain results have specifically addressed the aspects of distributed and parallel processing. For example, in [12], part of the responsibility for monitoring location-based queries has been delegated to the participating mobile entities, whereas in [19] the inherent parallelism of a shared-nothing computing environment for storing and indexing the spatiotemporal data was exploited. With a similar motivation [4, 8] have considered the efficient processing of spatio-temporal queries in wireless sensor networks. A load-shedding approach for processing/monitoring a variety of spatio-temporal queries in the settings in which the *(location, time)* data arrives in a streaming manner is presented in [27]. While all these works have a same objective – exploiting the parallelism when processing spatio-temporal queries, what separates our work is that we have specifically focused on exploiting the benefits of multicore architecture for the purpose of efficient processing of spatio-temporal range queries.

The results on developing parallel algorithms for multiprocessor and/or multicore architectures abound [20], and the ideas of range-based partitioning of the data for efficient query processing have been around [15]. In this work, we proposed and experimentally evaluated three heuristics for spatio-temporal range queries processing in multicore settings – two of which (H1 and H2) focused on partitioning the trajectories' data, and one (H3) which focused on partitioning based on the query region. In addition, we considered a Boolean conjunction of regions of interest for the query, towards which we proposed a modification of H3 – the H3m heuristic to cater for distributing the "geographical load" among the cores. Our experiments have demonstrated that each of the proposed heuristics has consistently outperformed the baseline (sequential) approach for processing of the respective queries, with H3 yielding highest speed-ups. In a similar spirit, H3m outperformed the sequential applications of H1 and H2 for the conjunctive range queries.

While we presented possible approaches to the problem of efficient processing of spatio-temporal range queries in multicore settings, there are several challenging directions for future investigations. An immediate extension that we are currently pursuing is to investigate whether some of the geometric techniques (e.g., using variants of ham-sandwich cut [1, 35] can further improve the benefits of H3m. Along these lines, we would like to further generalize the conjunctive range queries to an arbitrary Boolean combination including disjunctions and negations. Another goal is to further investigate the level of the "cache-friendliness" of our approaches. In addition to increasing the efficiency of the query processing, this may lead to a better understanding of the anomaly observed in part of our experiments (cf. Figure 7(a) and 7(b)).

From a practical perspective, we need to incorporate some known constraints (e.g., a motion restricted by a known road-network, density of the moving objects, etc.) into the policies for allocating tasks among the cores.

References

1. Abbott, T.G., Burr, M., Chan, T.M., Demaine, E.D., Demaine, M.L., Hugg, J., Kane, D.M., Langerman, S., Nelson, J., Rafalin, E., Seyboth, K., Yeung, V.: Dynamic ham-sandwich cuts in the plane. Comput. Geom. 42(5), 419–428 (2009)
2. Barroso, L.A., Gharachorloo, K., McNamara, R., Nowatzyk, A., Qadeer, S., Sano, B., Smith, S., Stets, R., Verghese, B.: Piranha: a scalable architecture based on single-chip multiprocessing. In: Proceedings of the 27th Annual International Symposium on Computer Architecture, pp. 282–293 (2000)
3. Benetis, R., Jensen, C.S., Karciauskas, G., Saltenis, S.: Nearest and reverse nearest neighbor queries for moving objects. VLDB J. 15(3), 229–249 (2006)
4. Bestehorn, M., Böhm, K., Bradley, P., Buchmann, E.: Deriving spatio-temporal query results in sensor networks. In: Gertz, M., Ludäscher, B. (eds.) SSDBM 2010. LNCS, vol. 6187, pp. 6–23. Springer, Heidelberg (2010)
5. Blazewicz, J., Ecker, K., Plateau, B. (eds.): Handbook on parallel and distributed processing. Springer, Heidelberg (2000)
6. Camp, T., Boleng, J., Davies, V.: A survey of mobility models for ad hoc network research. Wireless Communications and Mobile Computing 2(5) (2002)
7. Cheng, R., Kalashnikov, D., Prabhakar, S.: Evaluating probabilistic queries over imprecise data. In: SIGMOD (2003)
8. Coman, A., Nascimento, M.A., Sander, J.: A framework for spatio-temporal query processing over wireless sensor networks. In: DMSN, pp. 104–110 (2004)
9. Demiryurek, U., Pan, B., Kashani, F.B., Shahabi, C.: Towards modeling the traffic data on road networks. In: GIS-IWCTS (2009)
10. Ding, H., Trajcevski, G., Scheuermann, P.: Towards efficient maintenance of continuous queries for trajectories. GeoInformatica 12(3) (2008)
11. du Mouza, C., Rigaux, R.: Multi-scale classification of moving objects trajectories. In: SSDBM (2004)
12. Gedik, B., Liu, L.: Mobieyes: A distributed location monitoring service using moving location queries. IEEE Transactions on Mobile Computing 5(10) (2006)
13. Gedik, B., Liu, L.: Quality-aware distributed data delivery for continuous query services. In: SIGMOD Conference (2006)
14. George, B., Kim, S., Shekhar, S.: Spatio-temporal network databases and routing algorithms: A summary of results. In: Papadias, D., Zhang, D., Kollios, G. (eds.) SSTD 2007. LNCS, vol. 4605, pp. 460–477. Springer, Heidelberg (2007)
15. Ghandeharizadeh, S., DeWitt, D.J.: Hybrid-range partitioning strategy: A new declustering strategy for multiprocessor database machines. In: McLeod, D., Sacks-Davis, R., Schek, H.-J. (eds.) 16th International Conference on Very Large Data Bases, pp. 481–492 (1990)
16. Güting, R.H., de Almeida, V.T., Ding, Z.: Modeling and querying moving objects in networks. VLDB J. 15(2) (2006)
17. Güting, R.H., Schneider, M.: Moving Objects Databases. Morgan Kaufmann, San Francisco (2005)
18. Hadjieleftheriou, M., Kollios, G., Tsotras, V.J., Gunopulos, D.: Efficient indexing of spatiotemporal objects. In: Jensen, C.S., Jeffery, K., Pokorný, J., Šaltenis, S., Hwang, J., Böhm, K., Jarke, M. (eds.) EDBT 2002. LNCS, vol. 2287, p. 251. Springer, Heidelberg (2002)
19. Hadjieleftheriou, M., Kriakov, V., Tao, Y., Kollios, G., Delis, A., Tsotras, V.J.: Spatio-temporal data services in a shared-nothing environment. In: SSDBM, pp. 131–134 (2004)

20. Herlihy, M., Shavit, N.: The Art of Multiprocessor Programming. Morgan Kaufmann, San Francisco (2008)
21. Iwerks, G.S., Samet, H., Smith, K.P.: Maintenance of k-nn and spatial join queries on continuously moving points. ACM Transactions on Database Systems (TODS) 31(2) (2006)
22. Kongetira, P., Aingaran, K., Olukotun, K.: Niagara: A 32-way multithreaded sparc processor. IEEE Micro. 25, 21–29 (2005)
23. Koubarakis, M., Sellis, T., Frank, A.U., Grumbach, S., Güting, R.H., Jensen, C.S., Lorentzos, N., Manolopoulos, Y., Nardelli, E., Pernici, B., Scheck, H.-J., Scholl, M., Theodoulidis, B., Tryfona, N.: Spatio-Temporal Databases – the CHOROCHRONOS Approach. LNCS, vol. 2520. Springer, Heidelberg (2003)
24. Kujipers, B., Othman, W.: Trajectory databases: data models, uncertainty and complete query languages. Journal of Computer and System Sciences (2009), doi:10.1016/j.jcss.2009.10.002
25. Lema, J.A.C., Forlizzi, L., Güting, R.H., Nardelli, E., Schneider, M.: Algorithms for moving objects databases. Computing Journal 46(6) (2003)
26. Lin, D., Cui, B., Yang, D.: Optimizing moving queries over moving object data streams. In: Kotagiri, R., Radha Krishna, P., Mohania, M., Nantajeewarawat, E. (eds.) DASFAA 2007. LNCS, vol. 4443, pp. 563–575. Springer, Heidelberg (2007)
27. Mokbel, M.F., Aref, W.G.: SOLE: scalable on-line execution of continuous queries on spatio-temporal data streams. VLDB J. 17(5), 971–995 (2008)
28. Mokhtar, H., Su, J.: Questo: A query language for uncertain and exact spatio-temporal objects. In: Atzeni, P., Caplinskas, A., Jaakkola, H. (eds.) ADBIS 2008. LNCS, vol. 5207, pp. 184–198. Springer, Heidelberg (2008)
29. Mouratidis, K., Yiu, M.L., Papadias, D., Mamoulis, N.: Continuous nearest neighbor monitoring in road networks. In: VLDB, pp. 43–54 (2006)
30. Olukotun, K., Nayfeh, B.A., Hammond, L., Wilson, K., Chang, K.: The case for a single-chip multiprocessor. In: ASPLOS, pp. 2–11 (1996)
31. O'Rourke, J.: Computational Geometry in C. Cambridge University Press, Cambridge (2000)
32. Pelanis, M., Saltenis, S., Jensen, C.S.: Indexing the past, present, and anticipated future positions of moving objects. ACM TODS 31(1) (2006)
33. Schiller, J.H., Voisard, A. (eds.): Location-Based Services. Morgan Kaufmann, San Francisco (2004)
34. Shermer, T.C.: A linear algorithm for bisecting a polygon. Inf. Process. Lett. 41(3), 135–140 (1992)
35. Stojmenovic, I.: Bisections and ham-sandwich cuts of convex polygons and polyhedra. Inf. Process. Lett. 38(1), 15–21 (1991)
36. Tao, Y., Papadias, D., Sun, J.: The tpr*-tree: An optimized spatio-temporal access method for predictive queries. In: VLDB (2003)
37. Trajcevski, G., Wolfson, O., Hinrichs, K., Chamberlain, S.: Managing uncertainty in moving objects databases. ACM TODS 29(3) (2004)
38. Yoon, H., Shahabi, C.: Robust time-referenced segmentation of moving object trajectories. In: ICDM (2008)

Performance Comparison of xBR-trees and R*-trees for Single Dataset Spatial Queries

George Roumelis[1], Michael Vassilakopoulos[2,1], and Antonio Corral[3]

[1] Master in Information Systems,
Open University of Cyprus, Cyprus
george.roumelis@st.ouc.ac.cy
[2] Dept. of Computer Science and Biomedical Informatics,
University of Central Greece, Greece
mvasilako@ucg.gr
[3] Dept. of Languages and Computing,
University of Almeria, Spain
acorral@ual.es

Abstract. Processing of spatial queries has been studied extensively in the literature. In most cases, it is accomplished by indexing spatial data by an access method. For queries involving a single dataset, like the Point Location Query, the Window (Distance Range) Query, the (Constrained) K Nearest Neighbor Query, the R*-tree (a data-driven structure) is a very popular choice of such a method. In this paper, we compare the performance of the R*-tree for processing single dataset spatial queries to the performance of a disk based structure that belongs to the Quadtree family, the xBR-tree (a space-driven structure). We demonstrate performance results (I/O efficiency and execution time) of extensive experimentation that was based on real datasets, using these two index structures. The winner depends on several parameters and the results show that the xBR-tree is a promising alternative for these spatial operations.

Keywords: Spatial Access Methods, R-trees, Quadtrees, Query Processing.

1 Introduction

Due to the demanding need for efficient spatial access methods in many spatial database applications, significant research effort has been devoted to the development of new spatial index structures [13,16]. However, as shown in several previous comparative studies [9,10,11,12], there is no single index structure that works efficiently, in all cases, across a variety of modern applications, where a variety of Spatial Queries arise, like Point Location, Window, Distance Range and Nearest Neighbor Queries (involving one spatial dataset), or Distance Join, Closest Pair and All-Nearest Neighbor Queries (involving two spatial datasets).

In this paper, we implement the External Balanced Regular (xBR) tree [17], a secondary memory structure that belongs to the Quadtree family (widely used

J. Eder, M. Bielikova, and A.M. Tjoa (Eds.): ADBIS 2011, LNCS 6909, pp. 228–242, 2011.

in graphics and GIS applications [16]), which is suitable for storing and indexing multidimensional points (and, in extended versions, line segments, or other
spatial objects). Moreover, we compare it with the popular R*-tree index, using
important criteria: storage requirements and time needed for the tree construction and spatial query operations performance. We have chosen the R*-tree,
which is the most commonly employed spatial indexing structure in the database
community [13].

The R*-tree is a member of the family of R-trees that are characterized as
data-driven access methods: they organize spatial objects into a hierarchy of Minimum Bounding Rectangles (MBRs), with shape, size and position that depends
to the data distribution in space. On the contrary, the xBR-tree belongs to the
family of Quadtrees that are characterized as *space-driven access methods*: they
organize spatial objects into a hierarchy of (hyper-)rectangles, formed by subdi-
viding the current region (originally, the whole space) into four sub-quadrants for
2d space, eight sub-octants for 3d space, etc., independently to the data distribution in this region. However, the number of levels of this hierarchical subdivision,
and thus the size of the rectangular areas, depend on the distribution of data.
The books [13,16] provide excellent information sources for the interested reader
about R-trees and Quadtrees, respectively.

The contributions of this paper are the conclusions arising from the (real data
based) experimental comparison of these two spatial access methods regarding
I/O performance and execution time for

- Tree building,
- Point Location Queries (*PLQs*),
- Window Queries (*WQs*) and Distance Range Queries (*DRQs*), also called
  Distance Similarity Queries,
- K-Nearest Neighbor Queries (*K-NNQs*) and Constrained K-Nearest Neigh-
  bor Queries (*CK-NNQs*), also called Distance-based Range Nearest-Neighbor
  Queries.

This paper is organized as follows. In Section 2 we review Related Work on
comparing spatial access methods, regarding query processing and provide the
motivation for this report. In Section 3 (4), we briefly review R*-trees (xBR-
trees) and the algorithms for processing single dataset Spatial Queries. In Section
5, we present representative results of the extensive experimentation that we
have performed, using real datasets, for comparing the performance of the two
structures. Finally, in Section 6 we provide the conclusions arising from our
work and discuss related future work directions.

2   Related Work and Motivation

Several previous research efforts have focused on efficient spatial query algorithms using the most cited spatial access methods (R-trees and Quadtrees).
In [9] a qualitative comparative study is performed taking into account three
popular spatial indexes (R*-tree, R+-tree and PMR Quadtree), in the context

of processing spatial queries (point query, nearest line segment, window query, etc.) in large line segment databases. The conclusion reached was that the R$^+$-tree and PMR Quadtree are the best when the operations involve search, since they result in a disjoint decomposition of space. On the other hand, R*-tree is more compact than R$^+$-tree (and PMR Quadtree) but its performance is not as good as the R$^+$-tree, due to the non-disjointness of the decomposition induced by it.

In [10], various R-tree variants (R-tree, R*-tree and R$^+$-tree) and the PMR Quadtree have been compared for the traditional spatial overlap join operation. They showed that the R$^+$-tree and PMR Quadtree outperform the R-tree and R*-tree using 2D GIS spatial data. That is, with respect to the overlap join, the spatial data structures based on a disjoint decomposition of space (as R$^+$-tree and PMR Quadtree) outperform spatial data structures based on a non-disjoint decomposition such as the numerous variants of the R-tree including the R*-tree. Moreover, as the size of the output of the spatial join increases with respect to the larger of the two inputs, methods based on a disjoint regular decomposition (PMR Quadtree) perform significantly better.

Moreover, in [12] the R-tree and the Quadtree have been compared, using a variety of range and *NN* queries on spatial data arising in 2D Geographical Information Systems (GISs). It was shown that, in general, the R-tree outperforms the Quadtree. From this experimental comparison, Oracle, in general, recommends using R-trees over Quadtrees, due to higher tiling levels in the Quadtree that cause very expensive preprocessing and storage costs.

The All-Nearest Neighbor (ANN) operation takes as input two datasets of multidimensional data points and computes for each point in the first dataset the *NN* in the second one. For this operation, in [2], a new distance metric between two MBRs was proposed, called NXNDIST (the minimum MinMaxDist), that reduces the MaxMaxDist by allowing the use of the minimal MaxDist for exactly one dimension. In general, NXNDIST is based on the observation that at each side of an MBR, there must be exactly one data point contained which realizes the minimum distance. It is a distance bound for MBRs that is guaranteed to contain at least one *NN* to any query point in the query MBR (i.e. it is an effective pruning distance for ANN). Moreover, the MBA algorithm that traverses the index in a depth-first fashion and expands the candidate search node bi-directionally was proposed. Finally, they showed that for *ANN* queries, using a Quadtree index enhanced with MBR keys for the internal nodes (MBR-Quadtrees) is a much more efficient indexing structure than the R*-tree index.

Recently, in [11] an experimental study comparing the R*-tree and the Quadtree using various criteria, including *K-NNQs* and K Distance Join Queries and index construction methods (dynamic insertion and bulkloading algorithm), is presented. It was shown that when data are static (when a bulkloading algorithm is used for an index construction) and *K-NNQs* / K Distance Join Queries are processed the R*-tree shows the best performance. However, when data are dynamic (i.e. there are frequent updates), the Quadtree begins to outperform the R*-tree. This is due to, once the dynamic R*-tree algorithm is used, overlap among MBRs increases with increasing dataset sizes, and the R*-tree performance degrades.

xBR-trees have been presented in [17] and results related to the analysis of their performance have been presented in [5]. Using xBR-trees for processing *PLQs*, *WQs*, or *DRQs* is rather straightforward, due to the organization of the xBR-tree. However, algorithms for processing *K-NNQs* and *CK-NNQs* by using these trees have only recently been developed [14] and tested with real datasets, with promising performance. The main objective of this paper is to compare the xBR-tree performance against the performance of the most popular spatial access method, the R*-tree, considering the most representative spatial queries where a single index is involved and to highlight the performance winner, considering the characteristics of each query.

3   R*-tree and Single Dataset Query Processing

3.1   R*-tree

R-trees [6] are hierarchical, height balanced data structures, designed for use in secondary storage, derived from B-trees [3]. They are used for the organization of a collection of arbitrary spatial objects by representing them as Minimum Bounding d-dimensional Rectangles (MBRs). The MBR represents the smallest aligned rectangle in which the spatial objects are contained. A 2d MBR is determined by two 2-dimensional points that belong to its faces, one that has the minimum and one that has the maximum coordinates (these are the endpoints of one of the diagonals of the MBR). Each R-tree node corresponds to the MBR that contains its children. The tree leaves contain pointers to the actual spatial objects in the database, instead of pointers to children nodes. The nodes are implemented as disk pages. For more details about the R-tree structure, see [13].

Many variations of R-trees have appeared in the literature (exhaustive surveys can be found in [4,13]). One of the most popular and efficient variations of the R-tree is the R*-tree [1], which uses more sophisticated node insertion and splitting algorithm. In general terms, the R*-tree added two major enhancements to the original R-tree, when a node overflow is caused. First, rather than just considering the area, the node-splitting algorithm in R*-trees also minimizes the perimeter and overlap enlargement of the MBRs. Second, an overflowed node is not split immediately, but a portion of entries of the node is reinserted from the top of the R*-tree (forced reinsertion).

3.2   *PLQs*, *WQs*, *DRQs*, *K-NNQs* and *CK-NNQs* on R*-trees

In general terms, the definitions of these spatial queries are as follows. Given an index I and a query point q, the *PLQ* returns true if q belongs to I, false otherwise. Given an index I and a query rectangle r, the result of the WQ is the set of all points in I that are completely inside r. Given an index I, a query point q and a distance threshold $\delta \geq 0$, the *DRQ* returns all points of I, that are within the specified distance δ from q (according to a distance function). Given an index I, a query point q, and a value $K > 0$, the *K-NNQ* returns K points of I which are closest to q based on a distance function. Finally, given an index I, a query point q, a value $K > 0$ and a distance threshold $\delta \geq 0$, the *CK-NNQ* returns k closest points of I which are within the distance δ from q.

PLQs and *WQs* can be processed in a top-down manner on the R*-tree. The query point (or window) is tested first against each entry (MBR, Addr) in the root. If the query point is inside (or query window overlaps with) the MBR, then the search algorithm is applied recursively on the R*-tree node pointed by the *Addr*. This process stops after reaching leaf nodes of the R*-tree. The selected entries in leaves are used to retrieve the spatial objects associated with the *Oids*.

Based on the branch-and-bound paradigm, the distance-based query algorithms use several metrics to prune the search space [15]. The most important metric is *mindist*(q, M), which reports the minimum distance between q and any point in a MBR M. Another metric, *minmaxdist*(q, M), refers to the minimum distance from q within which a point in M is guaranteed to be found. Finally, *maxdist*(q, M) is the maximum distance between q and any point in M.

The first Nearest Neighbor Query (*NNQ*) algorithm for R-trees, proposed in [15], traverses recursively the tree in a Depth-First (DF) manner. Starting from the root, all entries are sorted according to their *mindist* from q, and the entry with the smallest *mindist* is visited first. The process is repeated recursively until the leaf level is reached, where a potential *NN* is found. During backtracking to the upper levels, the algorithm only visits entries whose *mindist* is smaller than or equal to the distance of the *NN* found so far. The generalization to find the K Nearest Neighbor (*K-NN*) is straightforward. We just need an additional data structure, a (based on the distance from the query point q) maximum binary heap, holding the K nearest points encountered so far.

A Best-First (BF) algorithm for *NNQ* was proposed in [7] for Quadtrees and in [8] for R-trees. BF keeps a minimum binary heap with the entries of the nodes visited so far. Initially the heap contains the entries of the root sorted according to their *mindist*. When the root of the heap is chosen for processing, it is removed from the heap and the entries of the R*-tree node pointed by *Addr* are added together with their *mindist*. The algorithm continues visiting the entry with the minimum *mindist* in the heap, until it becomes empty or the *mindist* value of the node entry located in the root of heap is larger than the distance value of the *NN* that has been found so far (i.e. the pruning distance). BF is I/O optimal because it only visits the nodes necessary for obtaining the *NN*.

For the *DRQ* ($\delta \geq 0$), we just need to extent the DF or BF algorithms for *NNQ* in a simple way. Starting from the root node, several tree nodes are traversed down to the leaves, depending on the result of whether *mindist* is less than or equal to δ. When the query algorithm reaches the leaf nodes, all the data points which distance with respect to the query point q smaller than or equal to δ are added to the answer set.

Finally, the *CK-NNQ* is a combination of *K-NNQ* and *DRQ*, where we can also extent the DF or BF algorithms for NNQ. Starting from the root node, several tree nodes are traversed down to the leaves, depending on the result of whether *mindist* is less than or equal to $\min\{\delta, z\}$, where z is the *NN* that has been found so far. When the query algorithm reaches the leaf nodes, all the data points (a maximum of k) with distance with respect to q smaller than or equal to δ are added to the query result.

BF algorithms require considerable amounts of main memory. In this paper, we consider that RAM is limited (due to several applications running in the hosting server; a rather common situation), therefore, we use the DF R*-tree algorithm against an analogous algorithm for the xBR-tree (described in the next section). For the same reason, in our experimentation we do not use buffering for storing tree nodes.

4 XBR-tree and Single Dataset Query Processing

Although xBR-trees [15] can be defined for various dimensions, for the ease of exposition in the rest of the paper, we assume 2 dimensions. For 2 dimensions the hierarchical decomposition of space is that of Quadtrees (the space is subdivided in 4 equal subquadrants, any of which may be further subdivided recursively in 4 subquadrants).

The space indexed by an xBR-tree is a square, expressed in a coordinate system of real numbers (not in a digitized space). The nodes of xBR-trees are disk pages and are distinguished in two kinds: leaves, which store the actual multidimensional data themselves and internal nodes, which provide a multiway indexing mechanism for these data.

4.1 Internal Nodes

As described in [17], internal nodes contain pairs of the form (address, pointer). During implementation and experimentation we concluded that more fields are needed for each entry: (shape, address, REG, pointer). An *address* is used to determine the region of a child node and is accompanied by the *pointer* to this child. Since *addresses* are of variable size, the number of entries fitting in each node is not predefined. Apparently, the space occupied by all entries within a node must not exceed the size of this node. The maximum size of an *address* is only limited by the node size and in practice it never reaches this limit. *Shape* is a flag that determines if the region of the child is a complete or non-complete square (the area remaining, after one or more splits; explained later in this subsection). This field will be used widely in queries. Finally, *REG* stores the coordinates of the region referenced by *address*. We measured the execution time for queries and we found that it is more expensive if we do not save this field, but calculate its value every time we need it.

Each *address* represents a subquadrant which has been produced by Quad-tree-like hierarchical subdivision of the current space. It consists of a number of directional digits that make up this subdivision. The NW, NE, SW and SE subquadrants of a quadrant are distinguished by the directional digits 0, 1, 2

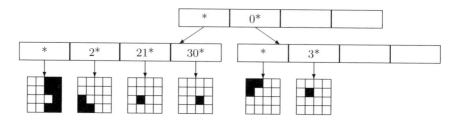

Fig. 1. An xBR-tree with two levels of internal nodes

and 3, respectively. For example, the *address* 1 represents the NE quadrant of the current space, while the *address* 10 the NW subquadrant of the NE quadrant of the current space.

However, the region of a child is, in general, the subquadrant of the related *address* minus a number of smaller subquadrants. The region of this child is the subquadrant determined by the *address* in its entry, minus the subquadrants corresponding to the next entries of the internal node (the entries in an internal node are saved sequentially, in preorder traversal of the Quadtree that corresponds to the internal node). For example, in Figure 1 an internal node (a root) that points to 2 internal nodes that point to 6 leaves is depicted. The region of the root is the original space, which is assumed to have a quadrangular shape. The region of the right (left) child is the NW quadrant of the original space (the whole space minus the region of the NW quadrant - a non complete square), depicted by the union of the black regions of the leaves of this child. The * symbol is used to denote the end of a variable size address. The *address* of the right child is 0*, since the region of this child is the NW quadrant of the original space. The *address* of the left child is * (has zero directional digits), since the region of the left child is the whole space minus the region of the right child.

Each of these *addresses* is expressed relatively to the minimal quadrant that covers the internal node (each *address* determines a subquadrant of this minimal quadrant). For example, in Figure 1, the *address* 2* is the SW subquadrant of the whole space (the minimal quadrant that covers the left right child of the root). During a search, or an insertion of a data element with specified coordinates, the appropriate leaf and its region is determined by descending the tree from the root. More details are given in the description of *PLQs*, in the following.

4.2   Leaf Nodes

External nodes (leaves) simply contain the data elements and have a predetermined capacity C. When C is exceeded, due to an insertion, the leaf is partitioned according to hierarchical (Quadtree like) decomposition, until each of the resulting two regions contains N data elements, where $N \leq xC, 0.5 < x < 1$. The choice of x affects the number of necessary subdivisions of an overflowed node and the size of addresses that result from a node split. A value closer to 0.5, in general, results in more subdivisions and larger addresses, since it is more difficult to partition the region of the leaf in subregions with almost equal numbers of elements. Of course, such a choice provides a better guarantee for the space occupancy of leaves. We used $x = 0.75$, which leads to a good compromise between size of addresses and leaf occupancy. Splitting of a leaf creates a new address that must be hosted by an internal node of the parent level. This can cause backtracking to the upper levels of the tree and may even cause an increase of its height.

4.3   Splitting of Internal Nodes

When an internal node overflows, it is split in two. The goal of this split is to achieve the best possible balance between the space use in the two nodes. The split is either based on existing quadrants or in ancestors of existing quadrants. First, a Quadtree is built that has as nodes the quadrants specified in the xBR-tree internal node. We use this tree for determining the best possible split of the internal node in two nodes (for details, see [15]). Although, in [15] we seek for a split in two nodes that have almost equal number of bits for storing addresses, by experimentation, we found that seeking for a split in two nodes that have almost equal number of addresses is equally effective and simpler.

4.4   *PLQs, WQs, DRQs, K-NNQs* and *CK-NNQs* on xBR-trees

PLQs can be processed in a top-down manner on the xBR, like the R*-tree. During a *PLQ* for a point with specified coordinates, the appropriate leaf and its region is determined by descending the tree from the root. Initially, the region under consideration is the whole space (the region of the root). As noted in subsection 4.1, the entries in an internal node are saved in preorder traversal of the Quadtree that corresponds to the internal node and are examined in reverse sequential order. So first we examine the last node of the Quadtree. If its subquadrant (specified by the address field of the entry) does not contain the query point, we continue with the next entry in reverse sequential order. The first subquadrant that hosts the query point determines the smallest region that hosts this point. Then we follow the pointer field to the related child at the next lower level, until we reach the leaf level. This way, we reach the unique leaf that may contain the query point. Unlike the R*-tree, in the xBR-tree a single path to the point we seek is followed.

Processing of *WQs* follows the same strategy to *PLQs*, regarding the way we examine regions/entries of an internal node. The decision about whether we are at a entry with a region likely to contain points inside the query window is the answer to the question: do the subquadrant of the current entry (specified by the address field of the entry) and the query window intersect? If yes, then we follow the pointer to the related child at the next lower level. We repeat until we have examined all entries of the internal node, or until the query window is completely inscribed inside the region of the entry that we examine (because none of the other, not examined, regions of the tree overlaps with this region).

DRQ follows the same strategy as *WQ*. At first, the querying circle is replaced from its MBR (the calculations are faster in this way) and if the answer about the intersection of the subquadrant of the current entry and the query MBR is positive, then we follow the pointer to the related child at the next lower level. If we reach a leaf with a region that intersects the query MBR, we select the leaf points that are inside the query circle.

For *K-NNQs* the search algorithm traverses recursively the tree in a DF manner, like proposed in [15]. Starting from the root (and for every node that is current) all node entries are sorted according to their *mindist* from the query point, and the entry with the smallest *mindist* is deleted from the list and visited first. The process is repeated recursively until the leaf level is reached, where a potential *NN* is found. Because of the xBR-tree structure, it is possible to reach a leaf (to which the query point may even belong), but the next *NN* may exist in a neighboring region (unlike the R*-tree, the regions to which the xBR-tree partitions space are not necessarily defined by points that fall on their boundaries). So we use a global (based on the distance to the query point) *max K-heap* and insert in it every point of this leaf that is nearest to the query point than the root of the *heap* (if the *heap* is full, storing K elements, the new entry replaces the root). When the *heap* is full and the next entry in *mindist* order of the current node is at a longer distance from the query point to the distance of the root of the *heap*, the search is stopped. Similarly, when the *heap* is full, this entry is a complete square, the query point falls within this square and the distance of the root of the *heap* is smaller than the minimum distance of the query point to the edges of the square, the search is stopped. More details about this algorithm appear in [14].

5   Experimentation

We designed and run a large set of experiments to discover advantages and disadvantages of xBR and R*-trees. We used 5 real datasets of different sizes. We used a real 2d dataset of California (CSN) that contains 98022 MBRs of streams (line-segments). We also used data from real spatial datasets of North America. Two datasets represent populated places (NApp) and cultural landmarks (NAcl) consisting of 24493 and 9203 points, respectively; railroads (NArr) of 191637 line-segments and, finally, roads (NArd) consisting of 569120 line-segments. To create 2d point datasets from non-point datasets, we used the centroids of the line-segment MBRs. The experiments were run on a Linux machine, with Intel core duo 2x2GHz processor and 3 GB of RAM. We run experiments for *PLQs*, *WQs*, *K-NNQs* and *CK-NNQs*, counting disk-page accesses (I/O) and total execution time for each index structure.

At first we built the xBR and R*-trees. We stored point coordinates as float numbers and constructed each tree for the following node (page) sizes: 512b, 1K, 2K, 4K and 8K. Results for the construction characteristics (Table 1) indicate that, in all cases, the xBR-tree uses less space (i.e. it is more compact) and time than the R*-tree (the R*-tree creation is slower, partially, due to the use of forced reinsertion that improves searching efficiency) and the difference in creation time is enlarged as the size of node increases. We show only one node size for each dataset, due to the limited space (other results were analogous). It must be noted that the R*-tree keeps points at leaf nodes by storing their MBR (four coordinates), while the xBR-tree keeps only two coordinates for each point.

Table 1. Tree construction characteristics

Dataset	Node size	Tree height		Tree size (bytes)		Creation time (secs)	
		R*	xBR	R*	xBR	R*	xBR
NAcl	512b	4	3	403284	130056	0.37	0.1
NApp	1K	3	2	1051846	334856	2.27	0.35
CSN	2K	3	2	3973953	1286152	45.02	2.04
NArr	4K	3	2	8032409	2576392	156.15	6.34
NArd	8K	3	2	24369063	7954440	1916.1	32.96

For the *PLQ*, for each dataset, we executed as many queries as the points in the dataset by searching for each of these points (we used the original datasets as query input). The results showed that the xBR-tree needs less disk read accesses and executes every query faster than the R*-tree. Results for the smallest dataset (NAcl) are shown in Figure 2. The results for the other datasets were analogous, always in favor of the xBR-tree.

We noticed that the xBR-tree needs a number of disk accesses equal to its tree height, while the R*-tree needs at least this number of access and, in most cases, even more. This finding can easily be explained from the analysis of the algorithms discussed above. When we use as input data points that did not exist in the database, the xBR-tree needs the same number of accesses while the R*-tree needs less. This is due to the structural difference of the two trees. Internal nodes of R*-trees contain information only about MBRs that include the data points. In this way, if the dataset has empty regions, the R*-tree does not build MBRs for these. However, the xBR-tree saves in internal nodes information about the regions in which the space is split, regardless if they contain the reference data point. It is important to mention that page access number becomes lower as the size of node increases, while execution time increases (for both trees).

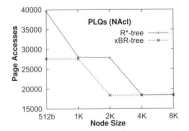
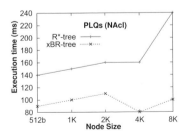

Fig. 2. Total disk accesses (left) and execution time (right) vs. node size for *PLQs*

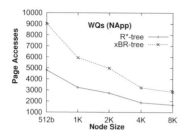 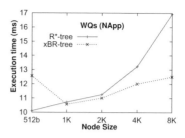

Fig. 3. Total disk accesses (left) and execution time (right) vs. node size for *WQs*

This may be surprising at first, but can be explained considering the fact that when the size of nodes increases, so does the time for main memory calculations and, consequently, the execution time. For large node sizes, the tree height, however, may decrease disproportionally.

For each of the other query types, we executed a large number of queries, as follows. For each dataset, we created rectangular query windows (and their inscribed circles) for studying *WQs* (*DRQs*) by splitting the whole space into $2^4, 2^6, \ldots, 2^{16}$ windows, in a row-order mapping manner (this sums up to executing the *WQ* and the *DRQ* 87376 times, for each node size, dataset and tree). The centroids of these windows were also used as query inputs for all the other queries (*DRQs*, *K-NNQs* and *CK-NNQs*). Especially, for *K-NNQs* and *CK-NNQs* we used the following set of K values: 1, 10, 30, 70, 100 (this sums up to executing the *K-NNQ* and the *CK-NNQ* 436880 times, for each node size, dataset and tree). Since the number of experiments performed was vast, we show only representative results, since results were analogous for each query category.

In Figure 3, for the WQ we depict the results for the second dataset (NApp), as one representative example. It is shown that the xBR-tree needs more accesses to find the population within the 1024 windows with which we scanned the whole space occupied from the 24493 data points of this dataset. As the size of node increases the I/O difference between the two trees becomes smaller. In both trees, a linear dependence of the number of accesses to the size of the node appears. This is due to reduction of tree height as the size of node increases. Note the reduction of the difference from the first size (512b) to the biggest size (8K).

In Figure 4, for the WQ we depict for the same dataset, but only for those query windows that were inhabited by points (non-empty windows). The significant improvement of the xBR-tree performance is obvious. It now becomes clearly faster (execution time) for all sizes of nodes. The explanation for this is again related to empty regions. The I/O efficiency of the two trees is closer now. R*-tree execution time is also improved, for non-empty windows, but the improvement of xBR-tree is larger. This leads us to the conclusion that main memory processing is simpler (and thus faster) for the xBR-tree.

For *DRQs* (1024 query circles, inscribed into the respective rectangular windows, δ value equal to $1/(2\sqrt{1024}) = 1/64$ of the space side length), the xBR-tree needs less disk accesses and is faster than the R*-tree, in all cases and for all datasets. In Figure 5, we show these results for the CSN dataset. The results were even better when the *DRQs* addressed only non-empty regions.

For the *NNQ*, the xBR-tree shows similar behavior to the *WQ*. The xBR-tree needs much more disk accesses for finding the NNs than the R*-tree. But the difference became smaller when the size of node (for the same dataset) increased. Regarding the execution time, the xBR-tree shows improved performance, in relation to its I/O difference from the R*-tree. In Figures 6 and 7, we show results for K=10 and the NArr dataset and for K=100 and the very large dataset (NArd), respectively. At this point, note the worse time performance of both trees, for larger node sizes (where the I/O cost is smaller). This is due to the fact that as the node size increases, the trees become very wide and very short. In this case, a node holds many elements to be processed and branching during tree descend plays a smaller role in restricting the search space. This leads us to the conclusion that the increase of the node size leads to many more calculations in main memory, which cancels the benefit of reducing I/O. For K=100 the I/O lines for the two trees are parallel and the relative performance difference is smaller. Diagrams not included in this report due to space limitations show that, for the 191637 points dataset and K>40, the xBR-tree is faster.

Finally, for the *CK-NNQs* we noticed that the xBR-tree is improved for both performance categories of our study. In Figure 8, we present the results of *CK-NNQs* for the large dataset (NArr) for all (1024 query points), setting K = 40 and δ value equal to $1/(2\sqrt{1024}) = 1/64$ of the space side length). Diagrams not included in this report, due to space limitations, show that, for non-empty regions, for the 98022 points dataset, the xBR-tree is faster, for K>10. In general, depending on the dataset, for non-empty regions the *CK-NNQ* time performance of the xBR-tree is almost the same to, or much better than the R*-tree.

In summary, the experimental comparison showed that

- The xBR-tree needs much less space and is built in much less time.
- The xBR-tree performance is higher for *PLQs* and *DRQs*.
- The R*-tree performance is higher for *WQs*, but when considering non-empty query windows only, the xBR-tree time-performance is higher.
- The R*-tree performance is higher for *K-NNQs* (the R*-tree excels to the xBR-tree more in I/O than in execution time).
- The R*-tree performance is higher for *CK-NNQs*, but when considering non-empty query results, the xBR-tree time-performance is in most cases higher.

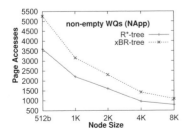 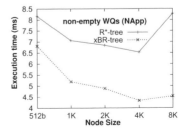

Fig. 4. Total disk accesses (left) and execution time (right) vs. node size for *WQs*

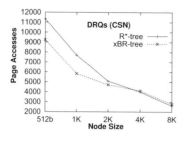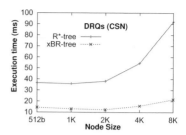

Fig. 5. Total disk accesses (left) and execution time (right) vs. node size *DRQs*

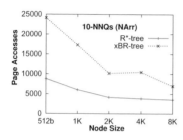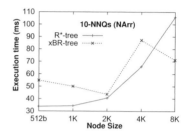

Fig. 6. Total disk accesses (left) and execution time (right) vs. node size for *K-NNQs*

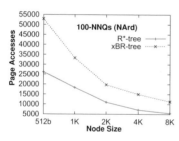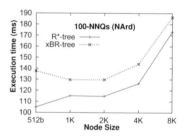

Fig. 7. Total disk accesses (left) and execution time (right) vs. node size for *K-NNQs*

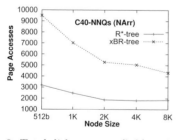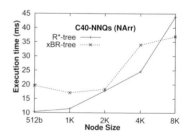

Fig. 8. Total disk accesses (left) and Execution time (right) vs. node size *CK-NNQs*

- The fact that xBR-trees do not model empty regions affects their performance for queries that do not return result points.
- Main memory processing of xBR-tree is simpler and faster.

6   Conclusions and Future Work

We performed an extensive (real data based) experimental comparison[1] of the xBR-tree I/O and execution time performance against the performance of the most popular spatial access method, the R*-tree, considering the most representative spatial queries where a single index is involved. The conclusions arising from this comparison show that the two structures are competitive. The xBR-tree is smaller (i.e. it is more compact) and is built faster than the R*-tree. The performance of the xBR-tree is higher for *PLQs* and *DRQs* and for *WQs* when the query window is non-empty, while the R*-tree is better for *K-NNQs* and needs less disk access for *CK-NNQs*. The execution time winner for *CK-NNQs* depends on whether the query returns result points (xBR-tree), or not (R*-tree).

Future work might include extending the xBR-tree for modelling empty regions too and studying the relative performance of the two trees for two dataset (join) queries. Moreover, studying the relative performance of the structures in the presence of buffering, or using memory consuming BF algorithms is another worthy target.

References

1. Beckmann, N., Kriegel, H.P., Schneider, R., Seeger, B.: The R*-tree: an Efficient and Robust Access Method for Points and Rectangles. In: SIGMOD Conference, pp. 322-331 (1990)
2. Chen, Y., Patel, J.M.: Efficient Evaluation of All-Nearest-Neighbor Queries. In: ICDE Conference, pp. 1056-1065 (2007)
3. Comer, D.: The Ubiquitous B-tree. ACM Computing Surveys 11(2), 121–137 (1979)
4. Gaede, V., Gunther, O.: Multidimensional Access Methods. ACM Computing Surveys 30(2), 170–231 (1998)
5. Gorawski, M., Bugdol, M.: New Trends in Data Warehousing and Data Analysis. In: Kozielski, S., Wrembel, R. (eds.) Cost Model for XBR-tree. Springer, Heidelberg (2009)
6. Guttman: R-trees: A Dynamic Index Structure for Spatial Searching. In: SIGMOD Conference, pp. 47-57 (1984)
7. Hjaltason, G.R., Samet, H.: Ranking in Spatial Databases. In: Egenhofer, M.J., Herring, J.R. (eds.) SSD 1995. LNCS, vol. 951, pp. 83–95. Springer, Heidelberg (1995)
8. Hjaltason, G.R., Samet, H.: Distance Browsing in Spatial Databases. ACM Transactions on Database Systems 24(2), 265–318 (1999)
9. Hoel, E.G., Samet, H.: A Qualitative Comparison Study of Data Structures for Large Line Segment Databases. In: SIGMOD Conference, pp. 205-214 (1992)
10. Hoel, E.G., Samet, H.: Benchmarking Spatial Join Operations with Spatial Output. In: VLDB Conference, pp. 606-618 (1995)
11. Kim, Y.J., Patel, J.: Performance Comparison of the R*-tree and the Quadtree for kNN and Distance Join Queries. IEEE Transactions on Knowledge and Data Engineering 22(7), 1014–1027 (2010)

---

[1] Due to space limitations, in this paper we present a small part of the experimental results. An extended set of experimental results is accessible from: http://delab.csd.auth.gr/~michalis/xBRsys/results.

12. Kothuri, R.K., Ravada, S., Abugov, D.: Quadtree and R-tree Indexes in Oracle Spatial: A Comparison using GIS Data. In: SIGMOD Conference, pp. 546–557 (2002)
13. Manolopoulos, Y., Nanopoulos, A., Papadopoulos, A., Theodoridis, Y.: R-Trees: Theory and Applications. Springer, Heidelberg (2006)
14. Roumelis, G., Vassilakopoulos, M., Corral, A.: Algorithms for processing Nearest Neighbor Queries using xBR-trees. In: 15th Panhellenic Conference on Informatics (PCI 2011) (to appear 2011)
15. Roussopoulos, N., Kelley, S., Vincent, F.: Nearest Neighbor Queries. In: SIGMOD Conference, pp.71-79 (1995)
16. Samet, H.: Applications of Spatial Data Structures: Computer Graphics, Image Processing, and GIS. Addison-Wesley, Reading (1990)
17. Vassilakopoulos, M., Manolopoulos, Y.: External Balanced Regular (x-BR) Trees: New Structures for Very Large Spatial Databases. In: Advances in Informatics: Proc. 7th Hellenic Conf. on Informatics (HCI 1999), pp. 324–333. World Scientific Publ. Co., Singapore (2000)

Efficient Detection of Minimal Failing Subqueries in a Fuzzy Querying Context

Olivier Pivert[1], Grégory Smits[2], Allel Hadjali[1], and Hélène Jaudoin[1]

[1] Irisa – Enssat, University of Rennes 1
Technopole Anticipa 22305 Lannion Cedex France
[2] Irisa – IUT Lannion, University of Rennes 1
Rue E. Branly 22300 Lannion Cedex France
{pivert,hadjali,jaudoin}@enssat.fr,
gregory.smits@univ-rennes1.fr

Abstract. This paper deals with conjunctive fuzzy queries that yield an empty or unsatisfactory answer set. We propose a cooperative answering approach which efficiently retrieves the minimal failing subqueries of the initial query (which can then be used to explain the failure). The detection of the minimal failing subqueries relies on a prior step of fuzzy cardinalities computation. The main advantage of this strategy is to imply a single scan of the database. Moreover, the storage of such knowledge about the data distributions easily fits in memory.

1 Introduction

The idea of introducing preferences into queries is gaining more and more attention in the database community. In this paper, we focus on the fuzzy-set-based approach to preference queries, which is founded on the use of fuzzy set membership functions that describe the preference profiles of the user on each attribute domain involved in the query.

With respect to Boolean queries, fuzzy queries reduce the risk of obtaining an empty set of answers since the use of a finer discrimination scale — $[0, 1]$ instead of $\{0, 1\}$ — increases the chance for an element to be considered somewhat satisfactory. Nevertheless, the situation may occur where none of the elements of the database satisfies the query even to a low degree.

In the context of fuzzy queries, beside the empty answer set (EAS) problem, another situation deserves attention: that where the answer set is not empty but only contains elements which satisfy to a *low degree* the preferences specified in the user query. We will show in this paper that a generic — and very efficient — type of approach, based on the use of fuzzy cardinalities, may be employed to provide explanations for both types of situations (empty or unsatisfactory answer set, the latter being denoted by UAS in the following). Minimal failing subqueries [1] constitute useful explanations about the conflicts in a failing query. These explanations may i) help the user revise or reformulate his/her initial query or ii) be used to set up an automatic and targeted relaxation strategy.

J. Eder, M. Bielikova, and A.M. Tjoa (Eds.): ADBIS 2011, LNCS 6909, pp. 243–256, 2011.
© Springer-Verlag Berlin Heidelberg 2011

The remainder of the paper is structured as follows. Section 2 consists of a reminder about fuzzy sets and fuzzy queries. In Section 3, we deal with the issue of explaining the causes of the failure (or near failure) of a query. More precisely, we propose an algorithm for efficiently computing the minimal failing subqueries associated with a failing (or almost failing) conjunctive fuzzy query. Section 4 presents an experimentation which illustrates the efficiency of the approach and the relevance of the explanations. Section 5 drafts the main lines of a repair strategy, whereas Section 6 discusses related work. Finally, Section 7 recalls the main contributions and outlines perspectives for future work.

2   Reminder about Fuzzy Sets and Fuzzy Queries

2.1   Basic Notions about Fuzzy Sets

Fuzzy set theory was introduced by Zadeh [2] for modeling classes or sets whose boundaries are not clear-cut. For such objects, the transition between full membership and full mismatch is gradual rather than crisp. Typical examples of such fuzzy classes are those described using adjectives of the natural language, such as *young, cheap, fast*, etc. Formally, a fuzzy set F on a referential U is characterized by a membership function $\mu_F : U \to [0,1]$ where $\mu_F(u)$ denotes the grade of membership of u in F. In particular, $\mu_F(u) = 1$ reflects full membership of u in F, while $\mu_F(u) = 0$ expresses absolute non-membership. When $0 < \mu_F(u) < 1$, one speaks of partial membership. Two crisp sets are of particular interest when defining a fuzzy set F:

- the core $C(F) = \{u \in U \mid \mu_F(u) = 1\}$, which gathers the *prototypes* of F,
- the support $S(F) = \{u \in U \mid \mu_F(u) > 0\}$.

In practice, the membership function associated with F is often of a trapezoidal shape. Then, F is expressed by the quadruplet (A, B, a, b) where $C(F) = [A, B]$ and $S(F) = [A - a, B + b]$, see Figure 1.

The α-cut of a fuzzy set F, denoted by F^α is an ordinary set of elements whose satisfaction degree is at least equal to α: $F^\alpha = \{u \in U \mid \mu_F(u) \geq \alpha\}$. Thus, $C(F)$ and $S(F)$ are two particular α-cuts of F where α is respectively equal to 1 and 0^+.

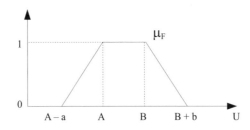

Fig. 1. Trapezoidal membership function

Let F and G be two fuzzy sets on the universe U, we say that $F \subseteq G$ iff $\mu_F(u) \leq \mu_G(u)$, $\forall u \in U$. The complement of F, denoted by F^c, is defined by $\mu_{F^c}(u) = 1 - \mu_F(u)$. Furthermore, $F \cap G$ (resp. $F \cup G$) is defined the following way: $\mu_{F \cap G} = min(\mu_F(u), \mu_G(u))$ (resp. $\mu_{F \cup G} = max(\mu_F(u), \mu_G(u))$).

As usual, the logical counterparts of the theoretical set operators $\cap$, $\cup$ and complementation operator correspond respectively to the conjunction $\wedge$, disjunction $\vee$ and negation $\neg$. See [3] for more details.

2.2   About SQLf

The language called SQLf described in [4] extends SQL so as to support fuzzy queries. The general principle consists in introducing gradual predicates wherever it makes sense. The three clauses *select*, *from* and *where* of the base block of SQL are kept in SQLf and the "from" clause remains unchanged. The principal differences affect mainly two aspects: i) the calibration of the result since it is made with discriminated elements, which can be achieved through a number of desired answers (k), a minimal level of satisfaction (α), or both, and ii) the nature of the authorized conditions as mentioned previously.

Therefore, the base block is expressed as:

select [**distinct**] $[k \mid \alpha \mid k, \alpha]$ attributes **from** relations **where** fuzzy-cond

where "fuzzy-cond" may involve both Boolean and fuzzy predicates. This expression is interpreted as:

- the fuzzy selection of the Cartesian product of the relations appearing in the "from" clause,
- a projection over the attributes of the "select" clause (duplicates are kept by default, and if "distinct" is specified the maximal degree is attached to the representative in the result),
- the calibration of the result (top k elements and/or those whose score is over the threshold α).

The operations from the relational algebra — on which SQLf is based — are extended to fuzzy relations by considering fuzzy relations as fuzzy sets on the one hand and by introducing gradual predicates in the appropriate operations (selections and joins especially) on the other hand. The definitions of these extended relational operators can be found in [5]. As an illustration, we give the definitions of the fuzzy selection and projection operators hereafter, where r denotes a fuzzy relation defined on the set of domains X.

- $\mu_{select(r, cond)}(t) = \top(\mu_r(t), \mu_{cond}(t))$ where *cond* is a fuzzy predicate and $\top$ is a triangular norm (most usually, *min* is used),
- $\mu_{project(r, Y)}(u) = max_{t \in r \mid t[Y]=u} \mu_r(t)$ where Y is a subset of X and u one of its values,

3   Explaining a Failure

3.1   About Minimal Failing and Unsatisfactory Subqueries

An empty set of answers associated with a fuzzy query $Q = P_1 \wedge P_2 \wedge \ldots \wedge P_n$ is necessarily due to an empty support (w.r.t. the current state of the database) for one at least of the *subqueries* of Q. The notion of an unsatisfactory set of answers generalizes this problem by considering an empty α-cut of Q where α is a user-defined qualitative threshold. As explained in Section 2, the support and the core of a fuzzy set are particular cases of α-cuts where α is respectively equal to 0^+ and 1. In the rest of the paper we only use the notion of an empty α-cut to refer to failing queries as well as unsatisfactory ones.

Thus, an extreme case of a failing query corresponds to an empty 1-cut for Q only. The opposite extreme is when one or several predicates P_i have an empty 0^+-cut. Between these two situations, it is of interest to detect the subqueries composed of more than one predicate and less than n predicates, which have an empty 0^+-cut. From an empty to an unsatisfactory set of answers, the problem defined above just has to be slightly revisited, where the condition of an empty 0^+-cut is transposed to α-cuts, where α is taken from an *a priori* defined scale of membership degrees $\mathcal{S} : 1 = \alpha_1 > \alpha_2 > \ldots > \alpha_f = 0^+$.

Definition 1. *Let Q be a query s.t. $Q = P_1 \wedge P_2 \wedge \ldots \wedge P_n$, let S and S' be two subsets of predicates s.t. $S' \subset S \subseteq \{P_1, P_2, \ldots, P_n\}$. A conjunction of the elements of S is a **subquery** of Q. A conjunction of the elements of S' is a **strict subquery** of Q.*

If one wants to explain why the result of the initial query is empty (resp. unsatisfactory), and/or weaken the query by identifying the subqueries whose α-cut is empty, one must naturally require that such subqueries be minimal: a subquery Q' of a query Q constitutes a minimal explanation if the considered α-cut is empty and if no (strict) subquery of Q' has an empty α-cut. This corresponds to a generalization of the concept of a Minimal Failing Subquery (MFS) [6].

Let us denote by Σ_Q^α the set of answers to the α-cut of a query Q against a given database D: $\Sigma_Q^\alpha = \{t \in D \mid \mu_Q(t) \geq \alpha\}$ (Σ_Q^α gathers those elements from D which satisfy Q with a degree of at least α).

Definition 2. *A **Minimal Failing Subquery** of a query $Q = P_1 \wedge P_2 \wedge \ldots \wedge P_n$ for a given α is any subquery Q' of Q such that $\Sigma_{Q'}^\alpha = \emptyset$ and for all strict subquery Q'' of Q', $\Sigma_{Q''}^\alpha \neq \emptyset$.*

When faced with an empty set of answers for a user-defined threshold α, the explanation process that we propose in this paper generates layered MFSs for different satisfaction degrees $\alpha_i, \alpha_i \in [\alpha, 1]$. This interval of satisfaction degrees is discretized using a scale $\mathcal{S} : 1 = \alpha_1 > \alpha_2 > \ldots > \alpha_f = 0^+$ of membership degrees. We will see in Section 3.2 that this discretization greatly facilitates the detection of gradual MFSs.

Obviously, due to the monotony of inclusion of α-cuts, one has $\Sigma_Q^{\alpha_i} \subseteq \Sigma_Q^{\alpha_j}$ if $\alpha_i \geq \alpha_j$. Therefore, a query Q that fails for a given α_j also fails for higher satisfaction degrees $\alpha_i > \alpha_j$. However, this property is not satisfied by minimal failing subqueries. Indeed, a subquery Q' can be an MFS of Q for a given α_j without being minimal for higher satisfaction degrees $\alpha_i > \alpha_j$ as a strict subquery of Q', say Q'', may fail for α_i and not for α_j.

During the layered MFS detection step (Section 3.2), when a subquery Q' of an initial failing or unsatisfactory query Q is detected for a degree α_j, one has to check for each higher level $\alpha_i > \alpha_j$ if Q' is also minimal according to previously identified MFSs before considering Q' as an MFS for level α_i.

3.2   Cardinality-Based MFS Detection

Our first contribution concerns the detection of layered MFSs of a conjunctive fuzzy query. We propose an efficient algorithm, which involves a single scan of the relation (or the join of relations) concerned during which fuzzy cardinalities are computed for each possible combination of the predicates specified in the initial query. A fuzzy cardinality for a given query Q and a given scale of membership degrees $\mathcal{S} : 1 = \alpha_1 > \alpha_2 > ... > \alpha_f = 0^+$ is represented as follows: $F_Q = \alpha_1/c_1 + \alpha_2/c_2 + ... + \alpha_f/c_f$ and expresses that c_1 tuples fully satisfy Q, c_2 tuples satisfy Q to a degree of at least α_2 and so on. For the examples and the experimentation we use the following scale $\mathcal{S} : 1 = \alpha_1 > \alpha_2 = 0.8 > \alpha_3 = 0.6 > \alpha_4 = 0.4 > \alpha_5 = 0.2 > \alpha_6 = 0^+$.

Let us consider a query $Q = P_1 \wedge ... \wedge P_n$. The preprocessing step — aimed at computing the fuzzy cardinalities associated with the results of all the possible subqueries — is as follows. One accesses each tuple t from the relation r (or the join of the relations) concerned and computes the satisfaction degrees of t for the predicates of the query. Then, one explores the possible conjunctions of those predicates and stops the development of a conjunction as soon as it returns an empty set of answer. During this process, one maintains a fuzzy cardinality for each combination in order to know how many tuples satisfy this combination for the different α_i's from $\mathcal{S}$. One just needs to maintain in the worst case $2^n - 1$ variables containing these diverse fuzzy cardinalities (which is not a problem in practice since in general $n \leq 10$). In the following, it is assumed that a single relation is concerned. Let us denote by V the vector of degrees $\langle \mu_{P_1}(t), \ldots, \mu_{P_n}(t) \rangle$ associated with the current tuple t. The algorithm that computes the set of fuzzy cardinalities is as follows:

for every tuple t of r **do**
    compute $\mu_{P_1}(t)$ and ... and $\mu_{P_n}(t)$;
    $V \leftarrow < \mu_{P_1}(t), ..., \mu_{P_n}(t) >$;
    update the fuzzy cardinalities for all parts of V;
done.

Remark. A possible heuristic for optimizing this fuzzy cardinality computation process is to consider the different predicates in increasing order of their size and to cut branches of the exploration tree as soon as a zero cardinality is found for an intermediate conjunction.

One may now detect the MFSs thanks to these fuzzy cardinalities for different empty α-cuts of Q, starting from a user-defined qualitative threshold to the highest satisfaction degree 1.

In the manner of Apriori [1], Algorithm 1 starts with atomic predicates and the first α_i-cut of interest, the one corresponding to the user-defined qualitative threshold α_i. To determine if an atomic predicate P_a is a failing subquery of Q, one just has to check the computed fuzzy cardinalities. If no tuple satisfies P_a at least with the degree α_i then P_a, as an atomic predicate, is by definition an MFS of Q and is also an MFS for $\alpha_j > \alpha_i$. Then, for the second round of the loop (line 1.7 of Algorithm 1), conjunctions containing two non failing predicates are generated and for each of them (line 1.11) one checks the fuzzy cardinalities so as to determine if it is an MFS. If one of these conjunctions, say $P_b \wedge P_c$, is an MFS for a degree α_i one tries to propagate it to higher satisfaction degrees (see Algorithm 2 where isMFS(L,$MFS_{\alpha_j}(Q)$) returns *true* if $L \in MFS_{\alpha_j}(Q)$, *false* otherwise). As the MFS property is not monotone with respect to α-cuts, one checks with Algorithm 2 for each $\alpha_j > \alpha_i$ if a subquery of $P_a \wedge P_b$ corresponds to a previously detected MFS for degree α_j; if it is not the case $P_a \wedge P_b$ is stored as an MFS of Q for α_j. Obviously, an atomic failing query is an MFS for all α-cuts. Then, the algorithm goes back to the loop (line 1.7) and conjunctions containing three predicates are generated for each considered satisfaction degree (line 1.8) taking care that these conjunctions do not contain an already identified MFS. This recursive process goes on until candidate conjunctions cannot be generated anymore.

Remark. In case of Boolean queries, Algorithm 1 can also be used without being reconsidered as one just needs to change the scale of satisfaction degrees for the singleton $\{0^+\}$.

The complexity of this algorithm is obviously exponential in the number of predicates involved in the failing query to explain, where the worst case corresponds to a single MFS Q for the maximal satisfaction degree of 1. In this case, the *foreach* loop (line 1.11) makes 2^n iterations where n is the number of predicates in Q. For a complete gradual explanation from $\alpha = 0^+$ to 1, the 2^n iterations are repeated f times, where f is the number of considered satisfaction degrees in $\mathcal{S} : 1 = \alpha_1 > \alpha_2 > ... > \alpha_f = 0^+$. Thus, the final complexity in the worst case is $f \times 2^n$ and more generally $\Theta(2^n)$.

But, as we said previously, this is not a problem in practice as the number of predicates specified by a user is rather low (≤ 10) in most applicative contexts. Therefore, this process remains tractable as we will show experimentally in Section 4.

Once the MFSs have been detected, it is possible to inform the user about the conflicts in his/her query, which should help him/her revise the selection condition of the failing query.

Input: a failing query $Q = P_1 \wedge \ldots \wedge P_n$; a scale of degrees
$\qquad$ $A = \alpha_f < \ldots < \alpha_2 < (\alpha_1 = 1)$; a user-defined qualitative
$\qquad$ threshold α_u;
Output: $MFS(Q)$ ordered sets of MFS's of Q, one set for each α-cut of
$\qquad$ Q.

1.1 **begin**
1.2 $\qquad$ **foreach** $\alpha_i \in A \mid \alpha_i \geq \alpha_u$ **do**
1.3 $\qquad\qquad$ $MFS_{\alpha_i}(Q) \leftarrow \emptyset$; $E_{\alpha_i} \leftarrow \{P_1, \ldots, P_n\}$;
1.4 $\qquad\qquad$ $Cand_{\alpha_i} \leftarrow E_{\alpha_i}$;
1.5 $\qquad$ **end**
1.6 $\qquad$ $nbPred \leftarrow 1$;
1.7 $\qquad$ **while** $Cand_{\alpha_1} \neq \emptyset$ **do**
1.8 $\qquad\qquad$ **foreach** $\alpha_i \in A \mid \alpha_i \geq \alpha_u$ **do**
1.9 $\qquad\qquad\qquad$ // generation of the candidates of size nbPred
1.10 $\qquad\qquad\qquad$ $Cand_{\alpha_i} \leftarrow \{M$ composed of $nbPred$ predicates present in E_{α_i}
$\qquad\qquad\qquad$ such that $\forall M' \subset M$, $M' \notin MFS_{\alpha_i}(Q)\}$;
1.11 $\qquad\qquad\qquad$ **foreach** L *in* $Cand_{\alpha_i}$ **do**
1.12 $\qquad\qquad\qquad\qquad$ **if** $card(L_{\alpha_i}) = 0$ **then**
1.13 $\qquad\qquad\qquad\qquad\qquad$ $MFS_{\alpha_i}(Q) \leftarrow MFS_{\alpha_i}(Q) \cup \{L\}$;
1.14 $\qquad\qquad\qquad\qquad\qquad$ //E_L contains the atomic predicates that compose L
1.15 $\qquad\qquad\qquad\qquad\qquad$ $E_{\alpha_i} \leftarrow E_{\alpha_i} - E_L$;
1.16 $\qquad\qquad\qquad\qquad\qquad$ //Propagate L to higher satisfaction degrees
1.17 $\qquad\qquad\qquad\qquad\qquad$ // $E = \cup_i E_{\alpha_i}$ and $MFS = \cup_i MFS_{\alpha_i}(Q)$
1.18 $\qquad\qquad\qquad\qquad\qquad$ propagate(α_i, A, L, MFS, E);
1.19 $\qquad\qquad\qquad\qquad$ **end**
1.20 $\qquad\qquad\qquad$ **end**
1.21 $\qquad\qquad$ **end**
1.22 $\qquad\qquad$ $nbPred \rightarrow nbPred + 1$;
1.23 $\qquad$ **end**
1.24 **end**

Algorithm 1. Gradual MFS computation

Example 1. Let us consider a real-estate database, and the failing query:

$$Q = big\ garden \wedge recent \wedge city\ center \wedge open\ view \wedge street\text{-}level$$

Let us assume that the satisfaction degrees associated with the predicates from Q are those represented in Table 1 (left). The computed fuzzy cardinalities appear in Table 1 (right). The minimal failing subqueries found are:

- for $\alpha = 0^+$: i) *recent*; ii) *big garden* $\wedge$ *city center*; iii) *city center* $\wedge$ *open view* $\wedge$ *street-level*.
- ...
- for $\alpha = 1$: i) *recent*; ii) *big garden*; iii) *open view*; iv) *city center* $\wedge$ *street-level* $\wedge$ *recent*; v) *city center* $\wedge$ *street-level* $\wedge$ *big garden*; vi) *city center* $\wedge$ *street-level* $\wedge$ *open view*.

Input: a satisfaction degree: α_i; a scale of degrees: A; detected MFS for α_i: L; a reference to the array of layered MFS: MFS; a reference to the array of predicates used for the generation of candidates: E;

2.1 **procedure** propagate(α_i, A, L, MFS, E) **begin**
2.2   **foreach** $\alpha_j \in A \mid \alpha_j \geq \alpha_i$ **do**
2.3   **if** $isAtomic(L)$ *or* $isMFS(L, MFS_{\alpha_j}(Q))$ **then**
2.4 $MFS_{\alpha_j}(Q) \leftarrow MFS_{\alpha_j}(Q) \cup \{L\}$;
2.5 $E_{\alpha_j} \leftarrow E_{\alpha_j} - E_L$;
2.6   **else**
2.7   break;
2.8   **end**
2.9   **end**
2.10 **end**
2.11 **end**

Algorithm 2. Procedure that propagates an MFS to higher satisfaction degrees

Then, the system provides the user with the following explanation: "no element is somewhat *recent (r)*, there is a conflict between *big garden (g)* and *city center (c)* on the one hand, and between *city center*, *open view (v)*, and *street-level (s)* on the other hand." Such explanations can be provided for any α-cut of the initial query, α being a user-defined threshold present in the query. ⋄

Table 1. Degrees associated with the predicates (left), and fuzzy cardinalities (right)

#id	(g)	(r)	(c)	(v)	(s)
t_1	0.9	0	0	0.4	1
t_2	0	0	0.7	0.2	0
t_3	0.8	0	0	0.9	1
t_4	0.7	0	0	0.8	0
t_5	0	0	1	0	1

combination	fuzzy cardinality
g	$1/0 + 0.8/2 + 0.6/3 + 0.4/3 + 0.2/3 + 0^+/3$
r	$1/0 + 0.8/0 + 0.6/0 + 0.4/0 + 0.2/0 + 0^+/0$
c	$1/1 + 0.8/1 + 0.6/2 + 0.4/2 + 0.2/2 + 0^+/2$
v	$1/0 + 0.8/2 + 0.6/2 + 0.4/3 + 0.2/4 + 0^+/4$
s	$1/3 + 0.8/3 + 0.6/3 + 0.4/3 + 0.2/3 + 0^+/3$
$g \wedge r$	$1/0 + 0.8/0 + 0.6/0 + 0.6/0 + 0.2/0 + 0^+/0$
$g \wedge c$	$1/0 + 0.8/0 + 0.6/0 + 0.4/0 + 0.2/0 + 0^+/0$
$g \wedge v$	$1/0 + 0.8/2 + 0.6/2 + 0.4/2 + 0.2/2 + 0^+/2$
...	
$g \wedge r \wedge c \wedge v$	$1/0 + 0.8/0 + 0.6/0 + 0.4/0 + 0.2/0 + 0^+/0$
$g \wedge r \wedge c \wedge v \wedge s$	$1/0 + 0.8/0 + 0.6/0 + 0.4/0 + 0.2/0 + 0^+/0$

Jannach [7] proposes an algorithm which is somewhat similar to ours, but which does not precompute the cardinalities (Table 1 (right)). Instead, it builds a binary matrix containing the satisfaction degrees obtained by each tuple for each atomic predicate, and combines these degrees in order to detect the MFS's. The main problem is that such a table can be very large, to the point of not fitting in memory and that a query is submitted for each atomic predicate on the whole dataset.

4   Experimentation

4.1   Context

We have implemented this cooperative approach as a research prototype over a database containing 10,479 ads about second hand cars respecting the following schema: {*idads*, *year*, *mileage*, *price*, *optionLevel*, *securityLevel*, *engineSize*, *horsePower*, *consumption*}. To provide a clear and user-friendly access interface, a vocabulary composed of 44 fuzzy predicates has been defined on the domains of eight of the attributes describing the different items, obviously the surrogate key is discarded. Figure 2 illustrates how this vocabulary is used to query the database and shows the explanations that are given in case of a failure.

Using this interface, we have submitted various conjunctive queries and analyzed the efficiency of this fuzzy cardinality-based explanation process for 100 failing queries. Figure 3 graphically shows the evolution of the execution time for

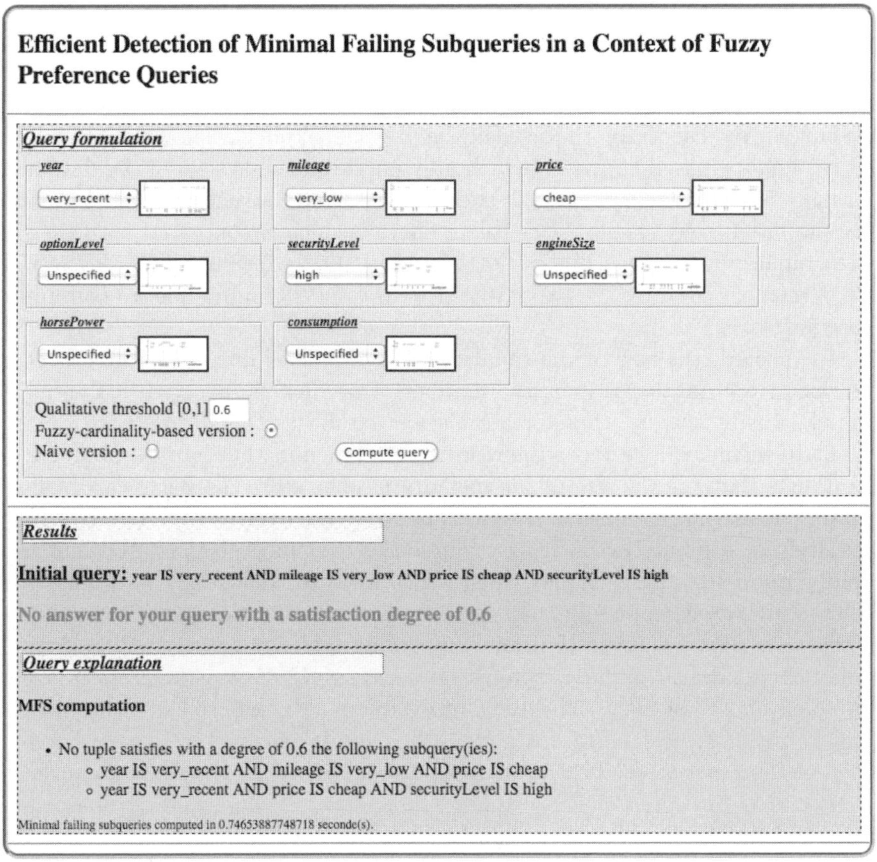

Fig. 2. Cooperative query interface

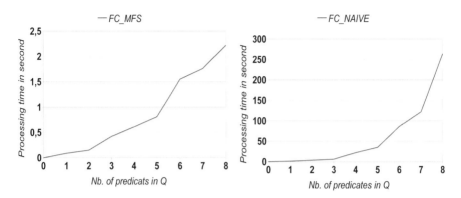

Fig. 3. Evolution of the computation time **Fig. 4.** Evolution of the computation time for FC_MFS   for the naive version

the fuzzy cardinality-based approach, where the measure FC_MFS includes the processing time of the two subsequent tasks: the fuzzy cardinalities computation step (FC_T) and the gradual MFS detection step (GM_T) (FC_MFS = FC_T + GM_T). Figure 4 illustrates the exponential execution time of a naive approach, where every subquery is executed on the database [8]. The comparison of these two curves, and especially the scale of the $y-axis$, explicitly shows the benefit of our cardinality-based approach which only implies a single scan of the database.

Figure 5 shows the evolution of the size of the data structure needed to store the computed fuzzy cardinalities (plain curve), which is compared with the size of the binary matrix used by [7] (dotted curve). These results reflect the average size of memory used by the two data structures for 25 failing queries containing 8 predicates.

As expected, the size of the binary matrices evolves linearly with respect to the size of the database times the number of predicates involved in the query. One can easily imagine that for large scale databases this data structure does not fit in memory. The most interesting phenomenon that can be observed in Figure 5 is that the the size of the memory used to store the fuzzy cardinalities is insignificant and increase in a logarithmic way according to the number of tuples. Indeed, the number of fuzzy cardinalities that have to be stored increases rapidly from 0 to 1,000 tuples, then very slowly to 6,000 and is stable from 6,000 to 10,000. This phenomenon was predictable and can easily be explained by the fact that whatever the number of tuples, the possible combinations of properties to describe them is finite and can quickly be enumerated. As an example, let us consider the failing fuzzy query Q composed of 8 predicates: *year IS recent AND mileage IS low AND price IS cheap AND optionLevel IS high AND securityLevel IS very_high AND engineSize IS big AND horsePower IS high AND consumption IS very_low.* On a subset of the initial database composed of 6,000 randomly selected tuples, 122 fuzzy cardinalities need to be stored to represent the data distribution over the 8 attributes. On the whole data set

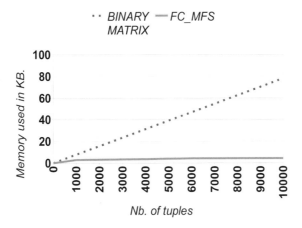

Fig. 5. Comparison of the fuzzy cardinalities-based approach and a naive one for queries containing 8 predicates

(10,479 tuples), 122 fuzzy cardinalities are needed too to represent the whole data distribution. The difference between the number of possible combinations of 8 predicates ($2^8 = 256$) and the observed number of useful combinations (122) is due to the presence of MFSs. Whatever the number of tuples in the database, some combinations of properties, corresponding to false presuppositions, are not observed, such as: *year IS recent AND mileage IS low AND price IS cheap, engineSize IS big AND horsePower IS high AND consumption IS very_low,* ...

So, one can legitimately expect that the size of the memory used to store the fuzzy cardinalities will not increase significantly. This phenomenon has been observed in the particular context of a database containing ads about second hand cars, but it obviously would occur for any database containing somewhat correlated attributes.

Concerning this prototype, it is worth noticing that it has been implemented with an interpreted language (PHP) and one may legitimately expect better execution times with a compiled language like C and parallel programming for the fuzzy cardinalities computation step. The efficiency of our approach relies on the fact that instead of depending on the size of the database, its complexity is related to the number of predicates involved in the query to explain. To illustrate the relevance of this strategy, we have analyzed the query interface of 12 web sites[1] proposing an access to ads about second hand cars. The maximum number of constraints (*i.e.* predicates) a user can specify through these interfaces varies from 5 to 12 with an average of 8.8 predicates. This observation confirms that in practice, the maximum number of predicates involved in queries is rather low (≈ 10).

---

[1] Some examples of web portals to databases containing ads about second hand cars: annoncesauto.com, paruvendu.fr, auroreflex.com, ebay.fr, lacentrale.fr, ...

5   Exploiting MFSs to Repair Failing Queries

When faced with a failing query, the explanations given by the layered MFSs help the user revise his/her initial query. Depending on the nature of the conflicts underlined in the MFSs, a user may:

- reconsider the qualitative threshold α specified in the query,
- remove one or several predicates involved in a conflict,
- replace one or several predicates involved in an MFS by predicates from the shared vocabulary that appear less conflicting,
- apply a repair step which aims at relaxing the definition of some predicates [9] or replace the conjunctive query Q by a fuzzy quantified statement of the type $Q^* = most(P_1, P_2, \ldots, P_n)$ [10].

Figure 2 of Section 4 illustrates a failing situation for an initial query *year IS very_recent AND mileage IS very_low AND price IS cheap AND securityLevel IS high* and a user-defined qualitative threshold $\alpha = 0.6$. The explanations related to this failure clearly point out that the predicate *price is CHEAP* is in conflict with the conjunctions of properties *year IS very_recent AND mileage IS very_low* on the one hand and *year IS very_recent AND securityLevel IS high* on the other hand. Guided with these explanations, one may revise the qualitative threshold and decrease it from 0.6 to 0.4 or replace the conflicting predicate *price is CHEAP* for a less demanding one like *price is MEDIUM* (Figure 6).

Results

Initial query: year IS very_recent AND mileage IS very_low AND price IS medium AND securityLevel IS high

-- Satisfaction degree of 1 --

brand	designation	year	mileage	OL	SL	ES	HP	price	cons.
fiat	Fiat 500 1.2l 8v 69 ch lounge...	2009	14800	1	3	1.2	94	11590	7
ford	Fiesta fun 1&virgule;4 TDCI année 2008 ...	2008	12000	1	3	1.4	68	7500	5
opel	Opel corsa 1.3 cdti ecoflex 75cv...	2009	19000	4	3	1.3	75	10500	9

Fig. 6. Example of an MFS-guided revision of an initial failing query

Figure 7 shows the gradual explanations given for another failing query concerning very cheap vintage cars:

$$Q = year\ is\ VINTAGE\ and\ price\ is\ VERY\_CHEAP.$$

Thanks to the gradual MFSs, the user knows that it is useless to expect answers with a maximum level of satisfaction if he/she keeps the predicate *year is VINTAGE* which constitutes an atomic MFS for $\alpha = 1$.

> **MFS computation**
>
> - No tuple satisfies with a degree of 0.2 the following subquery(ies):
>   - year IS vintage AND price IS very_cheap
> - No tuple satisfies with a degree of 1.0 the following subquery(ies):
>   - year IS vintage

Fig. 7. Gradual explanations of a failing query

6 Related Work

Compared with the binary matrix computed in the approach advocated by Jannach [7] that linearly increases with the size of database times the number of predicates, the size of the table of fuzzy cardinalities depends only on the number of predicates and we have experimentally shown that this size quickly converges and can easily fit in memory (contrary to the binary matrix used in [7]). Moreover, contrary to the approach described in [8,1], a single scan of the database is needed to compute the fuzzy cardinalities and then to detect gradual MFS. Finally, except the study done in [9] and to the best of our knowledge, there is no other work that has addressed the problem of MFS detection in the context of preference queries which induces a larger context of application that goes beyond failing queries *stricto sensu*. Indeed, it appears very useful for a user to know why his/her preferences are not satisfied or are only poorly satisfied.

With respect to the algorithm proposed in [6] which *processes* every query corresponding to a candidate MFS, the major interest of our approach is that, thanks to the precomputation of fuzzy cardinalities, the determination of the MFSs does not imply any additional query processing. Thus, the size of data linearly affects the complexity of our fuzzy cardinality-based approach to MFS detection.

7 Conclusion

In this paper, we first have generalized the problem of failing queries showing that this problem is a special case of poorly satisfied fuzzy queries. We then have proposed an efficient strategy for computing the MFSs of a failing or unsatisfactory query, which relies on the computation of fuzzy-cardinalities. The main benefits of this approach compared to related works are that: i) a single scan of the database is needed to compute the fuzzy-cardinalities, ii) the storage of these cardinalities is not costly and easily fits in memory. Thanks to these fuzzy cardinalities, MFSs can be detected for gradual satisfactions degrees of the user preferences. The MFSs provide interesting explanations of the failure or the dissatisfaction that the user can use to revise his/her initial query.

Among perspectives for future work, we intend to improve the computation of the fuzzy cardinalities using a compiled language and parallel programming.

References

1. McSherry, D.: Retrieval failure and recovery in recommender systems. Artif. Intell. Rev. 24(3-4) (2005)
2. Zadeh, L.A.: Fuzzy sets. Information and control 8(3), 338–353 (1965)
3. Dubois, D., Prade, H.: Fundamentals of fuzzy sets. The Handbooks of Fuzzy Sets, vol. 7. Kluwer Academic Pub., Netherlands (2000)
4. Bosc, P., Pivert, O.: SQLf: a relational database language for fuzzy querying. IEEE Transactions on Fuzzy Systems 3(1), 1–17 (1995)
5. Bosc, P., Buckles, B., Petry, F., Pivert, O.: Fuzzy databases. In: Bezdek, J., Dubois, D., Prade, H. (eds.) Fuzzy Sets in Approximate Reasoning and Information Systems. The Handbook of Fuzzy Sets Series, pp. 403–468. Kluwer Academic Publishers, Dordrecht (1999)
6. Godfrey, P.: Minimization in cooperative response to failing database queries. Int. J. Cooperative Inf. Syst. 6(2), 95–149 (1997)
7. Jannach, D.: Techniques for fast query relaxation in content-based recommender systems. In: Freksa, C., Kohlhase, M., Schill, K. (eds.) KI 2006. LNCS (LNAI), vol. 4314, pp. 49–63. Springer, Heidelberg (2007)
8. McSherry, D.: Incremental relaxation of unsuccessful queries. In: Funk, P., González Calero, P.A. (eds.) ECCBR 2004. LNCS (LNAI), vol. 3155, pp. 331–345. Springer, Heidelberg (2004)
9. Bosc, P., Hadjali, A., Pivert, O.: Incremental controlled relaxation of failing flexible queries. Journal of Intelligent Information Systems 33(3), 261–283 (2009)
10. Zadeh, L.: A computational approach to fuzzy quantifiers in natural languages. Computing and Mathematics with Applications 9, 149–183 (1983)

Rewriting Fuzzy Queries Using Imprecise Views

Hélène Jaudoin and Olivier Pivert

Irisa – Enssat, University of Rennes 1
Technopole Anticipa 22305 Lannion Cedex France
{jaudoin,pivert}@enssat.fr

Abstract. This paper proposes an approach to the tolerant rewriting of queries in terms of views when the views and the queries may involve *fuzzy* value constraints in the context of a Local-As-View mediation system. These constraints describe attribute values as a set of elements attached with a degree in [0, 1] that expresses the plausibility attached to a given element, i.e., attribute values more or less plausible/typical in the views, while in the queries, they denotes preferences, i.e., more or less desired values. The problem of rewriting queries is formalized in the setting of the description logic $\mathcal{FL}_0$ extended to fuzzy value constraints. We propose an algorithm of gradual and structural subsumption for this extended logic, that plays a key role in the query rewriting algorithm. Finally, we characterize the tolerant query rewriting forms and propose an algorithm to compute them.

Keywords: Data Integration System, LAV Approach, Fuzzy preferences, Imprecise views.

1 Introduction

Data integration systems provide a uniform querying interface to a set of distributed data sources, in the form of a global schema. The problem of answering queries in data integration systems has been much studied in the last decade [11]. Until recently, the global schema and the descriptions of the data sources of an integration system were assumed to be precise. In this context, it has been proven that the semantics of queries may be formalized in terms of *certain answers* [1]. A certain answer to a query Q expressed in terms of the global schema, according to the set of instances of the data sources, is an answer to Q for every database defined on this global schema which is consistent with the source instances. A technique for computing the certain answers to a query in a data integration system following an LAV (Local-As-View) approach, where the data sources are defined as queries, i.e, *views*, on the global schema, consists in reducing this problem into that of *query rewriting using views*. Given a query Q expressed on the global schema, the data sources which are relevant with respect to the resolution of the query are selected by means of a rewriting algorithm which reformulates Q into a query which is either equivalent to Q or maximally contained in Q, and whose definition only refers to the *views*. Every such rewriting must satisfy all of the constraints involved in Q in order to only return correct answers to Q.

J. Eder, M. Bielikova, and A.M. Tjoa (Eds.): ADBIS 2011, LNCS 6909, pp. 257–270, 2011.
© Springer-Verlag Berlin Heidelberg 2011

Classical data integration systems face the problem of empty/plethoric answers. Indeed they are not exception-tolerant, i.e., query rewritings supplying undesired data are discarded, even when these undesired data are not at all representative of the views involved (i.e., can be viewed as exceptions). They do not rank query rewritings according to the satisfaction level of the tuples they supply either. Moreover, to our knowledge, data integration systems do not allow vague description of views. However, in some integration contexts as well as in the setting of distributed multimedia information systems [10],[25], it may be the case that some source descriptions are *imprecise*, either because they have been subjectively defined by an expert, or because they have been obtained applying fuzzy rules, or because data are inherently vague. For example, in an oceanographic context, a biological station could list regions whose plausible fish species are $\{1/Tuna, 0.9/Mackerel, 0.5/Sardine\}$ in its study area, knowing some parameters as salinity or depth of the regions. An expert can also describe regions whose plausible fish species are only *big fish* where *big* is a fuzzy predicate. One way to overcome these disadvantages is to allow for the expression of preferences in the queries and imprecise descriptions in the views and then to use a tolerant mechanism of subsumption between queries and query rewritings.

Contributions. In this paper, we assume a data integration system based on the LAV approach, in which the views (assumed to be sound) and the queries involve fuzzy value constraints. We first express the query rewriting problem in this context using the formal framework of description logics [2], then we show how a *fuzzy pattern matching* technique can be exploited to determine whether a combination of views may constitute a *satisfactory* rewriting of a query, hence to define a new gradual and structural subsumption algorithm. This measure makes it possible to rewrite queries in the presence of *exceptions*. The semantics of answers is revisited in this context and we introduce the notion of an *α-certain answer*. Finally, we propose an algorithm to compute the rewritings satisfying a given query to a level at least equal to a threshold α.

Related Work. In this paper, the issue of rewriting fuzzy queries using imprecise views is formalized in the setting of description logics, which constitute a family of knowledge representation and reasoning formalisms based on first order logic. They make it possible to represent a knowledge domain by means of classes of elements and binary relationships between these classes. They are widely used in the context of the semantic web, notably through OWL which is a W3C standard for the definition of ontologies. Moreover, numerous fuzzy extensions of description logics (see [24], [23],[3],[17]) have been defined. To the best of our knowledge, none of these fuzzy extensions deals with the problem of rewriting queries using views. It is worth noting that the purpose of the proposed approach is not to define a new extension for description logics but rather to exploit an existing logics as a basis to our study. Recently, a few papers contributed to the development of data integration systems capable of taking into account imprecision or uncertainty. Most of the works along that line use probability theory in order to capture the form of uncertainty that stems from the schema definition process [6], [12], [16], [19], or that associated with the mere existence of data [5], or aim at modeling the approximate nature of the semantic links between the data sources and the mediated schema [7]. In this paper, the nature of the imprecision is different as it stems from the description of the data involved in the views and not from the view definition in terms of the global schema. In [13], the authors describe an approach that deals with query rewriting using views in the presence of value constraints. However, the constraints considered are not fuzzy and the semantics of the answers is that of certain answers. Consequently, the search for rewritings relies on a Boolean subsumption principle and the algorithms cannot be directly adapted to the context considered here. On the contrary, the work proposed here allows for generalizing the results of [13]. To our knowledge, none of the existing works in data integration systems has studied the problem of rewriting queries with preferences and imprecise descriptions of data. The work which relates the most to the approach presented here is [21] which describes an extension of object-oriented models to fuzzy classes in order to allow for the representation of imprecise data and the notion of exception. However, the type of reasoning that it considers mainly aims at classifying instances according to a hierarchy of fuzzy classes and is thus different from that involved in a query rewriting process. Finally, let us mention that the fuzzy descriptions that we use in this approach correspond to the concept of a *closed positive veristic variable* studied by R.R. Yager [20] in a different context.

The remainder of the paper is structured as follows. In Section 2, we give an example that illustrates our approach and we justify our modeling choices. In Section 3, we present the formal frameworks of fuzzy logic and description logics, and we focus on the logic $\mathcal{FL}_0(\mathcal{O}_F)$ used further. We also characterize a structural subsumption test for the logic $\mathcal{FL}_0(\mathcal{O}_F)$, suited to the rewriting problem. Section 4 addresses the issue of rewriting queries in the presence of fuzzy value constraints. Section 5 is devoted to the computation of α-certain answers whereas Section 6 concludes the paper and outlines perspectives for future work.

2   Running Example and Motivations

Let us introduce our running example and assume a query Q aimed at retrieving the *regions* whose *fish species* are $\{1/Tuna, 0.8/Mackerel, 0.5/Eel\}$. Four views are assumed available:

V_1 returns *regions*,
V_2 returns objects whose *species* are $\{1/Tuna\}$,
V_3 returns objects whose *species* are $\{1/Mackerel, 0.9/Tuna, 0.3/Sardine\}$ and,
V_4 returns objects whose *species* are $\{1/Mackerel, 0.8/Tuna, 0.6/Bass\}$.

The first two views may be considered precise whereas views V_3 and V_4 describe their data in an imprecise way. The degree attached to a value v in a fuzzy set F expresses the extent to which v matches the graded concept associated with F (here, the plausibility to be present in a certain region). Degrees do not correspond to probability degrees but rather give an order relationship between values. Some values are simply considered more likely than others.

Let us consider the problem of answering the query Q. It is possible to answer it using the semantics of certain answers by combining the data stemming from V_1 with those from V_2 but this cannot be done by combining V_1 with the views V_3 or V_4 alone. In a classical data integration framework, one can nevertheless combine the data from V_3 with those from V_4 so as to guarantee a specie in the set $\{Mackerel, Tuna\}$ and thus satisfy the criterion from the query. However, no indication about the satisfaction degree with respect to the preferences specified in the query can be given in this case. In a classical querying framework, V_3 would be discarded from the rewriting process. However, it is not very likely that it returns objects containing $Sardine$ according to its description ($Sardine$ is only 0.3-plausible). Furthermore, since the objects it returns are more likely to contain $Mackerel$ or $Tuna$, it would be interesting to select it, by making the rewriting process tolerant to exceptions. The fuzzy pattern matching technique, in particular the necessity measure that relies on Kleene-Dienes implication ($p \rightarrow_{KD} q = max(1 - p, q)$), constitutes an appropriate tool for expressing the extent to which view V_3 is a good rewriting candidate. This measure returns a degree between 0 and 1. Degree 1 means that the rewriting only returns answers which totally satisfy the query, i.e., $certain$ $answers$. Degree 0 means that the query rewriting is not at all subsumed by the query (there is a full exception, i.e., a value which is totally in the rewriting but which does not satisfy the query at all). On the other hand, a degree in $]0, 1[$ reflects the existence of partial exceptions, i.e., elements which are either unsatisfactory but not completely plausible, or plausible but weakly preferred. With our running example, this technique yields the following results. The truth value associated with the fact that any tuple obtained through the conjunction of V_1 and V_3 (resp. V_1 and V_4) be satisfactory w.r.t. the query equals $min(0.8, 1, 0.7) = 0.7$ (resp., $min(0.8, 1, 0.4) = 0.4$). If the user considers that a truth degree lower than 0.8 is not sufficient, then the combinations V_1 and V_3, as well as V_1 and V_4, cannot be considered satisfactory rewritings. On the other hand, the combination V_1 and V_3 and V_4 constitutes a satisfactory rewriting since it is included in Q to a degree 0.8 in the sense of Kleene-Dienes implication.

Note that in our running example, we consider a non-functional attribute (whose values are not mutually exclusive). If we had to deal with functional attributes (as for example the pollution degree of a zone), we could use the same theoretical framework to match user preferences against data source descriptions. The only difference is that the fuzzy sets used in the view descriptions would be interpreted as possibility distributions [26] with a disjunctive meaning. In the sequel, we will assume that a non-normalized intersection between two fuzzy sets describing non-disjoint geographical zones corresponds to an inconsistency. Notice that such an assumption makes sense in the context of a densely populated space (here, it corresponds to assuming that at least one species is fully compatible with every area of the ocean). Since the views are supposed to be sound, the possibility of such inconsistencies will be ignored.

3   Preliminaries

3.1   Basic Notions about Fuzzy Sets

Fuzzy set theory was introduced by Zadeh [15] for modeling classes or sets whose boundaries are not clear-cut. For such objects, the transition between full membership and full mismatch is gradual rather than crisp. Typical examples of such fuzzy classes are those described by means of adjectives of the natural language, such as *young, cheap, fast*, etc. Formally, a fuzzy set F on a referential U is characterized by a membership function $\mu_F : U \rightarrow [0,1]$ where $\mu_F(u)$ denotes the grade of membership of u in F. In particular, $\mu_F(u) = 1$ reflects full membership of u in F, while $\mu_F(u) = 0$ expresses absolute non-membership. When $0 < \mu_F(u) < 1$, one speaks of partial membership.

Two crisp sets are of particular interest when defining a fuzzy set F:

- the core $C(F) = \{u \in U \mid \mu_F(u) = 1\}$, which gathers the *prototypes* of F,
- the support $S(F) = \{u \in U \mid \mu_F(u) > 0\}$.

As usual, the logical counterparts of the theoretical set operators $\cap$, $\cup$ and complementation operator correspond respectively to the conjunction $\wedge$, disjunction $\vee$ and negation $\neg$ (see [9] for more details).

3.2   Logic $\mathcal{FL}_0(\mathcal{O}_F)$

Description logics [2] model a knowledge domain in terms of *concepts* (by means of unary predicates) which characterize subsets of elements (*individuals*) of a domain, and *roles* (by means of binary predicates) on this domain. Concepts are described by expressions formed using constructs. The various description logics differ according to the types of constructs they authorize. The concepts in $\mathcal{FL}_0$ are defined by means of constructs as follows:

$$B \rightarrow A \mid \forall R.C \mid \top$$

$$C \rightarrow B \mid C_1 \sqcap C_2$$

where A and R respectively denote an atomic (non decomposable) concept and an atomic role; B is a basic concept and may be either an atomic concept, or a restriction of role R to concept C, or the special concept Top ($\top$); C is a general concept which may be formed of the conjunction of two general concepts C_1 and C_2. For instance, the concept *Father* can be described by means of the conjunction *Parent* $\sqcap$ *Male* whereas the set of individuals whose children are only boys can be defined as $\forall child.Male$.

In this paper, on top of these constructs, we use a restriction of the One-Of ($\mathcal{O}$) operator [22,4] which makes it possible to restrict the range of roles to a set of *concrete values*, i.e., values different from those of the individuals of the domain [13]. By doing so, $\forall color.\{red, yellow, green\}$ is a concept which denotes those individuals whose color is necessarily *red, yellow* or *green*. In the following, the notion of a set of concrete values is extended to a fuzzy set somewhat in the manner of [3]. This construct defined as $\forall R_C.\{d_1/o_1, \ldots, d_n/o_n\}$ and denoted by $\mathcal{O}_F$ can be used for instance to express that the most plausible colors for a set of individuals are *red*, *yellow* and that *white* is a less plausible color: $\forall color.\{1/red, 1/yellow, 0.2/white\}$. As in [13,4], we will distinguish between two disjoint sets of roles: the roles from R_A whose range is a basic or general concept, and the concrete roles from R_C whose range is a fuzzy set of concrete values.

The knowledge base considered in this article only aims at giving an intensional description of an application domain. Therefore, it will be reduced to a terminology ($\mathcal{T}$), also known as an ontology or schema, and will only consist of a set of concept definitions $A \equiv C$ where A is a concept name and C is a concept expressed in $\mathcal{FL}_0(\mathcal{O}_F)$.

Semantics of $\mathcal{FL}_0(\mathcal{O}_F)$

The semantics of the concepts will be given by a fuzzy interpretation $\mathcal{I} = (\Delta^{\mathcal{I}}, .^{\mathcal{I}})$ associated with a concrete domain Δ_C which denotes the set of the concrete values. $\Delta^{\mathcal{I}}$ is a nonempty set of individuals called the *interpretation domain*, disjoint from Δ_C, and $.^{\mathcal{I}}$ is a fuzzy interpretation function which associates with each concrete value o_i an element of Δ_C such that $o_i \neq o_j$ implies $o_i^{\mathcal{I}} \neq o_j^{\mathcal{I}}$. Moreover, it associates

- with each C: a function $C^{\mathcal{I}} : \Delta^{\mathcal{I}} \to [0,1]$,
- with each abstract role R_a: a function of $R_a^{\mathcal{I}} : \Delta^{\mathcal{I}} \times \Delta^{\mathcal{I}} \to \{0,1\}$ and
- with each concrete role R_c: a function of $R_c^{\mathcal{I}} : \Delta^{\mathcal{I}} \times \Delta_C \to [0,1]$.

For $x \in \Delta^{\mathcal{I}}$, $C^{\mathcal{I}}(x)$ gives the degree to which x satisfies the fuzzy concept $C^{\mathcal{I}}$ under the fuzzy interpretation $\mathcal{I}$. Function $.^{\mathcal{I}}$ can be generalized to constructs the following way. Let $x \in \Delta^{\mathcal{I}}$ and E be a fuzzy set such that $E = \{d_1/o_1, \ldots, d_n/o_n\})$, then:

- $E^{\mathcal{I}}(v) = d_i$ if $v = o_i$,
- $\top^{\mathcal{I}}(x) = 1$
- $(C_1 \sqcap C_2)^{\mathcal{I}}(x) = C_1^{\mathcal{I}}(x) \wedge_{min} C_2^{\mathcal{I}}(x)$
- $(\forall R_a.C)^{\mathcal{I}}(x) = inf_{y \in \Delta^{\mathcal{I}}} \{R_c^{\mathcal{I}}(x,y) \to_{KD} C^{\mathcal{I}}(y)\}$
- $(\forall R_c.E)^{\mathcal{I}}(x) = R_c^{\mathcal{I}}(x,v) \to_{KD} E^{\mathcal{I}}(v)$ with $v \in \Delta_C$

where $\to_{KD}$ denotes Kleene-Dienes' fuzzy implication. It is worth noting that data are not uncertain but they belong with a certain degree to the fuzzy concepts involved in the descriptions of the data sources.

A concept C is *satisfiable* iff there exists a fuzzy interpretation $\mathcal{I}$ and an individual $x \in \Delta^{\mathcal{I}}$ such that $C^{\mathcal{I}}(x) > 0$. One then says that $\mathcal{I}$ is a model for C. A fuzzy interpretation $\mathcal{I}$ satisfies the definition $A \equiv C$ iff $A^{\mathcal{I}} = C^{\mathcal{I}}$. Lastly, a fuzzy interpretation satisfies a terminology $\mathcal{T}$, i.e., is a model for $\mathcal{T}$, iff it satisfies all of the concept definitions of $\mathcal{T}$.

The reasoning service which is at the basis of the query rewriting process is the notion of subsumption between two concepts. In this paper, the semantics of subsumption between two concepts C and D ($C \sqsubseteq D$) is based on the necessity measure defined in [8]. The necessity measure makes it possible to assess the degree of subsumption of C in D. Its evaluation relies on Kleene-Dienes's implication. Thus, concept subsumption is defined by the following expression:

$$(C \sqsubseteq D)^{\mathcal{I}} = inf_{x \in \Delta^{\mathcal{I}}} \{C^{\mathcal{I}}(x) \rightarrow_{KD} D^{\mathcal{I}}(x)\} \tag{1}$$

where $C^{\mathcal{I}}(x) \rightarrow_{KD} D^{\mathcal{I}}(x) = max(1 - C^{\mathcal{I}}(x), D^{\mathcal{I}}(x))$. An axiom of concept subsumption $\langle C \sqsubseteq D \geq n \rangle$ where $n \in]0,1]$ is said to be satisfiable by a fuzzy interpretation $.^{\mathcal{I}}$, denoted by $\mathcal{I} \models \langle C \sqsubseteq D \geq n \rangle$, iff $(C \sqsubseteq D)^{\mathcal{I}} \geq n$. We will say that concept C is subsumed by concept D to degree n, denoted by $\langle C \sqsubseteq D \geq n \rangle$, iff for every fuzzy interpretation $\mathcal{I}$, $(C \sqsubseteq D)^{\mathcal{I}} \geq n$.

Subsumption in $\mathcal{FL}_0(\mathcal{O}_F)$

Two types of algorithms have been proposed for testing concept subsumption: the structural approach and the semantic one [2], the latter being also known as the tableaux method [2]. The first approach relies on a comparison of the syntax of each of the concepts whereas the second one tries to build a valid interpretation for the axiom being tested. In the perspective of computing the rewritings of a query, given the large size of the search space, it is more convenient to have syntactical indications about the form of the rewritings in order to prune the search space. Therefore, we propose hereafter a characterization of subsumption in $\mathcal{FL}_0(\mathcal{O}_F)$ which makes it possible to devise a structural subsumption algorithm for $\mathcal{FL}_0(\mathcal{O}_F)$.

Characterizing the notion of subsumption as defined above entails expressing the concepts in their normal form, i.e., transforming them into an equivalent concept which makes its hidden related knowledge explicit. The normal form of a concept consists of a conjunction of atoms of the form $\forall m.P$ where $m = r_1.\ldots.r_n$ is a shortcut used to express a multilayered nesting of value restrictions $\forall r_1.\forall r_2.\ldots.\forall r_n.P$ and P is either an atomic concept (A) or a fuzzy set (E). $r_1.\ldots.r_{n-1}$ is a word in $\mathcal{R}_A^*$ and $r_n \in \mathcal{R}_A \cup \mathcal{R}_C$. If $n = 0$, then m is the empty word ϵ and $\forall m.P$ is equivalent to P.

Let C and C' be two classical concepts, and E and E' two fuzzy concepts. The normal form of a concept in $\mathcal{FL}_0(\mathcal{O}_F)$ is obtained by applying the normalization rules 1) et 2) as long as this is possible, then rule 3).

1. $\forall m.C \sqcap \forall m.C' \rightarrow \forall m.(C \sqcap C')$
2. $\forall m.(E_1 \sqcap E_2) \rightarrow \forall m.E$ where $E = E_1 \wedge_{min} E_2$
3. $\forall m.(C \sqcap C') \rightarrow \forall m.C \sqcap \forall m.C'$

Let C and D be two fuzzy concepts in $\mathcal{FL}_0(\mathcal{O}_F)$. The following algorithm computes the subsumption degree associated with $\langle C \sqsubseteq D \geq \alpha \rangle$:

1. if $D \equiv \top$ then $\alpha = 1$
2. otherwise
   (a) if there does not exist $\forall m_i.P_i'$ in C for a $\forall m_i.P_i$ of D, then $\alpha = 0$
   (b) otherwise, for all $\forall m_i.P_i$ of D, there exists $\forall m_i.P_i'$ in C such that
      - if $P_i = \top$, then $d_i = 1$
      - if $P_i = A$ and $P_i' = A$ then $d_i = 1$ (otherwise $d_i = 0$)

- if $P_i = E$ and $P'_i = E'$ then $d_i = (E' \rightarrow_{KD} E)$
and $\alpha = min_i\{d_i\}$.

The proof of the soundness and completeness of this algorithm mainly relies on a result from possibility theory which concerns the evaluation of the truth degree of a proposition [8]. This result states that for normalized possibility distributions (or fuzzy sets), the degree of inclusion (based on Kleene-Dienes' implication) between two conjunctive events $C = (C_1 \times \ldots \times C_n)$ and $D = (D_1 \times \ldots \times D_n)$ such that for each pair (C_i, D_i) both C_i and D_i are defined over the same domain, is equal to the smallest of the inclusion degrees $C_i \sqsubseteq D_i$.

4 Rewriting Fuzzy Queries Using Imprecise Views

4.1 Definition of the Data Integration System

Let us consider a data integration system based on a LAV approach. The schema and the views are defined by means of two terminologies $\mathcal{S}$ and $\mathcal{V}$ in $\mathcal{FL}_0(\mathcal{O}_F)$. The queries are made of concepts defined in $\mathcal{FL}_0(\mathcal{O}_F)$ in terms of $\mathcal{S}$.

Example 1. Let $\mathcal{S}$ be an ontology that partially describes shallow sea on Brittany coasts:

$ShallowSea \equiv WaterArea \sqcap \forall habitat.\{1/rock, 1/sand, 1/sediment\}$
$FishAreaOnShallowSea \equiv WaterArea \sqcap \forall fishPopulation.(Fish \sqcap$
$\qquad \forall size.\{1/small, 1/intermediate, 0.2/big\} \sqcap$
$\qquad \forall species.\{1/Mackerel, 1/Sardine, 0.5/Tuna\}).$

Let $\mathcal{V}$ be the terminology associated with the views, which gives the description of the result of the queries that can be processed on the sources:

$V_1 \equiv Area$
$V_2 \equiv \forall fishPopulation.\forall species.\{1/Tuna\}$
$V_3 \equiv$
$\qquad \forall fishPopulation.\forall species.\{1/Mackerel, 0.9/Tuna, 0.3/RedMullet, 0.2/eel\}$
$V_4 \equiv \forall fishPopulation.\forall species.\{1/Mackerel, 0.8/Tuna, 0.8/Bass, 0.3/eel\}$
$V_5 \equiv \forall fishPopulation.\forall size.\{1/small, 0.8/intermediate, 0.2/big\}$

An example of a query over $\mathcal{S}$ is

$Q \equiv Area \sqcap$
$\qquad \forall fishPopulation.\forall species.\{1/Tuna, 0.8/Mackerel, 0.5/Sardine\} \sqcap$
$\qquad \forall fishPopulation.\forall size.\{1/small, 1/intermediate\}.\diamond$

Let $(\mathcal{S}, \mathcal{V})$ be a mediation system in $\mathcal{FL}_0(\mathcal{O}_F)$. One assumes hereafter that $\mathcal{S}$ and $\mathcal{V}$ are acyclic, i.e., they do not contain any concept which refers to itself in its definition. Furthermore, descriptions in $\mathcal{S}$ and $\mathcal{V}$ are supposed to be extended, i.e., each defined concept A $(A \equiv D)$ is replaced by its definition D. The views are also assumed to be given in their normal form.

4.2   Computing the Rewritings

One now deals with the computation of the answers to Q involving preferences in the presence of imprecise views $\mathcal{V}$. Each query Q is associated with a threshold $\alpha \in [0,1]$ that specifies the minimal expected satisfaction degree attached with the answers. However, one aims at approximating this set of answers by authorizing exceptions, i.e., data whose plausibility degree w.r.t. to the description of their original source is less than $1 - \alpha$. In other words, an answer t, from a given source to a query Q, either has a low plausibility degree with respect to that source — hence can be seen as an *exception* —, or is has a high enough satisfaction degree with respect to Q. As an illustration, let us consider the Figure 1. The diamonds represent the query preferences, i.e., values 5, 8, 11, are preferred to 12, itself preferred to 13 then 10. If the threshold associated with the query is 0.6, then 10 and 13 are considered undesirable as any value of the domain that is not in the set $\{4, 7, 10, 11\}$. The only exceptions that are authorized are those values whose plausibility is less than 0.4, i.e., any value whose truth degree (w.r.t. to the description of the source) in under the green line. Therefore, the data plausibility given by triangles can provide an interesting approximation of the expected answers. The semantics of the answers in this context can be defined in terms of α-*certain* answers. Those are data stemming from the views for which it is certain to a degree $\geq \alpha$ that they satisfy the query. These answers can be computed by means of the α-certain rewritings of Q on the basis of Formula (1), as explained hereafter.

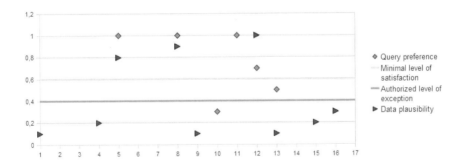

Fig. 1. Query semantics

Definition 1 (α-certain rewritings). *Let $(\mathcal{S}, \mathcal{V})$ be a mediation system in $\mathcal{FL}_0(\mathcal{O}_F)$, Q a query in terms of $\mathcal{S}$ and $\alpha \in]0,1]$. A concept Q' is an α-certain rewriting of Q in terms of $\mathcal{V}$ iff:*

i) Q' is a conjunction of views in terms of $\mathcal{V}$
ii) $\langle Q' \sqsubseteq_{(\mathcal{S},\mathcal{V})} Q, \beta \rangle$ and $\beta \geq \alpha$

Q' is a maximal rewriting of Q in terms of $\mathcal{V}$ if there does not exist any α-certain rewriting Q'' of Q such that $Q' \sqsubseteq Q''$.

Example 2. Let us come back to Example 1 and assume that the user wishes to obtain answers to Q with a satisfaction degree ≥ 0.6. Let us consider the concept $Q_1 \equiv V_1 \sqcap V_2 \sqcap V_5$ formed of the conjunction of the views V_1, V_2, and V_5. After extending Q_1, one gets

$Q_1 \equiv Area \sqcap \forall fishPopulation.\forall species.\{1/Tuna\} \sqcap$
$\qquad \forall fishPopulation.\forall species.\{1/small, 0.8/intermediate, 0.2/big\}.$

Then, using the subsumption algorithm previously defined, one deduces

$\langle V_1 \sqsubseteq Area, 1\rangle,$
$\langle V_2 \sqsubseteq \forall fishPopulation.\forall species.\{1/Tuna, 0.8/Mackerel, 0.5/RedMullet\}, 1\rangle$
$\langle V_5 \sqsubseteq \forall fishPopulation.\forall size.\{1/small, 1/intermediate\}, 0.8\rangle.$

Consequently, one has $\langle Q_1 \sqsubseteq_{(S,V)} Q, 0.8\rangle$. Q_1 is therefore an acceptable rewriting of $Q.\diamond$

In order to obtain as many answers as possible, one tries to compute all the maximal α-certain rewritings of Q. By definition, these rewritings contain a *minimal number of views*. For instance, let $Q' \equiv V_{i_1} \sqcap V_{i_2} \sqcap V_{i_3}$ and $Q'' \equiv V_{i_1} \sqcap V_{i_2}$ such that Q' and Q'' are two α-certain rewritings of Q; then Q' is not maximal since $Q' \sqsubseteq Q''$ (which is checked without extending the definitions of the views). Indeed, Q'' returns more answers than Q'. Let us first consider that Q is reduced to a sole atom $\forall m.C$. The following lemma, based on the subsumption algorithm defined in Subsection 3.2, gives the forms of its rewritings according to the properties of C as well as the maximal number of views needed to rewrite it.

Lemma 1 (Forms of the rewritings). *Let $Q \equiv \forall m.C$, l the cardinality of the largest fuzzy set in $\mathcal{V}$, and $Q' \equiv V_{i_1} \sqcap \ldots \sqcap V_{i_n}$, a conjunction of views from $\mathcal{V}$. Q' is a maximal α-certain rewriting of Q if it is formed of a minimal subset of views from $\mathcal{V}$ such that:*

- *if $C \equiv A$ (resp. $\top$), then $n = 1$ and the view from Q' contains the atom $\forall m.A$ (resp. any concept C) in its description,*
- *if $C \equiv E$, then the set $\{V_{i_1}, \ldots, V_{i_n}\}$ is such that i) every view V_{i_j} contains an atom $\forall m.E_{i_j}$, ii) $(\bigwedge_j E_{i_j} \rightarrow_{KD} E) \geq \alpha$ and iii) $n \leq l + 1$.*

This lemma generalizes the results of [13] to fuzzy sets. Note that the rewriting of fuzzy value constraints may require multiple views, and the worst case remains that of [13], i.e., the case where a total inclusion in E is expected ($\alpha = 1$) and there exists $l + 1$ views whose pairwise fuzzy sets contains l common values, disjoint from those in E. The intersection of $l + 1$ views products an empty set.

Example 3. Let us continue Example 2. The other candidate rewritings that we obtain are given hereafter:

$Q_2 \equiv V_1 \sqcap V_3 \sqcap V_5$ with degree 0.4,
$Q_3 \equiv V_1 \sqcap V_4 \sqcap V_5$ with degree 0.2
$Q_4 \equiv V_1 \sqcap V_3 \sqcap V_4 \sqcap V_5$ with degree 0.8.

Only Q_1 and Q_4 are maximal α-certain rewritings. Indeed, Q_2 and Q_3 do not reach the threshold (0.6) specified by the user, and the rewritings $Q_5 \equiv V_1 \sqcap V_2 \sqcap V_3 \sqcap V_5$, $Q_6 \equiv V_1 \sqcap V_2 \sqcap V_4 \sqcap V_5$ and $Q_7 \equiv V_1 \sqcap V_2 \sqcap V_3 \sqcap V_4 \sqcap V_5$ are not maximal since they are all subsumed by Q_1.$\diamond$

5   Algorithm for Computing α-Certain Query Rewritings

A classical approach to computing the global rewritings of a query, which can still be used here because we deal with unary predicates, consists in

1. computing, for each atom of the query (Algorithm 1, lines 2-6), a bucket which contains all of its α-certain rewritings,
2. building the global rewritings (Algorithm 1, lines 8-9) of the query by computing covers of query atoms from the elements in the buckets, and finally
3. discarding the rewritings which are not maximal (Algorithm 1, line 11).

Algorithm 1. ComputeRew

Require: $\mathcal{V} = \{V_1, ..., V_m\}$ a set of views, Q a query and α a threshold comprised between 0 and 1.
Ensure: $\mathcal{M}$ the set of maximally-contained rewriting of Q using $\mathcal{V}$
1: Let $Q \equiv \sqcap_{i=1}^n \forall w_i.P_i$
2: /* **Step 1: Buckets computation** */
3: **for all** conjunct $\forall w_i.P_i$ **do**
4:     $B(w_i, P_i) = $ BucketBuilding$(\mathcal{V}, \forall w_i.P_i)$
5:     /* *Pruning of inconsistent and non maximal rewritings* */
6:     $B(w_i, P_i) :=$ BucketPruning$(B(w_i, P_i))$
7: **end for**
8: /* **Step 2: Rewritings generation** */
9: $\mathcal{M} :=$ Cart_Prod$(B(w_i, P_i), i \in \{1, ..., n\})$
10: /* *Pruning of inconsistent and non maximal rewritings* */
11: $\mathcal{M} :=$ Pruning$(\mathcal{M})$
12: **return** $\mathcal{M}$

The computation of each query conjunct (BucketBuilding$(\mathcal{V}, \forall w_i.P_i)$) is performed by Algorithm 2 that lies on Lemma 1. One of the most costly step is that devoted to the retrieval of the rewritings of the conjuncts associated with elementary preferences (lines 7-9). In order to improve the step of generation and testing of preference rewritings made of multiple views, it would be interesting to envisage using data mining techniques in order to efficiently compute the rewritings, in the spirit of the approach proposed in [13]. Fortunately, the monotonicity of Kleene-Dienes implication makes this approach possible.

Let us consider the problem of rewriting the conjunct $\forall w.E$ (line 8 of Algorithm 2) and let us denote by $\mathcal{V}_{cand}$ the subset of views stemming from $\mathcal{V}$ that contain in their description a conjunct $\forall w.E_i$. The rewritings $Sol(w, E)$ of $\forall w.E$

Algorithm 2. ComputeBucket

Require: $\mathcal{V} = \{V_1, ..., V_m\}$ a set of views, $\forall w_i.P_i$ a subquery and α a threshold
Ensure: $B(w_i, P_i)$ the bucket associated with $\forall w_i.P_i$
1: **if** $P_i = A$ **then**
2:     $B(w_i, P_i) \leftarrow B(w_i, P_i) \cup \{\{V \in \mathcal{V}\}|V \text{ contains } \forall w_i.P_i\}$
3: **end if**
4: **if** $P_i = \top$ **then**
5:     $B(w_i, P_i) \leftarrow B(w_i, P_i) \cup \{\{V\}|V \in \mathcal{V} \text{ and } V \text{ contains } \forall w_i.AnyConcept\}$
6: **end if**
7: **if** $P_i = E$ **then**
8:     $B(w_i, P_i) \leftarrow B(w_i, P_i) \cup \{\mathcal{V}' \subseteq \mathcal{V}|\forall v_k \in \mathcal{V}', \ v_k \text{ contains } \forall w_i.E_k \text{ and }$
         $(\bigcap_k E_k \rightarrow_{KD} E) \geq \alpha \text{ and } \mathcal{V}' \text{ is minimal}\}$
9: **end if**
10: **return** $B(w_i, P_i)$

---

are conjunctions of the elements of minimal subsets of $\mathcal{V}_{cand}$ whose intersection of fuzzy sets E_i implies E to a certain degree α:

$$Sol(w, E) = Min_{\subseteq}\{ U \in 2^{\mathcal{V}_{cand}} \mid \bigcap_{V_k \in U} E_k \rightarrow_{KD} E) \geq \alpha\}$$

The problem of computing $Sol(w, E)$ can be reformulated as a problem of mining interesting patterns since it now has a set representation. For doing so, one has to set it in the theoretical framework of Mannila and Toivonen ([18]) that can be expressed as follows:

Let $\mathcal{D}$ be a database. Let $\mathcal{L}$ be a set of patterns and P a predicate that qualify the interesting patterns (Theory) with respect to $\mathcal{D}$. The problem is to generate all interesting patterns. However when a partial order ($\preceq$) exists among the patterns and if P is anti-monotonic, the interesting patterns in $\mathcal{L}$ can be represented by their positive and negative borders defined as follows:

$Bd^+ = \{X \in \mathcal{L} \mid P(X) \text{ is true and } \not\exists Y \in \mathcal{L} \text{ s.t. } X \preceq Y \text{ and } P(Y) \text{ is true}\}$
$Bd^- = \{X \in \mathcal{L} \mid P(X) \text{ is false and } \forall Y \in \mathcal{L} \text{ s.t. } Y \preceq X \text{ then } P(Y) \text{ is true}\}$

$Sol(w, E)$ can be set in this framework as follows. The pattern language is $\mathcal{L}_{Sol(w,E)} = \{U \subseteq \mathcal{V}_{cand}\}$ while the chosen predicate $P_{Sol(w,E)}(X)$ is true iff for all $V_i \in X$ and $\forall w.E_i \in V_i$, $(\bigcap_{V_i \in X} E_i \rightarrow_{KD} E) < \alpha$. The partial order is the set inclusion $\subseteq$.

Property 1. $P_{Sol(w,E)}(X)$ is anti-monotonic w.r.t. the set inclusion.

This yields the following result:

Theorem 1. $Bd^-(Theory(\mathcal{L}_{Sol(w,E)}, P_{Sol(w,E)}(X)))$ *is equal to* $Sol(w, E)$.

This makes it possible to use a levelwise approach as that underlying the APriori algorithm [18] to compute $Sol(w, E)$ and hence query rewritings in the presence of fuzzy value constraints. Preliminary experimentations show that it is possible to deal with $10,000$ views in $\mathcal{V}^{cand}$ (with at most 20 values in the fuzzy sets involved) to rewrite *each* fuzzy atom. In this case, there are 2^{10000} possible query rewritings for a given fuzzy atom. The second step of our algorithm requires the computation of covers of the query subgoals with minimal sets of views. Efficient implementations of this problem have been proposed, notably in [14]. An interesting perspective is then to adapt this algorithm in order to compute satisfactory query rewritings.

6 Conclusion

This paper constitutes a contribution to the definition of flexible data integration systems. It deals with query rewriting using views in the presence of fuzzy value constraints, which led us to introduce a new semantics of queries in integration systems, namely that of α-certain answers. Those are answers such that i) either their satisfaction degree w.r.t. the query is at least equal to a threshold $\alpha \in]0, 1]$ or ii) their plausibility w.r.t. the data sources is low ($\leq 1 - \alpha$). The problem was formalized in the setting of the fuzzy description logic $\mathcal{FL}_0(\mathcal{O}_F)$. In order to devise an algorithm aimed at computing the α-certain rewritings, we defined a test of structural subsumption in $\mathcal{FL}_0(\mathcal{O}_F)$, which is sound and complete. Lastly, we described the formats of the query rewritings in $\mathcal{FL}_0(\mathcal{O}_F)$ and showed that it is possible to adapt levelwise approaches to perform one of the most costly steps of our algorithm.

As a short-term perspective, we intend to implement the second step of our algorithm. Another perspective is to address the issue of extending the results reported here to a fuzzy hybrid language such as fuzzy-CARIN [17]. Finally, we also intend to investigate the potential application of this approach in geographical information systems.

References

1. Abiteboul, S., Duschka, O.M.: Complexity of answering queries using materialized views. In: PODS 1998, pp. 254–263 (1998)
2. Baader, F., Calvanese, D., McGuinness, D.L., Nardi, D., Patel-Schneider, P.F. (eds.): The Description Logic Handbook: Theory, Implementation, and Applications. Cambridge Univ. Press, Cambridge (2003)
3. Bobillo, F., Delgado, M., Gómez-Romero, J.: A crisp representation for fuzzy SHOIN with fuzzy nominals and general concept inclusions. In: URSW I pp. 174–188 (2008)
4. Borgida, A., Patel-Schneider, P.F.: A semantics and complete algorithm for subsumption in the classic description logic. JAIR 1, 277–308 (1994)
5. Dalvi, N., Suciu, D.: Answering queries from statistics and probabilistic views. In: VLDB, pp. 805–816 (2005)
6. Das Sarma, A., Dong, X., Halevy, A.: Bootstrapping pay-as-you-go data integration systems. In: SIGMOD, pp. 861–874. ACM, New York (2008)
7. Dong, X., Halevy, A., Yu, C.: Data integration with uncertainty. In: VLDB, pp. 687–698 (2007)
8. Dubois, D., Prade, H.: Possibility Theory. Plenum Press, New York (1988)

9. Dubois, D., Prade, H.: Fundamentals of fuzzy sets. The Handbooks of Fuzzy Sets, vol. 7. Kluwer Academic Pub., Netherlands (2000)
10. Fagin, R.: Combining fuzzy information from multiple systems. In: PODS, pp. 216–226 (1996)
11. Halevy, A.: Answering queries using views: A survey. VLDB Journal 10(4), 270–294 (2001)
12. Hayne, S., Ram, S.: Multi-user view integration system: An expert system for view integration. In: ICDE, Washington, DC, USA, pp. 402–409 (1990)
13. Jaudoin, H., Flouvat, F., Petit, J.M., Toumani, F.: Towards a scalable query rewriting algorithm in presence of value constraints. In: Spaccapietra, S. (ed.) JODS XII. LNCS, vol. 5480, pp. 37–65. Springer, Heidelberg (2009)
14. Kavvadias, D.J., Stavropoulos, E.C.: An efficient algorithm for the transversal hypergraph generation. Journal of Graph Algorithms and Applications 9 (2005)
15. Zadeh, L.A.: Fuzzy sets. Information and control 8(3), 338–353 (1965)
16. Magnani, M., Rizopoulos, N., Brien, P.M.C., Montesi, D.: Schema integration based on uncertain semantic mappings. In: Delcambre, L.M.L., Kop, C., Mayr, H.C., Mylopoulos, J., Pastor, Ó. (eds.) ER 2005. LNCS, vol. 3716, pp. 31–46. Springer, Heidelberg (2005)
17. Mailis, T., Stoilos, G., Stamou, G.: Expressive reasoning with horn rules and fuzzy description logics. In: Marchiori, M., Pan, J.Z., Marie, C.d.S. (eds.) RR 2007. LNCS, vol. 4524, pp. 43–57. Springer, Heidelberg (2007)
18. Mannila, H., Toivonen, H.: Levelwise search and borders of theories in knowledge discovery. DMKD 1(3), 241–258 (1997)
19. Nottelmann, H., Straccia, U.: splmap: A probabilistic approach to schema matching. In: Losada, D.E., Fernández-Luna, J.M. (eds.) ECIR 2005. LNCS, vol. 3408, pp. 81–95. Springer, Heidelberg (2005)
20. Yager, R.R.: Veristic Variables and Approximate Reasoning for Intelligent Semantic Web Systems. Studies in Fuzziness and Soft Computing, Preferences and Decisions Models and Applications 217, 231–249 (2007)
21. Rossazza, J.P.: Utilisation de hiérarchies de classes floues pour la représentation de connaissances imprécises et sujettes à exceptions: le système "SORCIER". Ph.D. thesis, Univ. Paul Sabatier de Toulouse (1990)
22. Schaerf, A.: Reasoning with individuals in concept languages. Data & Knowledge Engineering 13(2), 141–176 (1994)
23. Straccia, U.: Reasoning within fuzzy description logics. JAIR 14 (2001)
24. Tresp, C., Molitor, M.: A description logic for vague knowledge. In: ECAI, pp. 361–365 (1998)
25. Wimmers, E., Haas, H., Roth, X., Braendli, X.: Using fagin's algorithm for merging ranked results in multimedia distributed systems. In: Proc. of IFCIS 1999 (1999)
26. Zadeh, L.: Fuzzy sets as a basis for a theory of possibility. Fuzzy Sets and Systems 1, 3–28 (1978)

Personalizing Queries over Large Data Tables

Nicolas Spyratos, Tsuyoshi Sugibuchi, and Jitao Yang

Laboratoire de Recherche en Informatique, Université Paris-Sud 11, France
{Nicolas.Spyratos,Tsuyoshi.Sugibuchi,Jitao.Yang}@lri.fr

Abstract. We present a formal framework for the processing of preference queries over large data tables, in which user preferences are expressed as comparisons between attribute values (e.g. "I prefer *Red* to *Black*").The main contributions of the paper are as follows: (a) a formal framework for the statement of the problem, under no restrictions whatsoever on the preferences expressed by the user, (b) a rewriting algorithm that takes as input a preference query and returns a sequence of ordinary sub-queries whose evaluations construct the answer to the preference query, (c) a general definition of "skyline" and (d) a user-friendly interface supporting preference query formulation and incremental query evaluation with on-the-fly modification.

1 Introduction

Information personalization is an important aspect of the interaction between users and information systems. Its goal is to tailor, or customize the information returned to a user according to user needs and preferences. This paper focuses on a specific facet of information personalization, namely query personalization. The general idea is to allow users of information systems to express preferences online and take the preferences into account in presenting query results. Let us explain our approach informally using a very simple example.

Consider an internet company selling second hand cars through an e-catalogue T that can be thought of as a relational table (see Fig. 1). Each entry in T describes a car for sale, and T usually contains thousands or even tens of thousands of entries (or tuples). Potential customers access T to search for cars to buy. For example, consider a customer looking for a car which is either *Red* or *Black*. As the table T is often very large, the answer set is likely to contain hundreds of tuples. However, if the user can express a preference, say, "Red is preferred to Black", then the system can take this information into account in order to (a) compute the set T_{Red} of all tuples in T with *Red* as their Color attribute; and the set T_{Black} of all tuples in T with *Black* as their Color attribute and (b) present to the user first the tuples in T_{Red} and (if none of the cars in T_{Red} is of the user's liking) then present the tuples in T_{Black}.

If we consider the expressed preference P : "Red is preferred to Black" as a query, then the sequence $\langle T_{Red}, T_{Black} \rangle$ can be seen as the answer to P, in the sense that the set T_{Red} of red cars precedes the set T_{Black} of black cars (as required by P). The main advantage in viewing the preference P as a query

J. Eder, M. Bielikova, and A.M. Tjoa (Eds.): ADBIS 2011, LNCS 6909, pp. 271–284, 2011.
© Springer-Verlag Berlin Heidelberg 2011

and its answer as a sequence of sets of tuples is that the user will see the most preferred cars first (i.e. those in T_{Red}). Therefore it is quite likely that the user will like one of those cars, and that it will not be necessary to look for a car in the second set of tuples (i.e. the tuples in T_{Black}).

Note that if the preference is inverted (i.e. "Black is preferred to Red"), then the answer will be $\langle T_{Black}, T_{Red} \rangle$ that is, each of the sets T_{Black} and T_{Red} remains unchanged but the order in which these sets are presented to the user is inverted.

Now, it is not difficult to see that the sets of tuples T_{Black} and T_{Red} are the answers to the following ordinary queries: $Q_{Red} : (Color = Red)$ and $Q_{Black} : (Color = Black)$. Therefore we can rewrite the preference P as a sequence $\langle Q_{Red}, Q_{Black} \rangle$, where Q_{Red} and Q_{Black} are ordinary queries whose evaluations return the answer of P.

In the rest of the paper, we call *preference query* over T any set P of preferences, and we call *ordinary query* over T any boolean combination of elementary conditions of the form $\langle attribute \rangle = \langle value \rangle$; for example, $Q : (Color = Red) \land (Model = BMW)$ is an ordinary query over T. To simplify the presentation, we shall omit attribute names when writing down an ordinary query; for example, the previous query will be written as follows: $Q : Red \land BMW$ (the attributes Color and Model are understood from the values appearing in Q).

Note that what we call an ordinary query over T is just a usual SQL query, in which we omit the "select" clause (because we assume that each tuple is returned with all its attributes), and we also omit the "from" clause (because we assume only one table, namely the table T). In other words, an ordinary query is just the boolean combination of elementary conditions that appears in the "where" clause of a usual SQL query.

There are at least three advantages in performing the rewriting of a preference query P into a sequence of ordinary queries such as Q_{Black} and Q_{Red} above. First, the answer to P can be evaluated "incrementally" under user control. For example, if the user finds the car he is looking for in the result of Q_{Red} then he can ask the system to terminate the evaluation (i.e. not to evaluate Q_{Black}).

Second, the user can influence query evaluation "on the fly". For example, if the user doesn't find the car he is looking for in the result of Q_{Red} then it is reasonable to expect him to modify his preferences before evaluation of Q_{Black}, since Q_{Black} will return less preferred cars. For example he might ask the system to add a condition (such as $Model = BMW$, or $Price \leq 10000$, or order by "Mileage" etc.) before the system evaluates Q_{Black}.

Third, if the rewritng of the preference query and the presentation of its answer are done by an interface, then the whole process becomes transparent to the information system (i.e. the system evaluates just ordinary queries, submitted to it by the interface, and returns the answers).

The contributions of the paper can be summarized as follows:

- A formal framework for the statement of the problem under no assumptions whatsoever on the preferences expressed by the user.
- A rewriting algorithm that takes as input a preference query and returns a sequence of ordinary sub-queries whose evaluations construct the answer.

- A general definition of skyline with no restriction whatsoever over the preference relation producing it.
- A user-friendly interface, demonstrating the feasibility of the proposed approach.

In order to simplify the presentation, we shall assume that the information system is a relational database and that preference queries are expressed against a single table T, which is either present in the database or has been derived from the database tables using some SQL query. As mentioned earlier, our running example will be the table $T(Id, Model, Color, Price, Mileage)$ in which each entry describes a second hand car for sale (an instance of T is shown in Fig 1).

The paper is organized as follows. In section 2 we introduce the formal framework. In section 3 we present our rewriting algorithm assuming that the preferences are expressed over a single attribute. In section 4 we present rewriting algorithms in the case where preferences are declared over two or more attributes. In section 5 we present the basic ideas underlying our interface design. Section 6 concludes the paper with a discussion of related work, and perspectives.

Serial	Model	Color	Mileage	Price	Year
1	VW	red	35000	3800	2002
2	Clio	green	48000	4500	2001
3	VW	white	30000	3500	2003
...	...	...	...	...	...

Fig. 1. An instance of the table T of our running example

2   Basic Definitions

We begin this section by noting that there are various kinds of preferences and that they can be roughly classified in terms of their nature or in terms of their persistence in time. In terms of their nature preferences can be:

- *quantitative preferences* (or "absolute preferences"), which are expressed as a percentage capturing intensity of desire (e.g. I like BMW 80%, I like VW 30%, ...); however preferences of this kind are difficult to express by the casual user, albeit easy to compute by a machine (from query logs)
- *qualitative preferences* (or "relative preferences"), which are expressed by comparison and convey no intensity of desire (e.g. I like BMW more than VW); such preferences are easy to express by the casual user, and also easy to infer by a machine (from query logs)

In terms of their persistence in time preferences can be:

- *long term preferences*, which are either discovered by the system (unobtrusively, from query logs) or declared explicitly by the user (and in both cases stored in the so called "user profile")

- *short term preferences* (or "ephemeral preferences"), which are expressed explicitly by the user, online, and can be seen as queries.

We note that the nature of preferences and their persistence in time are orthogonal features. Long term preferences (whether qualitative or quantitative) are invoked during query evaluation and act as filters for delivering customized results. In contrast, short term preferences, are not stored in a profile; they are expressed on-line, explicitly, and override the user profile during query evaluation. We note that recommendation systems rely mostly on long term quantitative preferences stored in users' profiles.

In general, eliciting and managing user preferences is not a trivial task. Moreover, when preference elicitation is not performed over query logs but involves a dialogue with the user, a user friendly interface aiding the user to express preferences is indispensable.

In this paper, we focus on short term, qualitative preferences. Formally, qualitative preferences can be expressed as pairs of values from an attribute domain as defined below.

Definition 1 (Preference). *Let $T(A_1, \ldots, A_n)$ be a relational table; a preference over an attribute A of T is a pair (x, y), where x and y are values from the domain of A. The declaration of such a pair is interpreted as follows: "x is preferred to y", or "x precedes y". We shall call preference relation on A any finite set of preferences over A, and we shall denote it as $P.A$.*

For example, the following set of pairs is a preference relation over the attribute *Color* of our running example:

- $P.Color = \{(Red, White), (Red, Yellow), (Yellow, Blue)\}$

Each pair in this set represents a preference over the attribute "*Color*". As a preference relation is just a binary relation, we can represent it as a directed graph. This graph, called the *preference graph* of $P.A$, is defined as follows:

1. the value x is a node of the preference graph iff x appears in $P.A$.
2. $x \rightarrow y$ is an edge of the preference graph iff the pair (x, y) is in $P.A$.

A preference relation $P.A$ will be called (indifferently) a *preference relation*, a *preference graph* or a *preference query*; and a value x appearing in $P.A$ will be called (indifferently) a *value* or a *node*.

We stress here that, in our approach, no assumption is made on the shape of a preference relation. In particular, a preference relation might not be transitive and/or might contain cycles. In other words, the user can declare any set of preferences he wants. However, in defining the notion of rank below, we do make the assumption that the preference relation does not contain cycles (i.e. it is acyclic). Later on, in section 5, we shall see how this assumption can be relaxed.

Definition 2 (Rank). *Let $P.A$ be a preference relation whose preference graph is acyclic; then each node x of the preference graph can be associated to a nonnegative integer called the rank of x; the rank of x, denoted as $rank(x)$, is defined* *as follows: if x is a root then rank(x) = 0 else rank(x)= the maximal length of path among all paths going from a root to x.*

Note that not every value of A has necessarily a rank: a value a of A has a rank only if it appears in $P.A$.

The above definition of rank reflects the preferences expressed by the user in the following sense: (a) the roots are the most preferred values (as no value precedes a root) and (b) the further away from the roots a value is the less this value is preferred. Moreover, if two values have the same rank, then there is no edge between them; therefore they are non comparable, in the sense that they are equally distant from the roots and none of the two is preferred to the other.

The following algorithm takes as input an acyclic graph G and returns the sequence $B_0, B_1, \ldots, B_m$, where B_i is the set of all nodes of rank i and m is the maximal length of path among all paths starting from a root.

Ranking Algorithm

$Aux \leftarrow G, i \leftarrow 0$
while $Aux \neq \emptyset$ **do**
    $B_i \leftarrow \{r | r$ is a root of Aux$\}$
    $Aux \leftarrow Aux - \{$all the roots and the edges leaving the roots$\}$
    return B_i
    $i \leftarrow i + 1$
end while

The complexity of this algorithm is in the order of $n + e$, where n is the number of nodes and e the number of edges of G. Note that:

- this algorithm is a variant of the well known topological sorting algorithm.
- the sets (or "blocks") B_i form a partition of the set of nodes of G.
- there is no edge between any two nodes of B_i, $i = 0, \ldots, m$ (i.e. no node of B_i is comparable to any other node of B_i).
- for each node $y \in B_i$, there is a node $x \in B_{i-1}$ such that $x \rightarrow y$ is an edge of G, $i = 1, \ldots, m$.

3   Preferences over a Single Attribute

In this section we study preferences expressed over a single attribute, and in the following section we generalize the results to preferences over any number of attributes.

A basic assumption underlying our work is that a user declaring a preference relation $P.A$ is actually interested in *every* value a appearing in $P.A$ *but* in varying degrees; and in fact, the degree of interest in a value a is expressed by its rank[1]. This leads to the following definition of implied query:

---

[1] We also tacitly assume that a user declaring a preference query $P.A$ is *not* interested in any value *not* appearing in $P.A$ (i.e. we make a sort of Closed World Assumption, similar to the one for databases [18]).

Definition 3 (Implied Query). *Given a preference query $P.A$, let $a_1, a_2, \ldots, a_n$ be all the values of A appearing in $P.A$. The implied query of $P.A$, denoted $IQ(P.A)$, is the ordinary query defined as $IQ(P.A) = a_1 \lor a_2 \lor \ldots \lor a_n$.*

Fig. 2 shows a preference relation $P.Model$ and its implied query $IQ(P.Model)$. In order to define the answer of a preference query we need the notion of rank for a tuple.

Definition 4 (Rank of a Tuple). *Given a preference query $P.A$, the rank of a tuple t with respect to $P.A$ is defined as follows: if the value $t.A$ appears in $P.A$ then $rank(t) = rank(t.A))$ else $rank(t)$ is undefined.*

We can now define the answer to a preference query $P.A$ as follows.

Definition 5 (Answer of a Preference Query). *The answer to a preference query $P.A$ is defined to be the sequence $\langle T_0, T_1, \ldots T_m \rangle$, where T_i is the set of all tuples of rank i, and m is the maximal length of path among all paths starting from a root of the preference graph.*

To compute the answer of a preference query $P.A$ we can proceed in one of two ways, namely either by direct computation or by rewriting:

Direct Computation. We proceed in three steps as follows:

1. Compute the rank of each value appearing in $P.A$.
2. Compute the answer to the implied query $IQ(P.A)$.
3. Compute $T_i = \{t \in ans(IQ(P.A)) | rank(t) = i\}$, for $i = 0, 1, \ldots, m$.

Note that the higher the index i the less preferred the tuples of T_i are (i.e. T_0 contains the most preferred tuples, T_1 the next preferred tuples, and so on).

Rewriting. Let $B_0, B_1, \ldots, B_m$ be the sequence produced by the ranking algorithm applied to the preference graph, and let $B_i = \{a_{i1}, \ldots, a_{in_i}\}$, $i = 0, 1, \ldots, m$. Clearly, the set T_i can be obtained as the answer to the (ordinary) query $Q_i = a_{i1} \lor \ldots \lor a_{in_i}$, which is the disjunction of all values in B_i.

It follows that the preference query $P.A$ can be rewritten into a sequence of ordinary queries as follows:

1. Compute $B_0, B_1, \ldots, B_m$ (applying the ranking algorithm).
2. For $i = 0, 1, \ldots, m$, define $Q_i = a_{i1} \lor \ldots \lor a_{in_i}$.

Therefore the answer to $P.A$ is the sequence $ans(Q_0), ans(Q_1), \ldots, ans(Q_m)$, which is actually the sequence $T_0, T_1, \ldots, T_m$ of the direct computation *but* computed in a different way (i.e. $T_0 = ans(Q_0), \ldots, T_m = ans(Q_m)$).

Note that the rewriting of the query $P.A$ into the sequence $Q_0, Q_1, \ldots, Q_m$ is actually an ordered partition of the implied query $IQ(P.A)$ into m parts (sub-disjunctions) as shown in Fig. 2. Also note that one or more among the sets $T_i = ans(Q_i)$ might be empty. This is the case when none of the values $a_{i1}, \ldots, a_{in}$ of B_i appears in the current instance of T.

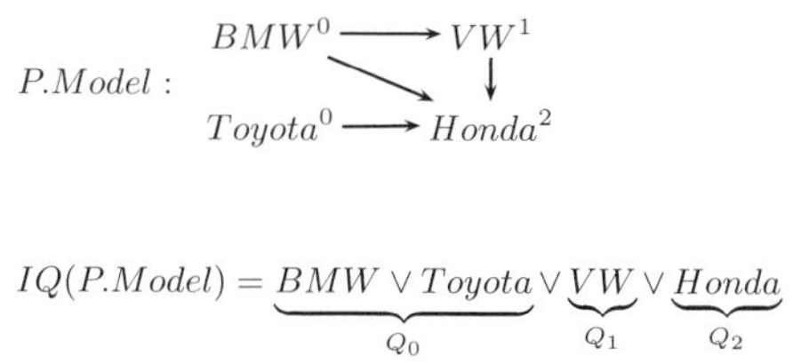

$P.Model$:
$$BMW^0 \longrightarrow VW^1$$
$$Toyota^0 \longrightarrow Honda^2$$

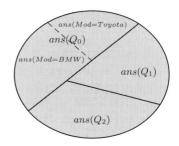

$$IQ(P.Model) = \underbrace{BMW \vee Toyota}_{Q_0} \vee \underbrace{VW}_{Q_1} \vee \underbrace{Honda}_{Q_2}$$

Fig. 2. The graph of $P.Model$ with its ranks and the implied query of $P.Model$

Fig. 3. The partitioned answer of implied query $IQ(P.Model)$

In principle, the answer to a preference query $P.A$ should be presented to the user in the form of pairs $(i, ans(Q_i))$, $i = 0, 1, \ldots, m$, even if $ans(Q_i)$ is empty for one or more values of i. Clearly, the first nonempty answer in the sequence $ans(Q_0)$, $\ldots$, $ans(Q_m)$ contains the "best" tuples *currently* available in the table. This first nonempty answer is called the *skyline answer*.

Definition 6 (Skyline). *Let $P.A$ be a preference query and let $T_0, T_1, \ldots, T_m$ be its answer (recall that $T_i = ans(Q_i)$). The skyline answer of $P.A$ (with respect to T) is defined to be the first non-empty T_i in the sequence; and if all T_is are empty then the skyline is defined to be empty.*

Hereafter, we shall simply say "skyline" instead of "skyline answer". Clearly, the skyline is nonempty iff at least one value appearing in $P.A$ also appears in T. Also, the skyline, say T_s, obviously depends on the current instance of T. Indeed, if one or more tuples are added in T, or removed from T, then the skyline might change in one of two ways: (a) T_s remains the first nonempty set in the answer of $P.A$ but contains different tuples than before the update of T, or (b) the skyline is not T_s any more but some T_r, where r is different than s.

The notion of skyline is an important notion. Indeed, when querying very large data sets (where answers are expected to be large as well) it is reasonable to expect that the user is interested only in the best tuples of the answer (where "best" is measured with respect to preferences expressed by the user).

The notion of skyline was first introduced in [4] for numeric attributes and has since been studied extensively in the context of data warehouses [14] [17]. It has generally been studied for numeric attributes, and to a lesser extent for categorical attributes [21].

Our approach to defining the skyline over categorical attributes generalizes that of [21], as we make no assumption on the preference relation $P.A$. In contrast, the approach in [21] makes the assumption that $P.A$ is a partial order over the domain of A.

4 Preferences over Two or More Attributes

Until now we have considered preferences over a single attribute A, and have defined a preference relation $P.A$ to be a binary relation over the domain of A.

Clearly, if the preference relation is a binary relation over tuples (e.g. over *Color and Model*), the approach remains the same except for one point: instead of having atomic values, such as "*Red*" or "*Yellow*" over the single attribute Color, we will now have tuples, such as "*Red* $\wedge$ *VW*" over the set of attributes $\{Color, Model\}$. In other words, if instead of having a preference relation $P.A$ we have a preference relation $P.\{A, B\}$, then the only difference is that each value appearing in $P.\{A, B\}$ is not an atomic value but a conjunction of two atomic values, one from the domain of A and the other from the domain of B. For example, the following is a preference relation over $\{Model, Color\}$.

$$P.\{Model, Color\}: \quad \begin{array}{c} Red \wedge VW^0 \\ \\ Black \wedge VW^0 \end{array} \longrightarrow Yellow \wedge BMW^1$$

Using the ranks we can rewrite $P.\{Model, Color\}$ as follows:

$$Q_0 : (Red \wedge VW) \vee (Black \wedge VW); \quad Q_1 : Yellow \wedge BMW$$

The answer to $P.\{Model, Color\}$ is now the sequence : $\langle ans(Q_0), ans(Q_1) \rangle$. However, things become more complex when the user wishes to express preferences over two or more attributes, but *independently* on each attribute. In other words, instead of expressing a single preference relation $P.\{A_1, \ldots, A_k\}$ over k attributes, the user wishes to express k preference relations $P.A_1, \ldots, P.A_k$, one over each of the k attributes. In this case, we adopt the following approach: first we generate a preference relation $P.\{A_1, \ldots, A_k\}$ from the given preference relations $P.A_1, \ldots, P.A_k$, and then we proceed as in the previous example of $P.\{Model, Color\}$. To do this we need to know whether each of $P.A_1, \ldots, P.A_k$ carries the same weight.

In this paper, we consider two cases: the case where all preference relations carry the same weight, known as the *Pareto* case (or "without priorities"); and the case where there is a priority over the k preference relations, known as the *prioritized* case (or "with priorities"). By *priority* over the attributes $A_1, \ldots, A_k$ we mean a total order over these attributes (and if such a priority is desired then it should be expressed by the user, together with the preference relations). The generated Pareto and prioritized preference relations over $A_1, \ldots, A_k$ are defined below, where P_a stands for "Pareto preference relation" and Pr stands for "Prioritized preference relation".

Note: We say that two tuples s and t of T are *equivalent* over $A_1, \ldots, A_k$. iff $s.A_i = t.A_i$ for all $i = 1, \ldots, k$ (i.e. iff s and t have the same value over each of $A_1, \ldots, A_k.$). In the following definitions all tuples are treated "up to equivalence".

Definition 7 (Pareto). *Let* $P.A_1, \ldots, P.A_k$ *be preference relations. The Pareto preference relation generated by* $P.A_1, \ldots, P.A_k$, *denoted by* $P_a(A_1, \ldots, A_k)$ *or simply by* P_a, *is defined as follows:* $(s, t) \in Pa$ *iff for all* $i = 1, \ldots, k$ *either* $s.A_i = t.A_i$ *or* $(s.A_i, t.A_i) \in P.A_i$, *where s and t are tuples of T which are not equivalent over* $\{A_1, \ldots, A_k\}$ *(i.e. their values differ over at least one* A_i).

Roughly speaking, s precedes t under P_a if $s.A_i$ either is equal to $t.A_i$ or precedes $t.A_i$ with respect to $P.A_i$, for all $i = 1, \ldots, k$.

In the example of Fig. 4, we see two preference relations $P.Model$ and $P.Color$ over the table T (hence $k = 2$ in this case). To derive the Pareto preference relation over $\{Model, Color\}$ we proceed as follows. First, we form all possible pairings (i.e. all possible 2-tuples) using one value appearing in $P.Model$ and one value appearing in $P.Color$; this gives four tuples over $\{Model, Color\}$, which are the nodes of the derived preference graph (see Fig. 4). To find the edges of the derived preference graph, we apply definition 7 to compare each node to all other nodes: if two nodes are comparable then we place an edge between them with the appropriate direction, otherwise no edge is placed between the two nodes. The resulting graph is the derived Pareto preference graph.

Definition 8 (Prioritized). *Let $P.A_1, \ldots, P.A_k$ be preference relations, and suppose (without loss of generality) that the priority over attributes is that of increasing index. The Prioritized preference relation over $A_1, \ldots, A_k$ generated by $P.A_1, \ldots, P.A_k$, denoted by $P_r(A_1, \ldots, A_k)$ or simply by P_r, is defined as follows: $(s, t) \in P_r$ iff*
either $(s.A_1, t.A_1) \in P.A_1$
or $[s.A_1 = t.A_1$ and $s.(A_2 \ldots A_k), t.(A_2 \ldots A_k) \in P_r(P.A_2, \ldots, P.A_k)]$
where s and t are tuples of T which are not equivalent over $\{A_1, \ldots, A_k\}$.

Roughly speaking, s precedes t under P_r if either $s.A_1$ precedes $t.A_1$ with respect to $P.A_1$ or (in case $s.A_1 = t.A_1$) $s.A_2 \ldots A_k$ precedes $t.A_2 \ldots A_k$ applying the previous rule recursively.

Continuing with the example of Fig. 4, in order to find the derived preference graph we proceed as in the Pareto case but this time applying definition 8. The result is shown in Fig. 4. Note that $Black \wedge VW$ and $Red \wedge Clio$ are non comparable under Pareto while they are comparable under Prioritized.

For a more detailed discussion on derived preference relations (with or without priorities), as well as on some variants of these derived relations, the interested reader is referred to [1][19].

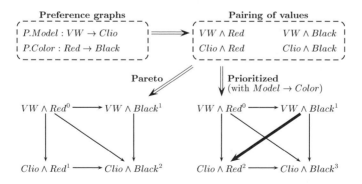

Fig. 4. Deriving the preference graph under Pareto and Prioritized

As the example of Fig. 4 shows, the number of nodes in the derived preference graph is the product of the number of nodes in the two given preference graphs. In general, the size of the derived preference graph grows rapidly with the number of given preference graphs and the number of nodes in each preference graph. On the other hand, as we have seen, the only reason why we use a preference graph is to compute the partition of its nodes into the blocks $B_0, ..., B_m$ that serve in the rewriting into a sequence of sub-queries. Hence the following question: is it possible to generate the partition $B_0, ..., B_m$ of the derived preference graph *without* constructing this graph, by simply combining the partitions of the given preference graphs? The answer is yes and the interested reader can find a complete account in the full version of this paper [22].

5   The User Interface

In our approach, query formulation and query evaluation are based on a dialogue between the user and the system. This dialogue is established through an interface which supports all the required steps, namely preference elicitation, rewriting, incremental evaluation with the possibility for on-the-fly modification, and presentation of the answer.

More precisely, when the user expresses a preference query P, the interface rewrites it into a sequence of sub-queries, as seen earlier. Actually the interface stores the sub-queries locally and passes control to the user for their evaluation (in a sense, the stored sequence of sub-queries constitutes a sort of "ephemeral" user profile). During the evaluation of P, the user can perform three main actions:

Action 1: Request the Skyline. In this case, the system returns the first pair $(i, ans(Q_i))$ for which $ans(Q_i)$ is nonempty. Optionally, the user can ask for the skyline *together with* a "order-by" and (eventually) a "top-k" clause. In this case the user will receive the top k tuples in the skyline, determined with respect to the attribute specified in the "order-by" clause. For example, if the user asks for the skyline answer together with the clauses "order-by increasing mileage" and "top-5", the interface will return the top 5 tuples from the skyline after the skyline has been sorted in increasing order of mileage.

Action 2: Request the Evaluation of the "Next" Sub-query. In this case (assuming that the user has already seen the answer to sub-query $Q_{(i-1)}$), the interface simply asks the system to evaluate the next sub-query, namely Q_i, and return the answer. Optionally, before the evaluation of Q_i takes place, the user can ask the interface to modify Q_i in one of the following ways:

1. Add a condition to Q_i. For example, let $Q_i = BMW \lor Toyota$ and suppose that the interface informs the user that the size of the answer to Q_i is in the order of hundreds. Then the user might decide to add the condition $Price \leq 8000$, in which case the interface will ask the system to evaluate the query: $Q_i \land (Price \leq 8000)$, instead of evaluating Q_i.[2]

---

[2] Estimating the size of the answer can be done using the frequency of appearance in T of the attribute values appearing in the query (and the frequencies of appearance in T can be estimated using probabilistic methods [16]).

2. Add a "order by" clause to Q_i and, optionally, a "top k" clause, as seen above.

Action 3: Terminate Evaluation. The user can terminate the evaluation of sub-queries, using a "next/stop" button. Typically, the "stop" button is used as soon as the user has found what he is looking for in the result of a sub-query.

Clearly, the above features allow the user to influence the evaluation of the query, inspect parts of the answer in decreasing order of preference, and stop evaluation at will. This way of evaluating the answer results in increased user convenience as well as in computational savings.

A last remaining point is the presence of cycles in the preference relation. As we mentioned earlier, the rewriting of a preference query into a sequence of sub-queries relies on the assumption that the preference graph contains no cycles. There are three approaches to cope with the creation of cycles during the the declaration of preferences:

- In the first approach, the user declares all his preferences, and if there is a cycle in the preference graph then the interface computes all (maximal) cycles, shows them to the user, and asks him to "break" them (by removing one or more of the declared preferences, as needed).
- In the second approach, the user declares all his preferences, and if there is a cycle in the preference graph then the interface transforms the preference graph into an acyclic graph by (a) considering all values in a cycle to be equivalent (i.e. of equal preference), and (b) coalescing all nodes of each maximal cycle into a single node (in this way, all cycles are "eliminated").
- In the third approach, the user declares his preferences one by one. After each preference is declared, the interface accepts the preference if its addition to the previously declared preferences creates no cycle, otherwise the interface follows one of two directions: refuse the declared preference or let the user declare all his preferences and then revert to one of the previous two approaches.

The main drawback of the first approach is that it requires quite sophisticated user intervention, namely inspecting all cycles and then "breaking" them; and a casual user might not have the desire or the time to do that.

The second approach does not require user intervention, and this is an advantage over the first approach. However, this approach introduces somewhat more complex semantics. Indeed, as we mentioned in the previous sections, our algorithms work "up to equivalence", where equivalence is understood as equality of tuples over the attributes used in declaring preferences; this kind of equivalence is quite natural to understand. However, if the second approach is adopted, then a new kind of equivalence is introduced (whereby two tuples are equivalent if they are on the same cycle). In presence of these two kinds of equivalence, the semantics of the answer becomes quite involved.

The third approach provides a sort of compromise in the following sense: if the user is prepared to accept refusal of some of his preferences then this approach becomes incremental (as to the declaration of preferences) and requires no user intervention (therefore more convenient for the user); otherwise this approach becomes similar to the first two approaches.

Note that the first two approaches need an algorithm to find all (maximal) cycles in a graph, and such algorithms do exist in the literature [13]; moreover, the preference graph is usually of small size, therefore computational time is not an issue.The third approach (in its simplest form) requires just an algorithm to check whether the addition of an edge to an acyclic graph creates a cycle; this can be easily checked by running a simple variant of the topological sorting algorithm (roughly, the addition of the new edge creates no cycle iff the result of applying topological sorting is the empty graph).

The current implementation of our interface follows the simplified form of the third approach, whereby a declared preference is accepted iff it does not create a cycle. Clearly, the choice of one of the approaches described above depends on the application environment and the user profiles, and its validation can only be done through experimentation.

6   Related Work and Concluding Remarks

Using preferences for ranking alternative choices has been around for over two centuries [2][7], and there is a huge body of literature in decision making and social choice theory since the 1950s (see [23] for a survey). However, the use of preferences in ranking query answers in information systems is quite recent and their embodiment in a query language presents a number of subtle problems [20]. Influential, recent papers on the subject include Andreka [1], Chomicki [5][6], and Kießling [11][12].

Probably the most important features that distinguish our approach from existing approaches are: (a) the definition of the answer to a preference query through rewriting, (b) incremental evaluation with on-the-fly modification of intermediate sub-queries and (c) a general definition of skyline and its computation by an ordinary query (with no restriction over the preference relation).

We note that the definition of the answer to a preference query through rewriting has also been considered in previous work by the first author of the present paper ([15] and [10]). However, the approach proposed in the present paper differs from those in [15] and [10] as follows:

– A preference query in [15] and [10] is defined to be a pair (Q, P), where Q is an SQL query and P a set of preferences over one or more attributes. However, the query Q is considered to have priority over the preferences P and therefore the semantics of the answer is different than in the present work.
– In [15] and [10] there is no possibility for on-the-fly modification before the evaluation of a sub-query, and no notion of skyline.
– In contrast to the present work, the approach in [10] assumes transitivity of the preference relation, an assumption which is debatable in the literature [9], and which limits the generality of the approach.

We are currently studying several aspects of the interface that we have implemented. The first concerns providing help to the user during preference declaration. Indeed, when a user declares preferences, he is not aware of the data values currently available in the table. Therefore we consider the possibility of adding a *data preview* facility, whereby the user can select an attribute of the table and have a look (by scrolling) at the current values of that attribute before declaring preferences. The second aspect concerns discretization of numeric attribute domains. For instance, referring to our running example, we would like to give the user the possibility to declare keywords of the sort "$LowPrice : Price \leq 5000$" and "$HighPrice : Price \geq 5000$", and use such keywords in the declaration of preferences. A third aspect that we consider is the organization of discrete domains into hierarchies (e.g. `dark`: {blue, black, green}, `light`: {white, yellow}) and its influence in preference declaration. For example, if the user declares a preference of `dark` over `light` how should we interpret it: "for every color x in `dark` and for every color y in `light` x is preferred to y"? or should it be: "for every color x in `dark` there is color y in `light` such that x is preferred to y"? and so on. A final aspect that we are studying is the usability of the interface. Indeed, in spite the choice of a rather simple mode of interaction, the combined activity of expressing queries and preferences at the same time might prove "cognitively" difficult to the casual user.

Regarding further work, there is one direction which is promising in our opinion. As pointed out in [3], most of the work done on preferences in the context of information systems takes into account the attributes over which preferences are declared but ignores completely the values of the attributes over which no preferences are declared. Therefore the authors of [3] propose a ceteris paribus assumption on these other attributes. We believe that what is actually needed is a preference specification language which allows the user to declare both, preferences and assumptions, on all attributes of the table. We are currently working towards that direction.

Acknowledgements. This work is partially supported by the following projects:

- European project ASSETS: Advanced Search Services and Enhanced Technological Solutions for the European Digital Library (CIP-ICT PSP-2009-3, Grant Agreement no 250527)
- French CNRS project NOVA: A Novel Data Model and Query Language for Digital Libraries (PICS 5220)

References

1. Andreka, H., Ryan, M., Schlobbens, P.-Y.: Operators and Laws for Combining Preferential Relations. Jrnl. of Logic and Computation 12(1), 13–53 (2002)
2. Borda, J-c.: Mémoire sur les élections au scrutin, Histoire de l'Académie Royale des Sciences, Paris (1781)

3. Boutilier, C., Brafman, R., Hoos, H., Poole, D.: Reasoning with conditional ceteris paribus preference statements. In: Proc. of Conf. on Uncertainty in Artificial (UAI) 1999, pp. 71–80 (1999)
4. Borzsonyi, S., Kossman, D., Stocker, K.: The Skyline Operator. In: Proc. of Conf. on Data Engineering (ICDE) 2001, pp. 421–430 (2001)
5. Chomicki, J.: Querying with Intrinsic Preferences. In: Jensen, C.S., Jeffery, K., Pokorný, J., Šaltenis, S., Hwang, J., Böhm, K., Jarke, M. (eds.) EDBT 2002. LNCS, vol. 2287, pp. 34–51. Springer, Heidelberg (2002)
6. Chomicki, J.: Preference formulas in relational queries. ACM Trans. Database Syst. 28(4), 427–466 (2003)
7. Condorcet, J.A.N.: Essai sur l'application de l'analyse à la probabilité des décisions rendues à la pluralité des voix. Kessinger Publishing, New York (1785)
8. Europeana - Homepage, http://www.europeana.eu/
9. Fishburn, P.C.: Nontransitive Preferences in Decision Theory. Jrnl. of Risk and Uncertainty 4(2), 113–134 (1991)
10. Georgiadis, P., Kapantaidakis, I., Christophides, V., Mamadou Nguer, E., Spyratos, N.: Efficient Rewriting Algorithms for Preference Queries. In: Proc. of Intl Conf. on Data Engineering (ICDE) 2008, pp. 1101–1110 (2008)
11. Kießling, W., Köstler, G.: Preference SQL - Design, Implementation, Experiences. In: Proc. of Very Large Data Bases (VLDB) 2002, pp. 990–1001 (2002)
12. Kießling, W.: Foundations of Preferences in Database Systems. In: Proc. of Very Large Data Bases (VLDB) 2002, pp. 311–322 (2002)
13. Mateti, P., Deo, N.: On algorithms for enumerating all circuits of a graph. SIAM J. Comput. 5, 90–99 (1976)
14. Morse, M., Patel, J.M., Jagadish, H.V.: Efficient skyline computation over low-cardinality domains. In: Proc. of Very Large Data Bases (VLDB) 2007, pp. 267–278 (2007)
15. Nguer, E.M., Spyratos, N.: A User-friendly Interface for Evaluating Preference Queries Over Tabular Data. In: Proc. of ACM Intl. Conf. on Design of Communication (SIGDOC) 2008, pp. 22–24 (2008)
16. Poosala, V., Ioannidis, Y.E.: Selectivity Estimation Without the Attribute Value Independence Assumption. In: Proc. of Very Large Data Bases (VLDB) 1997, pp. 486–495 (1997)
17. Papadias, D., Tao, Y., Fu, G., Seeger, B.: An optimal and progressive algorithm for skyline queries. In: Proc. of ACM SIGMOD Intl. Conf. on Management of Data 2003, pp. 467–478 (2003)
18. Reiter, R.: On closed world databases. In: Gallaire, H., Minker, J. (eds.) Logic and Data Bases, pp. 119–140. Plenum, New York (1978)
19. Spyratos, N., Meghini, C.: Combining Preference Relations: Completeness and Consistency. In: Proc. of Intl. Workshop on Personalized Access, Profile Management, and Context Awareness in Databases, PersDB 2010 (2010)
20. Stefanidis, K., Koutrika, G., Pitoura, E.: A Survey on Representation, Composition and Application of Preferences in Database Systems. ACM Trans. on Database Systems (TODS) 36(4) (to appear 2011)
21. Sacharidis, D., Papadopoulos, S., Papadias, D.: Topologically Sorted Skylines for Partially Ordered Domains. In: Proc. of IEEE Intl. Conf. on Data Engineering (ICDE) 2009, pp. 1072–1083 (2009)
22. http://www.lri.fr/spyratos/papers/adbis2011-full.pdf
23. Tsoukias, A.: From Decision Theory to Decision Aiding Methodology. European Jrnl. of Operational Research 187, 138–161 (2008)

An Analysis of the Structure and Dynamics of Large-Scale Q/A Communities

Daniel Schall and Florian Skopik

Distributed Systems Group, Vienna University of Technology
Argentinierstrasse 8/184-1, A-1040 Vienna, Austria
lastname@infosys.tuwien.ac.at,
http://www.infosys.tuwien.ac.at

Abstract. In recent years, the World Wide Web (WWW) has transformed to a gigantic social network where people interact and collaborate in diverse online communities. By using Web 2.0 tools, people contribute content and knowledge at a rapid pace. Knowledge-intensive social networks such as Q/A communities offer a great source of expertise for crowdsourcing applications. Companies desiring to outsource human tasks to the crowd, however, demand for certain guarantees such as quality that can be expected from returned tasks. We argue that the quality of crowd-sourced tasks greatly depends on incentives and the users' dynamically evolving expertise and interests. Here we propose expertise mining techniques that are applied in online social communities. Our approach recommends users by considering contextual properties of Q/A communities such as participation degree and topic-sensitive expertise. Furthermore, we discuss prediction mechanisms to estimate answering dynamics considering a person's interest and social preferences.

Keywords: Online communities, Expertise mining, Crowdsourcing.

1 Introduction

The collaboration landscape has changed dramatically over the last years by enabling users to shape the Web and availability of information. While in the past collaborations were bound to intra-organizational collaborations using company-specific platforms, and also limited to messaging tools such as email, it is nowadays possible to utilize the knowledge of an immense number of people participating in collaborations on the Web. The shift toward the Web 2.0 allows people to write blogs about their activities, share knowledge in forums, write Wiki pages, and utilize social platforms to stay in touch with other people. Task-based platforms for human computation and *crowdsourcing* enable access to the knowledge of thousands of people on demand by creating human tasks that are processed by the crowd. Platforms for human computation including Amazon Mechanical Turk (MTurk)[1] are examples for crowdsourcing applications associated to the business domain. Human tasks include activities such as

---

[1] http://www.mturk.com/

J. Eder, M. Bielikova, and A.M. Tjoa (Eds.): ADBIS 2011, LNCS 6909, pp. 285–301, 2011.
© Springer-Verlag Berlin Heidelberg 2011

designing, creating, and testing products, voting for best results, or organizing information.

In open and dynamic crowdsourcing environments it becomes essential to manage expertise profiles and reputation of people in an automated manner. Here we argue that the quality of crowd-sourced tasks depends to a large extent on the users' dynamically evolving expertise and interests. Companies desiring to outsource human tasks to the crowd typically demand for certain guarantees such as quality that can be expected from returned tasks. Therefore, we propose expertise mining techniques that are applied in online social communities. Somebody seeking help or advice on a specific problem or businesses issuing task requests using, for example, crowdsourcing platforms need to be able to find the right person who can assist by offering his or her expertise. Work in expert finding, see for example [4], has been addressing the search for persons with the right skill level by using ontologies and by combining diverse semantic information from user skill profiles. Since Web-scale collaborations involving a large amount of people does not only demand for scalable algorithms and ranking solutions, but in many cases it is also desirable to consider the global properties of a human interaction network to determine the importance of users.

In this paper, we present research performed in the context of Q/A communities and crowdsourcing with the following key contributions:

- We discuss the key characteristics of Q/A communities by studying the properties of a real-world dataset obtained from Yahoo! Answers [19]. Investigating interactions and Q/A behavior of people using Yahoo! Answers allows us to analyze community structures and evolution over time. Our analysis has important implications for the design of future crowdsourcing applications that demand for high-quality results of delivered crowd-sourced tasks.
- We highlight expertise mining techniques that are applied in online social communities. Our approach recommends users by considering contextual properties of Q/A communities such as participation degree and topic-sensitive expertise.
- We propose link analysis techniques derived from popular ranking and mining algorithms such as PageRank [14]. The presented approach accounts for *social dynamics* in Q/A communities. Furthermore, we propose prediction mechanisms to estimate answering dynamics considering a person's interest and social preferences.
- Our evaluations and discussions are based on the properties of a real Q/A community. Our experiments confirm that our proposed ranking approach is well suited for expertise mining and recommendations.

This paper is organized as follows: In Sect. 2 we present related work in the area of online communities and social network analysis techniques applied to expertise networks. Sect. 3 is concerned with the definition of interactions in Q/A systems and basic community characteristics. In Sect. 4, we present our expertise mining model that is based on fine-grained contextual answering behavior followed by our recommendation approach in Sect. 5. Finally, we conclude the paper in Sect. 6.

2   Background and Related Work

The availability of experts in emerging Q/A Web communities raises a number of research issues and provides unique opportunities for new business models. Recent crowdsourcing environments enable novel ways to utilize external expert knowledge. When focusing on large-scale networks, where thousands of individuals are interlinked through a social and collaborative network, the broad applicability of efficient and scalable algorithms becomes evident.

Discovery of experts in Q/A communities. Experts contribute their knowledge in various fields by providing answers, e.g., posting comments, in respective sub-communities [1]. However, since Q/A communities are typically open platforms (everyone can join), quality of provided content becomes a key issue [2]. One approach to cope with quality issues is to find authoritative users based on the link-structure in Q/A communities [8,11]. Well-established ranking models including HITS [12] and PageRank [14] have been applied and tested in online communities [21]. Our approach is to find experts with respect to different topics [15]. Most existing expertise mining approaches do not consider user posting behavior in different topics. An approach for calculating *personalized PageRank* scores was introduced in [9] to enable topic-sensitive search on the Web. In [18], topic-sensitive ranking in Twitter networks was discussed. The connectivity of users in different *contexts* (e.g., topics) may be very different based on their personal interaction preferences such as frequency of exchanged messages [6]. Thus, considering context and discussion topics may help to discover *relevant experts* in sub-communities.

Routing of requests within Q/A communities. The best candidate user for answering a given question may be the person with the highest reputation in a specific sub-community *and* with the highest number of trusted links. Aardvark [10], for example, is a social search engine that utilizes the social network structure of its users to route questions to be answered by the community. In such a system, it is essential to maintain an up-to-date view on users interests and trust relations. A wide range of computational trust models to control interactions have been proposed (e.g., [3]). Our approach is related to social trust [7,16,22] relying on previous behavior. In such Q/A communities, one of the key issues also is to provide recommendations to users whose questions they should answer based on their evolving interests.

Integration of crowds in enterprise systems. An emerging new business model is to integrate the capabilities of crowds in enterprise systems through crowdsourcing techniques [5]. The first commercially available platform was MTurk offering the crowd's capability via Web service-based interfaces. One of the key challenges in crowdsourcing is to estimate the quality that can be expected from a particular crowd member. This could be done based on the user's expertise. If the quality of returned task results is expected to be low (due to missing expertise), different actions can be taken such as issuing multiple instances of the same task and voting for best results. However, since open Web-based platforms are subject to frequent changes (members are joining and leaving), the user's expertise must be calculated automatically.

3   The Yahoo! Answers Community

The goal of this work is to study the characteristics of Q/A communities. To date, Yahoo! Answers (YA) [19] is one of the biggest and most successful Q/A community on the Web. YA was established in 2005 and attracts a large number of users to ask questions regarding different topics. The goal of our analysis of the YA community is twofold:

1. In this work, we analyze the basic properties of the popular YA community with the aim of providing insights in the structure and Q/A behavior of users. The results could be a valuable input when designing, for example, new rewarding or reputation schemes in online communities.
2. By understanding the basic properties of the community, we design a topic-sensitive expertise mining algorithm. Also, understanding the evolution of user interest plays a significant role when recommending questions or users in Q/A communities.

3.1   Interactions and Actor Relations

In the following (Fig. 1) we give an overview of the basic interactions as commonly found in Q/A communities such as YA.

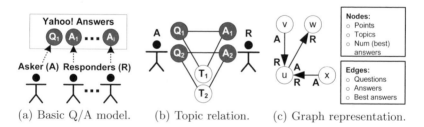

(a) Basic Q/A model.   (b) Topic relation.   (c) Graph representation.

Fig. 1. Representation of Q/A communities

Figure 1(a) shows the very basic actions taken by *Askers* (**A**) and *Responders* (**R**). Askers post questions that may be replied by answers of one or more responder(s). Typically, one answer is selected as the best answer, thereby setting the status of a question from open to closed. However, if the asker does not opt for a best answer, other community members may vote for a best answer and close a given question. Users get points for answering questions and more points if answers are selected as best answer (see [20] for details about the YA scoring system). The YA system has a predefined set of categories (hereafter refereed to as *topics*) in which questions are posted. Categories typically depict popular *question topics*. Thus, as shown in Fig. 1(b), each question and answer between

A and **R** is associated with a topic identifier. Of course, users may post and answer questions regarding different topics.

A central theme of our work is the automated analysis of expertise based on the users' answering behavior. The basic idea of our approach is to apply sound techniques from the information retrieval and social network analysis domain. An intuitive approach is to model the Q/A community as a directed graph $G(N, E)$ composed of the set of node N and the set of edges E. Nodes represent users and edges are established based on interactions that are derived from the underlying question/answering behavior. Given the YA system, each node has a set of properties such as the total collected points, set of topics the user is engaged in, number of answers as well as best answers, and so forth. Fig. 1(c) shows the essential graph model and the meaning of directed edges. In our approach, an edge points from, e.g., v to u, in standard graph notation written as $(v, u) \in E$, if u responded to v's question. Each edge between nodes holds further information such as the number of questions being asked and answered between a pair of nodes u and v.

3.2 Q/A Structure and Community Evolution

The presented analysis of the answers community is based on user interactions (questions and answers) that were posted from 2005 to 2008. We start our discussion of the YA community by analyzing basic statistics. In the following we show the amount of questions and answers posted over time.

Table 1. Q/A volume over time

	2005	2006	2007	2008
Number of questions $|Q|$	15886	76084	74501	50088
Number of answers $|A|$	40797	828282	790918	304138
Ratio A/Q	≈ 2.6	≈ 11	≈ 11	≈ 6
$|Q|_{year}/|Q|_{total}$	7%	35%	34%	23%
$|A|_{year}/|A|_{total}$	2%	42%	40%	15%

In Table 1, the number of questions and answers in consecutive years from 2005 to 2008 is shown. YA was established in 2005 so the number of questions $|Q|$ and answers $|A|$ is the lowest in this year. An answer per question ratio A/Q of approximately 11% can be observed (i.e., there is a high chance that a question receives multiple answers). Crawling of the dataset was performed until half of 2008. Thus, 2006 and 2007 represented the only years of full Q/A volume. In 2008, the A/Q ratio is lower due to a cutoff in the YA dataset.

The Q/A behavior in the YA community is best approximated as a power-law distribution, which is commonly used to describe the scaling laws of many naturally evolving social networks such as email communications or the access of Web resources [17]. We show the distributions of both *questions per user* and *answers per user* (see Fig. 2). The results show the distributions for the entire interval 2005-2008. The distributions in respective sub-figures can be modeled

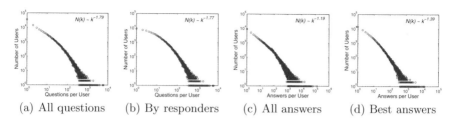

(a) All questions   (b) By responders   (c) All answers   (d) Best answers

Fig. 2. Question and answer distributions

as $N(k) \sim k^{-x}$ with exponent x. First, we analyzed the distribution of questions per user (see Fig. 2(a) with $x = -1.79$) and questions per user who also *respond to answers* (see Fig. 2(b) with $x = -1.77$). As in many Q/A communities, some users only ask questions (being askers only) and some users ask and answer questions. Second, we analyzed the community structure with respect to answers per users (see Fig. 2(c) with $x = -1.19$) and best answers per user (see Fig. 2(d) with $x = -1.39$).

In Fig. 3 we show the answering behavior per question. In Fig. 3(a), one can see that a large number of questions is answered by at least one response. Some questions are answered by a large number of responders. This is typically the case when users ask for opinions rather than factual answers. For example, in some cases the YA platform is used like a discussion forum to chat with other users about current news or celebrities [8]. This behavior is reflected by the tail of the distribution in Fig. 3(a).

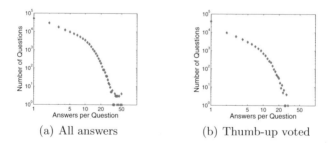

(a) All answers   (b) Thumb-up voted

Fig. 3. Answers per question

By selecting only those answers that were voted using the *thumb-up* feature of YA (a 'like' button), the maximum number of answers per question moves from about 50 to about 30 (see the tail of the distribution in Fig. 3(b)). Given that a large number of questions is answered by more than one answer, recommending users who are able to provide high-quality answers becomes important in Q/A communities. For example, a ranked list of answers could be presented to the asker based on the responders community reputation and experience.

The next step in this work is to analyze the role of askers and responders. The interesting aspect in YA is the role of *two-sided markets* [13]. In YA, users get

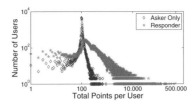

Fig. 4. Total points asker and responder

100 points by signing-up to the platform [20]. For each answer being provided, users get additional points (more points if the answer is selected as best answer). However, users lose points if they ask questions, thereby encouraging members to provide answers. Based on the rewarding scheme in YA, users tend to have either role – being asker or responder – instead of having both roles. This behavior is shown by Fig. 4 where we plot the total number of points and the count of users considering (i) askers only (see diamond shaped symbols) and (ii) responders (star-shaped symbols). One can see that users who only ask question have typically considerably less points than those who also answer questions (responders). Given those two sets of users, the highest scoring asker collected a total points value of 16.431 whereas the maximum value for responders is 425.225.

This observation means that many users who respond to questions also tend to use the YA platform over a longer period of time since more points are accumulated over time. On the contrary, many users just try using the platform by signing up to YA and asking a single question (see the spike at 100 points in Fig. 4). The details regarding user participation over time are shown in Table 2 by analyzing the overlap of community members (based on their Q/A behavior) in consecutive years and within the total duration from 2005 to 2008.

Table 2. User participation over time

	2005/2006	2006/2007	2007/2008	2006/2008
$sim(y_a, y_b)_A$	9%	4%	5%	1%
$sim(y_a, y_b)_R$	47%	31%	39%	12%
$sim(y_a, y_b)_{A \cap R}$	20%	5%	7%	1%

We use the following equations to obtain the entries in Table 2:

$$sim(y_a, y_b) = \frac{|U_{y_a} \cap U_{y_b}|}{|U_{y_a}|} \qquad (1)$$

The first equation (Eq. 1) calculates the similarity of sets of users U_{y_a} and U_{y_b} between two years y_a and y_b (these need not to be consecutive years). Thus, Eq. 1 calculates the overlap of how many users participating in year y_a also participated in year y_b in YA. In other words, Eq. 1 denotes the containment of y_a in y_b. The results are shown in Table 2 as $sim(y_a, y_b)_A$ (for askers only) and $sim(y_a, y_b)_R$ (responders only). One can observer that the number of responders continuing using the platform is much higher than the number of askers. This fact is even more drastic by comparing the overlap of users between the years 2006 and 2008. In particular, only 1% of askers who signed up in 2006 also continued to ask questions in 2008, whereas 12% of responders kept using the platform also in 2008.

Second, we calculate the similarity $sim(y_a, y_b)_{A \cap R}$ of users who are askers *and* responders as follows:

$$sim(y_a, y_b)_{A \cap R} = \frac{|U^A_{y_a} \cap U^R_{y_a} \cap U^A_{y_b} \cap U^R_{y_b}|}{|U^A_{y_a} \cap U^R_{y_b}|} \qquad (2)$$

The results are again shown in Table 2 (last row). The percentage of users who have both roles and continue using the platform in consecutive years is higher than for askers only but still much lower as compared to the set of responders. This is especially true for the comparison 2006/2008 where we observe equal values of $sim(y_a, y_b)_A$ and $sim(y_a, y_b)_{A \cap R}$. Thus, users who ask and answer questions do not typically use the platform over a longer period of time.

3.3 Community Topics

The YA community is structured in *topics* helping users to quickly access a set of relevant questions and answers. Before designing a topic-sensitive expertise mining approach, it is useful to understand the preference of users to post in one or more topics (i.e., whether users post in different topics at all). Figure 5 visualizes the number of topics per user.

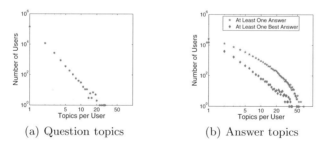

(a) Question topics   (b) Answer topics

Fig. 5. Number of topics per user

In Fig. 5(a), the number of topics by askers are shown. Since many users only ask one question (cf. previous section), the majority of askers is active in one topic only. The responders' behavior is shown in Fig. 5(b). We count the number of topics per user if the user provided at least one answer. Based on the results, we expect that a large fraction of responders typically post in 20-50 topics. However, by looking at those users who provided at least one best answer, about 20 topics by responder can be expected.

In the next step, we picked questions and answers in one particular year (2007) and analyzed the following properties: (i) which are the most discussed topics,

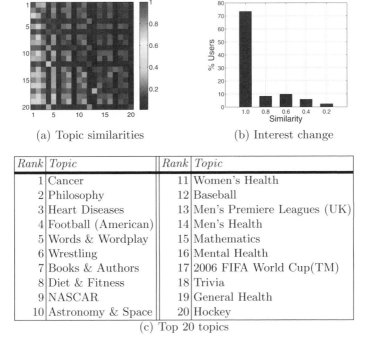

(a) Topic similarities   (b) Interest change

Rank	Topic	Rank	Topic
1	Cancer	11	Women's Health
2	Philosophy	12	Baseball
3	Heart Diseases	13	Men's Premiere Leagues (UK)
4	Football (American)	14	Men's Health
5	Words & Wordplay	15	Mathematics
6	Wrestling	16	Mental Health
7	Books & Authors	17	2006 FIFA World Cup(TM)
8	Diet & Fitness	18	Trivia
9	NASCAR	19	General Health
10	Astronomy & Space	20	Hockey

(c) Top 20 topics

Fig. 6. Top answer topics in YA (2007) and interest evolution

(ii) how much overlap is between the sub-communities in terms of users posting in multiple topics, and (iii) the evolution and change of user interest. The results are illustrated in Fig. 6. First, we rank each topic based on the number of answers, which is a good indicator for community interest (see Top 20 topics shown in Table 6(c)). For each topic in Table 6(c), we calculate how similar the user base is compared with another top-ranked topic using the formula $sim(T_a, T_b) = \frac{|U_{T_a} \cap U_{T_b}|}{|U_{T_a}|}$. In other words, similarity is expressed by how many users post in both topics T_a and T_b. This measure is useful when comparing the impact and results of topic-sensitive expertise mining. If topics (sub-communities) have virtually now overlap in their user base, it does not bring any benefit to rank experts in joint topics. The similarities of (ranked) topics are shown in Fig. 6(a). Here, the elements are interpreted as follows: given a top-ranked topic, say 'Cancer' (Rank 1), the similarity to other topics is shown from the left to the right were the index corresponds to the rank in Table 6(c). The color shade of each element in the topic comparison matrix corresponds to the numerical similarity value (see the colorbar on the right).

For all users who were active in 2007, we compared their interest evolution by analyzing user topics (in which topics users were active) in 2007 and 2008. We created a chart (Fig. 6(b)) that shows the percentage of users within the similarity intervals $[1, 0.8), [0.8, 0.6), [0.6, 0.4), [0.4, 0.2)$, and $[0.2, 0]$ according to the similarity value calculated using the definition of set overlap as discussed in the previous paragraph. Fig. 6(b) shows that 73% of users mostly post in the same topics they had posted in the past year. The remaining 27% have changing interests (to some degree) whereby a very low fraction of users have a high interest change (see most right category 0.2).

4   Topic-Sensitive Expertise Mining in Q/A Communities

Here we propose link-based expertise mining techniques. Open Web-based communities are governed by changes such as people joining and leaving the community. It is therefore important to automate aspects such as the management of interest and expertise profiles due to scale and temporary nature of communities. We propose the application of well-established ranking methods used by search engines to estimate users' experiences based on the link structure of Q/A communities. The goal of this work is to establish a *relative order* of users based on their participation degree. The top-ranked users (i.e., responders) are considered to be authoritative experts who are recommended for answering questions regarding a specific topic.

Specifically, we propose the PageRank method [14], which can be personalized for different topics as discussed in [9]. Consider a graph structure as previously introduced (cf. Fig. 1(c)), where edges point from askers $\mathbf{A}$ to responders $\mathbf{R}$. An edge is thus established whenever $\mathbf{R}$ responds to a question from $\mathbf{A}$, regardless whether the answer was selected as best answer. The PageRank $PR(u)$ of a node u is defined as follows:

$$PR(u) = (1-\alpha)p(u) + \alpha \sum_{(v,u) \in E} \frac{PR(v)}{\texttt{outdegree}(v)} \qquad (3)$$

The equation consists of two parts. The first part $p(u)$ is called *personalization vector* that is used to assign preferences to certain nodes (we shall discuss shortly how this is done). Without any preferences or bias towards a particular node, a personalization vector $p = [\frac{1}{|N|}]_{|N| \times 1}$ is assumed. The second part of Eq. 3 can be regarded as u's relative standing (reputation) within the entire network. The factor α is typically set to 0.85 because a random surfer on the Web is assumed to follow six links $(1-\alpha) = 1/6 = 0.15$ and 'teleports' with probability 0.15 to a randomly selected Web page.

Our approach is to assign preferences to users based on their answering behavior in a particular topic of interest. For example, if experts need to be discovered with regards to a topic T (see footnote[2]), then preferences are assigned to those users who have answered questions with respect to this topic T. Formally, the topic-sensitive personalization vector p is assigned as follows:

$$personalization_{topic} : (u, T) \mapsto \begin{cases} 0, & \text{if } u \text{ has not been active in } T \\ answers(u, T), & \text{otherwise} \end{cases}$$

---

[2] Notice, T could be a single topic or a joint topic with $T = \{T_a, T_b\}$.

The function $answers(u, T)$ returns the number of *all* answers of u in topic T. Answers may include best answers or even answers that were not useful at all (e.g., even thumb-down voted answers). Our hypothesis is that both answering behavior of users in a certain topic (i.e., personalization) and also a user's global (community wide) reputation are important.

Results. We performed three experiments in different topics and combinations thereof. All experiments are visualized in Fig. 7 (top-25 ranked users that are ordered based on their ranking scores). Ranking results have been obtained by using the personalized PageRank model as discussed before using the parameter $\alpha = 0.75$. If α was set to a higher value, ranking scores were biased too much towards the global reputation of a user. On the other hand, by setting α too low, also 'spammers' would be ranked high with, however, low community reputation. Each node in Fig. 7 is labeled by its position in the ranking result. The position of a node corresponds to an entry (# column) in Table 3. Only edges that connect nodes within the top-25 rankings results are shown. Also, to understand the quality of our results, we rank nodes according to (i) best answer count BAC (not topic sensitive) and topic-based best answer count BAC^T (the rank R is shown in parenthesis in Table 3).

This information is also encoded in the nodes's color and shape. The border of nodes is octagon-shaped and nodes have black fill color if nodes are ranked within the top-25 list by BAC and within the top-25 list by BAC^T. Nodes with fill color gray are only ranked in a high position by our method, but not by BAC or BAC^T. Nodes with dark borders and gray core are ranked high by our approach and also by BAC. Nodes with gray borders and a dark core are ranked high by our approach and also by BAC^T.

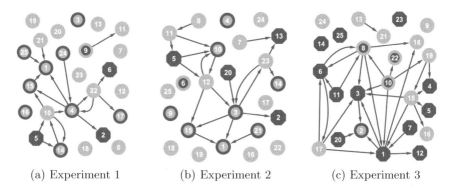

(a) Experiment 1   (b) Experiment 2   (c) Experiment 3

Fig. 7. Topic-sensitive expertise mining (users in top-25 results)

Further information shown in Table 3 includes: AC answer count, number of topics T the user posts answers in, and IL^T depicting the number of inbound neighbors (askers) given a specific topic. For each experiment, we create 5 rows depicting the properties of the top-10 ranked nodes (reduced from top-25 to top-10 due to space limits).

The **first experiment** (see Fig. 7(a) and first segment of rows in Table 3) shows the ranking results within the topic *General - Health*, which we selected because health related issues are of general interest in YA (a top ranked topic). The first ranked user at position 1 would have been ranked at position 1 based on BAC as well. However, the user never responded to a question in the specific topic (thus, topic related info is not shown in the table – n/a). In this case, the user's community-wide reputation was predominant over the personalization. By using a parameter $\alpha = 0.65$ (giving more preference towards topic-based personalization), the order of users ranked at position 1 and 2 would have been changed. However, we noticed that many other globally low ranked (by BAC) users would be ranked too high. The results of experiment 1 show that 8 of 10 have been active in the given topic, 4 of 10 would have been ranked top-10 by BAC^T in the topic, and 4 of 10 would have been ranked top-10 by BAC.

The **second experiment** (see Fig. 7(b) and second segment of rows in Table 3) shows the ranking results within the topics *General - Health* and *Diet & Fitness*, which we selected because both topics have a good overlap of users. The visualization shows that adding another overlapping topic results in more edges being added among the top-25 ranked nodes. The same node as in experiment 1 is ranked at position 1. However, the user previously ranked at position 4 is now ranked at position 3. Notice, the node with label 3 is now connected with a set of other nodes and was therefore ranked at a higher position (i.e., the community embedding of the user improved by adding another topic). In experiment 2, 8 of 10 have been active in the given topics, 3 of 10 would have been ranked top-10 by BAC^T given both topics, and 5 of 10 would have been ranked top-10 by BAC.

The **third experiment** (see Fig. 7(c) and third segment of rows in Table 3) shows the ranking results within the topics *General - Health* and *Heart Diseases*, to test the impact of combining the previously selected topic with the highest

Table 3. Experiment node properties

#	AC	BAC(R)	T	IL^T	AC^T	$BAC^T(R)$	#	AC	BAC(R)	T	IL^T	AC^T	$BAC^T(R)$
1	2072	1219 (1)	3	n/a	n/a	n/a	6	595	275 (6)	12	5	5	5(2)
2	396	257 (9)	22	29	29	26(1)	7	172	8 (386)	42	8	8	1(2)
3	505	118 (14)	18	18	18	0(589)	8	302	11 (266)	43	6	6	0(686)
4	1328	851 (2)	6	1	1	0(918)	9	58	14 (190)	12	7	7	2(8)
5	445	89 (17)	22	10	10	3(5)	10	126	63 (27)	2	n/a	n/a	n/a
1	2072	1219 (1)	3	n/a	n/a	n/a	6	35	5 (637)	2	30	30	5(6)
2	396	257 (9)	22	60	60	50(1)	7	370	33 (57)	14	19	19	1(496)
3	1328	851 (2)	6	1	1	0(4268)	8	97	3 (1359)	23	26	26	0(3895)
4	505	118 (14)	18	38	39	1(396)	9	845	316 (4)	10	1	1	0(2821)
5	445	89 (17)	22	29	29	4(9)	10	365	258 (8)	2	n/a	n/a	n/a
1	1328	851 (2)	6	1167	1300	842(1)	6	365	258 (8)	2	338	354	252(2)
2	2072	1219 (1)	3	n/a	n/a	n/a	7	491	128 (13)	7	456	482	124(6)
3	618	258 (7)	3	441	570	240(4)	8	126	62 (27)	2	121	125	62(11)
4	595	275 (6)	12	362	544	247(3)	9	393	15 (175)	9	360	380	15(39)
5	845	316 (4)	10	445	465	163(5)	10	286	59 (28)	2	270	285	59(12)

ranked topic. Using these topics, the top-ranked user is now also top-ranked by BAC^T and also high ranked by BAC. The combination with a high answering amount topic results in even more connections between the nodes. In experiment 3, 9 of 10 have been active in the given topics, 6 of 10 would have been ranked top-10 by BAC^T given both topics, and 6 of 10 would have been ranked top-10 by BAC.

To conclude the discussion on topic-sensitive expertise mining, personalization based on topic activity combined with community-wide reputation delivers very good results. Thus, the automatic discovery of experienced (authoritative) users with respect to topics is possible by using the **link-structure of Q/A communities**.

5 Personalized Recommendations

Due to the high number of questions that are typically asked in today's Q/A community, it becomes essential to provide recommendations (e.g., which questions should be answered). Our previous expertise ranking approach was based on the assumption that authoritative users also provide good answers, which we confirmed by comparing the rank of a user with the user's best answer count. Our recommendation approach works in a manner similar. Instead of analyzing questions, we recommend users that ask 'good' questions. This approach can be used to order, for example, askers in a responder's buddy list.

However, before actually designing a recommendation algorithm, the *peer answering behavior* of users must be analyzed. If responders rarely (or never) answer questions from the same asker, it is obviously not useful to provide recommendations because users would not prefer to interact repeatedly. The results are shown by Fig. 8.

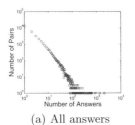

(a) All answers   (b) Best answers only

Fig. 8. Peer answering behavior

Here we take all edges in G and count the number of answers associated with $(v, u) \in E$. In Fig. 8, we show the number of asker/responder pairs and how often questions are typically answered by the same responder. On average, 6 answers are provided between the pair v, u (calculated as the weighted average). The average of best answers only is much lower (i.e., ≈ 0.4).

Notice, the objective is to recommend good askers to responders. To provide recommendations, we use essentially the same equation (cf. Eq. 3) as previously.

What we do, however, is to change the direction of edges. For example, $(v, u) \in E$ becomes $(u, v) \in E'$ where the link inverted graph is defined as $G'(N, E')$. This means that nodes (users) are ranked by their *question authority* instead of their *answer authority*. In particular, those users who ask good questions are recommended. However, not all responders should receive the same recommendations but rather based on their previous interactions. Thus, recommendations should be given from each responder's individual point of view. Thereby, recommendations are calculated using Eq. 3 based on the inverted graph $G'(N, E')$ with $n \in N$ and the following personalization of the preference vector:

$$recommendation : (u) \mapsto \begin{cases} 1, & \text{if } n = u \\ 0, & \text{otherwise} \end{cases}$$

This means that for each node $n \in N$ personalized PageRank scores are created by setting the preference to n with $p(n) \leftarrow 1$. To evaluate our approach we use the standard precision and recall metrics.

Recall: A measure of the ability of the system to present *all* relevant items.

$$recall = \frac{|\{rel\_nodes\} \cap \{rec\_nodes\}|}{|\{rel\_nodes\}|}$$

The set $\{rel\_nodes\}$ holds those nodes that are actually relevant and the set $\{rec\_nodes\}$ those nodes that were recommended.

Precision: A measure of the ability of the system to present *only* relevant items.

$$precision = \frac{|\{rel\_nodes\} \cap \{rec\_nodes\}|}{|\{rec\_nodes\}|}$$

Results. We performed recommendations for the top-200 ranked users based on their answer count AC (see Table 4). We calculated recall and precision for recommendations for a reduced list of 25, 50, 100, 150, and the full list of 200 users. Recommendations are performed as follows: based on the year 2007 we calculate recommendations as discussed before. So for example, a node u would be recommended a set of users s/he should also interact in the following year 2008 (i.e., $\{rec\_nodes\}$ based on previous interactions). Notice, the set $\{rec\_nodes\}$ would contain *all* users within YA since recommendations are available for all nodes in the graph G'. We reduced the number of users in $\{rec\_nodes\}$ by populating the set with users based on the ranked (recommended) list with a cutoff at IL in 2007 (i.e., the number of inbound (askers) neighbors in 2007). We compare recommendation results with the actual neighbors (asker) in year 2008 to check whether recommendations were accurate. The parameter α was set to 0.2. In PageRank nomenclature, the restart probability of a random walk on the graph G' is set to $(1 - \alpha) = 0.8$; thus with high probability a random walk would be restarted at node u. The simplest approach to provide recommendations, however, is to recommend the exact same asker neighborhood from 2007 in 2008

Table 4. Recall and precision in ranking results

	@25	@50	@100	@150	@200
Avg. IL (2007)	413	263	178	141	122
Avg. IL (2008)	426	284	199	163	143
Recall (noRank)	0.948	0.890	0.836	0.816	0.803
Precision (noRank)	1.000	1.000	1.000	1.000	1.000
Avg. RQ (noRank)	436	296	212	172	152
Avg. PQ (noRank)	426	284	199	163	143
Recall (rank)	0.944	0.883	0.827	0.804	0.789
Precision (rank)	0.995	0.992	0.988	0.984	0.980
Avg. RQ (rank)	437	297	214	174	153
Avg. PQ (rank)	427	285	200	164	145

again. These results are also calculated as *noRank* and compared against our approach *rank*.

The first two rows in Table 4 show the average IL neighborhood comparing the sets of users in 2007 and 2008. One can see a rising number IL which means that responders actually increased answering questions (more asker neighbors). This shows that recommendations are useful because responders increasingly answer questions of 'new' askers which they have not previously interacted with. In the next segment of Table 4, the average recall and precision of *noRank* recommendations are shown. These recommendations achieve surprisingly high accuracy in terms of precision and recall. This means that responders tend to interact with the same askers also in 2008 compared to 2007.

The next two rows show the average recall quality RQ and the average precision quality PQ. Instead of calculating recall and precision as a percentage value, we calculated a weighted average that takes the number of IL neighbors of askers into account. For example, recall and precision of responders who are very active in terms of answering questions are given more weights since they contribute more answers to the community. The following rows show the results in terms of recall and precision of our recommendation approach (*rank*). Our approach results are approximately as accurate as the *noRank* results. However, by comparing the average RQ and PQ, our approach outperforms *noRank* recommendations.

Final Remarks on Personalization and Performance Issues. Considering non-personalized recommendations using the standard PageRank model with G' as input, we obtained an *average recall value of 0.011* and *average precision value of 0.014*. This approach is clearly not usable for recommendations because personalization greatly increases the accuracy of results (i.e., recall and precision of the recommended neighborhood of askers).

Personalized recommendations and topic-sensitive expertise rankings using Eq. 3 are much faster computed due to faster convergence of the ranking algorithm towards stable ranking results (i.e., the iteratively computed ranking scores do not change the order of nodes in the ranked result list). Computing non-personalized recommendations or expertise rankings takes a magnitude longer. For example in graphs with $|N| = 59000$ computing personalized results takes a couple of minutes whereas computing non-personalized results takes several hours. This leads us to the conclusion that personalization is achievable also in larger graphs. Other approaches such as the PageRank linearity theorem as proposed in [9] could be applied to compute personalized PageRank results even faster.

6   Conclusion

In this paper, we provided a comprehensive analysis of the structure and dynamics of Q/A communities with the focus on Yahoo! Answers – one of the most popular Q/A platforms on the Web. Our analysis was based on fundamental properties such as Q/A answering behavior of users over time. Also, we studied context-based properties of Yahoo! Answers by analyzing topic-sensitive answering behavior. Our analysis provides important insights for the design of topic-based expertise mining algorithms. Furthermore, we discussed experiments focusing on recommending users to responders whose questions may be relevant to answer. Our research has important implications for future crowdsourcing environments in terms of designing incentive mechanisms for user participation as well as algorithms to assist question routing to relevant experts in open Web-based communities. In our future work, we will study different rewarding models in *two-sided markets* and crowdsourcing with the aim of encouraging crowd users to solve more complex tasks instead of just answering simple questions.

Acknowledgment. We thank Eugene Agichtein for providing us access to the Yahoo! Answers dataset. This work is supported by the European Union through the FP7-216256 project COIN.

References

1. Adamic, L.A., Zhang, J., Bakshy, E., Ackerman, M.S.: Knowledge sharing and yahoo answers: everyone knows something. In: WWW 2008, pp. 665–674. ACM, New York (2008)
2. Agichtein, E., Castillo, C., Donato, D., Gionis, A., Mishne, G.: Finding high-quality content in social media. In: WSDM 2008, pp. 183–194. ACM, New York (2008)
3. Artz, D., Gil, Y.: A survey of trust in computer science and the semantic web. J. Web Sem. 5(2), 58–71 (2007)
4. Becerra-Fernandez, I.: Searching for experts on the Web: A review of contemporary expertise locator systems. ACM Trans. Inter. Tech. 6(4), 333–355 (2006)
5. Brabham, D.: Crowdsourcing as a model for problem solving: An introduction and cases. Convergence 14(1), 75 (2008)
6. De Choudhury, M., Mason, W.A., Hofman, J.M., Watts, D.J.: Inferring relevant social networks from interpersonal communication. In: WWW 2010, pp. 301–310. ACM, New York (2010)

7. Golbeck, J.: Trust and nuanced profile similarity in online social networks. ACM Transactions on the Web 3(4), 1–33 (2009)
8. Gyongyi, Z., Koutrika, G., Pedersen, J., Garcia-Molina, H.: Questioning yahoo! answers. Technical Report 2007-35, Stanford InfoLab (2007)
9. Haveliwala, T.H.: Topic-sensitive pagerank. In: WWW 2002, pp. 517–526 (2002)
10. Horowitz, D., Kamvar, S.D.: The anatomy of a large-scale social search engine. In: WWW 2010, pp. 431–440. ACM, New York (2010)
11. Jurczyk, P., Agichtein, E.: Discovering authorities in question answer communities by using link analysis. In: CIKM 2007, pp. 919–922. ACM, New York (2007)
12. Kleinberg, J.M.: Authoritative sources in a hyperlinked environment. J. ACM 46(5), 604–632 (1999)
13. Kumar, R., Lifshits, Y., Tomkins, A.: Evolution of two-sided markets. In: WSDM 2010, pp. 311–320. ACM, New York (2010)
14. Page, L., Brin, S., Motwani, R., Winograd, T.: The PageRank Citation Ranking: Bringing Order to the Web. Tech. rep. (1998)
15. Schall, D., Dustdar, S.: Dynamic context-sensitive pagerank for expertise mining. In: Bolc, L., Makowski, M., Wierzbicki, A. (eds.) SocInfo 2010. LNCS, vol. 6430, pp. 160–175. Springer, Heidelberg (2010)
16. Skopik, F., Schall, D., Dustdar, S.: Modeling and mining of dynamic trust in complex service-oriented systems. Information Systems 35, 735–757 (2010)
17. Vázquez, A., Ao Gama Oliveira, J., Dezsö, Z., Goh, K.I., Kondor, I., Barabási, A.L.: Modeling bursts and heavy tails in human dynamics. Physical Review E 73(3), 36127+ (2006)
18. Weng, J., Lim, E.P., Jiang, J., He, Q.: Twitterrank: finding topic-sensitive influential twitterers. In: WSDM 2010, pp. 261–270. ACM, New York (2010)
19. Yahoo: Yahoo! answers - home, http://answers.yahoo.com (last access April 2011)
20. Yahoo: Yahoo! answers scoring system, http://answers.yahoo.com/info/scoring_system (last access April 2011)
21. Zhang, J., Ackerman, M.S., Adamic, L.: Expertise networks in online communities: structure and algorithms. In: WWW 2007, pp. 221–230. ACM, New York (2007)
22. Ziegler, C.N., Golbeck, J.: Investigating interactions of trust and interest similarity. Decision Support Systems 43(2), 460–475 (2007)

Forcasting Evolving Time Series of Energy Demand and Supply

Lars Dannecker[1], Matthias Böhm[2],
Wolfgang Lehner[2], and Gregor Hackenbroich[1]

[1] SAP Research Dresden, Chemnitzer Str. 48, 01187 Dresden, Germany
{lars.dannecker,gregor.hackenbroich}@sap.com
[2] Technische Universität Dresden, Database Technology Group
Nöthnitzer Str. 46, 01187 Dresden, Germany
{matthias.boehm,wolfgang.lehner}@tu-dresden.de

Abstract. Real-time balancing of energy demand and supply requires accurate
and efficient forecasting in order to take future consumption and production into
account. These balancing capabilities are reasoned by emerging energy market
developments, which also pose new challenges to forecasting in the energy domain not addressed so far: First, real-time balancing requires accurate forecasts
at any point in time. Second, the hierarchical market organization motivates forecasting in a distributed system environment. In this paper, we present an approach
that adapts forecasting to the hierarchical organization of today's energy markets.
Furthermore, we introduce a forecasting framework, which allows efficient forecasting and forecast model maintenance of time series that evolve due to continuous streams of measurements. This framework includes model evaluation and
adaptation techniques that enhance the model maintenance process by exploiting
context knowledge from previous model adaptations. With this approach (1) more
accurate forecasts can be produced within the same time budget, or (2) forecasts
with similar accuracy can be produced in less time.

Keywords: Forecasting, Energy, Hierarchy, Parameter Estimation.

1 Introduction

The energy market is changing from a day-ahead market to a continuous, intra-day trading that allows dynamic interactions between market participants. This new, liberalized
energy market in combination with emerging smart meter technologies requires fine-
grained planning capabilities. Also, the integration of more renewable energy sources
(RES, e.g. wind, solar) poses additional challenges. Unlike traditional energy sources,
the energy production of RES cannot be exactly planned, because the supply from RES
heavily depends on external factors like the weather. It is also very inefficient to store
energy produced from those sources, for what reason they have to be directly used when
they are available. As a result, it is necessary to balance energy demand and supply in
a fine-grained manner and establish possibilities of real-time balancing [1]. In addition,
real-time balancing in combination with the hierarchical organization of the energy markets with role-specific access to relevant demand and supply data lead to a distributed
data management architecture and therefore a distributed usage of forecasting models.

J. Eder, M. Bielikova, and A.M. Tjoa (Eds.): ADBIS 2011, LNCS 6909, pp. 302–315, 2011.
© Springer-Verlag Berlin Heidelberg 2011

Research projects like MIRABEL [2], MeRegio [3] and many more address the issues of real-time balancing and fine-grained scheduling of energy demand and supply. To do so a fundamental requirement is the availability of accurate predictions of future energy consumption and production. For this purpose we employ model-based forecast techniques, where a quantitative model is used to describe the characteristics and behavior of historic energy time series. Most forecast models involve a number of parameters, with each describing a specific aspect of the time series (e.g., seasonal patterns, energy output). The parameters are estimated on a training data set by minimizing the forecast error (i.e., difference between predicted and actual value) that is measured in terms of an error metric like (Symmetric) Mean Absolute Percentage Error [4]. The so created instances of forecast models are used to predict future values up to a defined horizon (e.g., one day). Important classes of forecast models are: autoregressive models [5], exponential smoothing models [6] and models that apply machine learning [7]. In most cases, forecast models from these classes are specifically adapted to the special characteristics of energy time series such as multi-seasonality and the dependence on exogenous factors. Thus, they produce more accurate forecasts compared to general purpose forecasting methods. However, the necessary real-time balancing capabilities pose new challenges to forecasting of energy demand and supply. Most importantly, accurate forecasts are necessary at any point in time to allow quick adaptations of the energy schedules. Fortunately, in the energy domain the current energy consumption and production can be measured constantly. This can be seen as a continuous stream of updates append to the time series in regular intervals. To ensure high accuracy in terms of exploiting this continuous feedback it is necessary to adapt the forecast model to the updates. However, the naïve adaptation strategy of re-estimating the forecast model after each update is not applicable, because typically the estimation of a forecast model is very time-consuming, as potentially a large number of parameters, spanning an exponential search space, have to be adjusted.

Our primary contribution is a forecast model maintenance framework, which continuously monitors the forecasting accuracy and exploits context knowledge from previous model adaptations and the hierarchical system to increase the model adaptation efficiency. The core idea bases on the assumption that forecast models, i.e. the values of the parameters, only gradually change over time, which typically can be observed in the energy domain. This framework enables existing forecasting models to work with evolving time series in the context of real-time energy balancing. Furthermore, we make the following more concrete contributions that also reflect the structure of this paper:

- We introduce a concept for forecasting in distributed systems in Section 2. Our approach synchronizes the forecast models instead of exchanging forecasts or measurements and therefore reduces the communication overhead.
- Subsequently, in Section 3, we present our novel maintenance approach that includes model evaluation techniques and a parameter estimation framework to ensure efficient model adaptations on single entities of the hierarchical system.
- We compare our estimation framework to other global optimization approaches in Section 4 and show the advantages of our solution.

Finally we conclude the paper with a summary and future work discussion in Section 5.

2 Distributed Forecasting

The European energy market is hierarchically organized. The lowest level comprises consumers and producers organized in balance groups, where a Balance Responsible Party (BRP) manages the energy consumption and production of each group. The BRPs represent the second level. The market operators that are responsible for the market balance areas represent the third level. Additional levels are possible (e.g., Neighborhood-Oriented Energy Balancing). This hierarchical organization of the energy market in combination with different roles and role-specific access to relevant demand and supply data motivates the use of a distributed data management architecture. This also requires a forecasting solution that works in this kind of architecture.

Related Work: Currently, no solution fully adapts forecasting to a distributed system architecture. Only partial aspects are solved like the distributed collection of data. Brabec et al., presented the nonlinear mixed effects model (NLME) implemented at the customer that provides information to a central system. Challa et al. described an approach for distributed sensor networks to incorporate values from different sources into one forecasting system [8]. To do so, they used a State-Space-Model based on Vector Auto Regression combined with the Interacting Multiple Model (IMM) estimation algorithm [9]. In addition, some work was done regarding hierarchical forecasting. For example the use of an AR-GARCH Model as suggested by Sohn et al. [10] or by combining hierarchically organized models using a special regression model as discussed by Hyndman et al. [11]. However, they mostly focused on aggregation strategies with regard to local settings and accuracy only rather than the distribution of the forecasting effort. In contrast to the aforementioned approaches, our solution addresses to provide efficient forecasting functionality to a distributed energy data management system.

2.1 Distributed System Architecture

A naïve approach for forecasting in a distributed system means the direct propagation of measurements or forecasts of lower level entities to the responsible next level entity. This entity then calculates a global forecast. This approach exhibits several drawbacks such as limited transmission granularity due to privacy restrictions that only allows data transmission every 15 minutes and a large communication overhead (e.g., 700 million customers in Europe sending data every 15 minutes).

For this purpose we introduce a more sophisticated forecasting approach for a distributed environment. The approach is based on the assumption that demand and supply measurements are available for all entities in the hierarchy, which motivates a more independent forecasting between hierarchy levels. Figure 1 presents an independent distributed forecasting approach that involves model synchronization. Each entity (S) calculates its own forecasting (F - forecast model) based on its own measurements. No measurements or predicted values are communicated. We rather suggest a synchronization of the forecast models between the different levels to still recognize local changes and to guarantee consistency between forecasts at different hierarchy layers. Model synchronization is conducted less frequently and by fewer entities compared to periodic measurement transmissions from all entities. Therefore, the communication effort

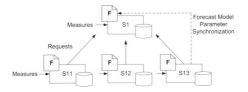

Fig. 1. Distributed Forecasting with Model Synchronization

between hierarchy levels is significantly reduced. In addition, privacy restrictions do not apply, because no measurements or predictions are transmitted. Thus, the forecasting can be based on data with a granularity lower than 15 min. Furthermore, model synchronization provides context knowledge about model adaptations on lower levels that can be used to enhance the model adaptation process on higher level entities. In the following, we describe the model synchronization in such a distributed architecture.

2.2 Model Synchronization

The goals of the model synchronization approach is to ensure the consistency of the forecasting results between the levels of the hierarchy, e.g., the aggregation of individual lower level forecasts should almost provide the same result as the forecast on the upper level. To reduce the communication efforts we assume that the impact of most single entities on the global forecasting is rather low, where the impact of an entity can be estimated by its share on the total consumption and production of its group. Therefore, it is not necessary to adapt the forecast model on higher levels for each model adaptation on a lower level entity. However, a large group of single entities could create a critical mass that is large enough to generate impact on the global forecasting. In addition, large customers, e.g., big companies, could have an impact that is sufficient to influence the forecast on upper level entities. We therefore, use a propagation strategy that involves selective change notifications from lower levels. This means that the change notification is only transmitted to the responsible next level entity. The model synchronization process works then as follows:

1. An entity adapts its forecast model (triggered by local model evaluation [as described in Section 3.2]).
2. A notification is sent to the responsible entity on the next level that includes a description of the model adaptation. The descriptions are transmitted as change vectors, containing the old and new forecast model parameters.
3. The next level entity collects the notifications until a critical mass with sufficient impact on its own forecast model is reached. The exact calculation of the critical mass is subject to future work.
4. Afterwards, an adaptation of the forecast model on the next level is performed using the change vectors of the lower level entities as input for the optimization.

This approach reduces the communication efforts by notifying responsible entities only when a forecast model adaptation is triggered. The information about the model adaptations can be used to enhance the model adaptation process on the next level entity.

For this purpose, before starting the model adaptation process the most recent changes on the child entities are considered, avoiding the usage of outdated information.

3   Forecast Model Maintenance

To ensure an efficient forecasting in the hierarchical system, it is important to also consider the forecasting processes on a single system entity. As mentioned before time series in the energy domain evolve over time. To guarantee up-to-date forecast models at any point in time a continuous model maintenance is needed. Also, additional energy domain specific particularities pose the challenge of causing unpredictable but gradual changes: (1) The energy consumption and the supply of renewable energy sources is influenced by uncertain exogenous factors like weather or temperature. (2) Real-time balancing capabilities allow market actors to constantly adapt energy consumption and production. These particularities require efficient model maintenance. In this section, we introduce the overall forecast model maintenance strategy for single system entities, which includes several model evaluation and adaptation techniques.

3.1   Maintenance Strategy Overview

Figure 2 illustrates our forecast model maintenance strategy that consists of three steps. First, for each new measurement we initiate an update of the local model. This step includes the incremental state adaptation (e.g., smoothing constant) of the forecast model and the persistent storage of measurements. This update is a simple insertion such that it is not the focus of this paper. Second, we continuously evaluate the accuracy of the forecast model upon evolution of the time series using different model evaluation techniques. Third, based on the outcome of the accuracy evaluation the adaptation of the forecast model to the new situation is triggered, which means a re-estimation of the parameters involved in the forecast model. When considering complex models that describe a lot of information like multiple seasons and exogenous information, the models involve a high number of parameters. Each parameter adds an additional dimension to the solution space, which increases the amount of solutions that have to be evaluated. For this reason the parameter re-estimation can be very time consuming. Triple Seasonal Holt Winters [12] for example involves five parameters, which leads to a number

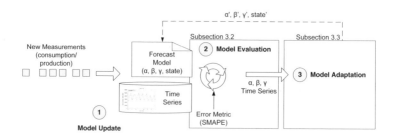

Fig. 2. Maintenance Strategy for Evolving Time Series

of x^5 possible parameter combinations, with x being the granularity of a parameter. The autoregressive multi-equation model EGRV [13] and its adaptations (e.g., Cottet et al. [14] and Dordonnat et al. [15]) model each hour as a separate model which leads to an even larger parameter space (e.g., EGRV: 24 times up to 31 parameters). As a result, parameter estimation approaches that efficiently find new optimal parameter combinations are needed to adapt even complex forecast models in reasonable time.

3.2 Forecast Model Evaluation

The second step in the maintenance process is the model evaluation. We distinguish two major groups of evaluation strategies: First, fixed interval model evaluation, where the model adaptation is triggered periodically. This strategy does not evaluate the forecast error. Therefore, it exhibits the problem of determining a reasonable model adaptation interval. Too short intervals mean unnecessary adaptations, whereas too long intervals pose the risk that arbitrary large errors may build up between the model adaptation intervals. The threshold-based model evaluation is the second strategy. It continuously evaluates the forecast model accuracy and triggers a model adaptation when the forecast error violates a previously defined error threshold. This enables quick adaptations of the forecast model to changes of the evolving time series. While this strategy guarantees that a certain forecast error is not exceeded, it exhibits a similar drawback like the fixed interval model adaptation as it also depends on the definition of suitable thresholds.

Due to the high influence of the adaptation criteria, we propose a heuristic approach that combines evaluation strategies. A combination weakens the disadvantages of the single techniques. There, a model adaptation is periodically triggered after a specified amount of time. Also, the model is continuously evaluated and adapted each time the error surpasses a defined threshold. The time counter is reset after a model adaptation was triggered by a threshold violation, to avoid unnecessary adaptations of the forecast model. The combination of model maintenance strategies reduces the dependence on single adaptation criteria, which makes it easier to determine suitable thresholds.

3.3 Enhanced Parameter Estimation

Once a model adaptation has been triggered, we try to re-estimate the parameters of the forecast model, which typically is a very time-consuming task. For this reason, a parameter estimation method that efficiently finds a new optimal parameter combination with regard to the forecast error is necessary.

Related Work: For the optimization of forecast model parameters several algorithms exist, which can be divided into two classes: (1) Algorithms that need a derivable function and (2) algorithms that can be used with arbitrary functions. In this paper, we only focus on the class of algorithms that do not require a derivable function, because they are more general and can be used with arbitrary forecasting models and error functions. These algorithms are classified into local and global optimization algorithms. Global optimization algorithms consider the whole solutions space with the goal of finding a solution that is the global optimum. Thus, in general they need more time to terminate.

In contrast, local optimization algorithms follow a directed approach and therefore converge faster to a solution but exhibit the risk of starvation in local suboptima. Also, they strongly depend on the position of a provided starting point that has to be close enough to the optimum to guarantee convergence. Examples for global optimization algorithms are: Simulated Annealing [16] or genetic algorithms [17]. Examples for local optimization are: Hook-Jeeves [18] and Nelder-Mead [19]. In addition to the described optimization algorithms, we can find parameters empirically using a naïve method called grid search that sequentially evaluates all solutions in a given granularity. However, its runtime exponentially increases with the number of parameters. Due to the limitations of the naïve method as well as local and global optimization algorithms, we enhance the parameter estimation process by introducing our parameter estimation framework.

Our core idea is to exploit context knowledge of previous model adaptations by determining starting points using information from the model synchronization within the hierarchical system and the previous parameter combination. The underlying assumption is that the combined parameter changes of the child entities approximately reflect the parameter changes of the forecast model at the parent entity. In addition, due to the continuous model evaluation we also assume that forecast model parameters will not change abruptly, but will be in the neighborhood of previous parameters. Our approach exploits both assumptions by combining the parameter changes suggested by the model synchronization with the previous parameter combination. Thus, we reduce the problem of finding a global optimum to the problem of finding local suboptima in the near surrounding of the determined starting point.

We supplement this local strategy with global coverage approaches. For this purpose, we introduce a parameter estimation framework for different search strategies that is illustrated in Figure 3: A starting point is determined by exploiting context knowledge of the continuous model maintenance and adaptation information of child entities in the hierarchical system. This starting point serves as input for a two-phase optimization process that refines the starting point to find a new, optimal parameter combination. The process consists of a local search for fast convergence and a global search to reduce the probability of getting stuck in local suboptima. In the following we describe (1) how to determine starting values for the parameter estimation process and (2) how to optimize this initial solution to find the global optimal parameter combination in detail.

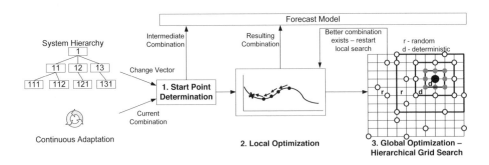

Fig. 3. Model Adaptation Process

Starting Point Determination. For local optimization approaches a good starting point is essential for finding a global optimal solution. Also, the runtime of such approaches is directly influenced by the position of the starting points. The closer the location of the starting point to the optimum, the lower the time needed to converge.

In our solution the forecast model is evaluated continuously and adapted regularly, which leads to the assumption of rather slight adaptations of the forecast model configuration. For the parameter estimation we therefore assume that the probability of finding the new global optimal parameter combination is highest in the near surrounding of the current combination. Given this assumption we set the current parameter combination as a starting point for any subsequent search strategy.

We further enhance the starting point determination by using the context knowledge of model maintenance from child entities in the hierarchy. The basis is the parameter change vector that is exchanged during model synchronization (compare Section 2.2). This vector contains the parameter combinations before and after the model adaptation of an entity, which means that the change vector represents the transition between forecast model configurations. These two parameter combinations can be used to estimate the direction of the parameter change by calculating their difference. We could also directly provide the difference of both vectors, but to allow more complex operations in the future we chose a more general solution in providing the parameter combinations. We can assume that the combination of all forecast model adaptations from entities on the lower level nearly represents the necessary changes to the model of the responsible parent entity. This is reasoned by the fact that the forecasting of the parent entity reflects the aggregated consumption or production of all connected child entities. To determine the starting point first, a global change vector is computed by combining the parameter change directions of the change vectors of multiple system nodes, weighting them according to the integral of consumed or produced energy. It is important to note that deviations from single entities are equalized due to the aggregation of many entities and the weighting. Subsequently, the new starting point is computed as the arithmetic mean of the old parameter combination and the global change vector.

In conclusion, for entities at the lowest level we exploit the continuous model adaptation with the assumption that the new global optimal solution is located in the near surrounding of the old parameter combination. On higher hierarchy levels, we use a combination of old parameter values and propagated changes of child entities to determine the start value. As a result, we compute start values that have a high probability of being close to the global optimal solution and thus, serve as good initial values.

Optimization Process. With a good starting point in place and assuming to find the optimal solution in the near surrounding of this starting point, we define a parameter estimation framework that allows fast convergence and that ensures stochastic properties of finding the global optimal solution likewise. The process comprises of the following two phases: First, we use the determined starting point as the input for a local optimization approach like the Nelder-Mead algorithm [19]. This simplex-based algorithm iteratively evaluates the neighborhood of the starting point until a local optimum is found. Due to the starting point determination the probability of the local optimum to be also the global optimal solution is fairly high. At the same time the local optimization with given starting points converges relatively fast. After the local optimum has been found,

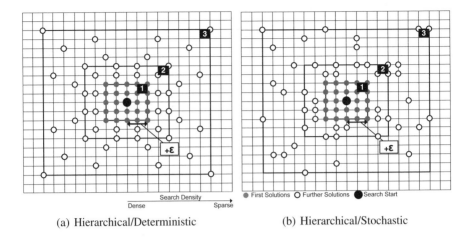

(a) Hierarchical/Deterministic   (b) Hierarchical/Stochastic

Fig. 4. Global Coverage: Enhanced Grid Search Variations

in the second phase we try to ensure its global optimality. For this purpose we start a hierarchical optimization that converges to the global optimal solution:

1. We span an area with a user defined margin $\pm\epsilon$ in all dimensions of the solution space around the local optimum.
2. Within this area, we sequentially evaluate all possible parameter combinations, which we did not evaluate in the first phase, in a predefined granularity.
3. We double the margin ϵ spanning a second area around the first one.
4. We repeat the search within this area. There, the search granularity is increased by considering the same amount of parameter combinations as in the first iteration, but distributed over an area that is 2^d times larger than the first one.

This process is repeated until the end of the grid is reached. If a solution is found that is better than the current best one, the local search starts again from this combination. It is important to note that for different start values during the continuous model adaptation process, the parameter combinations of the outer areas evaluated by this process also change. Thus, over time, the probability to find possible optima in other regions increases. An example of the deterministic approach is illustrated in Figure 4(a).

In addition to the standard deterministic search, a second possibility is to stochastically evaluate the solution space given a specific probability density function. It is faster due to the fact that no specific order of the results is required. Also, in contrast to the deterministic search, over time the stochastic search asymptotically evaluates all possible combinations in expectation. This leads to the following search strategy: In the ϵ-area all possible solutions are considered. In the further areas solutions are picked randomly, while the number of considered solutions always corresponds to the number of solutions in the core area. This leads to a more coarse-grained search resolution. Figure 4(b) illustrates an example of the hierarchical, stochastic search.

The size of the ϵ-area and the search resolution (res) can be defined freely. Nevertheless both directly influence the number of considered solutions. When considering complex forecast models (e.g., EGRV >30 parameters) the number of considered solutions in the ϵ-area increases exponentially with the number of parameters. We therefore define a maximal number of solutions in the ϵ-area, $maxPoints$, and therefore limit the maximal runtime of the enhanced grid search. To further reduce the run time of the global coverage, we can add some aspects of the well-known hill climbing optimization approach by terminating the grid search when in a subsequent area no better solutions is found than the current best. For example, if no better solution is found after the second expansion of the search area, the algorithm terminates. Essentially, the enhancements reduce the runtime, while sacrificing robustness.

4   Experimental Evaluation

In this section, we present our experiments that show the benefits of our parameter estimation framework. We demonstrate that precomputed start values enhanced by a local and a global search yield better results than solutions involving multiple random starting points. We validate our claims by comparing our parameter estimation framework to other global optimization algorithms.

4.1   Experimental Setting

We base our experiments on the publicly available data set from the UK National Grid organization. The data set contains metered electricity demand of the United Kingdom (UK) from April 1st 1971 to December 31st 2009. For our experiments we used the INDO[1] measure from January 1st 2002 to December 31st 2009 in a half hour granularity.

For the computation of the forecast, we chose the triple seasonal Holt Winters Exponential Smoothing (HWT). This model is tailor made for data from the energy domain and performs well on the above mentioned data set [12]. For the estimation process we split our data set as follows: The model was initialized with the years 2002 to 2008. We forecasted the year 2009 using a one-step ahead forecast to evaluate the applicability of the tested parameter combination. The forecasting error was calculated with the SMAPE error metric [4]. For our evaluation we used the following environment: Intel Core 2 Duo 2,0 GHz, 3 GB RAM, Windows 7 32bit operating system. All experiments were implemented using the C++ programming language.

4.2   Comparison of Efficiency and Accuracy

Parameter Estimation Framework versus Monte Carlo Grid Sampling: In the first experiment, we compared our parameter estimation framework to Monte Carlo grid sampling. Unfortunately, we did not have a hierarchical system with sufficient measurement data in place. For this reason, we simulated the start value for our framework by executing a coarse-grained grid search with a step size of 0.25 and subsequently modified the found point by 0.005 in random directions. The grid is configured with an ϵ-area of 0.1 and a search resolution of 0.05. Monte Carlo grid sampling chooses points from the solution space in a random fashion without using any further optimization. This method is

---

[1] **INDO** - Initial Demand Outturn based on operational generation metering." [20]

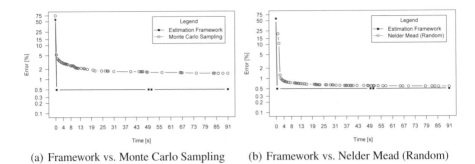

(a) Framework vs. Monte Carlo Sampling   (b) Framework vs. Nelder Mead (Random)

Fig. 5. Experiment: Comparing Accuracy and Efficiency

much faster in selecting possible solutions but it is totally stochastic with uniform PDF. Thus, we repeated the experiment 20 times and used the average of all attempts as the result. For both solutions, we tracked the development of the best-found solution.

Figure 5(a) illustrates the result of this experiment, which shows that at no point in time the Monte Carlo grid sampling exhibited a better development than the parameter estimation framework. The error progression is always below the curve of the Monte Carlo grid sampling with a minimal forecast error of 0.494%. In comparison the minimal forecast error of the Monte Carlo grid sampling was 1.446% using the same runtime as the estimation framework. The framework found its best solution already after 51 s, while the subsequent time was used for the global coverage. The framework quickly converged to a local optimum with a forecast error of 0.499% the subsequent grid search then found a better solution (forecast error 0.495%) that was then again enhanced by the Nelder-Meads algorithm to the global optimal solution. The Monte Carlo grid search started with a higher error value and converged slower than the parameter estimation framework. This clearly shows the benefit of using optimization algorithms instead of just randomly selecting points from the grid.

The overall time to converge is rather short with 91 s, however, when considering more complex models or longer time series the time frame needed to converge to the global optimum will increase considerably. When using the EGRV model with 22 parameters (22 instead of 31 due to missing weather information), where we limited the maximal considered solutions in the ϵ-area to 1000, the parameter estimation took 635 s.

Parameter Estimation Framework versus Local Search with Random Starting Points: In the second experiment we compared our parameter estimation framework to a repeated local optimization with random starting points. Since the local search is repeated with multiple starting points, it is equivalent to a global optimization. To ensure comparability we also chose Nelder-Mead for local optimization. Our estimation framework used the same configuration as in the first experiment.

The results are illustrated in Figure 5(b). For the local search with random starting point we exhibit a better error development compared to the Monte Carlo grid sampling. However, it still converged slower and did not reach an error as low as our estimation framework. The average error value of the local search with random starting point after 91 s was 0.566%. At any point in time we get a better solution with our parameter estimation framework. To find out the timeframe necessary for the local search with random starting point to converge to the same result as the parameter estimation framework, we removed the time limitation. The overall runtime was ten hours. The best results was found after 100 min with an error value of 0.499%. This also suggests that no better solution exists in the search space, which means that our estimation framework found the global optimal solution. In conclusion, the computation of a decent starting point in combination with our estimation framework is beneficial compared to just randomly selecting points as an input for the local optimization. We also showed that the subsequent global optimization is necessary to ensure that the global optimal solution is found.

Parameter Estimation Framework versus Nelder-Mead with Simulated Annealing: In this experiment, we compare the error development of our framework using enhanced grid search and simulated annealing for global coverage. The results show that first, both algorithms converge to the same intermediate solution of 0.499%, because both use the same local search algorithm. However, subsequently the enhanced grid search found a better solution that was again refined by the Nelder-Mead algorithm to 0.494%. In contrast, the simulated annealing algorithm did not find a better solution. An explanation is that our enhanced grid search considers solutions with minimal granularity in the near surrounding of the starting point, while the simulated annealing randomly selects points from the whole solution space.

Increasing Distance to the Starting Point: The starting point used for our framework was pre-computed via a coarse-grained grid search to simulate that the starting point is the last know parameter combination. To estimate the dependence of our framework to the starting point, in this experiment we increased the distance to the starting point from 0.005 to 0.1 and 0.35. The results are illustrated in Figure 6(a). They show that in all cases a similar error value is reached, but with a different convergence speed. For a distance of 0.1 the framework first converged slower to an intermediate solution that is tight above the result when using a distance of 0.005. The global optimal solution was reached after 194 s compared to 91 s. Using a distance of 0.35 the global optimal solution was not reached. The framework found another solution with an error value of 0.4943% that is almost as good as the global optimal solution with 0.4942%. As a result, when increasing the distance the timeframe to converge to the global optimal solution is increased and the chance of finding the global optimum is reduced.

Simulation of a Evolving Time Series: In this experiment, we simulated an evolving time series to demonstrate the knowledge exploitation of previous model adaptations. We used the data from the years 2002 to 2007 to initialize the model, the year 2008 to evaluate the initial parameters and the year 2009 to simulate the evolving time series. We continuously monitored the error and triggered a model adaptation when either the error threshold was violated (defined: 0.5% SMAPE) or once a day (48 data points). Our estimation framework was used with an ϵ-area of 0.1 and a search resolution of 0.1.

Figure 6(b) illustrates the error development over the year 2009. We observed the typical behavior of higher error values in the winter months, because there are more peak demands in the winter due to illumination and heating than in the summer. These peaks are harder to predict. Using a continuous evaluation and model adaptation, the error is always lower compared to a solution without continuous maintenance. This

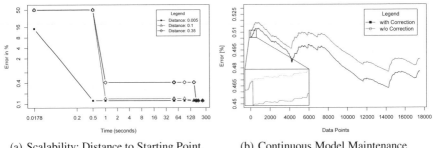

(a) Scalability: Distance to Starting Point   (b) Continuous Model Maintenance

Fig. 6. Experiments: Continuous Model Maintenance & Distance to Starting Point

shows the benefit of continuously adapting the forecast model to the current situation. In our evaluation the model adaptation based on previous parameters needed 28.23 s on average. The model adaptation was triggered 388 times and a better model was found 183 times. The small extract in Figure 6(b) illustrates the effect of a single parameter re-estimation. There, the error is reduced from 0.510% to 0.505%. In conclusion, our concept of exploiting knowledge from previous model adaptation is beneficial, because our framework finds parameter combinations that improve the forecast accuracy.

5   Conclusion

The liberalized energy market requires the balancing of energy demand and supply in real-time. This poses the challenge of forecasting in a distributed environment and the need for continuous model maintenance to provide reliable forecasts. In this paper, we introduced a forecasting approach for a hierarchical energy management system. This approach uses model synchronization instead of communicating measurements or forecasts to reduce the communication overhead. Thus, it allows an efficient forecasting in a distributed environment. We then investigated the maintenance of forecast models on a single entity and introduced our estimation framework that exploits context information from previous model adaptations and the hierarchical system to compute a suitable starting point for further optimization. The framework employs a combination of local and global search algorithms to find the global optimal solution in reasonable time, whereas arbitrary local and global algorithms can be used. Our evaluation shows the benefits of precomputed start values in combination with our estimation framework to increase the probability of finding the global optimal solution. In conclusion, we presented an efficient way to apply forecasting to distributed environments and to continuously maintain the forecast models on the entities of the system. With our approach efficient forecasting in distributed environments is possible. In addition, more accurate forecasts can be produced within the same time budget, or forecasts with similar accuracy can be produced in less time. In the future we will extensively evaluate the model synchronization approach to estimate its concrete potential in the application context.

Acknowledgment. The work presented in this paper has been carried out in the MIR-ACLE project funded by the EU under the grant agreement number 248195.

References

1. Belhomme, R., Asua, R.C.R.D., Valtorta, G., Paice, A., Bouffard, F., Rooth, R., Losi, A.: Address - active demand for the smart grids of the future. In: CIRED Seminar 2008: SmartGrids for Distribution (2010)
2. Project, M.: (2011), `http://www.miracle-project.eu`
3. MeRegio Project (2011), `http://www.meregio.de/en/`
4. Hyndman, R.J.: Another look at forecast-accuracy metrics for intermittent demand. Foresight: The International Journal of Applied Forecasting 4, 43–46 (2006)
5. Box, G.E.P., Jenkins, G.M., Reinsel, G.C.: Time Series Analysis: Forecasting and Control. John Wiley & Sons Inc., Chichester (2008)
6. Winters, P.R.: Forecasting sales by exponentially weighted moving averages. Management Science, 324–342 (April 1960)
7. Bunnoon, P., Chalermyanont, K., Limsakul, C.: A computing model of artificial intelligent approaches to mid-term load forecasting: a state-of-the-art- survey for the researcher. Int. Journal of Engineering and Technology 2(1), 94–100 (2010)
8. Challa, B., Challa, S., Chakravorty, R., Deshpande, S., Sharma, D.: A novel approach for electrical load forecasting using distributed sensor networks. In: Proceedings of the Third ICISIP, pp. 189–194 (2005)
9. Blom, H., Bar-Shalom, Y.: The interacting multiple model algorithm for systems with markovian switching coefficients. IEEE Transaction on Automatic Control 33, 780–783 (1988)
10. Sohn, S.Y., Lim, M.: Hierarchical forecasting based on ar-garch model in a coherent structure. European Journal of Operational Research 176, 1033–1040 (2005)
11. Hyndman, R.J., Ahmed, R.A., Athanasopoulos, G.: Optimal combination forecasts for hierarchical time series. Technical report, Monash University (2007)
12. Taylor, J.W.: Triple seasonal methods for short-term electricity demand forecasting. European Journal of Operational Research 204, 139–152 (2009)
13. Ramanathan, R., Engle, R., Granger, C.W., Vahid-Araghi, F., Brace, C.: Short-run forecasts of electricity loads and peaks. International Journal of Forecasting 13, 161–174 (1997)
14. Cottet, R., Smith, M.: Bayesian modeling and forecasting of intraday electricity load. Journal of the American Statistical Association 98, 839–849 (2003)
15. Dordonnat, V., Koopman, S., Ooms, M., Dessertaine, A., Collet, J.: An hourly periodic state space model for modelling french national electricity load. International Journal of Forecasting 24(4), 566–587 (2008)
16. Kirkpatrick, S., Gelatt, C.D., Vecchi, M.P.: Optimization by simulated annealing. Science New Series 220(4598), 671–680 (1983)
17. Fogel, D.B.: Evolutionary Computation: Towards a New Philosophy of Machine Intelligence. IEEE Press, Los Alamitos (2000)
18. Hooke, R., Jeeves, T.A.: Direct search solution of numerical and statistical problems. Journal of the ACM 8, 212–229 (1961)
19. Nelder, J., Mead, R.: A simplex method for function minimization. The Computer Journal 13, 308–313 (1965)
20. nationalgrid: nationalgrid UK - Metered half-hourly electricity demands (2010), `http://www.nationalgrid.com/uk/Electricity/Data/Demand+Data/`

The NestFlow Interpretation of Workflow Control-Flow Patterns

Carlo Combi, Mauro Gambini, and Sara Migliorini

Department of Computer Science – University of Verona
Strada Le Grazie, 15, 37134 Verona, Italy
{carlo.combi,mauro.gambini,sara.migliorini}@univr.it

Abstract. Business process models are designed using a set of control-flow and data-flow constructs provided by the chosen Business Process Modeling Language (BPML). As research confirms, the adoption of a structured control-flow is always desirable for enhancing model comprehensibility and reducing the presence of errors. However, existing BPMLs cannot promote a fully structured approach to control-flow design because any restriction imposed on the existing language constructs results in a loss of expressiveness in terms of definable models. This paper proposes a novel BPML called NestFlow, characterized by a small set of language constructs that together overcome the aforementioned limitation. NestFlow expressiveness is discussed in terms of supported Workflow Control-Flow Patterns (WCPs), showing how the right combination of control-flow and data-flow constructs allows one to express most of these patterns in a structured way.

Keywords: structured business process modeling languages, workflow control-flow patterns, process-aware information systems.

1 Introduction

A Business Process (BP) is a set of interrelated activities performed by a group of agents inside an organization in order to achieve a predefined goal. BPs are usually designed and specified using a graphical Business Process Modeling Language (BPML). The produced models can help to understand BPs and share knowledge about them among the different stakeholders. An executable BPML, like YAWL [1] and BPMN with WS-BPEL semantics [2], is used to design complete BP specifications that can be directly interpreted by a Process-Aware Information System (PAIS). Available BPMLs are usually *unstructured*, with a graph-oriented syntax and a token-based semantics: they allow a free composition of constructs without worrying much about type or position of the connected elements. The resulting models can be more or less structured, depending on the presence of properly nested sub-graphs with single entry and single exit points, in which correlated constructs have compatible types [3]. Research confirms that the use of structured forms leads to more comprehensible and modular models with less errors, as extensively discussed in [4,5,6,7]. However, existing

J. Eder, M. Bielikova, and A.M. Tjoa (Eds.): ADBIS 2011, LNCS 6909, pp. 316–332, 2011.

BPMLs are not able to support a fully structured control-flow design without losing expressiveness: to obtain a structured control-flow some syntactical restrictions have to be imposed on the existing constructs, reducing the number of representable models. Conversely, the NestFlow BPML introduced in this paper goes beyond the usual set of constructs and is able to support a structured control-flow design by means of well-formed nested control structures. NestFlow expressiveness is evaluated in terms of supported Workflow Control-Flow Patterns (WCPs) [1,8] showing how a structured BPML can represent the behavior of most WCPs with a small set of control-flow and data-flow constructs. The remainder of this paper is organized as follows: Sec. 2 discusses some related work that has inspired the NestFlow language. Sec. 3 presents the syntax and semantics of NestFlow with its essential properties. Sec. 4 shows how WCPs can be implemented with this language. Sec. 5 explains the method used to evaluate NestFlow expressiveness and its suitability for BP design. Evaluation parameters and results are exposed in Sec. 6. Finally, Sec. 7 summarizes the findings and how they are obtained.

2   Related Work

Structuredness is a recurring topic in BP modeling and in BPML design. A first notion of structured form can be found in [3], where Liu and Kumar introduce a taxonomy of unstructured forms and study which ones have an equivalent structured counterpart. In [9] Vanhatalo et al. introduce a parsing technique, called Refined Process Structured Tree (RPST), for detecting structured forms inside arbitrary models. The main idea of RPST is to extract the structure of a model to ease its mapping to a low-level executable language. The relation between structuredness and the probability to find errors inside a model has been empirically studied by Laue and Mendling in [4]: the authors conclude that structuredness is a key factor for enhancing the quality of BP models. Similarly, in [6] Reijers and Mendling experimentally measure the effect of modularity on model comprehension, determining the existence of a positive relationship between modularity and comprehensibility in large-scale BP models. On the basis of these and other empirical results, in [7] Mendling at al. define seven process modeling guidelines for enhancing BP quality. These guidelines have been prioritized on the basis of the opinion of several modeling experts and the two most important ones are: (1) design models as structured as possible and (2) decompose a model with more than 50 elements. The positive effects of structured forms for model comprehensibility have been also recognized in [9] where the authors suggest that RPST can help in analyzing and solving structural issues in BP models. However, as a parsing technique, RPST does not solve by itself problems that emerge with unstructured compositions: unstructured forms are ultimately mapped to unstructured constructs of the target executable language whenever possible. In [10] we discuss the main issues that can arise when an unstructured control-flow design is coupled with shared variables, parameter passing and message passing constructs, explaining why unstructured forms are mostly avoided during BP design.

3   The NestFlow Modeling Language

NestFlow is conceived to explore a particular BPML design solution in which structured control-flow constructs are tightly-coupled with asynchronous message passing connections. This is clearly a different approach from the established BPML solutions, in which unstructured control-flow constructs are coupled with parameter passing and shared task variables. In NestFlow control-flow constructs can be composed only in properly nested structures and task variables cannot be shared among concurrent entities. Furthermore, data-flow constructs are promoted as first-class citizens because they are invaluable for offering a uniform hierarchical decomposition mechanism that increases modularity. Conversely, data-flow constructs are a marginal feature in existing BPMLs that focus mainly on control-flow, in an attempt to design simpler languages. Actually, there is no evidence that a simple language simplifies the modeling activity: a simple language with few constructs often produces large and complex models. On the contrary, a fully-featured language can ease BP design by tackling the inherent complexity of the modeled reality through more expressive constructs. A questionable assumption about BPMLs is that control-flow relations are sufficient for expressing business logics and unstructured control-flow must be accepted for not reducing this expressiveness. Unfortunately, control-flow and data-flow concerns cannot be easily separated, and this is exceptionally true if the aim is to obtain executable specifications. The NestFlow rationale is to fuse control-flow and data-flow aspects for offering a structured control-flow without any loss of expressiveness and with positive effects on modularity.

3.1   NestFlow Syntax and Annotations

The concrete syntax of a programming language can be formally described through a context-free grammar encoded in one of the many extensions of the Backus-Naur-Form (BNF) meta-language. In similar way, the concrete syntax of NestFlow is given by the graphical BNF-like grammar depicted in Fig. 1, where $\langle P \rangle$ denotes the starting symbol, | is the usual BNF choice operator, $\langle A \rangle$ denotes non-terminal control-flow blocks and $\langle C \rangle$ denotes terminal command blocks.

A task declaration is essentially obtained by recursively substituting the non-terminal symbol $\langle A \rangle$ in $\langle P \rangle$ with the other blocks $\langle A \rangle$ and $\langle C \rangle$. Some minor restrictions are applied in this recursive composition: for instance, a `spawn` command can be placed only inside a `concurrent block`, while a `throw` command can be used only inside a proper `catch block` or a task that declares to raise the corresponding exception. These and other well-formedness properties can be statically checked and the most important ones are discussed in Sec. 3.2.

NestFlow distinguishes between a task type and its instances called *cases*. A task instance is denoted as $t : T$, where $t \in \mathcal{I}$ is an identifier chosen among the set of valid identifiers $\mathcal{I}$, and T is its type. In the graphical representation the textual identifier can be left implicit, because a task instance is uniquely identified by its place in the model, as further explained in the following sections.

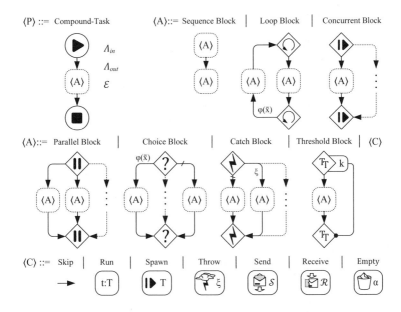

Fig. 1. The graphical concrete syntax of the NestFlow modeling language

In NestFlow data have a central role and their declaration is mandatory. However, many details are graphically shown in the diagram only when they are strictly necessary for understanding the expressed logic. From a preliminary modeling point of view, they can be seen as optional disambiguating annotations. The flow of objects among tasks is represented through *links* graphically denoted with dashed arrows, as in Fig. 2.

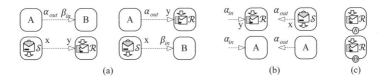

Fig. 2. Different combinations of NestFlow link notations

We can distinguish between internal and external links depicted in Fig. 2.a and Fig. 2.b, respectively. If necessary, links can be annotated with stream and variable identifiers. NestFlow links and their related commands can be hidden anywhere they can be subsumed by the control-flow. For instance, links with the same source and target can be grouped into a unique collapsed link and subsumed by a control-flow with the same direction; `receive` and `send` with only subsumed links can be hidden as well. The remaining links mostly describe interactions among concurrent entities.

3.2 NestFlow Semantics and Properties

In NestFlow a task can be atomic or compound: an *atomic task* can be implemented with a general-purpose language to provide some functionalities; for instance, it can provide a user-interface to support human activities or an adapter to drive external programs. The concept of *compound task* embodies the usual notions of sub-process and super-task. In this paper we do not consider all aspects of a BPML and a BP model simply coincides with a main process definition, obtained declaring a `compound-task` that has other tasks as components. Each `compound-task` has a main entry and exit point marked with a start arrow symbol and a stop square symbol, respectively. The *task logic* is expressed by expanding the non-terminal $\langle A \rangle$ between these two points with blocks in Fig. 1 and invoking instances of previously defined atomic or compound tasks: the recursive definition of tasks is essential for supporting uniform hierarchical decompositions. A task T can declare zero or more *task variables* with their own type; the set of all declared variables is called *store* and denoted as $\mathcal{M}(T)$. A variable may either contain an object of the declared type or may be *unbound*. Defining a complete type system is beyond the scope of this paper: we assume that a type X can be built starting from primitive types as in usual object-oriented languages, and we denote with $x : X$ an object x of type X. Variables are visible only inside the task where they have been declared and their scope does not extend to tasks contained in it. All task instances and variables inside the same `compound-task` have a unique identifier.

Any task T exposes a specific interface given by a set of streams $\Lambda(T) = \Lambda_{in}(T) \cup \Lambda_{out}(T)$, where $\Lambda_{in}(T) \cap \Lambda_{out}(T) = \emptyset$, and a set of *exceptions* $\mathcal{E}(T)$ that may be raised during its execution. We refer to $\Lambda(T)$ as the *stream interface* of T, $\Lambda_{in}(T)$ and $\Lambda_{out}(T)$ as sets of *input streams* and *output streams*, respectively. A *stream* is simply a queue of objects of a predefined type that can be used for modeling, not only the information flow, but also the set of objects needed and produced by a task. If an instance $t : T$ needs an object $a : A$ to proceed, it declares in its interface an input stream α_{in} of type A denoted with $\alpha_{in} : A$, such that $\alpha_{in} \in \Lambda_{in}(T)$. In similar way, if a task t may produce an object $b : B$ relevant for other tasks, it declares an output stream $\beta_{out} \in \Lambda_{out}(T)$ such that $\beta_{out} : B$. An instance of T can refer to one of its own interface stream $\alpha \in \Lambda(T)$ through the dot notation $self.\alpha$, where $self$ is a language keyword. Similarly, a stream α of an internal task instance $u : U$ is referred by $u.\alpha$, where u is the instance identifier and $\alpha \in \Lambda(U)$. The dot-notation ensures that all streams in a `compound-task` are uniquely identified and $self$ may be left implicit.

As previously mentioned, task logic is declared using $\langle A \rangle$ and $\langle C \rangle$ blocks in Fig. 1. The behavior of these elements is explained in the following, where $\bar{x} \subseteq \mathcal{M}(T)$ is a set of variables and $\varphi(\bar{x})$ is a condition over $\bar{x}$.

1. `sequence block` – It executes the specified blocks in sequence.
2. `loop block` – It executes the blocks contained in its branches multiple times. After the execution of the right branch in Fig. 1, the condition φ over a set of task variables $\bar{x}$ is evaluated. If the condition is false the loop exits, otherwise it executes the left and the right branch in sequence.

3. `concurrent block` – It is a dynamic `parallel block`. It initially executes $\langle A \rangle$ but one or more parallel branches can be added at run-time using a `spawn`; all threads join before exiting the block. A `spawn`/`concurrent block` pair provides a graphical representation of dynamically created instances by varying the model at run-time. BPMLs usually do not offer any representation of this dynamic behavior and concurrent instances are often left implicit.

4. `parallel block` – It executes the specified branches in parallel each one with its own thread of control. Threads are joined at the end of the block or reverted if an unhandled exception is thrown in one of the branches.

5. `choice block` – It evaluates the conditions $\varphi_i(\overline{x})$ associated to the i-th branch in sequence, as soon as one of these conditions is true the corresponding branch is executed, otherwise the default branch is chosen.

6. `catch block` – It executes the default branch and if an exception of type ξ is raised inside it, the execution is interrupted and resumed from the branch annotated with the corresponding exception to handle it. Exceptions may be raised by blocks and task instances inside the default branch or using an explicit `throw` command. When an exception of the specified type is raised, all blocks that contain it are recursively reverted until a proper `catch block` is reached, similarly to what happens in modern programming languages. A parallel branch exception, without a corresponding `catch block`, causes the raising of an asynchronous interruption in the remaining branches in order to revert the entire block. Different exceptions raised by parallel branches are grouped into a single exception before leaving the block.

7. `threshold block` – It limits to k the number of threads that can execute concurrently inside its body. The number of existing threads remains the same and corresponds to the number of declared parallel branches. If the number n of existing threads is greater then k, then $n - k$ threads will be suspended until the first k have completed their execution.

8. `skip` – It is useful for obtaining specific control-flow structures from generic ones; for instance, the usual while and repeat-until loops can be obtained replacing the right or the left branch of a `loop` with a `skip`, respectively.

9. `run` – It executes a task instance of the specified type. The thread of control is suspended until the task is completed or reverted with an exception.

10. `spawn` – It can be executed only inside a `concurrent block`. It creates a new task instance $t : T$ that is immediately executed into a new parallel branch added to the inner `concurrent block` containing the command.

11. `throw` – It raises an exception of the specified type, reverting recursively all blocks that contain it until a proper handler is reached.

12. `send` – It inserts the value of one or more variables into one or more corresponding output streams; the sending is asynchronous and the execution continues with the next block without waiting.

13. `receive` – It stores into one variable an object extracted from one of the available input streams. The `receive` temporally suspends the current thread of control until an object arrives or a timeout θ raises an exception. A `receive` with an associated timeout θ can be graphically annotated as in the second row of Fig. 2.c. A multiple `receive` stores the first arrived object

from a stream α_{in}^i into the corresponding variable x_i, resets the others to unbound and continues the execution: this behavior is called *or-receive*. A sequence of `receive` commands can be conceptually grouped into a unique *and-receive* which waits for an object from each connected stream before proceeding and is graphically depicted as in the first row of Fig. 2.c.

14. `empty` – It removes all objects in the specified stream α.

A `send` accepts only output streams and a `receive` only input streams: regardless of its name, a stream can have a different direction depending on the internal or external perspective. Furthermore, `send` and `receive` commands can be viewed as special task instances with their own identifiers: each variable x involved in a `send` $s:\mathcal{S}$ can be considered as an output stream $s.x_{out}$, while each variable y of a `receive` $r:\mathcal{R}$ can be considered as an input stream $r.y_{in}$.

Example 1. Fig. 3 depicts three NestFlow models that are useful for exemplifying several concepts about the notation and the semantics of some constructions. In order to keep at minimum the number of core constructs, NestFlow supports only message passing: this is not a limitation because parameter passing can be easily simulated by a `send`/`receive` pair. For instance, if task B in Fig. 3.a needs to store an object into the y variable at its completion, it can send such object through an output stream connected to the following `receive`. Such construction can be briefly denoted as in Fig. 3.b with $B \mid y$. In similar way, a `send` to a following task can be denoted as in Fig. 3.a with $x \mid D$. The model in Fig. 3.a can be further simplified by directly connecting B and D with a link, discarding the intermediate `send` and `receive` commands. Direct connections are essential for

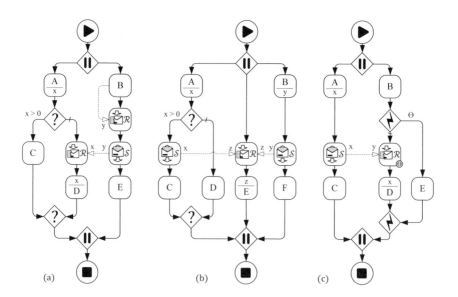

Fig. 3. (a) B produces an output that may be used by D. (b) E starts as soon as one input is made available. (c) The output of A is useful only if produced in time.

large-scale BP models; for instance, if B is a complex compound-task its output must be made available as soon as possible and not at completion. In similar way, D can run in parallel with B until the required input becomes strictly necessary. Not using an output produced by a task does not impact the model correctness: unused objects can be left in the case for successive executions, removed by an empty command or discarded when the case completes. For instance, Fig. 3.b depicts a model with an *or-receive* construction: task E is executed as soon as an object is stored in z by A or B. E needs only one input to run, the other one will be left in the case, if produced at all. As for the first model, the receive and the send after B can be substituted by a direct connection. In the model of Fig 3.c, D waits for the output of A at most θ units of time from the completion of B. In this case the output is useful only if produced in time: if the timeout expires an exception is raised and E is executed in place of D.

The NestFlow constructs description and the previous example should be sufficient for understanding the WCPs implementation in Sec. 4. A more detailed and formal description of the NestFlow semantics can be found in [11]. Two well-formedness properties are central in NestFlow models correctness: (1) a model T is valid only if its parallel branches do not share variables; (2) a model T is valid only if every task instance $a : A$ in T has a unique identifier $a \in \mathcal{I}$ and is executed in only one place. The first property avoids race conditions, while the second one guarantees a unique correspondence between graphical occurrences and task instances ensuring a unique graphical representation of senders and receivers: a task $a : A$ that sends an object to $b : B$ can be simply represented using an arrow from a to b and task identifiers can be left implicit.

4   NestFlow Representation of WCPs

The aim of this section is to give an idea on how WCPs can be obtained in NestFlow. To shorten the presentation, patterns are grouped by similarity and whenever possible they are described by difference with respect to the preceding ones. A more detailed description can be found in [12] where each WCP is recalled and explicitly implemented.

Sequence Patterns (G1) – Sequence (WCP-01) is represented in NestFlow through one or more sequence block. Interleaved Partial Order (P.O.) Routing (WCP-17) and Interleaved Routing (WCP-40) can be considered as special kinds of sequence, where only a partial order or no order between tasks is defined. In general, any finite partial order $R \subseteq S \times S$, among a set of tasks S, can be represented in NestFlow using parallel block, sequence block and link constructs, surrounded by a threshold block with $k = 1$. This block ensures that only one thread of control at time can execute inside the parallel block, thus only one task at time is running, while the other ones are suspended. A naive construction can be obtained with a single parallel block containing exactly one parallel branch for each task instance $t \in S$, then for each $(u, v) \in R$ a send s is added at the end of the branch containing u, a receive r is added at

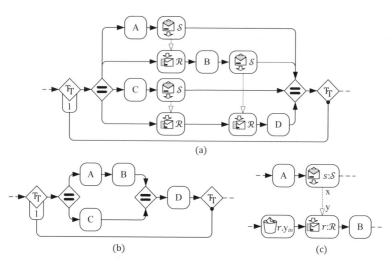

Fig. 4. (a) and (b) depict two possible representations of Interleaved Parallel Routing (WCP-17) of a set of four tasks. (c) depicts how to obtain a Milestone (WCP-18).

the beginning of the branch containing v and a link is added from s to r, as for A and B in Fig. 4.a. In Interleaved Routing (WCP-40) no order is defined and the tasks are distributed into a single **parallel block** wrapped by a **threshold block** with $k = 1$. This solution is used to prove the support of Interleaved P.O. Routing (WCP-17), but simpler constructions are possible, as exemplified in Fig. 4.b where all links have been substituted by control-flow constructs.

Critical Section (WCP-39) assumes the presence of a shared resource that has to be accessed in a mutually exclusive way. This resource can be managed by a single task instance, the other tasks can gain access to the underlying resource sending messages to it: stream serialization ensures the exclusive access to the resource. A shared resource can also be represented with a unique object exchanged by tasks using message passing. In Milestone (WCP-18) a task B can execute only when the process instance is in a specific state: for example, another task A is just concluded. This pattern can be represented as in Fig. 4.c: after A completion, an object is sent to the **receive** before B and if the thread of control in this branch is blocked in that **receive**, namely it is waiting that a specific state is reached, then the object is received and B is executed; otherwise, if the **receive** is performed some time after the completion of A, the **empty** command deletes all objects previously received in the stream $r.y_{in}$ and B is not executed.

Repetition Patterns (G2) – Structured Loop (WCP-21) is directly supported in NestFlow through the **loop block** that can also represent *repeat-until* and *while-do* loops by placing a **skip** in the first or second branch of the **loop block**, respectively. Recursive declarations of tasks (WCP-22) are also supported through lazy evaluation: an internal task instance will be created only when strictly necessary. As a design choice, arbitrary cycles (WCP-10) are not supported in NestFlow, because they drastically reduce modularity: this is not a severe limitation since any sequential composition of unstructured cycles can be transformed into an equivalent structured form [13], for instance duplicating some tasks or conditions.

Cancellation Patterns (G3) – The possibility to cancel a single activity (WCP-19) in any place of a model implies the ability to cancel an arbitrary region (WCP-25) or the entire case (WCP-20). We consider a generic behavior for WCP-19, because the original description does not clarify how to deal with compound tasks that are in an intermediate state, namely started but not completed. NestFlow supports cancellation through hierarchical exception handling that provides both a mechanism to manage task cancellation and the ability to specify clean-up actions to perform during the cancellation phase. Whenever required, a task has to be built cancellable for ensuring encapsulation, because only a task knows exactly how to exit. The cancellation of a task from a different thread of control can be obtained by sending a cancellation message to it.

When cancellation is not provided, a single or a block of activities $\langle B \rangle$ can be wrapped in a structure similar to the one in Fig. 5: if a cancellation message is sent to $\mathcal{R}$ during $\langle B \rangle$ execution, an interruption exception ξ is thrown on branch t_1 and the activities in this branch are cancelled; otherwise, after $\langle B \rangle$ completion a message is sent to $\mathcal{R}$ for terminating the other branch without raising an exception.

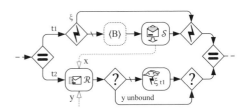

Fig. 5. A possible representation of Cancellation Patterns (G3)

Trigger Patterns (G4) – The activation of a task A through a signal is naturally supported by a link connected to A. In NestFlow streams are persistent (WCP-24): they retain objects until the receiver is able to consume them. A Transient Trigger (WCP-23) can be simulated by placing an `empty` command just before the `receive`, for canceling the content of its incoming stream.

Termination Patterns (G5) – In NestFlow there is no real difference between Implicit Termination (WCP-11) and Explicit Termination (WCP-43), because only one thread of control can enter a block and when it leaves the block, no other thread is left behind: only one thread reaches the stop place and the stop place is reached only when the last task instance completes. Besides the stop place, an explicit termination can be obtained by raising an exception that should be properly managed.

Branching Patterns (G6) – Branching patterns describe the divergence of thread of controls from a single point in the model. Parallel Split (WCP-02) is obtained with a single `parallel block`, while the Exclusive Choice (WCP-04) behavior is given by a single `choice block`. Conversely, there is no specific construct that supports Multi Choice (WCP-06) because its behavior can be obtained combining a `parallel block` with one `choice block` for each branch and an additional default branch. Undoubtedly, the introduction of such construct improves usability but redundancy shall be avoided in a *core* language.

Deferred Choice (WCP-16) is the most contrived WCP because any system has its own particular implementation: BPMN/WS-BPEL supports it with a `<pick/>` construct which suspends its thread of control and waits until an external event occurs, for instance receiving a message or a timeout expiration.

In YAWL the deferred choice places two or more tasks in the work-list; when one of these is chosen the other ones are instantaneously withdrawn. In NestFlow the first kind of behavior can be obtained with a single `receive` command that waits for the first incoming object from multiple streams or a timeout event. The second

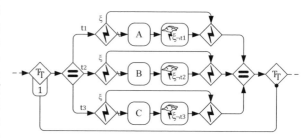

Fig. 6. One of the possible implementations of Deferred Choice (WCP-16)

kind of behavior can be also obtained, but here a slightly different solution is presented based on the `threshold block` suspension semantics. This solution is exemplified in Fig. 6, where a $\neg t_i$ in a `throw` command means "raise an exception to interrupt all parallel branches except t_i". The `threshold block` ensures that only one branch at a time is executed, while the other ones are suspended. Suspended tasks are not removed from the work-list, but they cannot be chosen by users. If the chosen task completes successfully, then the suspended tasks are definitively removed, otherwise they can become available again as alternatives of the failed one.

Synchronization Patterns (G7) – Synchronization (WCP-03) and Structured Synchronizing Merge (WCP-07) are characterized by only one divergency point and are directly supported by `parallel block` and `choice block`. Structured Discriminator (WCP-09), Blocking Discriminator (WCP-28) and Canceling Discriminator (WCP-29) are a specialization of Structured Partial Join (WCP-30), Blocking Partial Join (WCP-31) and Canceling Partial Join (WCP-32), respectively, when the number k of synchronized branches is one. In the same way, Structured Partial Join (WCP-30) and Blocking Partial Join (WCP-31) are a specialization of Canceling Partial Join (WCP-32) where the cancellation of the remaining activities is not necessary. Fig. 7.a shows the implementation of the most general Canceling Partial Join (WCP-32) from which the other patterns can be derived. After completion, each task involved in the synchronization sends a message to r which stores the first arrived object in the corresponding variable and continues the execution. The `receive` is performed k times for waiting the completion of exactly k tasks. When k tasks complete, multiple messages are sent to cancel the tasks that are still running. A cancellation message is sent to all involved task instances, but it does not affect the completed ones. The pattern can be executed multiple times preceding it with a reset phase. The blocking variation of partial joins (WCP-30, WCP-31) can be obtained by removing the send s after the `loop block` in Fig. 7.a. In the structured case all involved task instances belong to the same `parallel block`, while in the unstructured case they can belong to different execution paths. The discriminator patterns (WCP-29, WCP-09 and WCP-28) can be obtained in the same way of the corresponding partial join ones by removing the counting `loop block` around the receive r.

Generalized And-Join (WCP-33) can be obtained by replacing the bottom branch in Fig. 7.a, with a single and-receive followed by B. Acyclic Synchronizing Merge (WCP-37) and Generalized Synchronizing Merge (WCP-38) can be obtained using an and-receive, as in the previous case, and wrapping each task $A_1, ..., A_n$ into the branches of a `choice block` followed by a `send`. If the `choice block` condition $\varphi_j(\bar{x}_j)$ evaluates to true, then A_j executes, otherwise a `skip` is performed; in any case a message is sent to notify the and-receive.

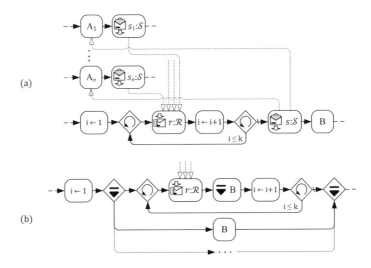

Fig. 7. (a) depicts the implementation of Canceling Partial Join (WCP-32). (b) represents the alternative branch for obtaining Multi Merge (WCP-08).

Simple Merge (WCP-05) can be obtained from the model in Fig. 7.a by removing the send s and placing B inside the `loop block`: this construction ensures that B is executed multiple times respecting the completion order of the involved task instances but without overlapping executions. Multi Merge (WCP-08) can be obtained by replacing the branch containing the loop in Fig. 7.a with the one in Fig. 7.b: any time one of the involved task instances completes a message is sent to r and the following `spawn` command creates a new instance of B which is immediately executed inside the `concurrent block`.

5   Evaluation Method

WCPs are a collection of behavioral patterns recurrently found in BP models. They are described in natural language and formalized as Colored Petri Nets (CPNs): a proper extension of Petri Nets enhanced with data, types and functional expressions. A particular WCP can be encoded as a CPN in several different ways; however, if the use of Petri nets constructs is preferred against the use of functional expressions, then the pattern complexity can be roughly quantified by the number of graphical constructs used for representing it. In particular, the WCPs encoding in [8] can be taken as a unit of comparison because it uses basic constructs and it is reasonable to assume that this encoding is nearly optimal, namely no simpler encoding can be given without exploiting advanced features.

WCPs has been used to analyze and classify a great number of BPMLs in order to expose their peculiarities and ease system comparison. In such analysis the commonly accepted scoring assigned to each WCP implementation is *fully supported* $(+)$, if there is a language construct that directly implements the pattern, *partially supported* $(\pm)$, if there is a construct that produces only a similar behavior to the one prescribed by the pattern, or *unsupported* $(-)$, if none of the provided constructs reproduces the required behavior.

This approach is questionable because it focuses only on language constructs at the expense of the actual behavior that can emerge when such constructs are put together. This leads to paradoxical conclusions: for instance, one should accept that CPNs, the language chosen to formalize all WCPs, supports only few of them, hence it is not suitable for BP design. The same conclusion holds even more for classical Petri nets that are strictly less expressive than CPNs, despite they are a widely accepted formal language for BP modeling and analysis.

In this paper the NestFlow evaluation is based on how many constructs are needed to reproduce a particular WCP with respect to the CPNs constructs used in [8]. CPNs seem a good reference for comparison because it is hard to conceive a graphical formal language with simpler constructs and equal expressiveness. Any language tailored for BP modeling should generally perform better in encoding WCPs than CPNs, otherwise something goes wrong with the language design.

At first glance, it seems easy to reach the maximum score in a WCPs analysis by conceiving a BPML that provides one construct for each pattern. Although this is ideally possible, one should explain what happens when two or more of these constructs are put together to design a BP model and how they can be used to express emerging patterns not considered in the current WCPs collection.

6   Evaluation of NestFlow WCPs Support

This section exposes the results of the NestFlow evaluation, giving a first evidence about its expressiveness and suitability for BP design. Multiple-instance patterns are not considered because they can be seen as a specialization of those presented here, where the involved task instances are always of the same type.

The result of this analysis is summarized in Table 1: the NestFlow support of each WCP is ranked excellent (★★★), good (★★☆), fair (★☆☆) and none

Table 1. A summary of NestFlow WCPs support. WCPs are grouped as in [1] and ordered on the basis of their similarities. *Code* is a unique pattern identifier defined in [8] and *Name* is its common name. Given the number of involved tasks n, *NF* is the number of NestFlow constructs used to represent the WCP in the *worst case* scenario, while *CPN* is the number of CPN constructs used in [8]; ρ is the ratio between the values in *NF* and *CPN* columns for large n, and γ is the ratio between the number of used links and the total number of NestFlow constructs. *Eval* is the overall evaluation.

	Code [8]	Name	NF	CPN	ρ	γ	Eval
	WCP-01	Sequence	$n+2$	$4n+2$	0.25	0.00	★★★
	WCP-17	Interleaved P.O. Routing	$4n+1^{(a)}$	$14n+6^{(a)}$	0.29	0.25	★★☆
G1	WCP-40	Interleaved Routing	$n+4$	$10n+10$	0.10	0.00	★★★
	WCP-39	Critical Section	$8n+9$	$14n+8$	0.57	0.37	★☆☆
	WCP-18	Milestone	10	18	0.56	0.10	★★★
	WCP-10	Arbitrary Cycles	—	29	—	—	☆☆☆
G2	WCP-21	Structured Loop	4	15	0.27	0.00	★★★
	WCP-22	Recursion	11	17	0.65	0.18	★★★
	WCP-19	Cancel Activity	14	30	0.47	0.14	★★★
G3	WCP-25	Cancel Region	$14n$	—	—	0.14	★★★
	WCP-20	Cancel Case	14	—	—	0.14	★★★
G4	WCP-23	Transient Trigger	6	16	0.37	0.17	★★★
	WCP-24	Persistent Trigger	5	9	0.55	0.20	★★★
G5	WCP-11	Implicit Termination	0	—	—	—	★★★
	WCP-43	Explicit Termination	1	—	—	0.00	★★★
	WCP-04	Exclusive Choice	$n+2$	$6n+4$	0.17	0.00	★★★
	WCP-16	Deferred Choice	$5n+4$	$4n+6^{(b)}$	1.25	0.00	★★★
G6	WCP-06	Multi Choice	$4n+1$	$7n+3$	0.57	0.00	★★★
	WCP-02	Parallel Split	$2n+1$	$7n+3$	0.29	0.00	★★★
	WCP-42	Thread Split	—	$n+5$	—	—	☆☆☆
	WCP-09	Structured Discriminator	$5n+2$	$5n+14$	1.00	0.20	★★☆
	WCP-28	Blocking Discriminator	$5n+2$	$9n+17$	0.56	0.20	★★☆
	WCP-29	Canceling Discriminator	$6n+4$	$10n+15$	0.60	0.33	★☆☆
	WCP-30	Structured Partial Join	$5n+6$	$5n+14$	1.00	0.20	★★☆
	WCP-31	Blocking Partial Join	$5n+6$	$9n+17$	0.56	0.20	★★☆
	WCP-32	Canceling Partial Join	$6n+7$	$10n+15$	0.60	0.33	★☆☆
G7	WCP-05	Simple Merge	$5n+8$	$5n+5$	1.00	0.20	★★☆
	WCP-08	Multi Merge	$7n+9$	$5n+5$	1.40	0.20	★☆☆
	WCP-07	Structured Synch. Merge	$4n+1$	$9n+6$	0.44	0.00	★★★
	WCP-37	Acyclic Synch. Merge	$7n+2$	$11n+6$	0.64	0.14	★★☆
	WCP-38	Generalized Synch. Merge	$7n+2$	—	—	0.14	★★☆
	WCP-03	Synchronization	$2n+1$	$7n+3$	0.14	0.00	★★★
	WCP-33	Generalized And-Join	$5n+2$	$7n+3$	0.71	0.20	★★☆
	WCP-41	Thread Merge	—	$n+5$	—	—	☆☆☆

G1 Sequence Patterns   G4 Trigger Patterns   G7 Synchronization Patterns
G2 Repetition Patterns   G5 Termination Patterns $^{(a)}$ Assuming P.O. graph size $n-1$
G3 Cancellation Patterns   G6 Branching Patterns $^{(b)}$ Considering the richer representation

($\star\star\star$) using the following parameters: (1) the ratio ρ between the number of NestFlow constructs used to encode the pattern in the worst case scenario and the number of CPNs constructs used in [8] for representing the same behavior, (2) the ratio γ between the number of links and the number of control-flow constructs used in the NestFlow interpretation and finally (3) the variety of involved NestFlow constructs. The *NF* column contains the total number of used NestFlow constructs, including the number of links and the maximum number of running threads. Similarly, the *CPN* column contains the number of CPNs constructs used in [8], including the number of transitions, places, arcs and the maximum number of involved tokens. For both languages the count does not include additional notations used to specify conditions and expressions. For estimating pattern complexity, a link can be considered a weak control-flow relation manipulated by more reliable control-flow structures; with this interpretation, γ gives a first clue about the level of unstructuredness of the pattern because links may cross the main control-flow structure; in practice structured forms are always preferred during design, and the number of links will be substantially below γ, which represents a worst case estimation.

Pattern support with worst case $\rho \leq 0.50$ and $\gamma \leq 0.20$ and an optimal use of constructs is ranked excellent ($\star\star\star$). We also accept in this category pattern implementations with $\rho > 0.50$ or a value of γ near to 0.20 when these ratios do not depend on the number of tasks n. Deferred Choice (Wcp-16) and Multi Choice (Wcp-06) are also ranked excellent even if $\rho > 0.50$, because Wcp-16 has an optimal implementation given by the `receive` construct and Wcp-06 can be captured by a single specialized construct [12]. Patterns with worst case $\rho \leq 1.00$ and $\gamma \leq 0.20$ and an adequate use of constructs are ranked good ($\star\star\star$). Interleaved P.O. Routing (Wcp-17) is also ranked good, because the worst case coefficients $\rho = 0.29$ and $\gamma = 0.25$ are related to an ideal worst case partial order. Patterns with $\rho > 1.00$ or $\gamma > 0.20$ are ranked fair ($\star\star\star$), while the unsupported patterns are ranked none ($\star\star\star$). Such criteria allows a more fine-grained and measurable evaluation of WCP support. Nevertheless, a pattern-based analysis remains a *qualitative* evaluation of expressiveness: a WCP specifies a system behavior that can be obtained in different ways, combining more or less sophisticated constructs. For instance, in modeling a process by CPNs we can obtain a particular behavior mainly using Petri nets constructs or alternatively using few graphical constructs, encoding most of the logic in functional arc expressions and transition guards. The considered language constructs are also important but a pattern-based evaluation that puts too much emphasis on constructs, instead of on the overall behavior, leads to paradoxical conclusions as explained in the previous section. The NestFlow evaluation is mainly based on the effort needed to replicate WCPs behavior: for each pattern such effort has been quantified by comparing the NestFlow implementation in the worst case scenario with the CPNs reference implementation [8]. The count of CPNs constructs is omitted for those patterns whose CPNs model is an ad-hoc construction that cannot be easily quantified, for instance because the number of constructs depends on the reachable states.

Any high-level BPML that wants to support WCPs will likely provide a more compact representation of these patterns, with more specific constructs than CPNs. Except for few patterns, the NestFlow ratio ρ is always less than one; more specifically, it usually needs half of the CPNs constructs for expressing the same behavior. In some cases the ratio ρ is greater than one or the pattern is not supported at all, as for Multi Merge (WCP-08), Arbitrary Cycles (WCP-10), Thread Merge (WCP-41) and Thread Split (WCP-42). The lack of support for these patterns is acceptable because consistent with the initial design intentions: NestFlow is built to be modular and these patterns hinder modularity.

7   Conclusion

Well established BP modeling practices and empirical research experiments suggested that structured control-flow forms are always desirable to enhance comprehensibility and modularity, and reduce the probability of introducing new errors in BP models. Unfortunately, existing BPMLs are not able to support a fully structured control-flow design without losing expressiveness. This paper introduces a novel structured BPML called NestFlow and shows how WCPs [8] can be implemented in a structured way. WCPs are a well accepted framework for evaluating BPMLs expressiveness and suitability; however, the generally adopted scoring method in WCPs-based analysis focuses on language constructs limiting its applicability. In this paper a more objective scoring method is introduced and the NestFlow WCPs support is evaluated on the basis of the used constructs with respect to the CPNs reference implementation proposed in [8], the number of links with respect to the total number of used constructs and the constructs variety. The analysis is performed, whenever possible, for a large number of involved tasks in the worst case scenario. NestFlow proves that structured BPMLs can be effectively built and are potentially more expressive than the existing ones especially when modularity is seriously taken into consideration.

References

1. ter Hofstede, A.H.M., van der Aalst, W.M.P., Adams, M., Russell, N.: Modern Business Process Automation: YAWL and its Support Environment. Springer, Heidelberg (2009)
2. Object Management Group (OMG). Business Process Modeling Notation (BPMN) 2.0 (Beta 1) (August 2009), http://www.omg.org/spec/BPMN/2.0/
3. Liu, R., Kumar, A.: An Analysis and Taxonomy of Unstructured Workflows. In: van der Aalst, W.M.P., Benatallah, B., Casati, F., Curbera, F. (eds.) BPM 2005. LNCS, vol. 3649, pp. 268–284. Springer, Heidelberg (2005)
4. Laue, R., Mendling, J.: Structuredness and Its Significance for Correctness of Process Models. In: Inf. Systems and E-Business Management, pp. 287–307 (2009)
5. Gruhn, V., Laue, R.: What Business Process Modelers Can Learn from Programmers. Science of Computer Programming 65(1), 4–13 (2007)
6. Reijers, H., Mendling, J.: Modularity in Process Models: Review and Effects. In: Dumas, M., Reichert, M., Shan, M.-C. (eds.) BPM 2008. LNCS, vol. 5240, pp. 20–35. Springer, Heidelberg (2008)

7. Mendling, J., Reijers, H.A., van der Aalst, W.M.P.: Seven Process Modeling Guidelines (7PMG). Information and Software Technology 52(2), 127–136 (2010)
8. Russell, N., ter Hofstede, A.H.M., van der Aalst, W.M.P., Mulyar, N.: Workflow Control-Flow Patterns: a Revised View. BPM Center Report BPM-06-22 (2006)
9. Vanhatalo, J., Völzer, H., Koehler, J.: The Refined Process Structure Tree. Data & Knowledge Engineering 68(9), 793–818 (2009)
10. Combi, C., Gambini, M.: Flaws in the Flow: the Weakness of Unstructured Business Process Modeling Languages Dealing with Data. In: Meersman, R., Dillon, T., Herrero, P. (eds.) OTM 2009. LNCS, vol. 5870, pp. 42–59. Springer, Heidelberg (2009)
11. Combi, C., Gambini, M., Migliorini, S.: The Nestflow Modeling Language Specification (2010), http://www.nestflow.org/downloads/nestflow_spec.pdf
12. Combi, C., Gambini, M., Migliorini, S.: Nestflow Workflow Control-Flow Patterns Support (2010), http://www.nestflow.org/downloads/nestflow_wcps.pdf
13. Oulsnam, G.: Unravelling Unstructured Programs. Comp. J. 25(3), 379–387 (1982)

On Simplifying Integrated Physical Database Design

Rima Bouchakri[1] and Ladjel Bellatreche[2]

[1] National High School of Computer Science Algiers, Algeria
r_bouchakri@esi.dz
[2] LISI/ENSMA Poitiers University Futuroscope, France
bellatreche@ensma.fr

Abstract. This paper deals with the problem of integrated physical database design involving two optimization techniques: horizontal data partitioning (HDP) and bitmap join indexes (BJI). These techniques compete for the same resource representing selection attributes. This competition incurs *attribute interchangeability* phenomena, where same attribute(s) may be used to select either HDP or BJI schemes. Existing studies dealing with integrated physical database design problem not consider this competition. We propose to study its contribution on simplifying the complexity of our problem. Instead of tackling it in an integrated way, we propose to start by *assigning* to each technique its own attributes and then it *launches* its own selection algorithm. This assignment is done using the K-Means method. Our design is compared with the state of the art work using APB1 benchmark. The results show that an *interchangeability attribute-aware database designer* can improve significantly query performance within the less space budget.

1 Introduction

Optimizing complex queries running on the top of very large database schemes such as data warehouses and scientific databases represents a *crucial performance issue* [19,21]. To reach this objective, several important tasks need to be performed by database administrators (DBA): **(1)** the choice of optimization techniques ($\mathcal{OT}$), **(2)** the choice of their selection mode (isolation or multiple), **(3)** the development of selection algorithms, **(4)** the generation of scripts corresponding to each selected $\mathcal{OT}$ and **(5)** the validation and deployment of the obtained solutions. Note that task **2** is the most important and more complex in the physical design phase, since it may impact the other tasks. Database community has demonstrated a great attention of studying the *individual selection (where only one technique is selected)* of $\mathcal{OT}$ (formalization, proposition of selection algorithms, development of advisors based in this mode, etc.) [2,3,7,8,13]. The individual selection is not sufficient to optimize the whole workload, since each $\mathcal{OT}$ is adapted to a *particular class of queries*. As consequence, the *multiple selection mode* has been introduced. Historically, studies related to this mode start with *two* $\mathcal{OT}$ [19], *three* [20] and now *four* [18,21]. Microsoft *AutoAdmin* project [19] is an example of work that deals with the multiple selection

J. Eder, M. Bielikova, and A.M. Tjoa (Eds.): ADBIS 2011, LNCS 6909, pp. 333–346, 2011.

of *materialized views* and *indexes* for a given workload. It has been extended by incorporating two others $\mathcal{OT}$: *vertical* and *horizontal partitioning* [18]. IBM database research group proposed a design advisor that supports four $\mathcal{OT}$: materialized views, partitioning, indexes and clustering. This mode is more complicated than the individual selection, since it requires the exploration of a large search space [21]. Three main implementations of this mode exist: *iterative, joint* (or *integrated*) and *hybrid*. The iterative solution is a naive way to implement the multiple selection, in which the selection of involved $\mathcal{OT}$ is done sequentially [20]. This solution *ignores* interaction between $\mathcal{OT}$. In the integrated implementation, a joint searching is performed directly in the *combined search space*, and heuristic rules are applied to limit the candidate sets being considered [21]. This implementation can better handle the interdependencies among different $\mathcal{OT}$. It has been used in *AutoAdmin* project to recommend indexes and materialized views by exploiting their interdependencies: both are redundant (they duplicate data), compete for the same resource representing storage space and cause update overhead [19]. The main drawback of this implementation is its extensibility and high complexity [21]. To reduce this complexity and to consider the interaction between $\mathcal{OT}$ during the selection process, a hybrid implementation has been proposed [6,21]. Its main idea is to analyze the interaction between $\mathcal{OT}$ and then imposes a selection order. Two main solutions of this implementation exist. The work done by [21], where two *dependency relations* between pairs of $\mathcal{OT}$ are identified: *strongly* and *weakly* dependencies. An $\mathcal{OT}$ ot_1 "strongly" depends on $\mathcal{OT}$ ot_2, if a change in selection of ot_2 often results in a change in that of ot_1 (example of materialized views and indexes). Otherwise, ot_1 "weakly" depends on ot_2. These relations are used to establish selection order. For instance, if only ot_1 strongly depends on ot_2, the authors propose to iteratively search ot_2 and ot_1, but make sure that ot_2 is searched before ot_1 so that ot_1 is properly influenced by ot_2. The second solution is given in [6], where the authors propose the use of horizontal data partitioning HDP (considered as a non redundant $\mathcal{OT}$) to *prune* the search space of bitmap join indexes BJI (a redundant $\mathcal{OT}$) selection problem. Their methodology starts by partitioning a data warehouse, identifies queries that will get benefit from this partitioning and selects BJI by considering only queries that do not get benefit from HDP (called *non profitable queries*). This work imposes a predefined selection order (HDP then BJI).

The proposed solutions dealing with the implementation of multiple selection do not analyze the body of $\mathcal{OT}$[1]. To illustrate this, let us consider two $\mathcal{OT}$: HDP and BJI, both are usually defined using the same set of selection attributes. If an attribute A_j is used to partition a given database and it gives satisfaction in optimizing queries, then *why we keep it for indexing selection process*. If, it is removed from the list of candidate attributes for indexing, the complexity of BJI selection problem may be reduced and a gain of storage and maintenance overhead may be guaranteed. The same reasoning is made if A_j is used to index the database. Removing this attribute from partitioning list may reduce the number of final fragments and facilitates the *manageability* of the partitioned database.

---

[1] The body of an $\mathcal{OT}$ describes table(s), attributes used in the definition of that $\mathcal{OT}$.

As consequence, the body analysis incurs a new phenomenon called *attribute interchangeability*. Recently, in *Coradd* project [14], another consequence of the body analysis of $\mathcal{OT}$ is identified and incurs *attribute correlation* phenomenon which has been exploited to select two redundant $\mathcal{OT}$: materialized views and indexes. The fact of ignoring body analysis of $\mathcal{OT}$ may affect seriously the dependency relations identified in [14,21]. For instance, the dependency relations between indexes and partitioning (partitioning and indexes) are considered as weak in [21]. But, if we consider *attribute interchangeability phenomenon they become mutually strong dependent*.

This paper studies the impact of body analysis on selecting HDP and BJI and proposes other solutions for the hybrid implementation to solve the multiple selection.

This paper is organized into seven sections. Section 2 presents background related to HDP and BJI. A genetic algorithm for selecting BJI is given in Section 3. Section 4 studies the impact of attribute interchangeability on $\mathcal{OT}$ selection. Section 5 describes an approach to assign attributes between HDP and BJI based on the K-Means method. Section 6 experimentally compares our proposal with existing studies. Section 7 concludes the paper summarizing the main findings of our research, and proposing directions for future work.

2   Background

To facilitate the understanding of attribute interchangeability, concepts and examples related to partitioning and BJI are given.

2.1   Horizontal Data Partitioning

HDP decomposes a table based on its instances. It is supported by most commercial and non commercial DBMS (Oracle, SQL Server, DB2, Postgres, etc.), where a native data definition language is proposed. Two main types of HDP exist: *mono table partitioning* and *table-dependent partitioning*. In the mono table partitioning, a table is partitioned using its own *attributes* used in selection operations. Several modes are proposed to implement this partitioning: *Range, List, Hash, Round Robin* (supported by Sybase), *Composite (List-List, Range-List, Range-Range, ...),* etc. Mono table partitioning optimizes selection operations, when partitioning keys match with selection attributes. In table-dependent partitioning, a table inherits the partitioning characteristics from other table. For example, a fact table of a given relational data warehouse schema may partitioned based on the fragmentation schemes of dimension tables[2] [9]. This partitioning is supported by Oracle11G (*known under the name referential partitioning*). It optimizes selections and joins simultaneously.

The problem of HDP has been formalized (*in isolation form*) in the context of relational data warehouses RDW with d dimension tables $\{D_1, ..., D_d\}$ and a fact

---

[2] A fragmentation schema is the result of partitioning process.

table $\mathcal{F}$ as follows [4,17]: given (i) a representative workload $\mathcal{Q} = \{Q_1, ..., Q_n\}$, where each query Q_i $(1 \leq i \leq n)$ has an access frequency f_i, defined on the RDW and (ii) a constraint (called maintenance bound $\mathcal{B}$ given by DBA) representing the maximum number of fact fragments that she/he wants. The problem of HDP consists in identifying dimension table(s) that could be used to partition the fact table $\mathcal{F}$ into N fragments, such that the overall execution cost of queries $(\sum_{Q_i \in \mathcal{Q}} f_i \times Cost(\mathcal{Q}, FS))$ is minimized and maintenance constraint is satisfied $(N \leq \mathcal{B})$, where FS represents the obtained fragmentation schema. This problem is known as NP-hard [4]. Several types of algorithms to find a near-optimal solution of this problem are proposed: *greedy, genetic, simulated annealing, data mining driven algorithms* [3,4,16].

2.2   Bitmap Join Indexes

Indexes considered in traditional databases and recommended by advisors usually concern only one table (B-tree, Bitmap, hash, etc.) [19,21]. A BJI is another form of indexes involving several tables. It computes the joins between the fact table and l $(l \geq 1)$ dimension tables using s $(s \geq 1)$ attributes used in selection operations. This join is materialized through a set of bit vectors built on the fact table based on dimension attribute(s) of *low cardinality*. The BJI are more efficient for count, and, or, not queries. The size of the binary index is proportional to the cardinality of the indexed attributes.

The formalization of the problem of selecting BJI *in isolation form* is quite similar to HDP problem, except it uses a storage capacity ($\mathcal{S}$) as a constraint [1]. The selected BJI shall minimize the query execution cost and satisfies $\mathcal{S}$. This problem is known as NP-hard [1]. To the best of our knowledge, only two classes of algorithms were proposed to deal this problem: *greedy heuristics* [6] and d*ata mining techniques* [1]. To give DBA a large spectrum of choices of BJI selection algorithms, we propose new genetic algorithm (GA) that selects BJI.

3   A Genetic Algorithm for Selecting BJI

GA have been used for a long time in the database physical design and proven to be a good for selecting optimization schemes [3,12]. Given a well-defined search space, they apply three different genetic search operations, namely, *selection, crossover* and *mutation*, to transform an initial population of chromosomes, with the objective to improve their quality.

For BJI selection, each chromosome is represented by an array of bits, where each cell corresponds to an indexable attribute. A cell value is set 1, if its corresponding attribute is used by a BJI, 0 otherwise. Figure 1 shows an example of chromosome involving five indexable attributes. This coding generates three different BJI defined on *City, Year* and *TypeProduct*. Note that the number of attributes candidates for BJI (or partitioning) is *very important. For instance, the star schema of photographic objects of Sloan Digital Sky Server Dat contains more than 400 attributes [11].*

Month	City	Year	Country	TypeProduct
0	1	1	0	1

Fig. 1. An example of chromosome

Note that each chromosome c_i of our GA represents a configuration of BJI. Let $Config_{c_i}$ and N_{c_i} be the set of selected indexes and its cardinal. To evaluate the quality of this configuration, two cost models are needed: one for estimating the cost needs to store $Config_{c_i}$ and the second to calculate the global query processing cost (in terms of inputs outputs) in the presence of $Config_{c_i}$. The storage cost required for a BJI bji_j of $Config_{c_i}$ defined on attribute A_k is given by [1,5]: $storage(bji_j) = (\frac{|A_k|}{8} + 16) \times |\mathcal{F}|$ bytes, where $|A_k|$ and $|\mathcal{F}|$ represent respectively, the cardinality of the attribute A_k and the number of instances of the fact table F. The cost of executing a query Q_i $(1 \leq i \leq n)$ in presence of bji_j is given by: $Cost(Q_i, bji_j) = log_m|A_k| - 1 + \frac{|A_k|}{m-1} + d \times \frac{||\mathcal{F}||}{8 \times PS} + ||\mathcal{F}|| \times (1 - e^{-\frac{N_r}{||\mathcal{F}||}})$ I/O, where $||\mathcal{F}||$, N_r, PS and d represent respectively the number of pages used by fact table $\mathcal{F}$, the number of tuples accessed by bji_j, the size of a disk page and the number of bitmaps used to evaluate the query Q_i. The global cost of executing all the n queries in the presence of the index configuration $Config_{c_i}$ is given by: $Cost(\mathcal{Q}, Config_{c_i}) = \sum_{i=1}^{n} \sum_{j=1}^{N_{c_i}} Cost(Q_i, bji_j)$ I/O.

To penalize a chromosome generating a configuration violating the storage constraint, a penalty value is introduced as a part of the fitness function. It is defined as follows: $Pen(Config_{c_i}) = \frac{storage(Config_{c_i})}{S}$ where $storage(Config_{c_i}) = \sum_{j=1}^{N_{c_i}} storage(bji_j)$. Our fitness function is defined as follows:

$$F(Config_{c_i}) = \begin{cases} Cost(\mathcal{Q}, Config_{c_i}) \times Pen(Config_{c_i}), & \text{if } Pen(Config_{c_i}) > 1 \\ Cost(\mathcal{Q}, Config_{c_i}), & \text{otherwise} \end{cases}$$

4   Attribute Interchangeability Impact on OT Selection

To facilitate the understanding the basic idea behind the attribute interchangeability and to show its utility, we consider the following motivating example.

4.1   Motivating Example

Let us assume that DBA decides to use HDP and BJI to optimise a set of parameterized queries[3], where each one has the following form:

```
Select count(*)
From Sales S, Customer C
Where C.Gender = 'F' And C.CID = S.CID
```

---

[3] A parameterized query is a query with selection operation defined on each different value of domain attribute.

She/he may execute algorithms for selecting relevant HDP and BJI. The obtained recommendations are: (1) the fact table *Sales* may be partitioned based on the fragmentation schema of the dimension table *Customer* also decomposed into two horizontal fragments based on the attribute *Gender* (that we call *partitioning attribute*). The obtained schemes are materialized by the following pseudo SQL statements (*Oracle Syntax*).

```
Create Table Customer (CID Number, Name Varchar2(20), Gender Char, Age Number)
Partition By List (Gender)
Partition Female Values ('F'), Partition Male Values ('M')

CREATE Table Sales (CID Number, PID Number, TID Number, ..., )
...
Partition by Reference  (CID)
```

(2) Defining a BJI on the fact table *Sales* using the attribute *Gender* (*indexable attribute*) of dimension table *Customer*.

```
Create Bitmap Index sales_cust_gender On Sales (Customer.Gender)
From Sales S, Customer C Where S.CID= C.CID
```

These two $\mathcal{OT}$ may give equivalent performance for the queries, as consequence, it is useless to consider them together (BJI require storage and maintenance costs, whereas HDP does not cause these overheads). Therefore, it will be better to assign attributes to $\mathcal{OT}$ before launching their selection algorithms, which reduces significantly DBA tasks. Note that the selection process is a task that requires much expertise from the DBA and consumes time and effort [10]. In order to capture the difficulty of the attribute assignment problem and to solve it, a formalisation of our integrated physical database design problem is recommended.

4.2   Formalisation

Our problem is formalized as follows: given a (i) workload $\mathcal{Q} = \{Q_1, Q_2, \cdots, Q_n\}$, where each query Q_j has an access frequency f_j, (ii) a set of restriction attributes $\mathcal{R}$ extracted from $\mathcal{Q}$, (iii) storage capacity $\mathcal{S}$ for BJI and (iv) a threshold $\mathcal{B}$ representing the maximum number of fact fragments. Our problem consists in selecting HDP and BJI schemes that reduce the query processing cost and satisfy the defined constraints ($\mathcal{B}$ and $\mathcal{S}$).

The resolution of this problem requires an exploration of the combined search space of HDP and BJI problems given by: $2^{Ins_{HDP}+Ins_{BJI}}$, where Ins_{HDP} and Ins_{BJI} represent respectively, the number of instances of both problems [21]. To simplify this resolution of this problem, we propose to first share attributes between these HDP and BJI and then to select each $\mathcal{OT}$ based on its assigned attributes using a selection algorithm.

To assign $\mathcal{R}$ to HDP and BJI, two main solutions are available for DBA: (i) she/he may use a manual assignment based on her/his experience. This solution is feasible if the cardinal of $\mathcal{R}$ is small. For extremely large databases with an important number of attributes, this solution is not useful. (ii) She/he may consider an exhaustive enumeration of all possible assignments (given by $2^{||\mathcal{R}||}$, where $||\mathcal{R}||$ represents the cardinal of $\mathcal{R}$). HDP and BJI selection algorithms are then executed for each assignment. Finally, the assignment offering a lower cost will be considered. This solution has high computational complexity which makes it unsuitable for practical applications and hence a low complexity suboptimal solution is proposed in the next section.

5   Clustering-Based Attribute Assignment by K-Means

Based on a deep analysis of each $\mathcal{OT}$ (BJI and HDP) [1,3,16,17,20], three clustering criteria are identified to guide attribute assignment process.

1. *Access frequency of a selection attribute*: this criterion represents the number of appearance of each attribute in queries. The existing works showed that HDP gives usually better performance when it is defined on most frequently attributes. Similarly, BJI defined on this type of attributes are efficient, especially for some classes of queries (e.g., count). If an attribute with high access frequency is concerned by HDP and BJI, we recommend using it for HDP due to its non redundancy nature.
2. *Attribute cardinality*: when the cardinalities of attributes are higher, the size of each fragment may increase, since HDP selection is based on splitting attribute's domains into sub domains [3]. This type of attribute may offer *a large number* of smallest fragments, especially when the maintenance cost ($\mathcal{B}$) is large enough (Section 2). When BJI is defined on higher cardinality, the storage cost increases dramatically. Based on this observation, the following assignment rule is established: a *high cardinality attribute is recommended for partitioning, whereas a lower cardinality attribute for BJI*[4].
3. *Selectivity factor defined on restriction predicates*: let A_i be an attribute used by k_i ($k_i \geq 0$) restriction predicates $\{p_1, \cdots, p_{k_i}\}$. If selectivity factors of these predicates are low then A_i is recommended for BJI. This is because to execute (*sum, avg, min, max,...*) queries, a small piece of fact table need to be loaded in the main memory. We define the *selectivity factor* (SF) of A_i as the average of selectivity factors of its predicates: $SF(A_i) = \frac{\sum_{i=1}^{k_i} Sel(P_i)}{k_i}$

To assign $\mathcal{R}$ to our $\mathcal{OT}$, clustering techniques such as K-Means [15], decision trees, etc. may be used. In this work, we use the K-Means method for the following reasons: a) it is well adapted to our assignment problem and b) it has been used to partition XML data warehouses in isolation way [16].

The K-Means method classifies a given data set $\mathcal{T}$, represented in $\Re^n$ space, through k clusters a priori fixed [15]. It defines k centroids $c_1, ..., c_k$, one for each cluster, and then assigns each point to one of the k clusters so as to minimize a measure of dispersion within the clusters. The algorithm is composed of the following steps: (1) place k initial points into the space represented by the data set $\mathcal{T}$; (2) assign each object x_i from $\mathcal{T}$ to the cluster that has the closest centroid c_j

---

[4] This rule follows partially the proposal in [20].

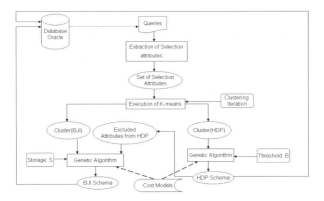

Fig. 2. Architecture of the HDP and BJI selection

(the proximity is often evaluated with the *Euclidean metric*); (3) re-compute the positions of the k centroids when all objects have been assigned; (4) repeat Steps 2 and 3 until the centroids no longer move. The best grouping is the partition of the data set $\mathcal{T}$ that minimizes the sum of squares of distances between data and the corresponding cluster centroid. To adapt the K-means method to our problem, correspondences are performed as follows:

- the data set of K-means represents our set of restriction attributes $\mathcal{R}$;
- K is equal to 2, since our combined selection problem concerns two $\mathcal{OT}$: HDP and BJI.
- The attributes are represented in $\Re^2$ space with coordinates (x, y) computed as follows: We define a classification weight for each restriction attribute A_i based on the three above criteria: $Weight(A_i) = Frc(A_i) + SF(A_i) + Card(A_i)$, where $Frc(A_i)$, $SF(A_i)$ and $Card(A_i)$ represent respectively, the frequency, selectivity factor and cardinality of A_i. During the development of the weights of attributes, we have noticed that the three criteria have different scales. To make the weight consistent, normalization is necessary. Once the weight is calculated, the coordinates in $\Re^2$ of each attribute A_i are specified as follows:

$$(x, y) = (position\ of\ attribute\ A_i, weight(A_i)) \tag{1}$$

Example 1. Let us consider a subset of queries[5] of our workload involving five attributes: $\mathcal{R} = \{Month, Year, City, Country, Class\}$. The weight of each attribute is given in Table 1. The coordinates of each attribute are given in Table 2. $NFrc$, NSF and $NCard$ represent respectively the normalized factor of frequency and selectivity factor and cardinality criteria.

Figure 3 shows a classification of restriction attributes into two subsets $Cluster_{BJI} = \{Country, Class\}$ and $Cluster_{HDP} = \{Year, Month, City\}$.

---

[5] http://www.lisi.ensma.fr/ftp/pub/documents/reports/2011/2011-LISI-.pdf

Table 1. Weight computation

Attribute	Frc	SF	Card	NFrc	NSF	NCard	Weight
Year	11	0.5	23	1.14	0.53	0.01	1.70
Month	5	0.33	12	0.26	1.41	-0.3	1.37
City	6	0.1	55	0.41	-0.13	0.94	1.22
Country	9	0.09	20	0.85	-0.2	-0.07	0.57
Class	3	0.02	62	0.02	-0.67	1.14	0.44

Table 2. Coordinates of restriction attributes

Attribute	Year	Month	City	Country	Class
Coordinates	[1, 1.70]	[2, 1.37]	[3, 1.22]	[4, 0.57]	[5, 0.44]

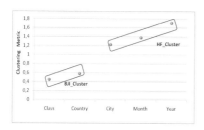

Fig. 3. Result of our classification   **Fig. 4.** Result of K-means

6   Performance Study

To validate our proposal, we conduct intensive experiments using data set of APB1 benchmark[6] and 47 queries[7] involving 11 restriction attributes. The schema of the used warehouse contains one fact table *Actvars* (24 786 000 tuples) and 4 dimension tables *Prodlevel* (9000 tuples), *Custlevel* (900 tuples), *Timelevel* (24 tuples) and *Chanlevel* (9 tuples). A Core 2 Duo machine with 2 GB of memory is used. To select our $\mathcal{OT}$, we use two different GA: one developed in [4] for HDP and another described in Section 3 for BJI. Our algorithms are implemented using Java Eclipse. Two API have been integrated: one named *SimpleK-Means*[8] that implements the K-Means method and the second, named, *JGAP* (Java Genetic Algorithms Package: http://jgap.sourceforge.net) used to implement our two GA. JGAP requires essentially coding of chromosome and fitness functions. To conduct our experiments, these steps are followed: (i) *classification of the 11 restriction attributes*: for 50 iterations, *K-Means* generates two

---

[6] http://www.olapcouncil.org/research/bmarkly.htm
[7] 47 queries are available at
http://www.lisi.ensma.fr/ftp/pub/documents/reports/2010/2010-LISI-.pdf
[8] http://weka.sourceforge.net/doc/weka/clusterers/SimpleKMeans.html

clusters (Figure 4): $C^{HDP} = \{Gender, Month, Year, All, Quarter, Group\}$ and $C^{BJI} = \{Family, Division, Class, City, Retailer\}$. (ii) *Selection of HDP and BJI schemes*: HDP GA is first executed and followed by BJI GA using C^{BJI} cluster and the attributes discarded by HDP. (iii) *Implementation of the obtained optimization schemes*: each obtained schema is implemented on a real data warehouse on Oracle11G using appropriate scripts. (iv) *Computation of the real Oracle cost of the queries*: to compute execution cost of a query running on a partitioned and indexed data warehouse, we developed a Java class, named *OracleCost* that calls *Explain plan* Oracle Optimizer tool (that displays execution plans) and accesses *Plan_Table* (a system table) to get its cost.

6.1   Tests and Results

In this section, we detail our obtained results. Unfortunately we could not compare our proposal with commercial and academic studies for two main reasons: referential partitioning is not well supported and the interaction between BJI and HDP is not clearly established. So, we compare three solutions of a hybrid implementations to select of HDP and BJI: (1) *OWC* (*Optimisation With Classification*), (2) *OPQ* (*Optimisation with Profitable Queries*) based on Boukhalfa et al.'s methodology that starts by partitioning the warehouses and indexes it by considering non profitable queries [6] (Section 1) and (3) *OWS* (*Optimisation With Simple selection*) that does not consider both attribute assignment and non profitable queries aspects (HDP is performed first on all restriction attributes, then BJI GA is executed on attributes not used in the final schema of HDP).

The first experiment aims at comparing the previous hybrid solutions using four optimization modes: (1) without optimisation, (2) only BJI, (3) only HDP and (4) multiple selection of HDP and BJI (*HDP&BJI*). For each mode, an instance of data warehouses is created and populated on Oracle11G and the cost of executing of 47 queries is computed. The storage space reserved for BJI is 500 MB and the maintenance constraint for HDP selection is $\mathcal{B} = 70$. Figure 5 and 6 give the cost in terms of inputs outputs (I/O) of executing the 47 queries and the rate of optimized queries offered by each mode. These results show that multiple selection outperforms the other methodologies, especially when attribute assignment is used (*OWC*). Indeed, the cost rises from 28.4 to

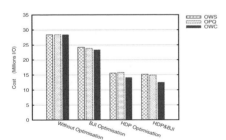

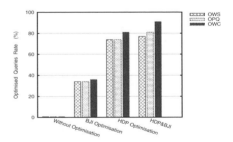

Fig. 5. Query cost with # modes   **Fig. 6.** Rate of optimized queries

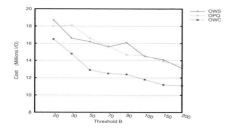

Fig. 7. Query cost vs. Threshold $\mathcal{B}$

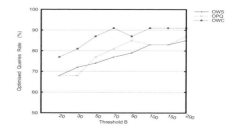

Fig. 8. Rate of optimized queries

12.5 million of I/0, which represents a reduction of 56 % of the total cost, and 91 % of queries are optimized. Lessons learned from these results are: (a) the multiple selection outperforms largely the individual ones; (b) the schema of HDP when OWC is used gives better results than that defined on all restriction attributes and (c) OWC slightly outperforms the other methodologies, especially for $HDP\&BJI$ mode. Indeed, OWC chooses for each $\mathcal{OT}$ selection the most appropriate attribute, contrary to the OPQ selection that only improves BJIs performance (BJI are selected to speed up the non profitable queries).

To study the impact of the maintenance bound $\mathcal{B}$ on query performance when using $HDP\&BJI$, we varied $\mathcal{B}$ while fixing the storage constraint $S = 500\ MB$. For each value of $\mathcal{B}$, we consider the above three solutions of the hybrid implementationn. The results illustrated in Figure 7 show that the best optimization is achieved when OWC is used, especially for $\mathcal{B} > 100$. Indeed, the cost is reduced from 58 % to 61 %, and 91% of 47 queries are optimized (Figure 8). When the maintenance bound $\mathcal{B}$ becomes larger, the probability that all attributes will be used in the partitioning process becomes higher.

We conduct other experiments to evaluate the impact of the storage constraint S on query performance. To do so, we vary S while fixing the value of maintenance bound $\mathcal{B} = 20$ (this value makes favourite indexing process). Figure 9 shows the obtained results. We note that, for $S < 900$ MB, OPQ gives better results than the simple selection OWS. Indeed, for these values of S, choosing restriction candidate attributes for indexing from a subset of queries (not profitable) reduces the complexity of BJI selection problem. But, when storage space increases, the costs corresponding to OPQ and OWS become linear. On

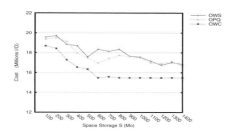

Fig. 9. Query Cost vs. Space Storage S

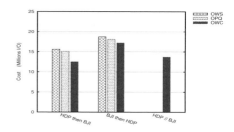

Fig. 10. Impact of selection order

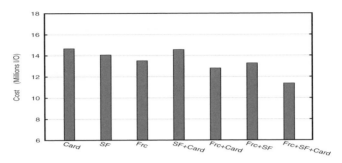

Fig. 11. Impact of Factors on assignment

the other hand, whatever the value of $\mathcal{S}$, our approach outperforms OPQ and OWS. In fact, the query processing cost for OWC is reduced to 15.8 million I/O, when $\mathcal{S} > 600MB$. This is due to the fact that our K-Means method assigns the right attributes for BJI.

Tests conducted so far select HDP then BJI (HDP then BJI). We conduct the same experiments as previously, but instead of considering (HDP then BJI) order, we consider two other selection orders: BJI then HDP and $HDP//BJI$ (the selection is done in a separate way, where each $\mathcal{OT}$ uses *only its own attributes*). This means that only OWC can be performed with this selection order. $\mathcal{S}$ and $\mathcal{B}$ are set to 500 MB and 70. Figure 10 summarizes the obtained results. The best performance of multiple selection is obtained when (HDP then BJI) is used with attribute assignment (12.5 millions of I/O). The second best approach is OWC with $HDP//BJI$ order (13.7 millions of I/O). The worst mode is attributed to OWS when BJI then HDP order is used (18.1 millions of IO). Several lessons can be learned from these results: (a) BJI then HDP gives worst performance. Indeed, the discarded attributes by BJI selection will be included to the attributes assigned to HDP, which may affect the choice of the final HDP schema. Contrary to the selection of BJI, where an attribute is either chosen or not, HDP selection may choose non appropriate attribute(s) that generate a small number of fragments, (b) the order $HDP//BJI$ is less beneficial than (HDP followed by BJI), since the two selections are done separately.

To investigate the relevance of our three assignment criteria, we conduct an experiment, where only OWC methodology is used, by changing the weight formula. This gives seven possible combinations. The results are shown in Figure 11. We note that the best optimization is obtained when all factors are used which confirms the relevancy of our criteria.

7   Conclusion

To meet the complex queries requirement, multiple selection of $\mathcal{OT}$ becomes a key solution. It is more complex than single selection due the large search space that should be explored. In addition to this complexity, interaction between certain $\mathcal{OT}$ shall be considered during the selection process. In this paper, we focused on hybrid implementation of this selection and we studied the contribution of *attribute interchangeability* on the selection process of two $\mathcal{OT}$: HDP (a non redundant structure) and BJI (a redundant structure). Similarities are identified between these $\mathcal{OT}$: both are defined on attributes of dimension tables and optimize restrictions and joins. These similarities are exploited to propose other solutions for the hybrid implementation that reduce the complexities of two sub problems. Instead of dealing with the joint problem exploring a large search space, we first assign restriction attributes to HDP and BJI and then each OT is selected based on its own attributes. This assignment is performed by a K-Means method with three criteria that we considered important: *attribute frequency, selectivity factor* and *cardinality of attributes*. A genetic algorithm is proposed for selecting BJI defined on a single attribute of a dimension table. Intensive experiments were conducted to compare our proposal with the existing state of art studies.

Two main issues need to be explored: (i) studying criteria when assigning the restriction attributes, such as storage and profiles of used queries, (ii) enriching our proposal by other optimisation techniques, such as materialized views and parallel processing and (iii) compare our proposal with more others selection approaches.

References

1. Aouiche, K., Boussaid, O., Bentayeb, F.: Automatic Selection of Bitmap Join Indexes in Data Warehouses. In: Tjoa, A.M., Trujillo, J. (eds.) DaWaK 2005. LNCS, vol. 3589, pp. 64–73. Springer, Heidelberg (2005)
2. Baralis, E., Paraboschi, S., Teniente, E.: Materialized view selection in a multidimensional database. In: VLDB, pp. 156–165 (August 1997)
3. Bellatreche, L., Boukhalfa, K., Abdalla, H.I.: Saga: A combination of genetic and simulated annealing algorithms for physical data warehouse design. In: Bell, D.A., Hong, J. (eds.) BNCOD 2006. LNCS, vol. 4042, pp. 212–219. Springer, Heidelberg (2006)
4. Bellatreche, L., Boukhalfa, K., Richard, P.: Referential horizontal partitioning selection problem in data warehouses: Hardness study and selection algorithms. International Journal of Data Warehousing and Mining 5(4), 1–23 (2009)
5. Bellatreche, L., Missaoui, R., Necir, H., Drias, H.: A data mining approach for selecting bitmap join indices. Journal of Computing Science and Engineering 2(1), 206–223 (2008)
6. Boukhalfa, K., Bellatreche, L., Alimazighi, Z.: Hp&bji: A combined selection of data partitioning and join indexes for improving olap performance. Annals of Information Systems 3, 179–2001 (2008)
7. Chaudhuri, S.: Index selection for databases: A hardness study and a principled heuristic solution. IEEE TKDE 16(11), 1313–1323 (2004)
8. Chaudhuri, S., Narasayya, V.: An efficient cost-driven index selection tool for microsoft sql server. In: VLDB, pp. 146–155 (August 1997)
9. Eadon, G., Chong, E.I., Shankar, S., Raghavan, A., Srinivasan, J., Das, S.: Supporting table partitioning by reference in oracle. In: SIGMOD, pp. 1111–1122 (2008)

10. Gebaly, K.E.L., Aboulnaga, A.: Robustness in automatic physical database design. In: 11th International Conference on Extending Database Technology (EDBT 2008), pp. 145–156 (2008)
11. Gray, J., Slutz, D.: Data mining the sdss skyserver database. Techreport Technical Report MSR-TR-2002-01, Microsoft Research (2002)
12. Ioannidis, Y., Kang, Y.: Randomized algorithms algorithms for optimizing large join queries. In: SIGMOD, pp. 9–22 (1990)
13. Karloff, H., Mihail, M.: On the complexity of the view-selection problem. pp. 167–173 (1999)
14. Kimura, H., Huo, G., Rasin, A., Madden, S., Zdonik, S.: Coradd: Correlation aware database designer for materialized views and indexes. PVLDB 3(1), 1103–1113 (2010)
15. MacQueen, J.B.: Some methods for classification and analysis of multivariate observations. In: Symposium on Mathematical Statistics and Probability, pp. 281–297 (1967)
16. Mahboubi, H., Darmont, J.: Data mining-based fragmentation of xml data warehouses. In: ACM DOLAP, pp. 9–16 (2008)
17. Papadomanolakis, S., Ailamaki, A.: Autopart: Automating schema design for large scientific databases using data partitioning. In: International Conference on Scientific and Statistical Database Management, pp. 383–392 (June 2004)
18. Sanjay, A., Narasayya, V.R., Yang, B.: Integrating vertical and horizontal partitioning into automated physical database design. In: SIGMOD, pp. 359–370 (June 2004)
19. Sanjay, A., Surajit, C., Narasayya, V.R.: Automated selection of materialized views and indexes in microsoft sql server. In: VLDB, pp. 496–505 (2000)
20. Stöhr, T., Märtens, H., Rahm, E.: Multi-dimensional database allocation for parallel data warehouses. In: VLDB, pp. 273–284 (2000)
21. Zilio, D.C., Rao, J., Lightstone, S., Lohman, G.M., Storm, A., Garcia-Arellano, C., Fadden, S.: Db2 design advisor: Integrated automatic physical database design. In: VLDB, pp. 1087–1097 (August 2004)

Generic Information System Architecture for Distributed Multimedia Indexation and Management

Mihaela Brut[1], Sébastien Laborie[2], Ana-Maria Manzat[1], and Florence Sèdes[1]

[1] Université de Toulouse – Université Paul Sabatier – IRIT UMR 5505
118 Route de Narbonne 31062 Toulouse Cedex 9, France
[2] T2i - LIUPPA - Universit de Pau et des Pays de l'Adour
2 Allée du Parc Montaury 64600 Anglet, France
{Mihaela.Brut,Ana-Maria.Manzat,Florence.Sedes}@irit.fr,
Sebastien.Laborie@iutbayonne.univ-pau.fr

Abstract. Effective and scalable distributed solutions for data indexation and management should provide solutions for three sensitive issues: the system's architecture that integrates the mentioned facilities, the management of the indexation techniques and the modeling of the metadata that describe the multimedia contents. This paper presents a generic and scalable distributed architectural solution that integrates a flexible indexation management technique and enables to deal with any metadata models. Our solution was developed in the context of the LINDO project, by capitalizing and enhancing the existing solutions for each issue while also respecting the project's requirements. The solution proposes an innovative indexing management technique, and includes the system's architecture and the metadata model that sustain it. In order to validate our proposal, several implementations are presented which are related to broadcast, video surveillance and archive activities.

Keywords: Multimedia Distributed Systems, Generic Architecture, Content Indexation Management, Implicit and explicit indexation.

1   Introduction

Nowadays, many multimedia contents are available in various domains, such as video surveillance, patient medical records, broadcast, personal information management. These contents are generally acquired and stored on different and heterogeneous locations. The huge quantity of multimedia contents, the increasing number of data transmissions between remote servers, as well as the indexation of the multimedia contents and the management of the generated metadata constitute the main scalability issues that have to be handled by a multimedia information management system. In this context, three sensitive problems occur in such a system [11], which are discussed in this paper:

- *Architectural solutions*: most of the current systems define a specific architecture that is strongly use case dependent;

J. Eder, M. Bielikova, and A.M. Tjoa (Eds.): ADBIS 2011, LNCS 6909, pp. 347–360, 2011.
© Springer-Verlag Berlin Heidelberg 2011

- *Indexation management techniques* are used to manage the indexation algorithms and the indexation process. In general, multimedia collections are indexed with a fixed and predefined set of indexing algorithms;
- *Metadata models* are used to organize and retrieve multimedia contents. Currently, a large palette of multimedia metadata formats exists, which specifies different vocabularies and structures, that are not necessary interoperable.

The paper exposes a generic framework developed in the context of LINDO project[1] (Large scale distributed INDexation of multimedia Objects) and intended to guide the formalization and the development of large scale distributed multimedia systems. Our focus was not to define yet another information indexing and retrieval model but rather to reuse many existing frameworks, by capitalizing and enhancing the existing solutions for each issue while also respecting the project's requirements. More precisely, the LINDO framework provides:

- an *architectural solution* that is applicable in multiple use cases, such as video surveillance, broadcast and archive systems;
- a distributed and dynamic *indexing management*, which has the advantage that any indexing algorithms could be integrated and deployed on demand on remote sites, depending on site's characteristics and on users queries; moreover, we developed the indexing management technique such as to minimize the system resources consumption while employing a dynamic algorithms selection before indexing, corresponding to each particular query;
- support for integrating different *metadata models* via translation facilities.

In the remainder, the paper presents first how existing projects address the three mentioned issues. Then, the LINDO generic architecture is detailed in Section 3, and exemplified in Section 4, including the presentation of its main workflows and a concrete use case implementation. Other possible instantiations of the framework are presented in Section 5 related to some specific use cases. Section 6 describes in more details the technical implementation of this architecture. Conclusions and further works are exposed in the end of the paper.

2   Related Work

We present in this section how existing projects related to large-scale distributed multimedia information systems address the three issues mentioned in Introduction. Some comparisons with the LINDO approach are then provided for each issue in order to emphasize its advantages.

2.1   The Employed Architectural Solutions

Two main architectural solutions are considered in the large-scale distributed multimedia information systems: peer-to-peer solutions and distributed architecture with a centralized management.

---

[1] http://www.lindo-itea.eu

In a peer-to-peer architecture each peer is in general equivalent in functionality and cooperates with other peers for retrieving some multimedia information [14]. Some multimedia information systems adopted a peer-to-peer architecture, such as [2] and SAPIR[2] project. However, this architecture involves many transactions between peers, while these operations are highly resource consuming. Moreover, a global view of the system is not available that could have enabled the selection of the most suitable indexing algorithms for a specified user query.

A distributed architecture with a centralized management is composed of many remote servers controlled by a central server, which is able to handle an overview of the entire distributed multimedia collection [16]. Such architecture is adopted by the distributed multimedia information systems that prefer to limit the number of transactions between servers, as the case of the following systems.

In the CANDELA project[3] (Content Analysis and Network DELivery Architectures) [12], a media distribution management component is included in the central server, which manages three data types: digital content, system control data, and the metadata that describes and controls access to the content. Its aim is to enable data storage and data exchange from content producers to distribution servers and all the way to end users devices, while an optimization of the data transfer or data indexing is not considered.

The VITALAS project[4] (Video & image Indexing and reTrievAl in the LArge Scale) [18] adopts the Web services solution for defining the architecture components in order to enable the uniform integration of the different partners indexing tools and infrastructures; with this respect, the architecture is generic and assures the centralized management of distributed multimedia servers.

The KLIMT project (Knowledge InterMediation Technologies) [8] proposes also a service-oriented architecture that allows the integration of heterogeneous multimedia indexing tools. A global centralized registry records all KLIMT services, while a controller manages the distribution of these services.

In the WebLab project[5], the indexing algorithms are also considered as Web Services, and are manually managed through a central facility [9].

In [13], a distributed image search engine based on mobile software agents is proposed. In order to deal with the metadata generated by the indexing algorithms, the authors propose two architectures: one centralizing the index and the another one distributing the indexes on the remote servers.

2.2 Information Indexation Management Techniques

The design of indexation workflow inside a distributed multimedia system could be set up in a multitude of ways, by developing solution for some important issues. The indexation process is managed through an indexation engine, which could includes a fix or variable set of indexing algorithms. The algorithms could

---

[2] http://www.sapir.eu

[3] http://www.hitech-projects.com/euprojects/candela

[4] http://vitalas.ercim.org

[5] http://weblab-project.org/

be executed over the entire multimedia collection or only over a filtered sub-collection. Moreover, the indexing algorithms set could be filtered or not before their effective execution. Also, the indexation could be accomplished in real time or off-line, at a central server level or at the remote servers level. As will be illustrated in Sections 3 and 4, the solutions for these issues developed in the LINDO framework were directed to the optimization of resources consumption by employing the indexation only when and where needed.

In this section we present the information indexation management approaches adopted by some multimedia distributed systems. As will be noticed, each of these systems raise some of the above mention issues and develop some corresponding solutions. Capitalizing their experience, LINDO framework integrated solution for all the issues, guided by the mentioned optimization principle.

CANDELA project, the indexation is uniformly accomplished in all remote servers, and managed at the central server level. Thus, time consumption is acquired because of the indexing the irrelevant information.

In order to develop large-scale search in audio-visual databases, Sapir project [1] adopts a centralized approach consisting in two steps: distributed multimedia information ingested at peers level is indexed in three specialized servers (one for images, one for textual documents and one for audio-visual content) and then it is fused by super-peers.

The indexation technique adopted by VITALAS project considers the integration (via web services) of partners' know-how cross-media indexing tools (text and visual features), as well as their complex combination based on support vector machine solutions managed by the central server.

The KLIMT project integrated different indexation algorithms for each media type, implemented as Web services. It provides also support for chaining and combining services for content indexation, thus the extracted information can be different in function of the use case scenario.

In the WebLab project the indexing algorithms are considered also as Web Services. The installation, the deployment and the execution of the available algorithms are manually accomplished, through an interface enabling as well to define certain algorithm chaining rules.

In [13] the indexation is accomplished with a fixed set of indexing algorithms that are implemented as mobile agents. These agents migrate from one site to another in order to extract different multimedia features. Thus the multimedia contents are not transfered over the network.

2.3   Multimedia Metadata Models

The multimedia metadata are available into a wide variety of standards, such as EXIF, DC, MXF. In order to avoid the interoperability problem, the systems that deal with multiple multimedia metadata formats must develop a solution for uniformly handle them. Such solutions are based on unified metadata models, which might be XML-based or RDF-based.

XML-based metadata models provide a uniform structure aiming to include the multimedia features into an exhaustive way:

- In CANDELA project the content labeling is based on MPEG-7, while the compliance with MPEG-4 and MPEG-21 is totally or partially covered.
- VITALAS project defined some concept dictionaries for expressing the cross-media indexing results, and developed an XML metadata repository including XML-based standard metadata formats.
- KLIMT project adopts an XML-based solution for managing multimedia metadata. The solution is developed on top of the SOAP messaging framework because the indexing algorithms are handled as Web services.

Current RDF-based metadata models are based on MPEG-7[17]. Some models were defined and/or adopted in some projects:

- The aceMedia Ontology Framework [4], developed inside the aceMedia project[6] define an integrated multimedia annotation framework.
- COMM (Common Ontology for Multimedia)[3] was developed by K-Space[7] and X-Media[8] projects in order to provide a formal semantics for MPEG-7.
- WebLab Reference Model was developed by WebLab project in order to provide support for annotating different segments of the multimedia content.

After presenting some important existing solutions for the three issues exposed in the Introduction, we provide in Table 1 an overview of these solutions.

Table 1. A comparative overview of some representative systems and approaches

System	Architecture	Indexation	Metadata
SAPIR	Peer-to-Peer	Uniform and fixed indexation at each peer	XML-based format, derived from MPEG-7
CANDELA	Generic distributed with central control	Uniform and fixed indexation at the same server as the content storage	XML-based format, derived from MPEG-7
VITALAS	Service oriented architecture	Variable set of indexing algorithms, running on some indexation servers	XML-based format
KLIMT	SOA with a central control	Realized at the query moment, with a variable set of IA, on dedicated servers	XML-based format
WebLab	SOA with central control	Realized at the acquisition moment, with a fixed set of IA, on dedicated servers	RDF-based format
[13]	Distributed architecture based on mobile agents	Accomplished by a fixed set of mobile IA on the content server	A set of feature vectors

After a careful study of solution proposed by the state of the art and an analysis of the concrete LINDO system needs, we decided to adopt a *distributed architecture with a centralized management* because it enables:

- To have a centralized management of indexing algorithms;

---

[6] http://www.acemedia.org/
[7] http://kspace.qmul.net:8080/kspace/
[8] http://www.x-media-project.org/

– To extract simultaneously multiple and diverse metadata by executing different indexation engines, in parallel, on different remote servers.
– To dispatch user queries only to some specific remote servers that might give some desired information.
– To process queries simultaneously on the central server and on remote servers.

The indexation process adopted in the LINDO project is managed at the central server level and intelligently accomplished at the remote servers levels:

– A generic interface was defined for indexing algorithms in order to uniformly handle them [7], and to enable the integration of new algorithms at any time into the LINDO architecture;
– There exist two indexation processes: (1) an *implicit indexation* that is executed over the multimedia contents from each remote server in the moment of acquisition; (2) an *explicit indexation* that could be executed, on demand, on a specific remote server, with a subset of indexing algorithms and only on a sub-collection of the multimedia contents.

Since LINDO project aims to *integrate any metadata models*, we defined an integrative metadata format [6], and included in LINDO architecture a *translation module* that assures the conversion between this format and other multimedia metadata vocabularies. In the following we detail the LINDO architecture.

3   The LINDO Generic Architecture

We have defined the LINDO generic architecture, illustrated in Figure 1, over two main components: (1) remote servers (§3.1) which acquire, index and store multimedia contents, and (2) a central server (§3.2) which has a global view of the overall system. Our architecture provides two advantages. First, each remote server is independent, i.e., it can perform uniform as well as differentiated indexations of multimedia contents. For instance, some remote servers may index in real time acquired multimedia contents, while others may proceed to an off-line indexation. Secondly, the central server can send relevant indexation routines or queries to relevant remote servers, while the system is running.

3.1   The Remote Servers Components

The remote servers in LINDO-based systems store and index all acquired multimedia contents, to provide answers to user queries. Hence, several modules have been defined and linked together in order to cover all these tasks:

The *Storage Manager* (SM) stores the acquired multimedia contents. Through the *Transcode* module, a multimedia content could be converted into several formats. Thus, an user can download different encodings of a desired content.

The *Access Manager* (AM) provides methods for accessing the multimedia contents stored into the SM. Apart from accessing a whole content, this module can select different parts of one multimedia content.

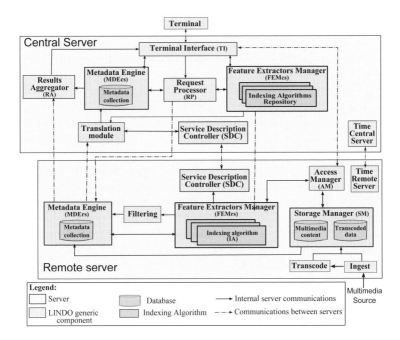

Fig. 1. The LINDO generic architecture

The *Feature Extractors Manager* (FEMrs) is in charge of managing and executing a set of indexing algorithms over the acquired multimedia contents. At any time, new algorithms can be uploaded into this module, while others can be removed or updated. It can permanently run the algorithms over all the acquired contents or it can execute them on demand only on certain multimedia contents.

The *Filtering module* module summarizes the metadata produced by indexing algorithms that may contain redundant or useless metadata.

The *Metadata Engine* (MDErs) collects and aggregates all extracted metadata about multimedia contents. Naturally, the metadata stored into this module can be queried in order to retrieve some desired information.

The *Time Client* handles time synchronization with the central server.

The *Service Description Controller* stores the description of the remote server, such as its location, its capacities, its configuration, all the indexing algorithms installed and their status, and the media acquisition context.

3.2   The Central Server Components

The central server controls the distributed remote servers by managing the remote indexation and query processes. One major difference between the central server and a remote server is that the central server does not index multimedia contents, but manages the indexation and includes a general knowledge about the indexed content. Thus, a central server is composed of:

The *Terminal Interface* (TI) enables a user to specify some queries and displays the query's results. Other functionalities are included in the TI, such as visualization of metadata collections and management of indexing algorithms.

The *Metadata Engine* (MDEcs) gives a global view of the system. It can contain some extracted metadata about multimedia contents, some contextual information about the system, the remote servers' descriptions, the descriptions of the available indexing algorithms, etc.

The *Service Description Controller* collects all remote server descriptions.

The *Feature Extractors Manager* (FEMcs) manages the entire set of indexing algorithms used in the system.This module communicates with its equivalent on the remote server side in order to install new indexing algorithms if it is necessary or to ask for the execution of a certain indexing algorithm on a multimedia content, or part of multimedia content.

The *Request Processor* (RP) treats some queries on the MDEcs or forwards them to specific remote server metadata engines. Moreover, through the FEMcs, it can decide to remotely deploy some indexing algorithms.

The *Results Aggregator* (RA) aggregates the results received from all the queried metadata engines and sends them to the TI, which displays them.

The *Translation module* homogenizes the data stored into the central server metadata engine. Indeed, many different models can be used by remote servers for indexing multimedia contents or describing their characteristics. Hence, this module unifies all descriptions in order to provide a global system view.

The *Time Server* provides a unique synchronization system time.

As could be noticed, the architecture's scalability is sustained by the opportunity to integrate at any moment new and heterogeneous remote servers. The central server considers this diversity when treating the user query. The provided solution is thus generic in terms of remote servers, multimedia data, as well as their associated metadata. Each remote server could employ a particular metadata format (e.g., XML or RDF based), while the Translation Module assures a uniform handling at the central server level.

Thus, the proposed scalable *generic architectural solution* offers an efficient *distributed and dynamic indexing algorithms management*, and enables to employ different *metadata models*. In order to illustrate how the global system is running, we present in the next section a concrete illustration of the system functioning.

4   A Concrete Illustration of the System Functioning

In this section, we will present the logic of our framework from two perspectives. First the chaining steps in the main workflows will be presented in §4.1 from a conceptual point of view. Further a concrete instantiation of the system for a particular use case will be detailed in §4.2.

4.1   The System's Workflows to Acquire Content Indexation Management

In order to save servers' resources consumption, multimedia content indexation can be accomplished at acquisition time (implicit indexation) and on demand (explicit indexation), that are explained in this section.

When a remote server acquires new multimedia content, the SM stores it and then the FEMrs starts its implicit indexation by executing a predefined set of indexing algorithms which is established according to the server particularities.

Once the execution of an indexing algorithm is achieved, the obtained metadata is forwarded to the Filtering module. The filtered metadata is then stored by the MDErs. In order to avoid the transmission of the whole collection of metadata computed on the remote servers, the MDErs only sends a summary of these metadata to the Translation Module on the central server[9]. Once translation is done, the metadata are sent to the MDEcs to be stored and further used in the querying process. Thus, the *implicit indexation* process is achieved.

The query process begins with the query specification through the TI. The user's query is sent to the RP module in order to be executed over the metadata collections. First, the RP processes the query in order to select the remote servers that could provide answers to the query and sends them the query for execution. Among the servers that were not selected at the first step there could be some servers that contain relevant information that has not been indexed with the right algorithms. For this reason, our solution detects such supplementary algorithms [5] and starts their execution (i.e., *explicit indexation*) on a sub-collection of multimedia contents. All the results obtained from the remote servers are sent to the RA, where they are combined and ranked in order to be displayed.

An important remark is that the two kinds of indexation can be mixed in the LINDO system, i.e., on some remote servers only the implicit indexation can be accomplished, while on others only the explicit indexation is done, while on others both indexation processes can be performed.

4.2   Implementation of a Video Surveillance Use Case

A first implementation of the LINDO architecture was realized for a specific video surveillance use case proposed by an industrial partner[10].

This use case concerns a public transportation company, that placed surveillance cameras in their buses, around the bus stations and the bus ticket machines. Each video stream is recorded with a real time association to a common time stamp and GPS location. A typical query may be to find all videos containing a specific event, such as a person with a red pullover, who took the bus in Place de la Concorde, Paris, on Monday 3rd May, between 6pm and 8pm.

The main identified issues are the extraction of metadata from real time videos, the storage of huge amounts of multimedia contents and their associated

---

[9] [10] proposes an algorithm which computes several metadata summaries based on RDF descriptions.

[10] http://www.thalesgroup.com/security-services

metadata, and the retrieval of the relevant content to users queries. The LINDO architecture provides a solution for these issues by associating several cameras to one remote server that stores and indexes the acquired video contents. Given the complexity of this particular use case, all the modules presented in Figure 1 are used. For example, the Time Server and the Time Client modules are important to synchronize the acquired videos. The SDCs are used to locate the cameras and to describe their acquisition contexts, e.g, street names, bus numbers.

At the central server level, the MDEcs stores the metadata received from the remote servers, that is enriched with additional semantic information (e.g., buses timetables). In order to semantically define contextual information, we have used an RDF-based representation for the MDEcs. Since different metadata models are used on the central server and on remote servers, the Translation module transforms the metadata through a modifiable set of transformation rules.

Concerning the workflows, the implicit and explicit indexation is done in real time, and thus the number of indexing algorithms executed must be minimum. From the specifications we can determine three categories of remote servers corresponding to cameras placed: (1) inside a bus, (2) nearby the bus ticket machines and (3) around a bus station. Thus, we establish different sets of implicit indexing algorithms to be installed on each type of server. For example, a crowd detection algorithm is associated with cameras placed around bus stations, while an algorithm for counting persons is associated with cameras placed inside the buses as well as nearby bus ticket machines.

In order to provide response to a particular user query, such as the one mentioned above, first the MDEcs is queried. If no results could be provided, the query is processed in two steps: (1) the spatio-temporal information are extracted [15], e.g., Paris, 3rd of May. These information are used to locate the relevant remote servers (e.g., related to cameras from the buses that stopped in Place de la Concorde in Paris) and the corresponding video fragments (e.g., the video fragments took on 3rd of May, between 6pm and 8pm). (2) the remained part of the query is used in order to determine what indexing algorithms are necessary to be executed over the video contents located at the first step. For our example, a color and a person detection algorithms are deployed and executed over the selected video content.

Through this indexing strategy, we avoid the execution of the entire set of indexing algorithms which is consuming all CPU resources and cannot be executed in real time. Thus, we limit resources consumption and we generate only the strict necessary metadata for a video surveillance system. If some other information is needed, the relevant indexing algorithms are applied, only on a part of the content. Hence, the size of the managed metadata collections is also limited.

5   Adopting the Framework in Other Use Cases

In order to illustrate that our architecture is generic enough to be adopted by various types of distributed multimedia information systems, we will present further two other use cases that involve different scenarios of utilization: broadcast (§5.1) and archive (§5.2) use cases. These use cases were proposed by the industrial partners involved in the LINDO project.

5.1 Broadcast Use Case

Our given broadcast use case[11] consists in the distribution of TV programs and movies diffused by a certain television during the last two months. A constraint specified for this use case is that the broadcaster has a unique access point and that only the low quality version of the multimedia contents are stored on a single remote server. Furthermore, all user queries are treated by the remote server. A typical query may be to find the comedies of Jim Carrey broadcasted last week.

Based on these requirements, the LINDO-based implementation is composed of a central server and a unique remote server. This simple architecture contains only the strict necessary modules. At the central server level only the creation of queries, their transmission to the remote server and the display of the results are needed. Thus, only the TI and the RP are necessary. Actually, the MDEcs and the RA are not used because all the metadata are stored on the unique remote server. At the remote server level, certain modules are not necessary, namely the Time Client and the SDC, because only one remote server is required. Furthermore, the requirements do not specify the need to filter the metadata obtained from the indexing process, thus the Filtering module is also skipped.

In this use case, the indexation process is accomplished off-line and there is no time constraint for the execution of the indexing algorithms. As a consequence, the explicit indexation is not needed. However, new indexing algorithms could be installed at any time on the remote server, thus improving the indexation.

Given the homogeneity of the multimedia contents, only one uniform metadata model is adopted. For the implementation of the MDErs we have used the Sinequa[12] engine because it treats uniformly the extracted metadata and the semantic knowledge associated to the multimedia contents.

The components used for both servers and the lack of the explicit indexation process simplify the query workflow. Actually, the RP gets the query from the TI and sends it directly to the MDErs. Because the RA is not used in this use case, the results obtained to a user query are directly sent to the TI.

5.2 Archive Use Case

Our given archive use case[13] consists in the management of patient medical records in a certain city. The system has to collect different information from all the hospitals. A constraint specified is that some of the information are kept only on the hospitals servers for privacy reasons. A typical query may be to find the medical history of a patient.

---

[11] http://www.sgt.eu/
[12] http://www.sinequa.com
[13] http://www.hi-stor.com/

The instantiation of this use case is a system composed of numerous remote servers, related to each hospital, and a central server. Moreover, almost all the modules presented in Figure 1 are used. For example, the SDCs are used to locate and describe the hospitals, e.g., the hospitals facilities.

The information used for the retrieval of documents are not very complex, e.g., the patient identification, the date of the visit. Consequently, the indexation is realized only at acquisition time and the explicit indexation is not needed. Furthermore, since many textual medical reports have to be acquired, we have decided to employ Lucene[14] which offers full-featured text search engine facilities.

At the central server level, the MDEcs stores some metadata received from remote servers in the same model as the one used in remote servers. Hence, the Translation module is skipped. Because some information are confidential, from the remote servers only secured metadata are transmitted to the central server.

In this context, the querying process is simplified: the central server only sends the query to the remote servers that might have some answers and collects the results from all the remote servers in order to display them to the user.

These three use cases and their instantiations demonstrate that our architecture responds to the three issues mentioned in the Introduction:

- an *architectural solution* that can be used for multiple use cases;
- a distributed and dynamic *indexing algorithms management* where the indexation can be realized at acqusition time (implicit indexation) and/or on demand (explicit indexation);
- support for integrating different *metadata models* via translation facilities encapsulated in the Translation module.

6   Experimental Tests and Results

Exploiting the specific competences of the project partners, we adopted the proposed architecture in developing a distributed system with a single central server, where remote servers were progressively integrated corresponding to different domains as video surveillance, broadcast and archive.

For implementation, we adopted the Java programming language for all modules, except the SM module that is implemented in C++. The communication between the different modules is realized through Java implementations of Web services and JMS (Java Message Service)[15] API. These implementations were employed and tested for both Unix and Windows platforms.

At the central server level, the RP employs some natural language processing techniques in order to treat the user query and to transform it into a machine understandable format (such as XQuery).

We proposed ourselves to integrate many indexing algorithms developed and/or currently used by the project partners[16],[17]. As a matter of fact, the ma-

---

[14] http://lucene.apache.org
[15] http://java.sun.com/products/jms/overview.html
[16] http://www.supelec.fr/
[17] http://www-list.cea.fr/

jority of these algorithms have their outputs encoded in an XML-based model. For this reason, we decided to employ at the remote server level an XML-based model for the metadata representation, such as the one proposed in [6] which encapsulates the most common metadata standards. We adopted Oracle Berkeley DB XML[18] to store and query our XML-based metadata in the MDErs.

The MDEcs of central server provides a global view of the existing information in the entire system. This is acquired with the support of Topic Maps[19] representation of the concise version of the metadata stored on the remote servers, as mentioned in §4.1. This general Topic Map includes also information that describes the characteristics and the context of each remote server. The unification of these disparate information inside the mentioned topic map is acquired through a data virtualization technology proposed by a project partner[20].

7   Conclusions and Perspectives

In this paper, we have presented a generic framework for indexing large scale distributed multimedia contents, developed in the LINDO project and tested for several use cases. We addressed the scalability issues by providing a flexible architecture which encapsulates distributed and dynamic indexation techniques and multiple metadata models. The indexation process comprises two steps: an implicit indexation and an explicit indexation that is applied when needed.

Currently, the relevance of a query result is established according to the indexation techniques used on each remote server. At the central server side, a general order of these relevancies is accomplished. In the future, we plan to study different merging techniques for improving the global results relevance. More precisely, different result rankings will be compared and evaluated in order to establish their real impact to the global results list.

Acknowledgment. This work is supported by the EUREKA project LINDO (ITEA2 – 06011).

References

1. Agosti, M., Di Buccio, E., Di Nunzio, G.M., Ferro, N., Melucci, M., Miotto, R., Orio, N.: Distributed information retrieval and automatic identification of music works in sapir. In: The 15th Italian Symposium on Advanced Database Systems, pp. 479–482 (2007)
2. Ardizzone, E., Gatani, L., La Cascia, M., Lo Re, G., Ortolani, M.: A P2P architecture for multimedia content retrieval (LNCS 4351). In: Cham, T.-J., Cai, J., Dorai, C., Rajan, D., Chua, T.-S., Chia, L.-T. (eds.) MMM 2007. LNCS, vol. 4351, pp. 462–474. Springer, Heidelberg (2006)

---

[18] http://www.oracle.com/technology/products/berkeley-db/xml/index.html
[19] http://www.topicmaps.org/
[20] http://www.denodo.com/

3. Arndt, R., Troncy, R., Staab, S., Hardman, L., Vacura, M.: COMM: Designing a Well-Founded Multimedia Ontology for the Web. In: Aberer, K., Choi, K.-S., Noy, N., Allemang, D., Lee, K.-I., Nixon, L.J.B., Golbeck, J., Mika, P., Maynard, D., Mizoguchi, R., Schreiber, G., Cudré-Mauroux, P. (eds.) ASWC 2007 and ISWC 2007. LNCS, vol. 4825, pp. 30–43. Springer, Heidelberg (2007)
4. Bloehdorn, S., Petridis, K., Saathoff, C., Simou, N., Tzouvaras, V., Avrithis, Y., Handschuh, S., Kompatsiaris, Y., Staab, S., Strintzis, M.G.: The 2nd European Semantic Web Conference
5. Brut, M., Laborie, S., Manzat, A.-M., Sèdes, F.: A Framework for Automatizing and Optimizing the Selection of Indexing Algorithms. In: The 3rd Conference on Metadata and Semantics Research, pp. 48–59 (2009)
6. Brut, M., Laborie, S., Manzat, A.-M., Sèdes, F.: A generic metadata framework for the indexation and the management of distributed multimedia contents. In: The 3rd Conference on New Technologies, Mobility and Security, pp. 1–5. IEEE Computer Society, Los Alamitos (2009)
7. Brut, M., Sèdes, F., Manzat, A.-M.: A web services orchestration solution for semantic multimedia indexing and retrieval. In: The 2nd Workshop on Frontiers in Complex, Intelligent and Software Intensive Systems, pp. 1187–1192. IEEE Computer Society, Los Alamitos (2009)
8. Conan, V., Ferran, I., Joly, P., Vasserot, C.: KLIMT: Intermediations Technologies and Multimedia Indexing. In: The 3rd Workshop on Content-Based Multimedia Indexing, pp. 11–18. INRIA (2003)
9. Giroux, P., Brunessaux, S., Brunessaux, S., Doucy, J., Dupont, G., Grilheres, B., Mombrun, Y., Saval, A.: Weblab: An integration infrastructure to ease the development of multimedia processing applications. In: The 21st Conf. on Software & Systems Engineering and their Applications (2008)
10. Laborie, S., Manzat, A.-M., Sèdes, F.: Managing and querying efficiently distributed semantic multimedia metadata collections. IEEE MultiMedia 16(4), 12–21 (2009)
11. Özsu, M.T., Valduriez, P.: Principles of distributed database systems, 3rd edn., Hardcover (2011)
12. Pietarila, P., Westermann, U., Järvinen, S., Korva, J., Lahti, J., Löthman, H.: CANDELA – storage, analysis, and retrieval of video content in distributed systems. In: The IEEE Conference on Multimedia and Expo, pp. 1557–1560 (2005)
13. Roth, V., Peters, J., Pinsdorf, U.: A distributed content-based search engine based on mobile code and web service technology. Scalable Computing: Practice and Experience 7(4), 101–117 (2006)
14. Schoder, D., Fischbach, K., Schmitt, C.: Core concepts in peer-to-peer networking. In: Peer to Peer Computing: The Evolution of a Disruptive Technology. ch.1, pp. 1–27. Idea Group Inc, USA (2005)
15. Strötgen, J., Gertz, M., Popov, P.: Extraction and exploration of spatio-temporal information in documents. In: The 6th Workshop on Geographic Information Retrieval, pp. 1–8. ACM, New York (2010)
16. Tanenbaum, A.S., Van Steen, M.: Distributed Systems: Principles and Paradigms. Prentice Hall PTR, Upper Saddle River (2001)
17. Troncy, R., Bailer, W., Hausenblas, M., Hofmair, P., Schlatte, R.: Enabling multimedia metadata interoperability by defining formal semantics of MPEG-7 profiles. In: The 1st Conference on Semantics And digital Media Technology, pp. 41–55 (2006)
18. Viaud, M.-L., Thièvre, J., Goëau, H., Saulnier, A., Buisson, O.: Interactive components for visual exploration of multimedia archives. In: Conf. on Content-based Image and Video Retrieval, pp. 609–616. ACM, New York (2008)

Automatic Physical Database Tuning Middleware for Web-Based Applications

Jozsef Patvarczki* and Neil T. Heffernan

Worcester Polytechnic Institute, Computer Science Department,
100 Institute Road, Worcester, Massachusetts, USA
{patvarcz,nth}@cs.wpi.edu
http://www.cs.wpi.edu/~patvarcz

Abstract. In this paper we conceptualize the database layout problem as a state space search problem. A state is a given assignment of tables to computer servers. We begin with a database and collect, for use as a workload input, a sequence of queries that were executed during normal usage of the database. The operators in the search are to fully replicate, horizontally partition, vertically partition, and de-normalize a table. We do a time intensive search over different table layouts, and at each iteration, physically create the configurations, and evaluate the total throughput of the system. We report our empirical results of two forms. First, we empirically validate as facts the heuristics that Database Administrators (DBAs) currently use as in doing this task manually: for tables that have a high ratio of update, delete, and insert to retrieval queries one should horizontally partition, but for a small ratio one should fully replicate a table. Such rules of thumb are reasonable, however we want to parameterize some common guidelines that DBAs can use. Our second empirical result is that we applied this search to our existing data test case and found a reliable increase in total system throughput. The search over layouts is very expensive, but we argue that our method is practical and useful, as entities trying to scale up their Web-based applications would be perfectly happy to spend a few weeks of CPU time to increase their system throughput (and potentially reduce the investment in hardware). To make this search more practical, we want to learn reasonable rules to guide the search to eliminate many layout configurations that are not very likely to succeed. The second aspect of our project (not reported here) is to use the created configurations as input into a machine learning system, to create general rules about when to use the different layout operators.

Keywords: Database tuning, partitioning, layout search, Web-based applications.

1 Introduction

Nowadays, Database Administrators have to be familiar with multiple available data stores to select the best fit for their Web-based applications. Upon a successful selection, scalability issues related to the database could be a possible

---

* For the full list of our funders please see http://www.webcitation.org/5xp605MwY.

J. Eder, M. Bielikova, and A.M. Tjoa (Eds.): ADBIS 2011, LNCS 6909, pp. 361–374, 2011.
© Springer-Verlag Berlin Heidelberg 2011

bottleneck of the system. Especially, if the requests are distributed among multiple database servers that could lead to slow response time or a possible system crash. Scalability issues and their possible solutions should be automatically addressed without spending enormous amount of time on investigating the database structure or investing into expensive hardware solutions. Our work is motivated by scalability issues of Web-based applications. Our novel assumption is that we can increase the total throughput of a Web-based application by automatically creating database configurations that are capable of answering all the incoming query templates using a single node. We use a second assumption that in the case of a Web-based application we can know all the incoming query templates beforehand because the user interacts through a possible Web interface such as web forms [9]. The application logic executes the same hard wired queries over and over again for the same web form request. We conceptualize the database layout problem as a state search problem where each state is a valid configuration of layouts across database nodes. We search for the best configuration that can maximize the total system throughput and increase the performance of the application. Our system is capable of connecting to the application database and it conducts an automated state space search over the database layouts based on the given workload. The workload needs to contain all the incoming query templates of the application to be able to create the join graph of the tables and determine the applicable operators. The join graph represents the connections between tables in all the queries. The system also examines the constraints graph of the tables of the database to validate the relationships of each involved tables. The constraint graph represents the relationships between tables that are involved in the queries such as one-to-one, one-to-many, and many-to-many. Based on the constructed graphs the system performs an analysis to determine the groups of possible applicable partitioning operators. These groups serve as an input to the state space search that can eliminate many invalid created layout configurations. A state S_n is considered to be valid if and only if the created layout configuration L_i correctly answers each and every query Q_i from the given workload W. To further increase the practical aspect of the state space search we propose to learn general guiding rules that can eliminate valid states with high precision. These valid states have no or have worse impacts on the overall throughput of the system. In Section 2 we describe the related work and in Section 3 we outline our proposed solution. Our solution for State Space Search, features selection, and Cut-off points detection is described in Section 4. Our middleware system is described in Section 5. Experimental results are discussed in Section 6, in Section 7 we summarize our contribution, and in Section 8 we conclude the work and discuss future directions.

2   Related Work

2.1   Parallel and Hybrid Solutions

MapReduce [7] is a programming model with an associated implementation to process and generate huge amounts of data in a large scale (hundreds and thousands nodes), heterogeneous shared-nothing environment. More recently, a marriage of MapReduce with DBMS Technologies led to HadoopDB [1] a hybrid architecture. HadoopDB targets the performance and scalability of a parallel database and the fault-tolerance feature of the flexible MapReduce to achieve better structured data processing. It uses Hadoop [22], the open source implementation of MapReduce, to parallelize the queries across nodes. HadoopDB uses PostgreSQL [10] as a database layer that processes the translated SQL queries. Parallel databases use shared-nothing infrastructures in a clustered environment and execute queries in parallel using multiple nodes. They can scale up well if the number of the involved nodes is small. In a heterogeneous environment the probability of a possible server failure is high and parallel databases are not designed for use in a large-scale heterogeneous environment. There is a conceptual difference between parallel database management systems (DBMS) and MapReduce based systems (Hadoop). In the case of DBMS, user can state what he/she wants in SQL language. In the case of MapReduce-like systems the user presents an algorithm to specify what he/she wants in a low-level programming language [17]. Analytic Database [27] utilizes cheap shared-nothing commodity hardware and it is designed for large scale data warehouses. It uses a column-store architecture where each column is independently stored on different nodes. It applies vertical partitioning on the original dataset to create multiple partitions that can be replicated across cluster nodes. It is used mostly for read intensive analytical applications where the system has to access a subset of columns. Verticas optimizer is designed to operate on this column-partitioned architecture to reduce I/O cost dramatically. The optimizer of HadoopDB does not have full support for joins nor cost-based optimization of the queries. Netezza's [6] parallel system is a two-tiered system capable of handling very large queries from multiple users. Teradata [25] supports single row manipulation, block manipulation and full table or sub-table manipulation as well. It distributes the data randomly utilizing all the nodes. Its built-in optimizer can handle sophisticated queries, ad-hoc queries, and complex joins. IBM DB2's data partitioning [21] tool suggests possible partitions based on the given workload and the frequency of each SQL statement occurrence. For IBM DB2's data partitioning selection strategy [29] the main goal is - for a given static database schema and workload characteristic - to minimize the overall response time of the workload in multiple nodes. There are several other parallel databases available like Exadata [15] (parallel database version of Oracle), MonetDB [26], ParAccel [12], InfoBright [24], Greenplum [8], NeoView [28], Dataupia [5], etc. that all combine different techniques to achieve better performance and reliability. According to our best knowledge, none of them conducts a state space search and involves full replication, horizontal and vertical partitioning, and de-normalization operators by answering each query using a single database node.

2.2   Automated Physical Design Solutions

While there has been work in the area of automating physical database design [21,29] we are not aware of any work that addresses the problem of incorporating the full range of common operators and can learn rules to better partition a given database with multiple database nodes. Papers [19,29,2,20] study full replication and data consistency using distributed heavy weighted transactions. For example, in [3] they automatically select an appropriate set of materialized views and indexes to optimize the physical design for a given database and query workload as a built in tool in Microsoft SQL server 2000 using a single node. The paper [3] has tackled the very related problem of how to know automatically which indexes the system should use and which materialized views should apply. In paper [2], they added a new operator, called horizontal partitioning, to the previous optimization goal in Microsoft SQL server 2005. Microsoft SQL server 2005 offers an automated database design tool that offers physical design recommendation for horizontal partitioning. IBM DB2 [21] examined the problem of laying out multiple nodes which use many of the same operators we propose. GlobeTP [9] exploits the same fact that our system does: the workload of the web application is composed of a small set of query templates. In [13] their approach describes two common properties of Web-based applications. According to their strong assumption, workload is (1) dominated by reads and (2) it consists of a small number of query and update templates (typically between 10 and 100). Using the second assumption, their system solves strong consistency management of the servers. DBProxy [4] observed that most applications issue template-based queries and these queries have the same structure that contains different string or numeric constraints. AutoPart [14] deals with large scientific databases where the continuous insertions limit the application of indexes and materialized views. For optimization purposes, their algorithm horizontally and vertically partitions the tables in the original large database according to a representative workload using a single node. Ganymed [18] uses a novel scheduling algorithm that separates update and read-only queries using multiple nodes. GlobeDB [23] offers a different approach for edge servers to handle data distribution. Their system automatically partitions and replicates the database through a wide area network using multiple nodes.

3 Proposed Solution

Our middleware architecture is also based on shared-nothing commodity hardware where each node has its own CPU, disk, RAM, and file system. We focused on a Web-based application where the workload consists of a fixed number of query templates. This means the system is not presented with ad-hoc and unexpected queries. According to our best knowledge, none of the existing systems specializes for Web-based applications and considers the database layout problem as a state space search problem with the assumption that all the incoming queries should be answered by a single node. Our system is also novel because the state based search applies four different operators: full replication, horizontal and vertical partitioning, and de-normalization. Since we know all the query templates beforehand our system can pre-partition the data using these operators and pre-determined heuristics. Comparing to HadoopDB, we do not need to re-partition the data since we do not have unexpected and ad-hoc queries. By characterizing the problem as a state space search over database layout configurations, we iteratively minimize the total cost of the workload creating different database layouts and increase the total system throughput. Moreover, the second aspect of the project is a system that has a built-in corpus with generalized machine learned rules to determine which operator is applicable and when. As soon as the layout is determined, the data is distributed across the server nodes. A central dispatcher similar to the HadoopDN catalog - maintains the statistics about the current layout (table descriptors, data part locations, etc.). All the joins are pre-computed in our middleware and the communication bottleneck is eliminated. Our central dispatcher can push each query into the database layer directly where the well-defined schemas support indexing. We support SELECT with joins, INSERT, UPDATE, and DELETE SQL statements natively. We do not have additional failure detection mechanism, but the system is easily expandable with a full copy of the original database. Furthermore, a possible integration with Hadoop grants an additional layer that is capable to scale up to thousands of nodes as needed.

4   State Space Search over Layouts

We consider the database layout problem as a state space search problem with the assumption that all incoming queries should be answered by a single node. We do a time intensive search over different layouts, and each time, physically create the configurations, and evaluate the total throughput of the system. A state is a given assignment of tables to computer servers. The operators in the search are to fully replicate, horizontally partition, vertically partition, and denormalize a table. The search over layouts is very expensive, but we argue that our method is practical and useful, as entities trying to scale up their Web-based applications would be perfectly happy to spend few weeks of CPU time to increase their system throughput. Figure 1 shows an initiated complete search. As a start state (state 0) we fully replicate all tables across all database nodes and measure the total throughput of the system using the given workload. The system provides breath-first-search and depth-first-search algorithms to traverse the search tree. The default search algorithm is depth-first-search. We traverse down an entire path (state 1, 2, 3, and 4) before backtracking to the next valid path (state 2, 5, and 6). A state S_n is considered to be valid if and only if the created layout configuration L_i correctly answers each and every query Q_i from the given workload W. As soon as a valid state is created the system measures the system throughput. As one of the guiding rules we backtrack to the next valid path if the throughput of a child is less than the throughput of its parent. For example, if the throughput of sate 18 is less than the throughput of state 17 then we will not explore states 19, 20, 21, and 22. To further increase the practical aspect of the state space search we propose to learn general guiding rules that can eliminate valid states that are not very likely to succeed because they have worse or no impact on the overall throughput of the system.

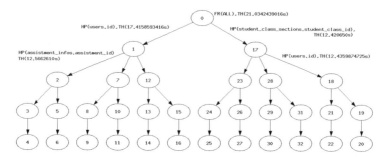

Fig. 1. State Space Search

Table 1. Illustrating Table, Query, Workload, and State features used by the system

Table Features	Query & Workload Features
Table Features	**Query & Workload Features**
Table size	UDI vs. R query ratio
Primary key	% of UDI and R queries for Table A
Foreign key	% of queries of Table A with "WHERE" condition
# of indexes	% of "JOINS" on Table A comparing to other JOIN queries
# of distinc values	% of queries that involves a JOIN and Table A comparing to all queries
# of columns	
Is schema heavy?	% of queries that involves Table A compared to other queries
# of INSERTS	
# of UPDATES	% of queries that involves Table A and other tables compared to all queries
# of DELETES	
# of table accesses	% of queries that involves Table A's columns compared to other queries
State Features	
Full Replication	
Horizontal Partitioning	% of queries that involves Table A's columns and other tables' columns compared to other queries
Vertical Partitioning	
De-normalization	
Total Throughput	The frequency of UDI and R queries

We parameterize common guidelines that DBAs use as heuristics in doing this task manually. Table 1 shows the features we applied to parameterize guidelines for the search. We divided the features into three categories: table-related, query- and workload-related, and state-related features. Table-related features like table size, distinct values, number of columns, etc. help to pre-select the applicable operators for a specific table. "Is schema heavy?" feature considers the table schema heavy if one of the table columns' type can generate high database memory and caching demand. If the column type is text, byte, xml, etc. and the frequency of the column specific retrieval query is high then one should help the database to share the memory requirements among multiple nodes. The detection of 'WHERE' conditions can help to identify a possible operator and a partitioning key. We do not consider horizontal partitioning operator for tables that are involved in queries without a 'WHERE' clause. For example 'SELECT * from table_a' can break our assumption that we can answer each query using a single node. Although we rule out applicable operators if specific query templates are involved, this assumption can reduce the intercommunication cost between nodes because each join is pre-computed and pushed into a single data node. Query and workload features are also very important to determine exact cut-off points. For example, a table with few tuples is not worth horizontally partition unless the number of table-related inserts versus the select queries' ratio is high enough. As another example: consider the frequency of the queries in the workload (for all queries and table specific ones as well) and determine a cut-off point to characterize the application as write intensive (high number of Update, Delete, and Insert queries) or as read intensive one (high number of Select queries). With the determination of cut-off points we can pre-determine heuristics, guide a parameterized state space search better, and eliminate valid states that otherwise would not increase the system throughput.

4.1   Cut-Off Points Detection

Cut-off points help the state space search to focus on creating layout configurations that could boost the performance of the application. Cut-off points reduce the size of the search space by eliminating valid table-operator keypairs because of their possible negative performance effect on the system. Figure 2 presents an example for the detection of the cut-off points. ASSISTments [11] is a Web-based Intelligent Tutoring System. In the system the number of sessions can be balanced among application servers but the continuous database retrieval queries, and update, delete and insert (UDI) queries decrease the system response time significantly. Currently, the ASSISTments system supports thousands of users over Massachusetts. It consists of multiple application servers, a load balancer, and a database server. We collected real-time queries during a week interval and constructed a workload that reflects a heavy tutoring day. The workload has all the query templates of the system (136) and 50,000 queries. The frequency of the queries reflects the real system's workload characteristics. We determined the Cut-off points for select vs. insert, select vs. update, and select vs. delete query ratios. We removed all the UIDs from the workload and replaced them with retrieval queries to maintain 50,000. The system fully replicated all the tables and created multiple partitions across three database nodes. We simulated three application servers each with the 50,000 queries and captured the total throughput of the system (Times[s]) averaging the results. We turned off the cache of the PostgreSQL servers to eliminate the effect of query caching. After the initial measurements we increased the UDI query ratios in the workload. We measured each state three times to be able to determine that the results are significantly different. We found significant differences in each case. Table 2 presents one of the two-tailed T-TEST results. We zoomed into the 0%-75% interval to make the Cut-off points precise. Based on the results, we can see that is worth fully

Table 2. Two-tailed T-TEST results for retrieval vs. update ratio

R/U ratio	Run 1[s]	Run 2[s]	Run 3 [s]	AVR[s]	TTEST p-value
75%/25%					
Full Replication	4080.551	4089.372	4085.493	4085.139	less than 0.01
Horizontal Part.	4624.573	4630.825	4632.409	4629.269	
81.25%/18.75%					
Full Replication	5416.133	5452.314	5421.186	5429.878	0.6394
Horizontal Part.	5501.191	5450.292	5393.945	5448.476	
87.5%/12.5%					
Full Replication	5169.829	5170.803	5176.965	5172.532	less than 0.01
Horizontal Part.	5195.582	5194.494	5206.076	5198.717	
93.75%/6.25%					
Full Replication	4913.791	4911.573	4925.373	4916.912	0.1231
Horizontal Part.	4914.215	4914.195	4927.423	4918.611	
100%/0%					
Full Replication	4401.551	4409.372	4402.493	4404.472	less than 0.01
Horizontal Part.	4673.159	4660.961	4675.263	4669.794	

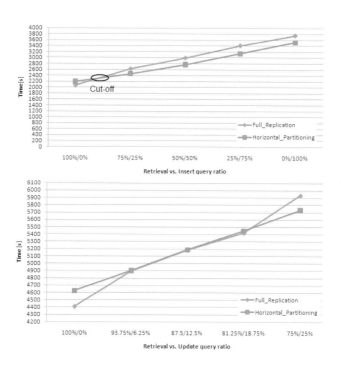

Fig. 2. Cut-off points of Horizontal Partitioning and Full Replication (upper: retrieval vs. insert query ratio with detected Cut-off point; lower: retrieval vs. update query ratio with zoomed Cut-off range

replicating a table if the percentage of the UDI vs. retrieval query ratio is less than or equal to 6%. We can also see that is worth considering horizontal partitioning if this ratio is greather than 18%. The average of the standard deviations is +/- 9.7. We also tried to change the total number of queries in the workload from 50,000 to 100,000 (system response time increased), but the Cut-off points did not change significantly.

5   The Middleware

In this section, we describe our middleware and the process of the state space search over different layout configurations. Figure 3 illustrates the high-level architecture of the system and its main inputs. The workload specifies the sequence of queries that were collected during normal usage of the database. It needs to contain all incoming query templates of the application. Constraints represent the relationships between tables. We considers one-to-one, one-to-many, and many-to-many relations. The placement algorithm determines the applicable operators, partitions, and partitioning keys based on the workload and constraints. The system also considers the application source database and the available database nodes for partitioning which applies the nodes' addresses and database connector strings. Figure 4 shows the modularized architecture of the system. The Data Placement Algorithm (DPA) [16] is responsible for determining the valid sets of tables and keys for each operator based on the given workload, constraints, and node information. As soon as the algorithm creates the sets it passes them to the State Space Search Module (SSSM) (Section 4). This module is the heart of the search. If the SSSM determines a valid state then it contacts the Layout Module (LM) to initiate the creation of the physical configuration with the selected operator. The LM stores information about the created configurations in the Layout Bank (LB). Therefore, it asks the LB for the required configuration. If the LB has no previous information about the requested layout then the LM starts the layout generation process. First, it connects to the source database to initialize the layout generation. Then it

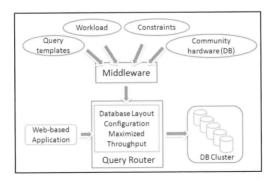

Fig. 3. High-level architecture of the Database Tuning Middleware

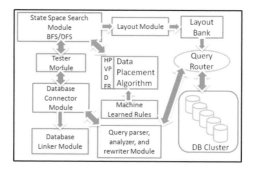

Fig. 4. Modules of the Database Tuning Middleware

collects information about the source table Users (e.g. column names and types, indices, triggers, etc.) and generates the new table schema (Table name + Applied operator + New Table identifier + Partitioning Key) utilizing the given database nodes (e.g. Users_HP_2abc4_id). In the next step the LM creates the new tables using the cloned table information on each node. With a pre-defined hash function, it distributes the tuples among multiple database nodes before sending the layout information to the LB. When the layout creation finishes, the SSSM initiates a test request to measure throughput of the created state. The Tester Module (TM) simulates the real world example with multiple application servers. Each simulated application server uses the workload to generate hundreds or thousands of requests for the back-end. The Database Connector (DCM) and Linker (DLM) modules are responsible for initializing and maintaining the database connections towards the available database nodes. The Query Analyzer Module (QAM) parses the queries in the workload and rewrites them to replace the original table names with the partitioned ones (e.g. replace table Users with Users_HP_2abc4_id). It uses the same hash function the LM applies to determine the correct database nodes for data retrieval. Once the data is laid out on the database servers, the LM updates the middleware's Query Router (QR) about the new changes in the layout configuration [16]. The QR maintains multiple connections to the database nodes, routes queries to the correct node, and transfers results back to the requestor. As soon as a new configuration is laid out, the Web-based application can connect to the QR without needing any code modification. The application detects the QR as a database node that manages and hides the configuration differences utilizing multiple databases.

6   Experimental Results

Our empirical result is determining a reliable increase in total system throughput by applying a hill climbing search to our existing data test case (the ASSISTments Intelligent Tutoring System). We understand that heuristics searches converge to local minimums but this is an acceptable compromise to achive a faster search. We collected real-time queries during a week interval and constructed a workload that reflects an average tutoring day. The workload had all the query templates of the system (136) and consisted of 1000 queries. The frequency of the queries reflected the real system's workload characteristics. Our three database nodes had the following configurations: node 1, node 2, and node 3 are an Intel Xeon 4 Core CPU with 8 GB RAM running FreeBSD 7.1 i386. The database software used on all nodes is PostgreSQL version 8.2.13. We turned off the cache of the PostgreSQL servers to eliminate the effect of query caching. We utilized our query router and used three simulated application servers, each issued 1000 queries respectively. The router was an Intel Pentium 4, 3 GHz machine with 4 GB RAM running Ubuntu 4.1.2 OS. The code for this middleware is written in Python version 2.6. The bandwidth between the query router and the database nodes is 100 Mbps, and the number of hops from the query router to the database servers are equal. Each simulated application server repeated the throughput measurement 4 times and the entire measurements was repeated 3 times. We took the average of the three repeated executions. The experiment was running for 48 hours. The state space search considered 168 valid states and eliminated 155 ones that had no effect on the total throughput of the system. Figure 5 shows the results. We had 11 distinct operators. Each state is significantly different from the initial state 0 where we fully replicated all the 44 tables across the nodes. We found that State 10 gave the most significant 10% throughput improvement of the system.

State	Table Name.Partitioning key	Operator	Parent	THP[s]	Gain[s]	TTEST_State0	TTEST_Parent	Run 1[s]	Run 2[s]	Run 3[s]
0	pending_parent_requests,assistments	FR	N/A	124.1669112	N/A	N/A	N/A	124.54332	123.46219	324.49523
	assistment_types,variables									
	sequences,student_class_sections									
	sequence_ownerships,item_difficulty_logs									
	class_files,enrollment_states									
	problem_to_skill_associations									
	class_assignment_types,assignment_logs									
	section_links,tags,comments									
	taggings,subjects,user_roles									
	student_classes,sections									
	tag_categories_tags,folders									
	assistment_infos,users,sessions									
	user_details,problems,answers									
	class_types,enrollments									
	progresses,hints									
	class_assignments,roles									
	assistment_ownerships,skills									
	teacher_classes,messages									
	transfer_models,item_difficulties									
	districts,frameworks									
	tutor_strategies									
	transfer_model_ownerships									
1	variables.assistment_id	HP	0	118.624273	-118.62427	0.015410998	0.015410998	119.14472	120.60649	116.12161
2	student_classes.id	HP	1	117.1130089	1.5112641	0.029413508	0.57479507	113.42944	117.22172	120.68786
3	student_class_sections.student_class_id	HP	2	116.56672	0.5462889	0.006421197	0.839449499	117.12638	113.88937	118.68443
4	assignment_logs.assignment_id	HP	3	115.2239789	1.3427411	0.019209459	0.648047138	114.91456	119.40695	111.35043
5	enrollment_states.name	HP	4	114.9910504	0.2329285	0.00137231	0.932366582	114.92414	113.11517	116.93385
6	sessions.session_id	HP	5	113.2827233	1.7083271	0.006672899	0.50785965	109.1325	115.18914	115.52653
7	class_assignments.student_class_id	HP	6	113.1958117	0.0869115	0.000508022	0.971784079	111.18961	114.00953	114.3883
8	item_difficulty_logs.problem_id	HP	7	112.3757076	0.8201041	0.021244595	0.818255744	116.43877	106.09194	114.59641
9	user_roles.user_id	HP	8	112.2583943	0.1193134	0.000123903	0.972651709	113.27337	112.6767	110.81911
10	comments.user_id	HP	9	111.6127518	0.6436425	0.000252626	0.623276083	112.72006	112.4167	109.70149
11	class_files.class_assignment_id	HP	5	114.2448575	0.7461929	8.53725E-05	0.571303821	114.51211	114.94886	113.27361
12	class_files.class_assignment_id	HP	6	112.6936737	0.502138	3.94697E-05	0.795715773	112.16215	113.62023	112.29866
13	comments.user_id	HP	7	112.0949376	0.28077	4.37215E-06	0.336668676	112.07477	112.07477	312.13527
Total throughput improvement[%]				MIN[s]	MAX[s]	DIFFERENT[s]	TTEST_State0			
10.11071252				111.6127518	124.16691	12.55415944	**0.000252626**			

Fig. 5. Experimental results

7   Contributions

The contribution of this paper is a methodology which can help Web-based applications that deal with scalability problems. The typical scaling bottleneck of an application is at the database side. We propose a simple solution to resolve this issue and we make certain assumptions about how to simplify the task. We propose a search methodology by which we search for better database layouts. One of the assumptions we actually impose upon ourselfs is that queries should be answerable by a single database node. By making this assumption we simplified the processing of individual queries to the databases. Of course, sometimes DBAs want to write queries that can go accross all database nodes involving multiple tables, e.g. for analytic purposes but these analytic queries can be executed as a background task by the DBAs. Furthermore, we proposed an established middleware that is general and it can be used by any Web-based application. We established two empirical results. First, we verified and parameterized the common assumption that one should horizontally partition tables or fully replicate them if the ratio of UDIs to the retrieval queries is greather than 18% or less than 6%. Second, we reported our experimental results using real application workload and performing the layout search over multiple database nodes. By conceptualizing the problem as a state space search problem and by doing a full state space search, we were able to physically create the layouts and evaluate the overall throughput of the system to parameterize the guiding rules. With our methodology we were able to increase the overal system throughput by 10%. We also identified other features that can be important to guide the search. We propose to use them as an input of a machine learning system to create general rules about when to use the different layout operators.

8   Future Work

We would like to use our middleware to actually find more interesting rules involving the proposed features of tables, queries, workload, and states. We propose to use them as rules of thumb of a machine learning system to recommend when to use the different layout operators. Our hypothesis is that we can learn more rules to capture human-like expertise and use these rules to better partition a given database. We also propose to evaluate our solution against many different web applications to illustrate the benefits of our approach.

References

1. Abouzeid, A., Bajda-Pawlikowski, K., Abadi, D., Silberschatz, A., Rasin, A.: Hadoopdb: an architectural hybrid of mapreduce and dbms technologies for analytical workloads. Proc. VLDB Endow. 2, 922–933 (2009)
2. Agrawal, S., Chaudhuri, S., Kollar, L., Marathe, A., Narasayya, V., Syamala, M.: Database tuning advisor for microsoft sql server 2005. In: Proceedings of VLDB, pp. 1110–1121 (2004)

3. Agrawal, S., Chaudhuri, S., Narasayya, V.R.: Automated selection of materialized views and indexes in sql databases. In: Proceedings of the 26th International Conference on Very Large Data Bases, VLDB 2000, pp. 496–505. Morgan Kaufmann Publishers Inc., San Francisco (2000)
4. Amiri, K., Park, S., Tewari, R., Padmanabhan, S.: DBProxy: A Dynamic Data Cache for Web Applications. In: IEEE Int'l Conference on Data Engineering (ICDE), Bangalore, India (March 2003)
5. Dataupia. Dataupia satori server (2008),
   http://www.dataupia.com/pdfs/productoverview/
   Dataupia%20Product%20Overview.pdf
6. Davidson, G.S., Boyack, K.W., Zacharski, R.A., Helmreich, S.C., Cowie, J.R.: Data-centric computing with the netezza architecture. In: Sandia Report, Unlimited Release (April 2006)
7. Dean, J., Ghemawat, S., Inc, G.: Mapreduce: simplified data processing on large clusters. In: OSDI 2004: Proceedings of the 6th Conference on Symposium on Opearting Systems Design and Implementation. USENIX Association (2004)
8. Greenplum. Greenplum database 3.2 administrator guide (2008),
   http://docs.huihoo.com/greenplum/GPDB-3.2-AdminGuide.pdf
9. Groothuyse, T., Sivasubramanian, S., Pierre, G.: GlobeTP: Template-Based Database Replication for Scalable Web Applications. In: Int'l World Wide Web Conf (WWW), Alberta, Canada (May 2007)
10. P. G. D. Group. Postgresql (2011), http://www.postgresql.org
11. Heffernan, N.T., Turner, T.E., Lourenco, A.L.N., Macasek, M.A., Nuzzo-Jones, G., Koedinger, K.R.: The ASSISTment Builder: Towards an Analysis of Cost Effectiveness of ITS creation. In: FLAIRS, Florida, USA (2006)
12. MacFarland, A.: The speed of paraccels data warehousing solution changes the economics of business insight (2010),
   http://www.clipper.com/research/TCG2010008.pdf
13. Olston, C., Manjhi, A., Garrod, C., Ailamaki, A., Maggs, B.M., Mowry, T.C.: A scalability service for dynamic web applications. In: Proc. CIDR, pp. 56–69 (2005)
14. Papadomanolakis, E., Ailamaki, A.: Autopart: Automating schema design for large scientific databases using data partitioning. In: Proceedings of the 16th International Conference on Scientific and Statistical Database Management, pp. 383–392. IEEE Computer Society, Los Alamitos (2004)
15. A. O. W. Paper. A technical overview of the sun oracle database machine and exadata storage server (2010),
   http://www.oracle.com/technetwork/database/exadata/
   exadata-technical-whitepaper-134575.pdf
16. Patvarczki, J., Mani, M., Heffernan, N.: Performance driven database design for scalable web applications. In: Grundspenkis, J., Morzy, T., Vossen, G. (eds.) ADBIS 2009. LNCS, vol. 5739, pp. 43–58. Springer, Heidelberg (2009)
17. Pavlo, A., Paulson, E., Rasin, A., Abadi, D.J., DeWitt, D.J., Madden, S., Stonebraker, M.: A comparison of approaches to large-scale data analysis. In: SIGMOD 2009: Proceedings of the 35th SIGMOD International Conference on Management of Data, pp. 165–178. ACM, New York (2009)
18. Plattner, C., Alonso, G.: Ganymed: Scalable replication for transactional web applications. In: Jacobsen, H.-A. (ed.) Middleware 2004. LNCS, vol. 3231, pp. 155–174. Springer, Heidelberg (2004)
19. Plattner, C., Alonso, G., Özsu, M.T.: DbFarm: A scalable cluster for multiple databases. In: van Steen, M., Henning, M. (eds.) Middleware 2006. LNCS, vol. 4290, pp. 180–200. Springer, Heidelberg (2006)

20. Ramamurthy, R., DeWitt, D.J., Su, Q.: A case for fractured mirrors. In: International Conference on Very Large Databases, pp. 430–441. Morgan Kaufmann Publishers, San Francisco (2002)

21. Rao, J., Zhang, C., Lohman, G., Megiddo, N., Rao, J., Zhang, C., Megiddo, N.: Automating physical database design in a parallel database. In: Proc. 2002 ACM SIGMOD, pp. 558–569 (2002)

22. Shafer, J., Rixner, S., Cox, A.L.: The hadoop distributed filesystem: Balancing portability and performance (April 2010)

23. Sivasubramanian, S., Pierre, G., van Steen, M.: GlobeDB: Autonomic Data Replication for Web Applications. In: Int'l World Wide Web Conf (WWW), Chiba, Japan (May 2005)

24. Ślezak, D., Eastwood, V.: Data warehouse technology by infobright. In: Proceedings of the 35th SIGMOD International Conference on Management of Data SIGMOD 2009, pp. 841–846. ACM, New York (2009)

25. Sue Clarke, T.: Butler group research paper (October 2000), http://www.teradata.com/library/pdf/butler_100101.pdf

26. Vermeij, M., Quak, W., Kersten, M., Nes, N.: Monetdb, a novel spatial column-store dbms. In: Netterberg, I., Coetzee, S. (eds.) Academic Proceedings of the 2008 Free and Open Source for Geospatial (FOSS4G) Conference, OSGeo, pp. 193–199 (2008)

27. Vertica. The vertica analytic database–introducing a new era in dbms performance and efficiency (2009), http://www.redhat.com/solutions/intelligence/collateral/vertica_new_era_in_dbms_performance.pdf

28. Winter, R.: Hp neoview architecture and performance (2009), http://h20195.www2.hp.com/v2/GetPDF.aspx/4AA2-6924ENW.pdf

29. Zilio, D.C., Jhingran, A., Padmanabhan, S.: Partitioning key selection for a shared-nothing parallel database system. IBM Research Report RC (1994)

XML Data Transformations as Schema Evolves*

Jakub Malý, Irena Mlýnková, and Martin Nečaský

XML Research Group, Charles University in Prague, Czech Republic
{maly,mlynkova,necasky}@ksi.mff.cuni.cz

Abstract. One of the key characteristics of XML applications is their dynamic nature. When a system grows and evolves, old user requirements change and/or new requirements accumulate. Apart from changes in the interface, it is also necessary to modify the existing documents with each new version, so they are valid against the new specification. The approach presented in this paper extends an existing XML conceptual model with the support for multiple versions of the model. Thanks to this extension, it is possible to define a set of changes between two versions of a schema. This work contains an outline of an algorithm that compares two versions of a schema and produces a revalidation script in XSL.

Keywords: XML schema, conceptual modeling, evolution, revalidation.

1 Introduction

The eXtensible Markup Language (XML) [16] has become a standard for data representation and manipulation and, hence, invoked a boom of so-called *XML applications* that exploit a whole family of XML technologies. A typical current XML application usually consists of a set of sub-applications, each being responsible for a particular logical execution part. The life-cycle of a such an *XML system of applications* is similar to a life-cycle of a single application, however the complexity is much higher. First of all we need to design of numerous data structures, i.e. XML schemas, that are exchanged and processed by business processes of the system. What is more, they are usually mutually related or overlayed. In other words, each application of the system utilizes several *views* of a common problem domain represented by XML schemas. Hence, they cannot be designed separately, but as a whole complex system. In addition, sooner or later the user requirements of the applications change and, hence, the data structures they input, process and output must be modified respectively – we usually speak about the problem of *XML schema evolution*.

In our previous work [15] we have proposed a five-level XML evolution framework that enables one to face the described issues. It utilizes the concepts of MDA (*Model-Driven Architecture*) [12] hierarchy of conceptual models that enable a user to abstract from specifics of a particular XML format, enables one to model a whole set of related XML applications concurrently and preserves the

---

* Supported by GAČR grant no. P202/10/0573, the grant no. 201/09/P364 and the grant no. P202/11/P455.

J. Eder, M. Bielikova, and A.M. Tjoa (Eds.): ADBIS 2011, LNCS 6909, pp. 375–388, 2011.

respective relations between them. Consequently, it naturally supports evolution management.

In this paper we focus on one particular aspect of the complex framework – propagation of changes between multiple versions of an XML schema to the respective instances, i.e. XML documents. We speak about the process of *revalidation*. We describe a unique approach that enables one to output so-called *revalidation script*, i.e. an XSLT [6] script that, when applied on the given set of XML documents valid against an old version of an XML schema, outputs a set of XML documents valid against the new version of the schema. The approach enables to reduce manual and, hence, error-prone tasks via cutting down the user interaction to the necessary minimum.

The paper is structured as follows: In Section 2 we provide an overview of the related work. In Section 3 we describe the conceptual model we utilize for the purpose of change specification. In Section 4 we define the set of changes that can be performed over the models and in Section 5 we provide the algorithm for generation of the XSLT revalidation script. In Section 6 we describe implementation of our approach and provide a complete illustrative example. Finally, in Section 7 we conclude and outline future work.

2   Related Work

For the goal of determining whether the set of documents was invalidated with the newly coming version of the schema, the system must recognize and analyze the differences between them. There are two possible ways to recognize changes – *recording of the changes* as they are conducted during the design process and *comparing the two versions* of the schema.

An evolution system that utilizes recording of the changes (e.g. [7]) usually provides some kind of a command that initiates the recording and after issuing this command all operations carried out by a user over the schema are recorded. When the desired schema is reached, the user finishes recording and the system has all the information about the changes made – the sequence of performed operations. When the recording is finished, the system can normalize the sequence for example by eliminating operations that cancel each other or by replacing groups of operations by other groups that lead to the same result but in a more straight way. These normalizing rules must be defined in the system.

An alternative approach is to base the change detection on comparison of the two versions. The user can work with both schemas independently until (s)he is satisfied with them. Before detecting changes, the mapping between the two schemas must be found, which requires some degree of user interaction ([10] uses a visualization tool for mapping editing). The change detection algorithm then takes the two schemas as an input and compares them. The result of the comparison is a list of differences between the schemas.

On the contrary, systems *X-Evolution* [3] and *XEM* [4], built upon graphical editor for creating schemas in the XML Schema [18] or DTD [19] respectively, use an *incremental validation*. Each single evolution operation executed upon the schema is propagated to valid documents. In addition, neither of the systems uses any conceptual model for the schemas, so the user must cope with all technical details of the languages. The authors of [2] also chose this method, only the evolution operations are executed upon UML model, from which they are propagated to XML schemas and documents.

None of the existing approaches meets the requirements for a full-fledged evolution framework. Moving elements/attributes is not supported in [3,4], [7] does not support cardinalities entirely, [4], [7] and [10] are strongly dependent on the chosen implementation language. Except for [2], none of the aforementioned approaches utilizes a conceptual model or any higher level of abstraction. The drawback of [2] is the lack of control over the resulting XML schema (common constructs like `choice` or `set` have no counterpart in UML). All the incremental approaches are not suitable when XML documents are stored in a relational database supporting schema evolution [1], because in this scenario, a complete revalidation script is required for the revalidation cycle.

3   Conceptual Model

Our approach does not work directly at the low level of XML schemas. Instead, we proposed a two-layered framework called *XSEM* [14] for schema modeling and revalidation. XSEM is a model for XML data and it exploits two layers of the MDA – *platform-independent model* (PIM) and *platform-specific model* (PSM). A schema in a platform-specific model (PSM) models how a part of the reality modeled by the PIM schema is represented in a particular data model. In our case, we consider XML as the data model and our PSM schema specifies representation in a particular XML format. Separation of the PIM allows to easily model a whole family of related schemas sharing the same problem domain. Moreover, the layered model is highly useful for integration of heterogenous systems sharing a problem domain, as we have shown in [9], and especially for evolving the family of schemas as a whole (where a one conceptual change must be reflected in several parts of the system). In the rest of this paper, we will use following notation:

- $\mathcal{L}$ to denote an infinite set of string *labels*, λ will be used for empty string,
- $\mathcal{D}$ to denote a finite set of *data types* (a data type $D \in \mathcal{D}$ is a possibly infinite set of data values, e.g. an infinite set of all positive integers, etc.),
- $\mathcal{D}^*$ to denote the union of all considered data types ($\mathcal{D}^* = \bigcup_{D \in \mathcal{D}} D$),
- $\mathcal{C} = \{m..n : m \in \mathbb{N}_0 \wedge n \in (\mathbb{N}_0 \cup \{*\}) \wedge (m \leq n \vee n = *)\}$ to denote an infinite set of *cardinality constraints* where $\mathbb{N}_0$ denotes the set of natural numbers including 0. We will use auxiliary functions low and $upp : c = \{m..n\} \in \mathcal{C} \to low(c) = m \wedge upp(c) = n$.
- $2^{\mathcal{X}}$ and $2^{(\mathcal{X})}$ to denote the set of all subsets and ordered subsequences of a set $\mathcal{X}$, respectively. For $x \in L, L \in 2^{(\mathcal{X})}$ the function $position(x, L)$ returns position of x in the ordered subsequence L.

The full specification of both PIM and PSM for XSEM can be found in [13], its prototype implementation *XCase* was presented in [8]. We use (simplified) UML class diagrams on the PIM level (class, attributes, binary associations). PIM contains exactly one schema (unlike PSM). For this paper, we will use a slightly simplified definition of PSM. Def. 1 formally introduces PSM schemas. A PSM schema 'shapes' PIM concepts to model a particular XML format. The separate PSM layer allows reuse of shared concepts from the PIM and thus reduce the manual work of the schema designer. A PSM schema can be automatically translated to an XSD [13].

Fig. 1 shows an XML format for a purchase in an e-commerce system, each purchase document has an identification of the customer, customer's address and the list of purchased products.

Definition 1. *A* platform-specific schema (PSM schema) *is a 5-tuple* $\mathcal{S}' = (\mathcal{S}'_c, \mathcal{S}'_a, \mathcal{S}'_r, \mathcal{S}'_m, \mathcal{C}'_{\mathcal{S}'})$ *of disjoint sets of* classes, attributes, associations, *and* content models, *respectively, and one specific class* $\mathcal{C}'_{\mathcal{S}'} \in \mathcal{S}'_c$ *called* schema class. $\mathcal{S}'$ *must be a forest with one of its trees rooted in* $\mathcal{C}'_{\mathcal{S}'}$. *In addition, PSM schema defines functions* name', card', type', xform', attributes', cmtype', content' *:*

- Class $C' \in \mathcal{S}'_c$ *has a name* $\in \mathcal{L}$ *assigned by function* name'(C').
- Attribute $A' \in \mathcal{S}'_a$ *has a name* $\in \mathcal{L}$, *data type* $\in \mathcal{D}$, *cardinality* $\in \mathcal{C}$ *and XML form* $\in \{e, a\}$, *assigned by function* name'(A'), type'(A'), card'(A') *and* xform'(A') *respectively. Function* attributes'$(C') : \mathcal{S}'_c \rightarrow 2^{(\mathcal{S}'_a)}$ *assigns a sequence of attributes for each class.*
- Association $R' \in \mathcal{S}'_r$ *is a pair* $R' = (C'_1, C'_2)$ *where* C'_1 *and* C'_2 *are called* parent *and* child *of R and denoted* parent'(R') *and* child'(R'). *R' has a name* $\in \mathcal{L}$ *and cardinality* $\in \mathcal{C}$ *assigned by function* name'(R') *and* card'(R').
- Content model $M' \in \mathcal{S}'_m$ *has a content model type* $\in \{$sequence, choice, set$\}$ *assigned by function* cmtype'(M') .
- $\mathcal{N}' = \mathcal{S}'_c \cup \mathcal{S}'_m$, *function* content' $: \mathcal{N}' \rightarrow 2^{(\mathcal{S}'_r)}$ *assigns to each node an ordered sequence of association starting in that node.*

The graph $\widehat{\mathcal{S}}' = (\mathcal{N}', \mathcal{S}'_r)$ *with classes and content models as nodes and associations as ordered edges is a tree rooted in the schema class* $C'_{\mathcal{S}'}$. *Members of* $\mathcal{S}'_c$, $\mathcal{S}'_a$, $\mathcal{S}'_r$, *and* $\mathcal{S}'_m$ *are called* components *of* $\mathcal{S}'$.

We will also use auxiliary functions *inAssociation'*, *parentNode'* and *childNodes'*:

$$inAssociation'(N') = R' \Leftrightarrow \exists R' \in \mathcal{S}'_r \text{ s.t. } child'(R') = N'$$
$$childNodes'(N') = (N_1', \dots, N_n') \Leftrightarrow content'(N') = (R_1', \dots, R_n') \wedge$$
$$N_1' = child'(R_1') \wedge \dots \wedge N_n' = child'(R_n')$$

Tab. 1 shows how a specific XML format is modeled by a PSM schema.

4   Changes between Versions

Before revalidation, we need to find out what changes were made during schema evolution, i.e. what operation the user performed when evolving a PSM schema

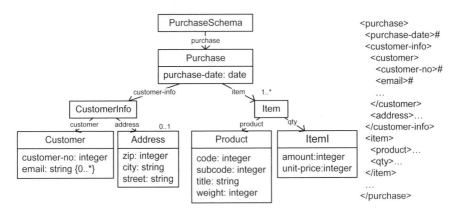

Fig. 1. Sample PSM schema (and a rough structure of a modeled XML document)

Table 1. XML attributes and XML elements modeled by PSM constructs

Construct	Modeled XML Construct
$C' \in \mathcal{S}'_c$	A complex content which is a sequence of XML attributes and XML elements modeled by attributes in $attributes'(C')$ followed by XML attributes and XML elements modeled by associations in $content'(C')$
$A' \in \mathcal{S}'_a$, s.t. $xform'(A') = a$	An XML attribute with name $name(A')$, data type $type(A')$ and cardinality $card(A')$
$A' \in \mathcal{S}'_a$, s.t. $xform'(A') = e$	An XML element with name $name(A')$, simple content with data type $type(A')$ and cardinality $card(A')$
$R' \in \mathcal{S}'_r$, s.t. $name(R') \neq \lambda$	An XML element with name $name(R')$, complex content modeled by $child'(R')$ and cardinality $card(R')$
$R' \in \mathcal{S}'_r$, s.t. $name(R') = \lambda$	Complex content modeled by $child'(R')$
$M' \in \mathcal{S}'_m$	A complex content which is a sequence (or choice or set, respectively) of XML attributes and XML elements modeled by associations in $content'(C')$

$\mathcal{S}'$ to its new version $\widetilde{\mathcal{S}'}$ (we use tilde mark $\tilde{c}$ to denote the evolved schema/construct c). The detected changes can then be propagated to the XML documents. The algorithm for change detection takes $\mathcal{S}'$ and $\widetilde{\mathcal{S}'}$ as an input and outputs the set of detected changes. Tab. 2 lists types of changes that can occur grouped by their scope (class, attribute, association, content model). We also further categorize classes into four groups: addition, removal, migratory (e.g. *classAdded*, *classRemoved*, *classMoved*, *classRenamed*) and sedentary.

Due to space limitations, we will not explain in detail the algorithm of detecting changes, the in-depth description can be found in [11]. Here we cover only the core principles. The change detection combines the comparison and recording approaches to gain the benefits of both. The main principle is to compare the two versions, but during the user's evolution operations, each construct stays linked to

Table 2. Changes identified in XSEM between two versions of the same schema

Change	Description
classAdded	A new class was added to the schema
classRemoved	A class was removed from the schema
classRenamed	An existing class was renamed
classMoved	A class is moved to a new location in the tree
attributeAdded/Removed/Renamed	An attribute was added/removed/renamed
attributeTypeChanged	An attribute type changed
attributeIndexChanged	Attributes were reordered in the class
attributeCardinalityChanged	A cardinality of an attribute was changed
attributeXFormChanged	An xform of an attribute was changed
attributeMoved	An attribute was moved from a class to another
associationAdded/Removed/Renamed	An association was added/removed/renamed
associationIndexChanged	Associations were reordered in the class
associationCardinalityChanged	A cardinality of an attribute was changed
associationMoved	An association is moved in the PSM tree
contentModelAdded/Removed	A content model was added/removed
contentModelMoved	A content model is moved to the PSM tree
contentModelTypeChanged	A type of content model changed

its other versions. This eliminates possible misinterpretations of changes, where e.g. some 'rename' operation is incorrectly interpreted as 'remove & add'.

According to the categories of changes, the algorithm divides classes, attributes and content models in the schema into disjoint sets $\mathcal{K}_a$, $\mathcal{K}_r$, $\mathcal{K}_m$ and $\mathcal{K}_s$ (of added, removed, moved and sedentary constructs) and also classifies nodes $\widetilde{\mathcal{N}'} \cup \widetilde{\mathcal{S}'_a}$ in schema $\widetilde{\mathcal{S}'}$ the tree into three disjoint groups:

- *red nodes* – the nodes that were changed + old and new parent nodes of all the migrated and renamed nodes/associations + classes that contain changed attributes + classes from which attributes were moved
- *blue nodes* – nodes that are not red, but contain a red node in their subtrees
- *green nodes* – other nodes

During revalidation, each of these groups is treated differently. Treatment of green nodes is straightforward – they can be skipped/copied to the result (with their whole subtree). Blue nodes were not modified, but the subtree cannot be copied as in the case of green nodes, because it contains at least one red node that needs to be revalidated. And finally, each red node needs to be processed separately.

5   Revalidation

The change detection algorithm outputs the set of changes between the schemas. Having the set of changes, we can now describe the algorithm for producing a *revalidation script* that outputs document $\widetilde{D'}$ revalidated against $\widetilde{S'}$ when applied on XML document D' valid against S'. The sequence of changes made over a PSM schema can be converted to an expression in any implementation language, e.g. *XQuery Update Facility* [21] or XSLT [6]. Assuming that the *XQuery Update Facility* is the implementation language, each change would be translated to an XQuery Update command(s):

- addition changes to `insert` commands
- removal changes to `delete` commands
- migratory changes would generate first `insert` command referencing some part of the document and thus copying the content and `delete` command to remove the content from its old location
- sedentary changes would generate `rename` command or again `insert` or `delete` commands

Each command would then be executed upon the revalidated document. The procedure when using *DOM API* [20] would be analogous. However, our approach, called *XSEM-Evo*, uses XSL stylesheets as implementation language due to the wide support for XSL among the tools working with XML data and especially the database systems supporting XML Schema [18] evolution. But, the previous procedure of translating each change into one command, is not directly applicable when using XSLT – for the following reasons:

1. *No Removal*: XSL does not have any means of explicit removing a content from a document. Removal is achieved by not putting the particular part of content to the output so that the processor never reaches the particular part of content, or by letting the processor go through the content without sending anything to the output.
2. *Processing of Unchanged Content*: XSL must process all content that should be sent to the output, not only content modified by some of the changes.
3. *Output Definitiveness*: When XSLT processor sends a content to the output, it can not be changed during the same transformation, the changes have to be grouped and conducted together.

5.1   Revalidation Script Overview

In XSL, stylesheets producing the same output can be written in several forms. To keep it transparent, comprehensible and easily modifiable, the revalidation stylesheet $\mathcal{F}$ generated by XSEM takes the following form:

- It is a one-pass stylesheet.
- It follows the *navigational stylesheet* pattern described in [5]. It relies on a detailed knowledge of the input document. XPath expressions used for `match` attributes of all top-level templates are always absolute.
- A top-level template is created for each red node.
- Each top-level template describes attributes and direct subelements of the processed red node.

- One common top-level template is added to process all green nodes and another to process all blue nodes.
- Implicit XSLT templates are never used, because they do not serve the desired purpose.
- The stylesheet grows (counting the number of top-level templates) with the amount of changes made in the schema, not with the complexity of the schema.

5.2   Generation of the XSLT Revalidation Script

Due to space limitations, we will show how the algorithm processes only the core constructs (classes, attributes with $xform' = e$ and associations with $name' \neq \lambda$). For the full description of the algorithm, see [11]. We start by showing the templates that process the blue and green nodes and then we show how templates processing the red nodes are constructed. For processing blue and green nodes, $\mathcal{F}$ contains templates depicted in Listing 1.1. The first template copies an element with its attributes and instructs the processor to continue with its subelements (where at least one element corresponding to a red node exists, which must be revalidated). The second template copies the element with its whole subtree to the output. Since all red and green nodes are processed by these two templates, the complexity of $\mathcal{F}$ does not grow with the size of the schemas, but with the amount of changes mad between $\mathcal{S}'$ and $\widetilde{\mathcal{S}'}$.

```
<xsl:template match="{blue-nodes-paths}">
  <xsl:copy>
    <xsl:copy-of select='@*' />
    <xsl:apply-templates select='*' />
  </xsl:copy>
</xsl:template>
<xsl:template match='{green-nodes-paths}'>
  <xsl:copy-of select='.' />
</xsl:template>
```

Listing 1.1. Green and blue nodes template

For each red node $\widetilde{N'}$ the algorithm generates one template in $\mathcal{F}$. During the process, the algorithm keeps a track of the currently processed node in the source schema (available through a variable `processedPath`). The algorithm can compute the XPath expression that selects instances of a given node in the input document from the processed node via function $relativeXPath(X, processedPath)$. For example:

processedPath	Path to node X	Relative path
/Purchase	/Purchase/Item/@amount	Item/@amount
/customer-info/Customer	/customer-info/Address/city	../Address/city

Listing 1.2 shows the basic structure of the template. If the processed node is an added node (i.e. $\widetilde{N'} \in \mathcal{K}_a$), it will be a named template (an auxiliary function `suggestName` returns a unique, but human-friendly name for the template).

Otherwise, it will be a template with **match** attribute. The template copies the element corresponding to the node via **xsl:copy**, then **processConstruct** subroutine is called for processing each attribute (if $\widetilde{N}'$ is a class and not a content model) and the same subroutine is called also for each child of the processed node.

If the node is an attribute, **xsl:value-of** is used to retrieve its value and copy it to the result. If the type of the attribute changed (*attributeTypeChanged* is detected, in that case let D' be the old type of the attribute and $\widetilde{D}'$ the new one), a conversion function must be called. In the pseudocode, this call is represented by the function $conv_{\widetilde{N}'} : D' \rightarrow \widetilde{D}'$. If the type did not change, the call can be omitted ($conv_{\widetilde{N}'} = identity$), similarly in the case when $D' \subseteq \widetilde{D}'$ (which is guaranteed e.g. when D' is a subtype of $\widetilde{D}'$). Subroutine **processConstruct** examines the state (whether the construct belongs to the set of added nodes) and also its cardinality.

```
<xsl:template
    { if  N̄' ∈ 𝒦ₐ } match='{N̄'.XPath}'  {else} name='{suggestName(N̄')}'>
    <{name'(N̄')}>
    { if  N̄' ∈ 𝒮'_c then foreach  Ã ∈ attributes'(N̄')
        processConstruct (Ã, card'(Ã))
      if  N̄' ∉ 𝒮'_a foreach  C̃ ∈ childNodes'(N̄')
        processConstruct (C̃, card'(inAssociation'(C̃)))
      else }  // attribute(leaf) → add the value
        <xsl:value-of select='{conv_N̄'}{(relativeXPath(N̄',processedPath))}' />
    <{name'(N̄')}>
</xsl:template>

procedure processConstruct
parameter  N̄' ∈ 𝒮'_a ∪ 𝒮'_c // processed attribute or class
parameter  card ∈ 𝒞 // cardinality
{
    case  N̄' ∈ 𝒦ₐ ∧ low(card) = 0 :
      exit ; // added optional element can be skipped
    case  N̄' ∈ 𝒦ₐ ∧ low(card) > 0 :
    case  N̄' ∈ 𝒦ₛ ∪ 𝒦ₘ ∧ cardinalityChanged(N̄') :
        generateElementCardinalityReference (N̄', card)
    otherwise : // added with card = 1 or card unchanged
        generateElementSingleReference (N̄')
}
```

Listing 1.2. Red nodes template – basic structure

Function **cardinalityChanged** looks up *associationCardinalityChanged/attributeCardinalityChanged* change (if there is one). There are two variants of *reference generating* subroutine – *single* (not dealing with cardinalities) and *cardinality* (designed to revalidate changes in cardinality). The first one is depicted in the first part of Listing 1.3. If the processed node is added, call of **instanceGenerator** template for the node is added to $\mathcal{F}$. This function must be supplied by the user after $\mathcal{F}$ is generated and should create the subtree for the added node. If the process node is not added, **xsl:apply-templates** is

```
procedure generateElementSingleReference
parameter: Ñ' ∈ Ñ' // referenced node
parameter: condition: XPath expression optional
{ if Ñ' ∈ 𝒦ₐ }
    <xsl:call-template name='{suggestName(Ñ')}' />
{ else }
    { var xpath ← relativeXPath(Ñ', processedPath) }
    { if condition is set }
        <xsl:apply-templates select='{xpath}[{condition}]' />
    { else }
        <xsl:apply-templates select='{xpath}' />
    { end if }
{ end if }

procedure generateElementCardinalityReference
parameter: Ñ' ∈ Ñ' // referenced node
parameter card ∈ 𝒞 // cardinality
/* routine called either when cardinality of element N changed
   or N' was added with lower cardinality > 1 */
{ if Ñ' ∈ 𝒦ₛ ∪ 𝒦ₘ // existing node
   // cardinality of N' changed, deal with existing nodes
    if ¬upper cardinality of Ñ' decreased
        generateElementSingleReference(Ñ')
    else
        generateElementSingleReference(Ñ', condition = 'position() ≤ '.upp(card))
    end if
  end if
  if Ñ' ∈ 𝒦ₐ ∨ lower cardinality of Ñ' increased
    // new nodes need to be created
    var countExpr
    var lower ← low(card)
    if (N' ∈ 𝒦ₐ)
        countExpr ← lower
    else
        var existing ← relativeXPath(Ñ', processedPath)
        countExpr ← lower.' - .count('.existing.')' }
        <xsl:call-template name='{instanceGenerator_Ñ'}'>
            <xsl:with-param name='count' select='{countExpr}' />
        </xsl:call-template>
    { end if }
{ end if }
```

Listing 1.3. Generating element reference

outputted (with possible *condition* – a parameter that is used when the *single* variant is called from the *cardinality* variant.

Finally, the *cardinality* variant of reference generating is depicted in the second part of Listing 1.3. There are two parts of the template. The first part concentrates on instances already present in the document (and is therefore skipped for *added* elements). Existing instances are processed again by the *single reference* subroutine – either all existing instances (when the upper cardinality of node $\widetilde{N'}$ was not decreased i.e. all existing instances can remain in the document) or the first k instances, where k is the new upper cardinality. The condition parameter of *single* variant with built-in XPath function position is utilized to restrict the number of instances processed. The purpose of the second part is to add new instances of N' to the document. Adding several instances may be

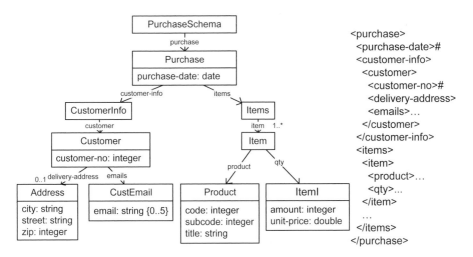

Fig. 2. Sample PSM schema – evolved version of the schema from Figure 1

needed for two reasons: either $\widetilde{N'}$ is an node with lower cardinality > 1 or the lower cardinality of $\widetilde{N'}$ was increased. Again, `instanceGenerator` template is made responsible for creating new instances.

As we stated in the beginning of 5.2, we will not show detailed pseudocode for revalidating content models. In brief, sequence model is revalidated similarly as class, choice and set models introduce branches to the generated stylesheets.

6   Implementation and Example

To prove our concepts we are continuously working on the implementation of the proposed models and algorithms. The first tool which utilizes our XML conceptual modeling framework XSEM is called *XCase*. XCase is available to the community as a free open-source software[1] and contains a full-fledged UML editor for creating and editing the PIM schema and deriving PSM schemas, support for maintaining an arbitrary amount of versions of schema and generating revalidation scripts. In XCase, the restrictions mentioned in 5.2 of the model are not enforced; content models, attributes and also our method for type reuse are included in this implementation.

We conclude the description of our approach by showing a concrete revalidation script that the algorithm generates for the schema depicted in Fig. 1 and a new version of the schema depicted in Fig. 2 – see Listing 1.4.

In the new version, association `address` was moved from `CustomerInfo` to `Customer` and renamed to `delivery-address`. New classes `Items` and `CustEmail` were added. Attribute `email` was moved from `Customer` to `CustEmail` and its cardinality was restricted to 0..5. Attributes of `Address` class were reordered and attribute `weight` was removed from the schema.

---

[1] http://xcase.codeplex.com/

```xml
<xsl:template match='/purchase'>
  <purchase>
    <xsl:apply-templates select='purchase-date'/>
    <xsl:apply-templates select='customer-info'/>
    <xsl:call-template name='purchase-items'/>
  </purchase>
</xsl:template>
<xsl:template match='/purchase/customer-info'>
  <customer-info>
    <xsl:apply-templates select='customer'/>
  </customer-info>
</xsl:template>
<xsl:template match='/purchase/customer-info/customer'>
  <customer>
    <xsl:apply-templates select='customer-no'/>
    <xsl:apply-templates select='../address'/>
    <xsl:call-template name='emails'/>
  </customer>
</xsl:template>

<xsl:template match='/purchase/customer-info/address'>
  <delivery-address>
    <xsl:apply-templates select='city'/>
    <xsl:apply-templates select='street'/>
      <xsl:apply-templates select='zip'/>
  </delivery-address>
</xsl:template>

<xsl:template name='emails'>
  <emails><xsl:copy-of select='email[position() &lt;= 5]'/></emails>
</xsl:template>
<xsl:template name='purchase-items'>
  <items><xsl:apply-templates select='item'/></items>
</xsl:template>
<xsl:template match='/purchase/item/product'>
  <product><xsl:apply-templates select='code|subcode|title'/></product>
</xsl:template>

<!-- blue nodes template -->
<xsl:template match='/purchase/item'>
  <xsl:copy>
    <xsl:copy-of select='@*'/>
    <xsl:apply-templates select='*'/>
  </xsl:copy>
</xsl:template>
<!-- green nodes template -->
<xsl:template match='/purchase/purchase-date
  | /purchase/customer-info/customer/*
  | /purchase/item/product/*[.= ../code|../subcode|../title]
  | /purchase/item/qty/* | /purchase/customer-info/address/*'>
    <xsl:copy-of select='.'/>
</xsl:template>
```

Listing 1.4. Purchase revalidation example

7   Conclusion and Open Problems

In this paper we proposed an algorithm for automatic revalidation of XML documents according to changes in respective XML schema. The revalidation script can deal with structural modifications automatically, user input is required only where necessary (e.g. when a new content must be added during revalidation). Our approach expects that there may exists several versions (which can be edited separately) in the system and can produce revalidation script between any selected pair (including translation from a new version to the old one). It is effective – the change detection step can decide, which changes does not invalidate existing documents and skip the revalidation in those cases, the document is processed only once and the revalidation script grows linearly with the amount of changes made. The algorithm is able to correctly distinguish moving operations from adding/deleting, correctly handle renaming, reordering and complex composite changes in the structure of the document.

Using our approach, the designer does not have to create XSLT scripts manually, instead of working at the low level of XML schemas, XML documents and XSLT (which would require him to study both the old and new version of the schema, spot the differences and write and debug the lengthy revalidation scripts manually), he can make the changes at a conceptual level. He can also generate revalidation scripts for any two versions in the system. A script, that transforms from version A to C can be simpler and more effective then the pair A $\rightarrow$ B, B $\rightarrow$ C (some operations may cancel each other, etc.). Combining them manually can be very difficult when the scripts get more complex.

The algorithm in its current version deals mainly with revalidation of (1) structure and (2) data already present in the document. Since new data are often required for new versions, we will focus our future work on obtaining this data for the revalidated documents. For this purpose, we will utilize the existing connection between PIM and PSM and a new similar connection between PIM and the model of a data storage (e.g. an ER schema [17]).

How to obtain the list of changes between the versions of the schema is not described in this paper. It is created automatically when the new version is obtained via editing the old version in the tool. However, when both old and new version of the schema is imported to the system, the mapping needs to be defined. We plan to propose heuristics for finding the mapping semi-automatically.

Finally, a complex system, besides a precise definition of how the data are structured, requires a support for modeling and checking integrity constraints (ICs). ICs also change as system evolves and can also be used to describe evolution operations in greater detail. Support for ICs will further enhance capabilities and applicability of the algorithm.

References

1. Oracle XML DB Developer's Guide – XML Schema Evolution.,
   http://download-uk.oracle.com/docs/cd/B28359_01/appdev.111/
   b28369/xdb07evo.htm#BCGFEEBB
2. Domínguez, E., Lloret, J., Rubio, A.L., Zapata, M.A.: Evolving xml schemas and documents using uml class diagrams. In: Andersen, K.V., Debenham, J., Wagner, R. (eds.) DEXA 2005. LNCS, vol. 3588, pp. 343–352. Springer, Heidelberg (2005)
3. Guerrini, G., Mesiti, M., Sorrenti, M.A.: Xml schema evolution: Incremental validation and efficient document adaptation. In: Barbosa, D., Bonifati, A., Bellahsène, Z., Hunt, E., Unland, R. (eds.) XSym 2007. LNCS, vol. 4704, pp. 92–106. Springer, Heidelberg (2007),
   http://dblp.uni-trier.de/db/conf/xsym/xsym2007.html#GuerriniMS07

4. Su, H., Kramer, D.K., Rundensteiner, E.A.: XEM: XML Evolution Management, Technical Report WPI-CS-TR-02-09 (2002)
5. Kay, M.: XSLT 2.0 and XPath 2.0, 4th edn. Wrox (2008)
6. Kay, M.: XSL Transformations (XSLT) Version 2.0. W3C (January 2007), http://www.w3.org/TR/xslt20/
7. Klettke, M.: Conceptual xml schema evolution — the codex approach for design and redesign. In: Workshop Proceedings Datenbanksysteme in Business, Technologie und Web (BTW 2007), Aachen, Germany, pp. 53–63 (March 2007)
8. Klímek, J., Kopenec, L., Loupal, P., Malý, J.: XCase – A Tool for Conceptual XML Data Modeling. In: Grundspenkis, J., Kirikova, M., Manolopoulos, Y., Novickis, L. (eds.) ADBIS 2009. LNCS, vol. 5968, pp. 96–103. Springer, Heidelberg (2010)
9. Klímek, J., Nečaský, M.: Semi-automatic integration of web service interfaces. In: IEEE International Conference on Web Services, pp. 307–314 (2010)
10. Kwietniewski, M., Gryz, J., Hazlewood, S., Van Run, P.: Transforming xml documents as schemas evolve. Proc. VLDB Endow. 33, 1577–1580 (2010)
11. Malý, J.: XML Schema Evolution. Master Thesis, Charles University in Prague, Czech Republic (2010)
12. Miller, J., Mukerji, J.: MDA Guide Version 1.0.1. Object Management Group (2003), http://www.omg.org/docs/omg/03-06-01.pdf
13. Nečaský, M., Mlýnková, I.: When conceptual model meets grammar: A formal approach to semi-structured data modeling. In: Chen, L., Triantafillou, P., Suel, T. (eds.) WISE 2010. LNCS, vol. 6488, pp. 279–293. Springer, Heidelberg (2010)
14. Nečaský, M.: Conceptual Modeling for XML. Dissertations in Database and Information Systems Series, vol. 99. IOS Press/AKA Verlag (January 2009)
15. Nečaský, M., Mlýnková, I.: On different perspectives of xml schema evolution. In: FlexDBIST 2009: Proceedings of the 5th International Workshop on Flexible Database and Information System Technology, Linz, Austria. IEEE Computer Society, Los Alamitos (2009)
16. Bray, T., Paoli, J., Sperberg-McQueen, C.M., Maler, E., Yergeau, F.: Extensible Markup Language (XML) 1.0 (5th edn.) W3C (November 2008), http://www.w3.org/TR/REC-xml/
17. Thalheim, B.: Entity-Relationship Modeling: Foundations of Database Technology. Springer, Berlin (2000)
18. Thompson, H.S., Beech, D., Maloney, M., Mendelsohn, N.: XML Schema Part 1: Structures (2nd edn.) W3C (October 2004), http://www.w3.org/TR/xmlschema-1/
19. Tim Bray, C.M.S.-M., Paoli, J.: Document type declaration (2000)
20. W3C. Document Object Model (DOM) specification, http://www.w3.org/DOM/
21. W3C. XQuery Update Facility 1.0 specification, http://www.w3.org/TR/xquery-update-10/

Partial Repairs That Tolerate Inconsistency

Hendrik Decker[*]

Instituto Tecnológico de Informática, Valencia, Spain

Abstract. The consistency of databases can be supported by enforcing integrity constraints on the stored data. Constraints that are violated should be repaired by eliminating the causes of the violations. Traditionally, repairs are conceived to be total. However, it may be unfeasible to eliminate all violations. We show that it is possible to get by with partial repairs that tolerate extant inconsistencies. They may not eliminate all causes of integrity violations but preserve the consistent parts of the database. Remaining violations can be controlled by measuring inconsistency, and further reduced by inconsistency-tolerant integrity checking.

1 Introduction

The semantic consistency of stored data can be modeled by integrity constraint conditions to be enforced on the database.

Integrity can be enforced in essentially two ways, that complement each other. One is to check each update for preserving the conditions imposed by the constraints. Another is to eliminate the causes of extant inconsistencies that have manifested themselves as violations of the integrity constraints.

In spite of a variety of possible preventive measures, such as integrity checking or careful transaction modeling, the accumulation of integrity violations in databases is commonplace and eventually unavoidable. That may be due to negligence (e.g., integrity checking had been switched off for database uploading a backup and not switched on again afterwards), efficiency considerations (e.g., integrity is not taken care of sufficiently in real-time applications), architectural impediments (e.g., the lack of integrity support in distributed databases) or other circumstances (e.g., altered constraints are not checked against legacy data, or locally consistent data fail to comply with global constraints after federation). Thus, the accumulation of inconsistency in databases is inevitable and commonplace. Hence, some approach to cope with extant inconsistencies is needed.

A radical way to deal with inconsistency is to eliminate it. That is the objective of database repairing. Traditionally, repairs are conceived to be *total*, i.e., all causes of violations are supposed to be eliminated, so that none of them persists and no new violation is caused. Although several solutions for total repairs have been proposed in the literature, for limited classes of constraints, the general case of eliminating all integrity violations is intractable in theory and may be infeasible in practice [15,6,3], even for propositional integrity constraints [4].

---

[*] Partially supported by FEDER and the Spanish grants TIN2009-14460-C03 and TIN2010-17139.

J. Eder, M. Bielikova, and A.M. Tjoa (Eds.): ADBIS 2011, LNCS 6909, pp. 389–400, 2011.

In this paper, we relax the ambitious radicality of a total elimination of inconsistency. That relaxation is achieved by *partial* repairs that are *inconsistency-tolerant*. Such repairs procure, for a database D and an integrity theory IC, the elimination of some subset S of all constraint violations while tolerating the persistence of violations in the complement of S, for an indefinite amount of time. Remaining violations can be controlled by inconsistency-tolerant integrity checking. The latter may be based on some violation metric, and may also serve to further reduce inconsistency.

The main problem with partial repairs is that they may have the unwanted side effect to induce new violations of constraints that are not in the repaired subset S. Our solution to this problem is the concept of *inconsistency-tolerant* repairs that are also *integrity-preserving*. These are partial repairs which ensure that the total amount of integrity violations before the repair is not increased after the repair. In other words, integrity-preserving repairs are updates which assure that, over time, inconsistency decreases, while, at any time, any amount of integrity violations can be tolerated.

As we are going to see, an easy way to decide if a repair is integrity-preserving is to examine the repairing update by a method for integrity checking that is inconsistency-tolerant [11]. In [11], we have shown that most, though not all known integrity checking methods are inconsistency-tolerant. Without incurring any additional cost, such methods can simply waive the usual but gratuitous requirement that all constraints must be satisfied before an update can be checked efficiently for consistency preservation.

In Section 2, we sketch the formal framework for the remainder. In Section 3, we define repairs in general, and partial as well as integrity-preserving repairs in particular. Both partial and integrity-preserving repairs tolerate inconsistency, but only the latter guarantee the preservation of consistency. The theme of Section 4 is integrity-preserving repair management. Its goal is to show how to compute partial integrity-preserving repairs, in 4.3. For that purpose, we recapitulate inconsistency-tolerant integrity checking in 2.3, and integrity-preserving updating in 4.2. In Section 5, related work is addressed. In Section 6, we conclude with an outlook to further work.

2   The Formal Framework

In subsection 2.1, we first outline some elementary preliminaries. Then, we recapitulate the notion of 'cases', i.e., instances of constraints that are going to be of use for three objectives: simplified integrity checking, quantifying constraint violations and tolerating inconsistency. In subsection 2.2, we axiomatize violation metrics for measuring inconsistency. Based on violation metrics, we characterize inconsistency-tolerant integrity checking in subsection 2.3. Unless specified otherwise, we use notations and terminology that are common for datalog [24] and first-order predicate logic.

2.1   Databases, Updates, Constraints, Cases

A *database clause* is a universally closed formula of the form $A \leftarrow B$, where the *head* A is an atom and the *body* B is a possibly empty conjunction of literals. A *database* is a finite set of database clauses.

By overloading, we use $=$ as the identity predicate, as assignment in substitutions, or meta-level equality; $\neq$ is the respective negation. The symbol $\Rightarrow$ denotes meta-implication.

An *update* is a finite set of database clauses to be inserted or deleted. For an update U of a database state D, we denote the database in which all inserts in U are added to D and all deletes in U are removed from D, by D^U.

An *integrity constraint* (in short, *constraint*) is a first-order predicate logic sentence which, w.l.o.g, we assume to be always represented by a *denial clause*, i.e., a universally closed formula of the form $\leftarrow B$, where the body B is a conjunction of literals that asserts what should not hold in any state of the database. An *integrity theory* is a finite set of constraints.

As usual, we assume that each variable is *range-restricted* (i.e., for each clause F, each variable in F occurs in a positive literal in the body of F), and has a finite universal domain, the elements of which we represent by natural numbers.

From now on, let D, IC, I, U and adornments thereof always stand for a database, an integrity theory, a constraint and, resp., an update.

For each sentence F, we write $D(F) = true$ (resp., $D(F) = false$) if F evaluates to *true* (resp., *false*) in D. Similarly, we write $D(I) = true$ (resp., $D(I) = false$) if I is satisfied (resp., violated) in D, and $D(IC) = true$ (resp., $D(IC) = false$) if all constraints in IC are satisfied in D (resp., at least one constraint in IC is violated in D). A *case* of I is an instance of I obtained by substituting the variables in I with terms of the underlying language.

Let $\mathsf{Cas}(IC)$ denote the set of all cases of all $I \in IC$. Further, let $\mathsf{SatCas}(D, IC)$, resp., $\mathsf{VioCas}(D, IC)$ be the set of all $C \in \mathsf{Cas}(IC)$ such that $D(C) = true$, resp., *false*. The use of cases for simplified integrity checking is illustrated in Example 1.

Example 1. A constraint which requires that each person's *ID* be unique, by asserting that no two persons with the same identifier x may have different names y_1, y_2, nor different birth dates z_1, z_2, is represented by the denial $I = \leftarrow p(x, y_1, z_1) \wedge p(x, y_2, z_2) \wedge (y_1 \neq y_2 \vee z_1 \neq z_2)$. For the insertion of a new person, e.g., $p(999, joe, 1\text{-}1\text{-}11)$, methods for simplified integrity checking do not evaluate I in its full generality, but just the relevant (non-ground) case $\leftarrow p(999, joe, 1\text{-}1\text{-}11) \wedge p(999, y_2, z_2) \wedge (y_1 \neq y_2 \vee z_1 \neq z_2)$.

2.2   Violation Metrics

Let $\preccurlyeq$ symbolize an ordering that is antisymmetric, reflexive and transitive. For expressions E, E', let $E \prec E'$ denote that $E \preccurlyeq E'$ and $E \neq E'$. Further, for two elements A, B in a lattice, let $A \oplus B$ denote their least upper bound.

Definition 1. We say that $(\mu, \preccurlyeq)$ is a *violation metric* (in short, a *metric*) if μ maps pairs (D, IC) to some lattice that is partially ordered by $\preccurlyeq$, and, for each pair (D, IC) and each pair (D', IC'), the following properties (1) – (4) hold.

$$\text{If}\quad D(IC) = true \text{ and } D'(IC') = false \text{ then } \mu(D, IC) \prec \mu(D', IC') \quad (1)$$

$$\text{If}\quad D(IC) = true \text{ then } \mu(D, IC) \preccurlyeq \mu(D', IC') \quad (2)$$

$$\mu(D, IC \cup IC') \preccurlyeq \mu(D, IC) \oplus \mu(D, IC') \quad (3)$$

$$\mu(D, IC) \preccurlyeq \mu(D, IC \cup IC') \quad (4)$$

Property (1), called *violation is bad* in [10], ensures that the measured amount of inconsistency in any pair (D, IC) for which integrity is satisfied is always smaller than what is measured for any pair (D', IC') for which integrity is violated. Property (2), which could be called *satisfaction is best*, ensures that inconsistency is lowest, and hence integrity is always highest, in any database that totally satisfies its integrity theory. Property (3) is a triangle inequality which states that the inconsistency of a composed element (i.e., the union of (D, IC) and (D, IC')) is never greater than the least upper bound of the inconsistency of the components. Property (4) requires that the values of μ grow monotonically with growing integrity theories.

Occasionally, we may identify a metric $(\mu, \preccurlyeq)$ with μ, if $\preccurlyeq$ is understood.

Example 2. A simple example of a coarse, binary violation metric β is provided by the equation $\beta(D, IC) = D(IC)$, with the natural ordering $true \prec false$ of the range of β, i.e., integrity satisfaction $(D(IC) = true)$ means lower inconsistency than integrity violation $(D(IC) = false)$.

Other examples of violation metrics are given by $(\mathsf{VioCas}, \subseteq)$ and by the cardinalities of sets $\mathsf{VioCas}(D, IC)$, with $\preccurlyeq \, = \, \leq$. Similar to cases, another kind of metrics can be defined by sets of causes or by counting causes; roughly, causes are minimal extracts of the database that explain why a constraint is violated [9]. Other violation metrics are discussed in [10].

Violation metrics can be used to decide if an update preserves integrity, i.e., does not cause any integrity violation that did not exist before the update, according to the following definition. Intuitively, an update preserves integrity if it does not increase the measured violation.

Definition 2. For a metric $(\mu, \preccurlyeq)$, an update U of a database D with integrity theory IC is said to be *integrity-preserving* (or, synonymously, to *preserve integrity*) *with regard to* $(\mu, \preccurlyeq)$ if $\mu(D^U, IC) \preccurlyeq \mu(D, IC)$.

2.3   Inconsistency-Tolerant Integrity Checking

Due to the possibly complex quantification of constraints, integrity checking tends to be prohibitively expensive, unless some simplification method is used [7]. Simplification theory traditionally requires that, for each update U, the state to be updated by U must satisfy all constraints. However, that requirement is unnecessary for inconsistency-tolerant integrity checking, as shown in [11].

The definition below significantly generalizes the one in [11], since the latter is based on sets of cases, which is a special instance of a metric, as seen in 2.2. Inconsistency-tolerant integrity checking methods (in short, methods) are abstractly defined by their i/o behaviour. Each method $\mathcal{M}$ maps triples (D, IC, U) to $\{ok, ko\}$. Intuitively, ok means that U does not increase the amount of inconsistency given by some metric, and ko that it may.

Definition 3. (*Inconsistency-tolerant Integrity Checking*)
An *integrity checking method* maps triples (D, IC, U) to $\{ok, ko\}$. For a metric $(\mu, \preccurlyeq)$, a method $\mathcal{M}$ is called *sound (complete) for μ-based inconsistency-tolerant integrity checking* if, for each (D, IC, U), (1) (resp., (2)) holds.

$$\mathcal{M}(D, IC, U) = ok \ \Rightarrow \ \mu(D^U, IC) \preccurlyeq \mu(D, IC) \tag{5}$$

$$\mu(D^U, IC) \preccurlyeq \mu(D, IC) \ \Rightarrow \ \mathcal{M}(D, IC, U) = ok \tag{6}$$

Intuitively, (5) says: a method is sound if, whenever it outputs ok, the amount of violation of IC in D as measured by μ is not increased by the update U. Conversely, (6) says: a method is complete if it outputs ok whenever U does not increase the amount of integrity violation.

Essentially, the only difference between conventional integrity checking and inconsistency-tolerant checking is that the former additionally requires total integrity before the update, i.e., that $D(IC) = true$ in the premise of Definition 3. The metric μ used for traditional integrity checking has a binary range: $\mu(D, U) = true$ means that IC is satisfied in D, and $\mu(D, U) = false$ that it is violated.

Example 3. Let I and $U = insert\ p(999, joe, 1\text{-}1\text{-}11)$ be as in Example 1. Then, VioCas-based inconsistency-tolerant integrity checking methods evaluate the case $C = \leftarrow p(999, joe, 1\text{-}1\text{-}11) \wedge p(999, y_2, z_2) \wedge (y_1 \neq y_2 \vee z_1 \neq z_2)$ in order to check U for integrity preservation. If $D^U(C) = true$, there is no other person with ID 999. Thus, U does not increase the set of violated cases of I, no matter if any such violations exist or not. Hence, that check is inconsistency-tolerant, while guaranteeing that all consistent parts of the database remain consistent.

3 Repairs

Roughly, repairing is to compute and execute updates to databases in order to eliminate extant integrity violations. As already mentioned, repairing can be intractably costly. Thus, it should be a reasonable heuristic to curtail inconsistency by not repairing *all*, but only *some* violations, particularly in large databases with hidden or unknown inconsistencies.

The definition below distinguishes between total repairs, which eliminate all inconsistencies, and partial repairs, which repair only a fragment of the database. Partial repairs tolerate inconsistency, since violated constraints in the complement of the repaired set may persist.

Definition 4. (*Repair*) [11]
Let D be a database, IC an integrity theory and S a subset of $\mathsf{Cas}(IC)$ such that $D(S) = false$. An update U is called a *repair* of S in D if $D^U(S) = true$. If $D^U(IC) = false$, U is also called a *partial repair* of IC in D. Otherwise, if $D^U(IC) = true$, U is called a *total repair* of IC in D.

In the literature, repairs usually are required to be total and minimal. Mostly, subset-minimality is opted for, but several other notions of minimality exist [6] or can be imagined. Note that Definition 4 does not involve any particular variant of minimality. However, Example 4 features subset-minimal repairs.

Example 4. Let $D = \{p(a,b,c), p(b,b,c), p(c,b,c), q(a,c), q(c,b), q(c,c)\}$ and $IC = \{\leftarrow p(x,y,z) \wedge \sim q(x,z), \leftarrow q(x,x)\}$. Clearly, the violated cases of IC in D are $\leftarrow p(b,b,c) \wedge \sim q(b,c)$ and $\leftarrow q(c,c)$. There are exactly two minimal total repairs of IC in D, viz. $\{delete\,q(c,c), \ delete\,p(b,b,c), \ delete\,p(c,b,c)\}$ and $\{delete\,q(c,c), \ insert\,q(b,c), \ delete\,p(c,b,c)\}$. Each of $U_1 = \{delete\,p(b,b,c)\}$ and $U_2 = \{insert\,q(b,c)\}$ is a minimal repair of $\{\leftarrow p(b,b,c) \wedge \sim q(b,c)\}$ in D and a partial repair of IC in D. Both tolerate the persistence of the violation of $\leftarrow q(c,c)$. Similarly, $U_3 = \{delete\,q(c,c)\}$ is a minimal repair of $\{\leftarrow q(c,c)\}$ in D and a partial repair of IC, which tolerates the violation of $\leftarrow p(b,b,c) \wedge \sim q(b,c)$.

A significant problem with partial repairs is that they may not preserve integrity, i.e., they may cause the violation of some constraint that is not in the repaired set, as shown by the following example.

Example 5. Consider again D and IC in Example 4. As opposed to U_1 and U_2, U_3 causes the violation of a case in the updated state that is satisfied before the update. That case is $\leftarrow p(c,b,c) \wedge \sim q(c,c)$; it is satisfied in D but not in D^{U_3}. Thus, the non-minimal partial repair $U_4 = \{delete\,q(c,c); \ delete\,p(c,b,c)\}$ is needed to eliminate the violation of $\leftarrow q(c,c)$ in D without causing a violation that did not exist before the partial repair. Indeed, all cases in $\mathsf{SatCas}(D, IC)$ remain satisfied in D^{U_4}.

The enlargement of U_3 to U_4, i.e., deleting also $p(c,b,c)$, fortunately does not induce any similar side effect as produced by deleting $q(c,c)$ alone. In general, iterations such as the one from U_3 to U_4 may possibly continue indefinitely, due to iterative side effects. The termination of such iterations is unpredictable, in general, as is known from repairing by triggers [5]. However, such iterations can be avoided by checking if a given repair is an integrity-preserving update, according to Definition 2.

Example 6. As seen in Example 5, both U_1 and U_2, and also U_4, preserve integrity since all cases in $\mathsf{SatCas}(D, IC)$ remain satisfied in the updated state. According to Definitions 2 and 4, U_4 is a minimal integrity-preserving repair of $\{\leftarrow q(x,x)\}$, although U_4 is not a mere minimal repair of $\{\leftarrow q(x,x)\}$, since the minimal repair U_3 of $\{\leftarrow q(x,x)\}$ is a proper subset of U_4. However, U_4 is preferable to U_3 since U_4 preserves integrity, while U_3 does not, as seen in Example 5. In general, each total repair (e.g., the two total repairs in Example 4) trivially preserves integrity, since no violations remain after total repairs.

4   Integrity-Preserving Repair Management

In Section 3, we have distinguished desirable (partial) and preferable (integrity-preserving) repairs. However, all we have so far are definitions and examples, while a method to compute such repairs is still missing. The goal of this section is to close that void.

Fortunately, the main building blocks of the technology to compute partial and inconsistency-preserving repairs already exist. They recur on inconsistency-tolerant integrity checking (from now on, in short, *ITIC*), as outlined in 2.3, and methods for computing integrity-preserving updates for satisfying given update requests, as discussed in [11], for the special case of $(\mu, \preceq) = (\mathsf{VioCas}, \subseteq)$. In 4.1, we show how to check if repairs are integrity-preserving or not. In 4.2, we then recapitulate update computation. In 4.3, we finally show how update computation plus ITIC can compute partial and integrity-preserving repairs.

4.1   Checking Repairs for Integrity Preservation

Clearly, each integrity-preserving update, hence each integrity-preserving repair, is inconsistency-tolerant, in the sense that there may be arbitrarily many constraint violations in D that persist in D^U. Thus, the following result is an immediate consequence of Definitions 2 and 3.

Theorem 1. For each triple (D, IC, U) and each inconsistency-tolerant integrity checking method $\mathcal{M}$, U is integrity-preserving if $\mathcal{M}(D, IC, U) = ok$.

In general, the only-if version of Theorem 1 does not hold. A counter-example is provided by each method that is incomplete for inconsistency-tolerant integrity checking in the sense of Definition 3 (e.g., those in [22,25] have been shown to be incomplete in [11]). However, it is easy to see that it does hold for methods that are complete for inconsistency-tolerant integrity checking e.g., the well-known method in [23] is complete with regard to the metric $(\mathsf{VioCas}, \subseteq)$, as shown in [11]).

Thus, theorem 1 is important for the following reason: For each partial repair U, each inconsistency-tolerant integrity checking method can be used to check if U is integrity-preserving, and each complete inconsistency-tolerant method is a procedure for deciding if U is integrity-preserving or not.

4.2   Integrity-Preserving Update Methods

We are going to define update methods as algorithms that take as input an update request and compute candidate updates as their output.

Definition 5
a) An *update request* in a database D is a first-order sentence R that is to be made *true* by some integrity-preserving update U, i.e., $D^U(R) = true$ is requested to hold.

b) An update U is said to *satisfy* an update request R if $D^U(R) = true$ and U preserves integrity. Clearly, view update requests are a well-known special kind of update requests.

c) An *update method* is an algorithm that, for each database D and each update request R, computes candidate updates $U_1, \ldots, U_n$ ($n \geq 0$) such that $D^{U_i}(R) = true$ ($1 \leq i \leq n$).

A well-known special case of update requests are view update requests. Essentially, a view update request is expressed by a literal whose predicate is not a base relation but a database view predicate. It is to be satisfied, i.e., to be made *true*, by an update of the base relations by which the view predicate is defined. Thus, the class of methods for computing view update requests is a special case of update methods.

Note that, according to Definition 5c, an update method is impartial with regard to any integrity violation that may be caused by any of the U_i. As opposed to that, Definition 6, below, is going to take such undesirable side effects into account.

To avoid that updates cause new integrity violations, many of the known update methods in the literature (e.g., [8,16,19]) postulate the total satisfaction of all constraints in the state before the update, in analogy to the total integrity premise of traditional integrity checking, as mentioned in 2.3. However, that requirement is as superfluous for satisfying update requests as for integrity checking, for the class of update methods defined next.

Definition 6. (*Integrity-preserving Update Method*)
An update method $\mathcal{UM}$ is *integrity-preserving* if each update computed by $\mathcal{UM}$ preserves integrity.

For an update request R and a database D, several update methods in the literature work by two separate phases. First, a candidate update U such that $D^U(R) = true$ is computed. Then, U is checked for integrity preservation by some integrity checking method. If that check is positive, U is accepted. Else, U is rejected and another candidate update, if any, is computed and checked. Hence, Theorem 2, below, follows from the definitions above.

Theorem 2. Each update method that uses an inconsistency-tolerant method to check its computed candidate updates is integrity-preserving.

Theorem 2 serves to identify several known update methods as integrity-preserving, since they use inconsistency-tolerant integrity checking. Among them are the update methods described in [8] and [16,17]. Several other known update methods are abductive e.g., [19,20,12]. They interleave the two phases as addressed above. Most of them are also integrity-preserving, as has been shown in [11] for the method in [19].

The following example illustrates the usefulness of integrity-preserving update methods, by featuring what can go wrong if an update method that is not integrity-preserving is used.

Example 7

Let $D = \{q(x) \leftarrow r(x) \wedge s(x); \; p(a, a)\}$, $IC = \{\leftarrow p(x, x); \; \leftarrow p(a, y) \wedge q(y)\}$ and R the update request to make $q(a)$ *true*. To satisfy R, most update methods compute the candidate update $U = \{insert\ r(a); \; insert\ s(a)\}$. To check if U preserves integrity, most methods compute the simplification $\leftarrow p(a, a)$ of the second constraint in IC. Rather than accessing the p relation for evaluating $\leftarrow p(a, a)$, integrity checking methods that are not inconsistency-tolerant (e.g., those in [18,21]) may be mislead to use the invalid premise that $D(IC) = true$, by reasoning as follows.

The constraint $\leftarrow p(x, x)$ in IC is not affected by U and subsumes $\leftarrow p(a, a)$; hence, both constraints remain satisfied in D^U. Thus, such methods wrongly conclude that U preserves integrity, since the case $\leftarrow p(a, y) \wedge q(y)$ is satisfied in D but violated in D^U. By contrast, each inconsistency-tolerant method rejects U and computes the update $U' = U \cup \{delete\ p(a, a)\}$ for satisfying R. Clearly, U' preserves integrity. Note that, incidentally, U' even removes the violated case $\leftarrow p(a, a)$.

In fact, the reduction of the amount of inconsistency in Example 7 is not entirely coincidental. In general, as long as inconsistency-tolerant integrity checking is applied for each update, the number of violated cases is not only prevented from increasing, but also is likely to decrease over time, since each update, be it accidentally or on purpose, may repair some or all inconsistencies.

4.3 How to Compute Integrity-Preserving Repairs

The following example illustrates a general approach of how partial repairs can be computed by update methods off the shelve.

Example 8. Let $S = \{\leftarrow B_1, \ldots, \leftarrow B_n\}$ $(n \geq 0)$ be a set of cases of constraints in an integrity theory IC of a database D. Thus, $D(S) = false$ if and only if $D(B_i) = true$ for some i. Further, suppose that there is a case in $IC \setminus S$ that is violated in D. Hence, a partial repair can be computed by each update method, simply by issuing the update request $\sim vio_S$, where vio_S be defined by the clauses $vio_S \leftarrow B_i$ $(1 \leq i \leq n)$.

Now we recall from Section 3 that partial repairs may not preserve integrity. That problem is solved by the following consequence of Theorems 1 and 2. It says that the integrity preservation of partial repairs can be checked by inconsistency-tolerant integrity checking (part a), and that integrity-preserving repairs can be computed by integrity-preserving update methods (part b).

Theorem 3

a) For each tuple (D, IC), each partial repair U of IC in D and each inconsistency-tolerant method $\mathcal{M}$ such that $\mathcal{M}(D, IC, U) = ok$, U is integrity-preserving.

b) Each partial repair computed as in Example 8 with an integrity-preserving update method is integrity-preserving.

5   Related Work

Traditionally, concepts of repair in the literature (e.g., in [2,15,13] only deal with total repairs. To the best of the author's knowledge, partial repairs have never been addressed elsewhere, except in [11]. In [14], null values and a 3-valued semantics are used to "summarize" total repairs. Since integrity preservation is a trivial issue for total repairs, there is also no notion of integrity-preserving updates or repairs in the literature.

Total repairs can be exceedingly costly, and so can partial repairs, in general. However, by comparison, partial repairs are more feasible than total repairs, simply because the violations of some integrity constraints may be hidden, unknown or not resolvable, while the repair of the violation of others may be fairly straightforward. Moreover, the application of our definitions and results is not compromised by any limitation with regard to the syntax of integrity constraints, while severe syntactical restrictions are typical in the literature on repairs.

A broadly discussed issue in the literature about repairs is repair checking, i.e., algorithms for deciding if a given update is a repair or not. Analogous to similar definitions in [6,1], the problem of *integrity-preserving partial repair checking* can be defined as the check if a given update is an integrity-preserving repair. Thus, Theorem 3a entails that each inconsistency-tolerant integrity checking method is an implementation of inconsistency-tolerant repair checking.

Probably the most widely discussed topic related to repairs is consistent query answering (CQA) [2]. It defines an answer to be consistent in (D, IC) if it is *true* in each minimal repair of IC in D. CQA suffers from its dependence on the chosen notion of minimality, of which our definitions are steered clear. Moreover, CQA usually is not computed by computing each repair, but by techniques of semantic query optimization or disjunctive logic programming. It should be interesting to devise a new way of computing CQA by computing partial instead of total repairs, since, in general, not all violated constraints are relevant with regard to the given query.

6   Conclusion

The evolution of a database typically involves fallacious updates and other events that may compromise the integrity of the stored data, e.g., during down- and uploads, migrations, changes in the schema, system failures, etc. In particular, it is hard to avoid that some violations of integrity constraints occur and persist. Thus, the need for a systematic control of the integrity of the stored data arises. One way to meet that challenge is to eliminate extant violations of integrity constraints. Since a total elimination of all inconsistencies is intractable, in general, the need to tolerate inconsistency imposes itself as well.

In this paper, we have presented an approach to reconcile the conflict between eliminating and living with integrity violations in databases. It consists in possibly partial repairs, instead of total repairs. Partial repairs are inconsistency-tolerant, in the sense that only some but not all causes of integrity violations are eliminated, while violations of constraints not included in the repaired subset may remain inconsistent. With regard to inconsistency tolerance, partial repairs are not only useful, but there may even be no better choice, since some integrity violations may be hidden or unknown. As illustrated by a paradigmatic example, partial repairs can be computed by any method for view updating.

A severe problem with partial repairs is that they may have the unpleasant side effect of increasing the amount of inconsistency in the fragment of the database that is not repaired. In order to avoid that problem, inconsistency-tolerant updates that are integrity-preserving need to be filtered out of the set of candidate partial repairs. To do that, the updates associated to partial repairs should be checked for integrity preservation.

Traditionally, integrity checking methods had been believed to be not applicable for checking updates for inconsistency-tolerant integrity preservation. They all have insisted on the requirement of total consistency, which cannot be complied with whenever repairs are partial. Fortunately, however, many known integrity checking methods could be shown to be inconsistency-tolerant [11], and hence applicable to check partial repairs for integrity preservation.

Future work is concerned with replacing the notion of cases by a similar but more basic notion of causes, for explaining the reasons for integrity violations. Causes provide a uniform basis for an alternative concept of inconsistency tolerance and, at the same time, of 'answers that have integrity' (AHI) [9]. The latter is not provided by case-based ITIC. Based on causes, AHI is, by intents and purposes, similar to CQA, and, as argued in [9], compares favorably to CQA.

Replacing repairs of violated cases by repairs of the actual causes of integrity violation is going to be elaborated in a follow-up version of this paper.

References

1. Afrati, F., Kolaitis, P.: Repair checking in inconsistent databases: algorithms and complexity. In: 12th ICDT, pp. 31–41. ACM Press, New York (2009)
2. Arenas, M., Bertossi, L., Chomicki, J.: Consistent query answers in inconsistent databases. In: PODS 1999, pp. 68–79. ACM Press, New York (1999)
3. Caroprese, L., Greco, S., Zumpano, E.: Active Integrity Constraints for Database Consistency Maintenance. IEEE TKDE 21(7), 1042–1057 (2009)
4. Caroprese, L., Truszczynski, M.: Active Integrity Constraints and Revision Programming. To appear in TPLP (2010), http://arxiv.org/abs/1009.2270
5. Ceri, S., Cochrane, R., Widom, J.: Practical Applications of Triggers and Constraints: Success and Lingering Issues (10-Year Award). In: Proc. 26th VLDB, pp. 254–262. Morgan Kaufmann, San Francisco (2000)
6. Chomicki, J.: Consistent query answering: Five easy pieces. In: Schwentick, T., Suciu, D. (eds.) ICDT 2007. LNCS, vol. 4353, pp. 1–17. Springer, Heidelberg (2006)
7. Christiansen, H., Martinenghi, D.: On simplification of database integrity constraints. Fundamenta Informaticae 71(4), 371–417 (2006)
8. Decker, H.: Drawing Updates From Derivations. In: Kanellakis, P.C., Abiteboul, S. (eds.) ICDT 1990. LNCS, vol. 470, pp. 437–451. Springer, Heidelberg (1990)

9. Decker, H.: Toward a Uniform Cause-based Approach to Inconsistency-tolerant Database Semantics. In: Meersman, R., Dillon, T., Herrero, P. (eds.) OTM 2010. LNCS, vol. 6427, pp. 983–998. Springer, Heidelberg (2010)

10. Decker, H., Martinenghi, D.: Modeling, Measuring and Monitoring the Quality of Information. In: Heuser, C.A., Pernul, G. (eds.) ER 2009. LNCS, vol. 5833, pp. 212–221. Springer, Heidelberg (2009)

11. Decker, H., Martinenghi, D.: Inconsistency-tolerant Integrity Checking. IEEE TKDE 23(2), 218–234 (2011)

12. Dung, P.M., Kowalski, R., Toni, F.: Dialectic proof procedures for assumption-based admissible argumentation. Artificial Intelligence 170(2), 114–159 (2006)

13. Eiter, T., Fink, M., Greco, G., Lembo, D.: Repair localization for query answering from inconsistent databases. ACM TODS 33(2) (2008)

14. Furfaro, F., Greco, S., Molinaro, C.: A three-valued semantics for querying and repairing inconsistent databases. Ann. Math. Artif. Intell. 51(2-4), 167–193 (2007)

15. Greco, G., Greco, S., Zumpano, E.: A logical framework for querying and repairing inconsistent databases. IEEE TKDE 15(6), 1389–1408 (2003)

16. Guessoum, A., Lloyd, J.: Updating knowledge bases. New Generation Computing 8(1), 71–89 (1990)

17. Guessoum, A., Lloyd, J.: Updating knowledge bases II. New Generation Computing 10(1), 73–100 (1991)

18. Gupta, A., Sagiv, Y., Ullman, J., Widom, J.: Constraint checking with partial information. In: Proc. PODS 1994, pp. 45–55. ACM Press, New York (1994)

19. Kakas, A., Mancarella, P.: Database updates through abduction. In: Proc. 16th VLDB, pp. 650–661. Morgan Kaufmann, San Francisco (1990)

20. Kakas, A., Kowalski, R., Toni, F.: The role of Abduction in Logic Programming. In: Handbook in Artificial Intelligence and Logic Programming, pp. 235–324 (1998)

21. Lee, S.Y., Ling, T.W.: Further improvements on integrity constraint checking for stratifiable deductive databases. In: Proc. VLDB 1996, pp. 495–505. Morgan Kaufmann, San Francisco (1996)

22. Lloyd, J., Sonenberg, L., Topor, R.: Integrity constraint checking in stratified databases. J. Logic Programming 4(4), 331–343 (1987)

23. Nicolas, J.M.: Logic for improving integrity checking in relational data bases. Acta Informatica 18, 227–253 (1982)

24. Ramakrishnan, R., Gehrke, J.: Database Management Systems. McGraw-Hill, New York (2003)

25. Sadri, F., Kowalski, R.: A theorem-proving approach to database integrity. In: Foundations of Deductive Databases and Logic Programming, pp. 313–362. Morgan Kaufmann, San Francisco (1988)

Modularisation in Maude of Parametrized RBAC for Row Level Access Control

Ścibor Sobieski and Bartosz Zieliński

Department of Theoretical Physics and Informatics
Faculty of Physics and Applied Informatics
University of Łódź
ul. Pomorska nr 149/153, 90-236 Łódź, Poland
{scibor,bzielinski}@uni.lodz.pl

Abstract. We formalize a Parametrized Role-Based Access Control in the language Maude. We demonstrate how this formalization can be used to specify a row level access control policy in a database and how module algebra capabilities of Maude assist in modularization of such specification.

1   Introduction

Ever growing demands on sensitive data and privacy protection made crude table level access control, now standard in database management systems, insufficient for many applications. This has led to the creation of row level access control frameworks, such as SE-PostgreSQL, Oracle Virtual Private Database [1], and some in-house solutions (see e.g. [18]).

Unfortunately, fine grained access control increases greatly the burden of access rights administration. Roles [22], [11], [21] are now the standard administrative help directly supported by majority of databases. For row level access control, however, simple RBAC is not sufficient because of proliferation of roles which differ only by a partition of data they are associated with. For instance, there may be separate accounting role for each division of an enterprise. Even more demanding would be to allow each of the employees to have access to their own personal data and not to the data of other people, unless specifically authorised. Creating a separate role for each user is certainly not what RBAC is about!

A natural solution is to add attributes to roles. Parametrized RBAC [2], [15], [12] introduces role templates (constructors), the arguments of which define some partition of data. Instantiation of such templates with actual parameters creates the usual roles. For example, consider the role template

```
accountant(d : department)
```

with single argument of sort `department`. Instantiation `accountant("Phys")` creates the role of (presumably) an accountant in the Department of Physics at the university. The advantage of this approach is that one can use quantification

J. Eder, M. Bielikova, and A.M. Tjoa (Eds.): ADBIS 2011, LNCS 6909, pp. 401–414, 2011.

with respect to role parameters when defining the permissions associated with roles, e.g.:

> For all departments `d`, allow `accountant(d)` to select financial data of the employees of the department `d`.

Even with the help of role parametrization, specifying roles for large system containing hundreds of tables and numerous ways of partitioning data within the tables can be very daunting without some support for modularisation and creating abstractions, just like in the case of writing large procedural programs. In particular it should be possible to first define partitioning of data into semantic categories independently of its physical division into tables and to define roles in terms of accesses to those semantic categories. Only on the lower level, those semantic categories should be associated with particular (parts of) tables.

This work examines the possibility of using Maude system [8] to specify in a modular way the parametrized RBAC policy for a row level access control system in a database. In our approach, while the specification is defined outside the database system, it is designed specifically to allow easy automatic generation of the database code implementing the policy. Our only essential assumption is that the row level access control is implemented by some variant of query modification [24], [20], and therefore our work should be aplicable to many systems (like those using Oracle VPD [1]). Our solution is, however, ultimately designed with a specific system [18] in mind, our original purpose being the creation of a high level policy description language for the HR database at the University of Łódź.

Note that the current database systems lack high level, modularized policy description languages, apart from the support for (unparametrized) roles.

A theory of access control policy modularisation was extensively studied (see e.g. [14], [10]), especially for the needs of heterogenous access control policies.

Mathematical basis of Maude system are term rewriting logic [16] and, as its subset, membership equational logic [17], [5]. Using term rewriting and algebraic methods to specify security policies is already well established in the literature (see e.g. [19], [3], [10], [6]). Also Maude itself was used in the specification of security protocols (see e.g. [9]) and the analysis of access control policies (see e.g. [23]).

We were unable to find any published work describing term rewriting formalism for specification of access control policy, which would be geared towards the needs of row level access control in relational databases and which would allow easy, automatic translation into implementation code. To the best of our knowledge, at least in the context of term rewriting and relational databases, the idea of using high level abstract predicates to describe data to which we grant access is also new.

2   Access Control by Query Modification

The basic idea of access control by query rewriting [24], [20] is to modify user query by adding appropriate `WHERE` clauses to each query referring to protected tables. For example, the query

```
SELECT * FROM Employees WHERE salary>3000;
```

sent by the user allowed to see only data of employees from a certain department might get rewritten to

```
SELECT * FROM Employees WHERE salary>3000 and department_id=10;
```

Oracle VPD does it internally, dynamically appending to each query the `VARCHAR` value returned by special user-defined functions associated with protected objects and types of access. Also updatable views can be used as a mechanism for implementing such query rewriting which does not require modifying the query compiler. For large systems, with additional requirements such as controlling access with respect to sessions, using views directly is impractical, and some additional framework for access control is needed, such as the one described in [18].

Access control systems which work by rewriting user queries are sometimes called *Truman systems* [20]. Their advantage is the simplicity of implementation. Their main disadvantage, is that while a true access control system should either accept or reject a query, Truman systems silently change the query semantics, which can be especially dangerous for aggregate queries. Therefore *non-Truman* systems were proposed [20], in which user query is either proven to be equivalent to the modified one, and then executed unchanged, or rejected otherwise. There are unfortunately no simple ways of implementing such systems. The formalism developed in this paper, however, should work equally well for Truman and non-Truman systems.

3   Equational Membership Logic and Maude

This section will give very basic overview of mathematics behind Maude and algebraic specifications. The readers should be aware that some of the material is oversimplified. We will use only functional modules and theories of Maude, which can be understood as describing an order sorted algebra with the equational membership logic used to specify the properties of algebra operators. The simplifications of algebra expressions are Maude computations, and the posible simplifications are given by equations understood as rewritings.

3.1   Order Sorted Algebras

Let $(S, \leq)$ be an ordered set, called the set of sorts (type names). *An S-sorted set A* is a collection of sets $\{A_r\}_{r \in S}$. Denote by S^* the set of all finite strings in the alphabet S, including the empty string ϵ. The order on S extends to S^* by the formula:

$$r_1 \ldots r_n \leq s_1 \ldots s_m \quad \equiv \quad n = m \wedge r_1 \leq s_1 \wedge \ldots \wedge r_n \leq s_n.$$

An order sorted signature (see e.g. [13]) (S, Σ) is an $S^* \times S$-sorted set Σ of function symbols. A function symbol $f \in \Sigma_{w,r}$ can be also written as $f : w \to r$

to avoid ambiguity as sets $\Sigma_{w,r}$ and $\Sigma_{w',r'}$ are not assumed to be disjoint for $(w,r) \neq (w',r')$ (we allow overloading of symbols). Function symbols $c : \epsilon \to r$ are constant symbols of sort r. The signature can be understood as a syntax of a program.

An order sorted (S, Σ)-algebra A (see e.g. [13]) consists of an S-sorted set $|A|$ and a collection of functions $\{|A|_{f:w\to r} : |A|_w \to |A|_r\}_{w \in S^*, r \in S, f \in \Sigma_{w,r}}$, where $|A|_{r_1 \ldots r_n} = |A|_{r_1} \times \ldots \times |A|_{r_n}$, satisfying

1. $r \leq r' \;\;\Rightarrow\;\; |A|_r \subseteq |A|_{r'}$,
2. $r \leq r' \wedge w \leq w' \wedge f \in \Sigma_{w,r} \cap \Sigma_{w',r'} \;\;\Rightarrow\;\; |A|_{f:w'\to r'}|_{|A|_w} = |A|_{f:w\to r}$.

As sorts r corespond to types $|A|_r$, the two conditions above mean that an order on sorts correspond to type hierarchy, and function overloading works well with type hierarchies. The algebra of a signature can be understood as a semantics of a program – it gives meaning to the function symbols defined by the signature.

A homomorphism h of (S, Σ)-order sorted algebras (see e.g. [13]) A and B is a collection of functions $\{h : |A|_r \to |B|_r\}_{r \in S}$ which preserve constants and such that for all function symbols $f : w \to r$ in the signature Σ we have $h_r \circ |A|_{f:w\to r} = |B|_{f:w\to r} \circ h_w$, where $h_{w_1 \ldots w_n} = h_{w_1} \times \ldots \times h_{w_n}$.

Let $\mathcal{V}$ be an S-sorted set of variables, where we might denote by $x : r$ the element of $\mathcal{V}_r$ to emphasize the sort (though we do not allow overloading for simplicty). The term algebra $T_{\Sigma,S}(\mathcal{V})$ (see e.g. [13]) is defined recursively as:

1. $x : r \in |T_{\Sigma,S}(\mathcal{V})|_{r'}$ for all $r \leq r'$.
2. $f : \epsilon \to r \in |T_{\Sigma,S}(\mathcal{V})|_{r'}$ for all $r \leq r'$.
3. If $r_1 \ldots r_n \leq w$ and $t_i \in |T_{\Sigma,S}(\mathcal{V})|_{r_i}$ for all $1 \leq i \leq n$, then $f(t_1, \ldots, t_n) \in |T_{\Sigma,S}(\mathcal{V})|_{r'}$ for all $f : w \to r$ and $r \leq r'$.

Hence the term algebra is just the set of all type-correct formulas. The algebra operations are defined in an obvious way, i.e., $|T_{\Sigma,S}(\mathcal{V})|_f(t_1, \ldots, t_n) = f(t_1, \ldots, t_n)$. Terms which are variable free are called *ground terms*. We denote the sub-algebra of ground terms by $T_{\Sigma,S}$.

When a signature satisfies the preregularity condition [13], which means that all terms in the term algebra can be assigned a unique least sort, then the ground term algebra $T_{\Sigma,S}$ is *initial*. This means that for all order sorted (S, Σ)-algebras there exists a unique algebra homomorphism $h : T_{\Sigma,S} \to A$.

3.2   Conditional Equations

Any sort preserving map $v : \mathcal{V} \to T_{\Sigma,S}(\mathcal{V})$ such that $v(x) = x$ for all but a finite number of elements of $\mathcal{V}$ is called a *variable substitution*. This map extends in an obvious way to an algebra endomorphism $\hat{v} : T_{\Sigma,S}(\mathcal{V}) \to T_{\Sigma,S}(\mathcal{V})$.

Usualy we define algebras as quotients of ground term algebra by an equivalence relation defined by a family of equations. In Equational Membership Logic [17] those equations can be conditional, and in the most general case they have the form

$$p = q \quad \text{if} \quad p_1 = q_1 \wedge \ldots \wedge p_n = q_n \wedge t_1 : r_1 \wedge \ldots \wedge t_m : r_m$$

where p, q, p_i's and q_i's are terms in $T_{\Sigma,S}(\mathcal{V})$, and the r_i's are sorts in S. Expression $t : r$ is true if and only if term t has sort r. The equations are understood to be implicitly universally quantified with respect to all posible substitutions, i.e., for all substitutions $v : \mathcal{V} \to T_{\Sigma,S}(\mathcal{V})$ such that

$$\hat{v}(p_1) = \hat{v}(q_1) \wedge \ldots \wedge \hat{v}(p_n) = \hat{v}(q_n) \wedge \hat{v}(t_1) : r_1 \wedge \ldots \wedge \hat{v}(t_m) : r_m$$

also $\hat{v}(p) = \hat{v}(q)$ holds.

Maude computes the simplifications of terms using equations as rewrite rules from left to right [7]. Hence it requires that the rewriting system defined by equations is terminating and Church-Rosser (i.e., all paths of computation lead to a unique normal form).

Some properties of operations like commutativity and associativity are difficult or impossible to express using equations interpreted as terminating and Church Rosser rewritings. Therefore Maude allows to define associativity and commutativity of algebra operations using special attributes (`assoc` and `comm`).

Functional modules in Maude have the initial semantics, i.e., equations are assumed to fully define the result of operations in algebra. Maude also allows defining functional theories which have the loose semantics, i.e., the equations are assumed to define properties of a class of algebras.

Algebras defined in modules are declared to satisfy a theory through the use of views.

4   Parametrized RBAC for Row Level Access Control

In this section we present a specification in Maude of a significant part of flat parametrized RBAC model [21] and we outline the method of generating the implementation of the policy in the database from the specification. In Role Based Access Control access permissions are assigned to roles, instead of being assigned directly to subjects. Each subject is viewed by a security system as a part of some session. Roles are assigned to sessions, and accesses are granted to sessions if they are permitted by some role belonging to a given session.

Sessions are created on behalf of the users. For each session there is a unique user on behalf of whom the session was created. For accounting reasons, one usually requires one to one correspondence between users and physical persons. Each user has a pool of roles and the session created on behalf of a given user can have only the subset of the user roles. We omit the discussion of formalization of user session relations.

Some explanations of Maude syntax accompany most of the code listings. Readers are referred to [7] for more details about the language.

4.1   Specification in Maude

First, possible access types need to be defined in order to allow different treatment of, say, selects and updates. In order to separate various aspects of the system as much as possible we do it in a separate, very simple module:

```
fmod ACCESSTYPES is
  sort AccessType .
  ops select update insert delete : -> AccessType [ctor] .
endfm
```

which defines only one sort `AccessType` and four constants of this sort. The attribute `ctor` means that the constants are constructors, and are not supposed to be simplified any further.

Next, we need to describe basic access decisions. We also need an operation (or) to combine access decisions, for example from different roles active in the same session. Here for simplicity, we assume that the access policy cannot contain negative access rights, and therefore, it is enough for the access to be granted that the session has one role for which the access decision is positive. Hence, in the module below, we introduce only two constants, `permit` and `notpermit`, with the rule that `permit` combined with anything yields `permit`:

```
fmod ACCESSDECS is
  sorts Decision .
  ops permit notpermit : -> Decision [ctor] .
  op _ or _ : Decision Decision -> Decision [assoc comm]   .
  var D : Decision .
  eq (permit or D) = permit . eq (notpermit or D) = D .
endfm
```

Note that the infix operator `or` was defined as associative and commutative (attributes `assoc` and `comm`) and that its identity is `notpermit`. Note also that the property that `permit or` anything else is `permit` and the identity property were expressed using a (universaly quantified) variable D instead of using explicit values. As a consequence, when we extend the module `ACCESSDECS` with more irreducible terms of sort `Decision`, the `permit` constant preserves its dominant property and `notpermit` is still an identity.

One of the key ideas of this paper is the introduction of "unfinished" decisions (i.e., additional, not always reducible terms of sort `Decision`), which essentially correspond to `WHERE` clauses which the database system should add to user query. In short, instead of letting Maude do the whole job, we leave some of it to the database. We cannot define such unfinished decisions in the general `ACCESSDECS` module because they must refer to the data sorts. This time we create a theory instead of a module to allow a parametrization of further layers of our access control system by a different relational schemas, or even other data models.

```
fth DATATH is
  including ACCESSDECS .
  sort DataItem .
  op [_ when _] : DataItem Bool -> Decision .
  var DI : DataItem . vars P Q : Bool .
  eq [DI when P] or [DI when Q] = [DI when P or Q] .
  eq [DI when true] = permit .
```

```
  eq [DI when false] = notpermit .
endfth
```

Here `DataItem` is the most general sort corresponding to the data objects we want to deal with in our security policy. The partial "mixfix" decision operator `[_ when _]` has two arguments – one is a data item argument, the other one is a boolean term. Usually the boolean term will be some predicate on data item. For example, the request of access to a row `t` in a relational table `EMPLOYEES`, by someone entitled only to seeing data on employees in a particular department, might return a partial decision term

```
[t when dept-id(t)==10]
```

Whether the above term gets reduced any further or not, depends on the opaqueness of the term `t`. When the term `t` is, say, a full row expression, it might allow a full evaluation of the predicate `dept-id(t)==10` into `true` or `false` constants and hence the full reduction of `[t when dept-id(t)==10]` into `permit` or `notpermit` terms. On the other hand, when the term `t` is completely opaque – it means only "some row of table `EMPLOYEES`", then the unreduced decision term `[t when dept-id(t)==10]` might be used to generate appropriate `WHERE`-clause for the query rewriting engine in the database.

Next, we create a theory for roles, which includes the theory for data items and the module for access types. These are necessary to define the signature for the operation `request`:

```
fth ROLESTH is
 including DATATH . protecting ACCESSTYPES .
 sort Role .
 op request : Role AccessType DataItem -> Decision .
endfth
```

Modules implementing the above theory of roles will have to define constructors of sort `Role` corresponding to (parametrized) roles. Each reduction rule for the `request` operator will correspond to adding some access right to the role.

Each session will have a set of roles attached. In Maude one can use a standard module `SET` parametrized by the trivial theory `TRIV` of elements. The `TRIV` theory contains only the definition of sort `Elt` of elements. In order to use sets of terms of sort `Role` we need a view from `TRIV` theory into `ROLETH` theory:

```
view RolesT from TRIV to ROLESTH is
  sort Elt to  Role .
endv
```

and later also views from `TRIV` theory into modules implementing `ROLETH` theory.

Our module specifying basic session operations will be parametrized with the `ROLETH` theory. Note that one still needs to extend the instatiation of this module with definitions of session constants and sets of roles associated with them.

We start with declaring the theory parameter `X` and importing instantiation of the `SET` module (see [7] for details):

```
fmod SESSIONS{X :: ROLESTH} is
   protecting SET{RolesT}{X}
                        * (sort Set{RolesT}{X} to RoleSet{X}) .
```

Note the module algebra operator * used to rename the sort Set{RolesT}{X}.
Next, we define the sort name Session{X}, and signatures of session operations:

```
   sort Session{X} .
   op roles : Session{X} -> RoleSet{X} .
   op request : Session{X} AccessType X$DataItem -> Decision .
   op nrsrequest : NeSet{RolesT}{X} AccessType X$DataItem
                                                    -> Decision .
```

Operations roles and request are required by the RBAC model. The function
roles returns the session roles, and the function request gives the access decision based on the roles associated with a session. The operation nrsrequest is
an auxiliary one needed for recursive implementation of the request operation.
Note that the Session{X} sort name is parametrized by the module parameter.
Different instantiations of the role theory implementations will create different
session types. This might be useful if we want to model accesses to different
databases or different parts of a database. Note also that the sort names from
the ROLESTH theory are prefixed by X$ ([7]).

Next we declare variables:

```
   var R : X$Role . var NRS : NeSet{RolesT}{X} .
   var DI : X$DataItem . var AT : AccessType .
   var  S : Session{X} .
```

needed in recursive implementation of request:

```
   ceq request(S,AT,DI) = notpermit if roles(S) == empty .
   ceq request(S,AT,DI) = nrsrequest(roles(S),AT,DI)
                             if roles(S) =/= empty .
   eq nrsrequest(R,AT,DI) = request(R,AT,DI) .
   eq nrsrequest((R,NRS),AT,DI) = request(R,AT,DI) or
                                   nrsrequest(NRS,AT,DI) .
endfm
```

which ends the module. The operation request works as follows. It returns
notpermit if the session has no roles. Otherwise it reduces to the auxiliary
operation nrsrequest called with the set of roles associated to a session as a
first argument. Then nrsrequest reduces (recursively) to the or of the request's
for roles, for each role in the set. Note that for this operation to be well defined,
it was essential to declare or as commutative and associative. The intended
semantics of requests for sessions is that the operation is allowed for a session if
and only if at least one of the session's roles allows it.

Now we can start implementing concrete data and role descriptions. We start
with the module BASICDATA, the contents of which is just copy and paste from
the contents of the theory DATATH:

```
fmod BASICDATA is
  including ACCESSDECS .
  sort DataItem .
  op [_ when _] : DataItem Bool -> Decision .
  var DI : DataItem . vars P Q : Bool .
  eq [DI when P] or [DI when Q] = [DI when P or Q] .
  eq [DI when true] = permit . eq [DI when false] = notpermit .
endfm
```

Now we can extend this module, creating the high level description of partitions of data, the use of which is another of the leading ideas of this paper. For example the following module

```
fmod DATAPREDICATES is
  including BASICDATA . protecting STRING .
  sorts Dept Person .
  op persinfo : DataItem Person -> Bool .
  op fininfo : DataItem -> Bool .
  op deptinfo : DataItem Dept -> Bool .
  op dept : String -> Dept [ctor] .
  op person : String -> Person [ctor] .
endfm
```

declares parametrized predicates persinfo, fininfo, and deptinfo the intended meaning of which is to define the division of all data into:

- Personal information of person p: the data items I for which persinfo(I,p) reduces to true.
- Financial information: the data items I for which fininfo(I) reduces to true.
- Departmental information for the department d: the data items I for which deptinfo(I,d) reduces to true.

In addition, functions dept and person allow to construct terms of sort Dept and Person, respectively, from character strings, which might represent the person and department identifiers in the database.

The next module will declare the database relations as opaque tables. These declarations will then be used to define the predicates from the DATAPREDICATES module:

```
fmod RELVARS is
  including DATAPREDICATES .
  sorts SalaryD PersonD PersNSD DeptD  .
  subsorts SalaryD PersonD PersNSD DeptD < DataItem .
  subsort PersNSD < PersonD .
  op deptd   : DeptD Dept -> Bool [ctor] .
  op salaryd : SalaryD Dept -> Bool [ctor] .
  op salaryd : SalaryD Person -> Bool [ctor] .
```

```
  op persond : PersonD Dept -> Bool [ctor] .
  op persond : PersonD Person -> Bool [ctor] .
  op salaries : -> SalaryD [ctor] .
  op people : -> PersonD [ctor] .
  op peopleNS : -> PersNSD [ctor] .
  op departments : -> DeptD [ctor] .
endfm
```

Here the subsorts `SalaryD`, `PersonD`, `DeptD` of the `DataItem` sort correspond to the (sets of) the rows of relational tables `salaries`, `people`, `departments`. Subsort `PersNSD` corresponds to the rows of the view `peopleNS` of `people` table which contains only "nonsensitive" subset of columns, such as name and family name, but not, say, passport number or national ID. Predicates `deptd`, `salaryd` and `persond` choose a part of the respective table associated to a given person or department. For example, `persond(I,d)` should reduce to `true` only for those rows `I` of the table `people` which describe employees of the department `d`.

Now we are ready to implement parametrized predicates `persinfo`, `fininfo`, and `deptinfo` from the module `DATAPREDICATES`:

```
fmod PREDSIMPL is
  including RELVARS .
  var P : PersonD . var PNS :PersNSD . var S : SalaryD .
  var DP : DeptD . var pP : Person . var pD : Dept .
  var DI : DataItem .
  eq persinfo(P,pP) = persond(P,pP) .
  eq persinfo(S,pP) = salaryd(S,pP) .
  eq persinfo(DI,pP) = false [owise] .
  eq fininfo(DP) = true .
  ceq fininfo(P) = true if (P :: PersNSD) .
  ceq fininfo(P) = false if not (P :: PersNSD) .
  eq fininfo(DI) = false [owise] .
  eq deptinfo(P,pD) = persond(P,pD) .
  eq deptinfo(DP,pD) = deptd(DP,pD) .
  eq deptinfo(S,pD) = salaryd(S,pD) .
endfm
```

Note that the personal data for a person p, as defined by the predicate `persinfo`, consists of rows from tables `people` and `salaries` (sorts `PersonD` and `SalaryD`) associated with person p and nothing else (attribute `owise` means "otherwise").

Note also the conditional equations defining the `fininfo` predicate for rows from `Personal` table. The intended meaning is that only the non-sensitive part (sort `PersNSD`) of `people` table is contained in the financial data. Here we use the `Bool` valued, built in function `_:: PersNSD`. Such a function is defined for each sort.

We are now ready to define example role constructors in terms of the just defined predicates `persinfo`, `fininfo`, and `deptinfo`. First, we declare the sort `Role` and the signature for the `request` operation and role constructors named persrole and accountant parametrized by the person and department, respectively:

```
fmod ROLES is
  including PREDSIMPL . protecting ACCESSTYPES .
  sort Role .
  op request : Role AccessType DataItem -> Decision .
  op persrole : Person -> Role [ctor] .
  op accountant : Dept -> Role [ctor] .
```

Next we declare variables needed for the request operation definition:

```
  var pP : Person . var pD : Dept .
  var DI : DataItem . var AC : AccessType .
```

Then we assign permissions to the role persrole(pP) to read (but not modify) the data associated to a person pP.

```
  eq request(persrole(pP),select,DI) = [DI when persinfo(DI,pP)] .
  eq request(persrole(pP),AC,DI) = notpermit [owise] .
```

Note that here we assign the access permissions to the roles in terms of high level predicates describing meaning of data, such as the predicate persinfo, and not in terms of tables. Similarly, we define permissions for the accountant(pD) role:

```
  eq request(accountant(pD),AC,DI)
          = [DI when fininfo(DI) and deptinfo(DI,pD)] .
  eq request(accountant(pD),AC,DI) = notpermit [owise] .
endfm
```

Here one can observe how one can combine in a natural way the high level predicates, in order to express the property, that an accountant should have access to financial data, but only in his own department.

In order to define specific sessions by extending and instantiating module SESSIONS, we first need to define view from the role theory ROLETH to the module ROLES:

```
view RolesV from ROLESTH to ROLES is endv
```

The view is trivial, as we took care to name the sorts and operations identically in the module as in the theory. Finally we define specific session named session:

```
fmod EXAMPLESESSION is
  including ROLES . including SESSIONS{RolesV} .
  op session : -> Session{RolesV} .
  eq roles(session) = (persrole(person("A")),
                       accountant(dept("B"))) .
endfm
```

The session session has two roles: persrole(person("A")) and accountant(dept("B")). With all the definitions above, the term

```
request(session, select, people)
```

should reduce to

```
[people when persond(people, person("A"))]
```

Finally we present how the actual database policy can be generated from Maude specification. We need to generate WHERE conditions which will be added to user queries (for instance by using them in the definition of security views), for each combination of the role, access type and protected relational variable. In the module MAUDETOSQL we define a function permtostr which converts terms of sort Decision to strings using the auxillary function condtostr:

```
fmod MAUDETOSQL is
  including PREDSIMPL . including ROWCONSTRUCTORS .
  op condtostr : Bool -> String .
  op permtostr : Decision -> String .
  var di : DataItem . var b : Bool .
  eq permtostr(permit) = "1=1" .
  eq permtostr(notpermit) = "1<>1" .
  eq permtostr([di when b]) = condtostr(b) .
```

Hence condtostr does most of the job and it is defined as follows:

```
  var D : DeptD . var S : SalaryD .
  var P : PersonD . var St : String .
  eq condtostr(true) = "1=1" . eq condtostr(false) = "1<>1" .
  eq condtostr(deptd(D,dept(St))) = "Item.DeptId='" + St + "'" .
  eq condtostr(salaryd(S,dept(St))) = "'" + St + "'=(
      SELECT q.DeptID FROM People q WHERE q.Id = Item.PersId" .
  eq condtostr(salaryd(S,person(St)))
                = "Item.PersId='" + St + "'" .
  eq condtostr(persond(P,dept(St)))
                = "Item.DeptId = '" + St + "'" .
  eq condtostr(persond(P,person(St))) = "Item.Id = '" + St + "'" .
endfm
```

Lines above represent actual implementation of predicates such as salaryd. While most of them are defined as simple comparison of a column with a value, in many cases the SQL implementation involves complicated joins, like in the example above, where the DeptID for comparison had to be provided by a sub-query. After loading the modules MAUDETOSQL and ROLES:

```
fmod POLICYGEN is including MAUDETOSQL . including ROLES . endfm
```

one can start generating the where clasuses. For instance the term

```
permtostr(request(persrole(person("John")),select,salaries))
```

gets reduced to the string "Item.PersId = 'John'".

5   Conclusion and Future Work

We have shown how to specify (a significant part of) a flat parametrized RBAC in Maude, in a way which is particularly well suited for row level access control systems, implemented by query rewriting methods, and which allows easy automatic code generation from the specification. We have also shown how assignment of access premissions to (parametrized) roles might be carried in terms of high level predicates defining semantic partitioning of data. And finally, we have demonstrated how sophisticated module algebra capabilities of Maude assist in structuring such a specification. This is the begining of a research project, and the study described in this paper can be extended in several directions:

- We have specified only flat RBAC. It would be interesting to add role hierarchies and separation of duty constraints [21] as well.
- A lot of research (see e.g. [14], [10]) goes into cooperation between different kinds of security policies. It would be interesting to define in our formalism cooperation between, say, parametrized RBAC and Bell-LaPadula [4] policies.
- Finally, our formalism works well with the rows coming from single tables (views). However a more general solution, would require implementing some part of relational algebra in Maude, which is an interesting project in itself.

References

1. The Virtual Private Database in Oracle9ir2. An Oracle White Paper (2002)
2. Abdallah, A., Khayat, E.: A Formal Model for Parameterized Role-Based Access Control. In: Dimitrakos, T., Martinelli, F. (eds.) Formal Aspects in Security and Trust, IFIP, vol. 173, pp. 233–246. Springer, Boston (2005)
3. Barker, S., Fernandez, M.: Term Rewriting for Access Control. In: Damiani, E., Liu, P. (eds.) DBSec 2006. LNCS, vol. 4127, pp. 179–193. Springer, Heidelberg (2006)
4. Bell, D., LaPadula, L.: Secure Computer Systems: Mathematical Foundations and Model. The MITRE Corporation Technical Report M74-244 (May 1973)
5. Bouhoula, A., Jouannaud, J.P., Meseguer, J.: Specification and Proof in Membership Equational Logic. Tech. rep., SRI International (1988)
6. Bourdier, T., Cirstea, H., Jaume, M., Kirchner, H.: On Formal Specification and Analysis of Security Policies, preprint inria-0042924
7. Clavel, M., Durán, F., Eker, S., Lincoln, P., Marti-Oliet, N., Meseguer, J., Talcott, C.: Maude Manual, Version 2.6 (2011)
8. Clavel, M., Durán, F., Eker, S., Lincoln, P., Martí-Oliet, N., Meseguer, J., Talcott, C.: The Maude 2.0 System. In: Nieuwenhuis, R. (ed.) RTA 2003. LNCS, vol. 2706, pp. 76–87. Springer, Heidelberg (2003)
9. Denker, G., Meseguer, J., Talcott, C.: Protocol Specification and Analysis in Maude. In: Workshop on Formal Methods and Security Protocols (1998)
10. Dougherty, D.J., Kirchner, C., Kirchner, H., De, A.S.: Modular Access Control via Strategic Rewriting. In: Biskup, J., López, J. (eds.) ESORICS 2007. LNCS, vol. 4734, pp. 578–593. Springer, Heidelberg (2007)

11. Ferraiolo, D., Kuhn, D., Chandramouli, R.: Role-Based Access Control. Artech House computer security series. Artech House, Boston (2003)
12. Ge, M., Osborn, S.: A Design for Parameterized Roles. In: Farkas, C., Samarati, P. (eds.) DBSec. IFIP, vol. 144, pp. 251–264, Kluver (2004)
13. Goguen, J.A., Meseguer, J.: Order-Sorted Algebra i: Equational Deduction for Multiple Inheritance, Overloading, Exceptions and Partial Operations. Theor. Comput. Sci. 105, 217–273 (1992)
14. Jajodia, S., Samarati, P., Sapino, M.L., Subrahmanian, V.S.: Flexible Support for Multiple Access Control Policies. ACM Trans. Database Syst. 26(2), 214–260 (2001)
15. Kuhn, D.R., Coyne, E.J., Weil, T.R.: Adding Attributes to Role-Based Access Control. IEEE Computer 43(6), 79–81 (2010)
16. Martí-Oliet, N., Meseguer, J.: Rewriting Logic as a Logical and Semantic Framework. In: Meseguer, J. (ed.) Electronic Notes in Theoretical Computer Science, vol. 4. Elsevier Science Publishers, Amsterdam (2000)
17. Meseguer, J.: Membership Algebra as a Logical Framework for Equational Specification. In: Parisi-Presicce, F. (ed.) WADT 1997. LNCS, vol. 1376, pp. 18–61. Springer, Heidelberg (1998)
18. Miodek, K., Pychowski, J.: Elastyczny System Uprawnień Użytkowników w Systemie Zarządzania Bazą Danych PostgreSQL. In: Bazy Danych - Modele, Technologie, Narzedzia, pp. 309–314. WKL Gliwice (2006)
19. de Oliveira, A.S.: Rewriting-Based Access Control Policies. Electr. Notes Theor. Comput. Sci. 171(4), 59–72 (2007)
20. Rizvi, S., Mendelzon, A., Sudarshan, S., Roy, P.: Extending Query Rewriting Techniques for Fine-Grained Access Control. In: Proceedings of the 2004 ACM SIGMOD International Conference on Management of Data, pp. 551–562 (2004)
21. Sandhu, R., Ferraiolo, D., Kuhn, R.: The NIST Model for Role-Based Access Control: Towards A Unified Standard. In: Proceedings of the Fifth ACM Workshop on Role-based Access Control, pp. 47–63 (2000)
22. Sandhu, R.S., Coyne, E.J., Feinstein, H.L., Youman, C.E.: Role-Based Access Control Models. IEEE Computer 29(2), 38–47 (1996)
23. Stoller, S.D., Yang, P., Gofman, M.I., Ramakrishnan, C.: Symbolic Reachability Analysis for Parameterized Administrative Role-Based Access Control. Computers & Security 30(2-3), 148–164 (2011)
24. Stonebraker, M., Wong, E.: Access Control in a Relational Database Management System by Query Modification. In: Proceedings of the 1974 Annual Conference ACM 1974, vol. 1, pp. 180–186. ACM, New York (1974)

A Clustering-Based Approach for Large-Scale Ontology Matching

Alsayed Algergawy, Sabine Massmann, and Erhard Rahm

Department of Computer Science, University of Leipzig
{algergawy,massmann,rahm}@informatik.uni-leipzig.de

Abstract. Schema and ontology matching have attracted a great deal of interest among researchers. Despite the advances achieved, the large matching problem still presents a real challenge, such as it is a time-consuming and memory-intensive process. We therefore propose a scalable, clustering-based matching approach that breaks up the large matching problem into smaller matching problems. In particular, we first introduce a structure-based clustering approach to partition each schema graph into a set of disjoint subgraphs (clusters). Then, we propose a new measure that efficiently determines similar clusters between every two sets of clusters to obtain a set of small matching tasks. Finally, we adopt the matching prototype COMA++ to solve individual matching tasks and combine their results. The experimental analysis reveals that the proposed method permits encouraging and significant improvements.

1 Introduction

There is a proliferation of schema- and ontology-based web data sources using models and languages, such as XML, RDF, and OWL [1]. Identifying semantic correspondences among such heterogeneous data sources and their metadata models (schemas and ontologies) is the biggest obstacle for making these data sources interoperable. The process of identifying these correspondences across different metadata models is called *schema matching* or *ontology matching*.

For its importance, a myriad of matching algorithms has been proposed and a large number of matching systems have been developed (see e.g., [17,4,2] for surveys). Unfortunately, most of these systems severely lack performance when dealing with large matching problems. The results of previous OAEI contests[1] show that more than half of the matching systems couldn't match large ontologies in less than one hour [12]. Consequently, several approaches have been proposed to address the problem of matching two large schemas, such as *MOM* [20], *COMA++* [6] and *Falcon* [11]. As we will further discuss in Section 2, the current approaches to partition-based matching have several limitations and the design space for such solutions has not yet sufficiently been explored.

In this paper we address two of these limitations. The first issue is *partition identification*. Some solutions, such as Falcon, are specific to certain ontology

---

[1] http://www.ontologymatching.org

J. Eder, M. Bielikova, and A.M. Tjoa (Eds.): ADBIS 2011, LNCS 6909, pp. 415–428, 2011.

languages and cannot be applied to other data models. Other solutions, such as COMA++, use relatively simple heuristic rules to partition the input schemas resulting often in too few or too many partitions. The second issue is *determination of similar partitions*. Some solutions, such as COMA++, only use limited information about the partition (only the root node of the partition) to determine the similarity between partitions of the input schemas, which results in less matching quality. Other solutions such as Falcon, fully evaluate the input ontologies to assess the partition similarity that leads to higher response time.

To cope with these challenges and limitations, we therefore propose and evaluate a new, more efficient, partition-based matching strategy. The proposed approach shares the general procedure to match large ontologies with existing matching systems. However, the approach introduces new methodologies to overcome the observed limitations. In particular, we make the following contributions:

- We propose a new clustering-based approach to cope with the large matching problem. The approach is generic. Similar to the current implementation of COMA++ [6], we first represent input schemas and ontologies as directed acyclic graphs, called schema graphs. We further apply a structure-based clustering algorithm to partition each input ontology into a set of disjoint sub-graphs. We thus achieve that elements that are structurally similar are in the same cluster, while elements in different clusters are dissimilar.
- Given the two cluster sets of the input ontologies, we apply a light-weight similarity measure to efficiently assess the similarity between cluster pairs. To this end, we represent each cluster as a *cluster document* and use of both the Vector Space Model and TF-IDF to determine the similarity between cluster documents. Having similar clusters, we adopt a standard match tool such as COMA++ to fully match elements inside similar clusters.
- We experimentally evaluate the efficiency and match quality of the proposed approach for different real-world schemas and ontologies. The resulting insights should be helpful for the development and evaluation of future match systems.

Section 2 discusses related work. We then introduce the basic definitions in Section 3. Sections 4 and 5 present the new approaches for structure-based clustering and identifying similar pairs of clusters. Section 6 presents experimental results. We conclude in Section 7.

2   Related Work

To cope with matching two large ontologies, several techniques can be used, such as reduction of search space, parallel matching, and self-tuning [16]. Reducing the match search space aims to limit the number of element comparisons either by early pruning dissimilar element pairs [7,14] or by partitioning the two ontologies [18,6,11,19,10]. *Quick ontology mapping* [7] was one of the first approaches that considers both matching quality and run-time complexity. It first

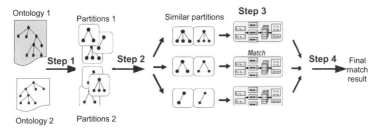

Fig. 1. Steps of matching two large ontologies

determines match candidates based on element labels and evaluates structure properties only for the most similar pairs from the first step. The approach proposed in [14] uses a set of filters within the matching process to prune dissimilar element pairs from intermediate match results.

Similar to our approach, partition-based matching aims at partitioning input ontologies/schemas in such a way that each partition of the first ontology has to be matched only with a subset of the second ontology. As shown in Fig. 1, the skeleton of partition-based matching involves four main steps. Step 1, *partition identification*, partitions the input schemas into a set of disjoint clusters. The second step, *determination of similar partitions*, is devoted to identifying similar partitions. Once settling on similar partitions (called fragments in COMA++ [6] or blocks in Falcon [11]), in Step 3, normal matching algorithms can be used to determine local correspondences between similar partitions. Finally, from these local correspondences, Step 4 is to construct the final match result. COMA++ implements one of the first approaches for partition-based matching. Its approach called *fragment matching* [6] has the four general steps similar of Fig. 1. Fragment matching first partitions two input schemas into two set of fragments, which are then compared with each other to identify the most similar fragments in the two sets worth to be fully matched later. Both *Falcon* [11] and *Anchor-Flood* [19] focus on matching (OWL) ontologies but do not support matching of XML schemas or relational schemas. The Falcon system uses a specific structure-based clustering technique to partition entities of ontology into blocks. Matching is then applied to the most similar blocks from the two ontologies. The Anchor-Flood system follows a dynamic partitioning technique. It starts off with an anchor, a pair of look-alike concepts from each ontology, gradually exploring concepts by collecting neighboring concepts until no further matches are found or all concepts are processed. The partitions are located around the anchors and their size depends on the continued success of finding match partners of the considered concepts. Further details about different techniques of large-scale matching can be found in [16].

3   Preliminaries

We first present definitions and basic concepts used throughout the paper.

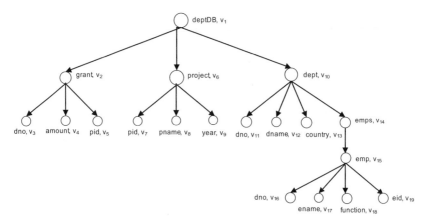

Fig. 2. Schema graph representation

Schema Graph. In order to make the proposed approach generic, we represent input schemas (e.g., XML schemas) and ontologies as labeled directed acyclic graphs, called *schema graphs* (SG).

Definition 1. *A schema graph is a rooted node-labeled directed acyclic graph. It is represented as a 3-tuple (V, E, Lab_v), where: $V = \{r, v_2, ..., v_n\}$ is a finite set of nodes, each of them is uniquely identified by an object identifier (OID), where r is the schema graph root node. $E = \{(v_i, v_j)|v_i, v_j \in V\}$ is a finite set of edges. Lab_v is a finite set of node labels. These labels are strings for describing the properties of the element and attribute nodes, such as name and data type.*

Fig. 2 represents the schema graph representation of an XML schema taken from [3]. *DeptDB* represents information about departments with their employees and grants, as well as the projects for which grants are awarded. The figure shows that each node is associated with the node name and the node identifier. For example, the node v_1 has the name *deptDB*.

Node Context. The context of a node in a schema graph is represented by its descendants, ancestors, and siblings. The descendants of the node include both its immediate children and the leaves of the subgraphs rooted at the node. The immediate children reflect its basic structure, while the leaves reflect the node's content. Without loss of generality, to construct the context of a node, we consider descendants and ancestors of the node up to one level, i.e., the parents and the children elements, as well as the node itself. Formally, we introduce the definition of the node context ($\mathcal{C}$) as follows:

Definition 2. *Given a schema graph $SG = (V, E, Lab_v)$, the context of a node $v \in V$ is given by $\mathcal{C}(v) = \{v_i|(v, v_i) \in E \cup (v_i, v) \in E \cup v\}$*

For the schema graph in Fig. 2, $\mathcal{C}(v_6) = \{v_1, v_6, v_7, v_8, v_9\}$. We claim that the more contexts two nodes share, the higher their structural similarity is. We therefore define and use the following context-based similarity measure.

Definition 3. *Given two nodes v_i and $v_j \in SG$, the context similarity, σ, between them is computed using the node contexts as follows:*

$$\sigma(v_i, v_j) = \frac{|\mathcal{C}(v_i) \cap \mathcal{C}(v_j)|}{\sqrt{|\mathcal{C}(v_i)|.|\mathcal{C}(v_j)|}} \tag{1}$$

$|\mathcal{C}(v_i) \cap \mathcal{C}(v_j)|$ represents the number of common nodes between their contexts and $\sqrt{|\mathcal{C}(v_i)|.|\mathcal{C}(v_j)|}$ is the geometric mean of the two contexts'size used to normalize the value of the structure similarity. In fact, Eq.1 guarantees that the more common nodes the two nodes share, the higher context similarity they have. Furthermore, the equation shows that context similarity has several properties. Among them are: it is normalized, $0 \leq \sigma(v_i, v_j) \leq 1$, and symmetric, $\sigma(v_i, v_j) = \sigma(v_j, v_i)$.

Example 1. The node contexts of nodes v_2, v_4 and v_6 are as follows: $\mathcal{C}(v_2) = \{v_1, v_2, v_3, v_4, v_5\}$, $\mathcal{C}(v_4) = \{v_2, v_4\}$ and $\mathcal{C}(v_6) = \{v_1, v_6, v_7, v_8, v_9\}$, respectively. The structure similarity between these nodes can be computed as follows: $\sigma(v_2, v_4) = 0.63$, $\sigma(v_2, v_6) = 0.2$, and $\sigma(v_4, v_6) = 0$.

4   Structure-Based Clustering

Our goal is to divide the schema graph into disjoint subgraphs in order to facilitate matching large ontologies represented as schema graphs. Clustering is a useful technique for grouping nodes such that nodes within a single cluster are structurally similar, while nodes in different groups are dissimilar. In the following, we present a clustering algorithm based on the introduced context similarity so that structurally similar nodes are placed in the same cluster while the nodes of different clusters are structurally dissimilar. We first describe how to use the computed structure similarity to construct so-called *links*. After this we introduce the proposed clustering algorithm.

   To avoid the repeated calculation of intra-ontology element similarities for clustering, we predetermine and store the structural similarity between selected node elements as so-called links. In particular, we are interested in the following set of element pairs for which the context similarity exceeds a predefined threshold, *th*:

$$links = \{L_i | L_i = (v_i, v_j, \sigma(v_i, v_j))\ s.t.\ \sigma(v_i, v_j) \geq th, v_i, v_j \in SG\} \tag{2}$$

Using this set of items (*links*) we construct a *links* hash table. Given a schema graph SG with n nodes, the worst case each node may be compared with $n - 1$ other nodes resulting in quadratic number of comparisons. However, as shown in Example 1, $\sigma(v_4, v_6) = 0$ since the two nodes have no common nodes in their contexts. Therefore, we limit the comparison of a node with the set of neighboring nodes to achieve a linear number of comparisons. By using a threshold value greater than 0 we can dramatically reduce the number of entries in the *links*

Algorithm 1. Clustering algorithm

Require: A schema graph, $SG = (V, E, Lab_v)$
Ensure: A dendrogram, a hierarchy of clusters
    {**Stage 1: Preparation**}
  1: $ClusterSet \Leftarrow \phi, Dendro \Leftarrow \phi$;
  2: $Nodes[] \Leftarrow SG.getNodes()$;
  3: $links[] \Leftarrow constructLink(SG)$;
    {**Stage 2: Cluster initialization**}
  4: **for** $n_i \in Nodes[]$ **do**
  5:    $Cluster\, C \Leftarrow new\, Cluster(n_i)$;
  6:    $ClusterSet.add(C)$;
  7: **end for**
  8: $Dendro.addLevel(ClusterSet)$;
    {**Stage 3: Cluster hierarchy construction**}
  9: $dist \Leftarrow 1$;
10: **while** $ClusterSet.size() > 1$ **do**
11:    $k \Leftarrow ClusterSet.size()$;
12:    $ClusterSet \Leftarrow mergeCluster(dist)$
13:    **if** $k > ClusterSet.size()$ **then**
14:        $Dendro.addLevel(ClusterSet)$;
15:        $computeIntraSim(ClusterSet)$;
16:        $k \Leftarrow ClusterSet.size()$;
17:    **end if**
18:    **if** $noMoreMerge()$ **then**
19:        $break$;
20:    **end if**
21:    $dist \Leftarrow dist + \delta$;
22: **end while**
    {**Stage 4: Best cluster set selection**}
23: return $Dendro.getBestCluster(BestLevel)$;

---

hash table. It should be noted that the similarity is assumed to be 0 if there is no pre-computed link.

The clustering algorithm presented in this paper is an agglomerative hierarchical algorithm mainly extended from the SCAN approach [21], which is a very scalable algorithm in the area of network clustering. The algorithm produces a tree representing the hierarchy of clusters in a bottom-up fashion, called *dendrogram*. Initially, each node represents its own single-member cluster. The algorithm iteratively merges nodes of a schema graph in descending order of structure similarity to build the hierarchy. As shown in Algorithm 1, the proposed clustering algorithm proceeds in four stages as follows.

- *Preparation.* The algorithm accepts the schema graph, *SG*, to be clustered and prepares it for the next stages. The stage starts by initializing the output set of clusters (*ClusterSet*) and the cluster hierarchy (*Dendro*), *line* 1. Then, the algorithm proceeds to extract schema graph nodes, elements to be clustered, *line* 2, and constructs the *links* hash table.
- *Cluster initialization.* The initialization stage constructs the bottom level of the cluster hierarchy. Each node represents its own cluster resulting into n clusters in the cluster set (*ClusterSet*), *lines* 4 *to* 7. Once getting the initial cluster set, the bottom level of the hierarchy is added to the dendrogram, *line* 8.

- *Cluster hierarchy construction.* This is the main stage of the clustering algorithm and is dedicated to construct the cluster hierarchy. It first initializes the distance between levels of hierarchy with 1, *line* 9. The algorithm iteratively merges clusters at a certain level until either the number of clusters reaches 1 or there is no possibility to merge more clusters. We keep the current size of the cluster set in variable k, *lines* 11 & 16. If the number of clusters after merging is changed, *line* 13, the new cluster set is added to the cluster hierarchy at the specified level. Furthermore, as we will explain later, the intra-cluster similarity is computed and the k value is updated. After that the algorithm checks if there is a possibility to further merge clusters and finally updates the distance for the current hierarchy level.
- *Best cluster set selection.* The task of the final stage is to select the best cluster set. Each level in the dendrogram is associated with a value that represents the average value of intra-cluster similarities of clusters at that level. Therefore, the algorithm returns the cluster set at the level with the best value, *line* 23.

In the following we give more details considering the two main operations in the clustering algorithm: *cluster merging* and *intra-cluster similarity computation*.

Cluster Merging. Once obtaining the first (bottom) level of the cluster hierarchy (*line* 8, Algorithm 1), we need to merge nodes into groups such that nodes in the same group are structurally similar while nodes in different groups are dissimilar. To this end, we call for a measure that quantifies relationship between individual clusters as well as a condition that should be satisfied to decide that nodes in two clusters have to be merged into one. To quantify the relationship between clusters, we rely on the pre-computed links. Having two clusters C_1 and C_2 containing k_1 and k_2 nodes (elements) respectively, the similarity between them can be expressed as the average context similarity of their elements. It can be represented as follows [9]:

$$Sim(C1, C2) = \frac{\sum_{i=1}^{k_1} \sum_{j=1}^{k_2} \sigma(v_{1i}, v_{2j})}{k_1 + k_2}. \tag{3}$$

where $\sigma(v_{1i}, v_{2j})$ is the context similarity between nodes $v_{1i} \in C_1$ and $v_{2j} \in C_2$ computed by Eq.1. Having this similarity between every cluster pair, a condition is required to decide if elements in the cluster pair should be merged. This condition has to reflect the level of the cluster hierarchy at which elements come together. Therefore, the introduced distance variable is used (*dist, line* 9). Elements of every cluster pair are combined when the similarity between the two clusters exceeds the predefined level similarity threshold ($1/dist$). The value of the level distance is then updated to reflect the nature of the next level (*line* 21, Algorithm 1).

It is worth noting that we add another condition in order to limit the merging process. Once two nodes in two different clusters have been merged into a new cluster, their links in the *links* hash table have been removed. The merging process stops when no more links are in the table, (*lines* 18 & 19).

Intra-Clustering Similarity. The proposed clustering algorithm produces a cluster hierarchy (dendrogram) in a bottom-up fashion. The cluster solution does not give information regarding the cut-off level. Cutting off the hierarchical tree requires the selection of a suitable level. To select the best level, we compute intra-clustering similarity at each level (*line* 15, Algorithm 1).

The intra-clustering similarity measures the cohesion within a cluster, how similar nodes within a cluster are. This is computed by measuring the similarity between each pair of data within a cluster, and the intra-clustering similarity of a clustering solution is determined by averaging all computed similarities taking into account the number of nodes. Let at a certain level of the cluster hierarchy, L, be a number of clusters K of the n nodes of a schema graph. The intra-clustering similarity, $IntraSim$, at this level can be computed from the following formula:

$$IntraSim_L = \frac{\sum_{i=1}^{K} IntraSim(C_i)}{n}. \tag{4}$$

where $IntraSim(C_i)$ is the intra-clustering similarity for the cluster C_i. In general, the larger the values of intra-clustering similarity ($IntraSim$), the better the clustering solution is.

Example 2. Applying Algorithm 1 to the schema graph represented in Fig. 2, the cluster solution consists of two clusters $C_1 = \{v_1, v_2, v_3, v_4, v_5, v_6, v_7, v_8, v_9\}$ and $C_2 = \{v_{10}, v_{11}, v_{12}, v_{13}, v_{14}, v_{15}, v_{16}, v_{17}, v_{18}, v_{19}, \}$[2]. It should be noted that this cluster solution represents a reasonable solution in the sense that C_1 includes information about projects and grants for these projects, while C_2 represents information about departments and their employees.

Complexity Analysis of Clustering Algorithm. Given a schema graph with n nodes, the algorithm contains four main stages. The worst case total cost of the preparation stage is $O(n(n-1)) = O(n^2)$ if every node in the schema graph has to be compared with all other $n-1$ nodes. However, on average, each node can only be compared with a set of nodes in the context of the node. With a typically constant average context size, this results in an average cost $O(n)$. The worst case total cost of the initialization stage is $O(n)$, and the time complexity of the final stage is $O(1)$. The cluster hierarchy construction stage initially iterates over n clusters, and then $n/2$ until either the number of clusters reduces to one or the merge condition, *line* 18, is satisfied. This results in an average time complexity of $O(n)$. Therefore, the time complexity of the clustering algorithm is $O(n) + O(n) + O(n) + O(1) = O(n)$. Results reported in the evaluation section verify and confirm this complexity.

---

[2] It should be noted that the cluster solution is based on the state of the schema graph. The state of schema graph represented in Fig. 2 is reduced. More information can be found in [6].

5   Determination of Similar Clusters

The goal of this step is to identify partitions (clusters) of the two schema graphs that are sufficiently similar to be worth matching in more detail. This aims at reducing the match overhead by not trying to find correspondences between unrelated partitions. The approach determines a *cluster document* per partition and makes use of the Vector Space Model (VSM) for computing the similarity between cluster documents.

To determine the similarity between clusters of different ontologies we can utilize different features of cluster elements, such as name, data type, cardinality constraints, etc. It has been verified that the node name is the most dominant feature [2]. Therefore, we construct a so-called cluster document based on the node name.

Definition 4. *Given a cluster J, the text document that contains the names of cluster elements is called a cluster document, CD_J.*

Adopting VSM provides the possibility to model document terms as elements of a vector space. Let $W_1, W_2, ..., W_t$ be the words (terms) in a cluster document. Let us suppose that there exists a unit length vector in the space corresponding to each word. We therefore can express each cluster document (CD_J) as a vector in terms of words as follows:

$$V_J = (W_{1J}, W_{2J}, ..., W_{tJ}) \tag{5}$$

where W_{iJ}s are real numbers reflecting the importance of word i in CD_J. Given a vector V_J representing the cluster document CD_J containing t words, the values of the vector elements can be computed using the $W_{iJ} = TF * IDF$ equation [5], where TF is the term frequency and IDF is the inverse document frequency.

To determine the cluster similarity, we propose the use of a light-weight similarity function based on the vector representation of the cluster document. Given two vectors V_{1I} and V_{2J} representing two cluster documents from two different ontologies, the cluster document similarity, $CDSim$, can be defined as the inner product (cosine) of the two vectors. It can be expressed as:

$$CDSim(CD_{1I}, CD_{2J}) = cos(V_{1I}, V_{2J}) = \frac{\sum_i^t W_{i1I}.W_{i2J}}{\sqrt{\sum_i^t (W_{i1I})^2 . \sum_i^t (W_{i2J})^2}} \tag{6}$$

where t is the size of the vectors. It should be noted that the equation yields a value of 0 when the elements of the two clusters do not have names in common, however, the similarity value becomes 1 when the elements of the two clusters have the same names. It is worth noting that representing cluster documents as vectors provides the possibility to utilize efficient similarity measures, such as the used one.

Now we are in a position to state the problem of similar clusters determination. Given two schema graphs SG_1 and SG_2 with n and m elements, and K and K' clusters, respectively. The problem is to identify the similar clusters between the

Table 1. Data set specification

Series	Tested schemas	No. matching tasks	min./max. No. of elements
1	PO (5 XDR)	10	27/74
	Spicy (4 XSD)	2	20/125
2	Webdirectory (4 OWL)	6	418/1132
3	Anatomy (2 OWL)	1	2746/3306

two sets. The computed similarities between cluster pairs of the two ontologies are used to construct a so-called cluster similarity matrix. Due to uncertainty inherent in ontology/schema matching, the best matching can actually be an unsuccessful choice [8]. To overcome this problem, the elements of the matrix are ranked according to their similarity to each other and the *top-k*[3] elements are selected from the ranked list.

Once settling on the similar clusters of the two ontologies, the next step is to fully match similar clusters to obtain the correspondences between their elements. Each pair of the similar clusters represents an individual match task that is independently solved. Match results of these individual tasks are then combined to a single mapping, which represents the final match result.

6   Experimental Evaluation

In order to evaluate the performance of the clustering-based matching approach, we conducted a set of experiments utilizing real-world ontologies of different sizes. We aim to evaluate both the quality and the efficiency of the proposed approach. We ran all our experiments on a 2.66 GHz Intel(R) Xeon(R) processor with 4 GB RAM running Windows 7.

6.1   Data Sets and Evaluation Criteria

Table 1 shows the characteristics of the test ontologies. In Series 1, we use five XML schemas for purchase orders (PO) taken from the COMA++ evaluation [6] and four XML schemas from [15]. In Series 2, we match four ontologies taken from the Web directory domain [13]. Series 3 contains a single match task taken from the OAEI initiative to match two large anatomy ontologies (AdultMouseAnatomy with 2,746 concepts vs. the anatomical part of the NCI Thesaurus with 3,306 concepts)[4]. We choose these data sets to demonstrate the applicability of our approach to different data sources represented in different models and having different characteristics. We performed the required matching tasks between schemas/ontologies within the similar domains with a total of 22 different matching tasks. More details about data sets in Table 1 can be found in [6,13].

---

[3] k may equal 1, 2, or 3 based on the similarity value between clusters.
[4] http://www.ontologymatching.org/

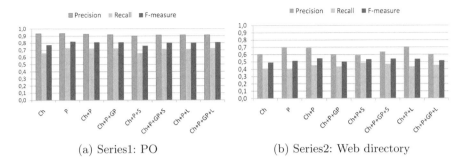

(a) Series1: PO   (b) Series2: Web directory

Fig. 3. Matching quality

To match elements within similar clusters, we used the *COMA++'s Allcontext* (a combination of Name, Path, Leaves, and Parents matchers) for match tasks of Series 1, the *Context* strategy (a Path matcher) for match tasks of Series 2, while the name matcher (without using synonyms) is used to perform the anatomy matching task. The threshold (*th*) used to construct *links* hash table is set to 0.15.

To measure the matching quality, we use the same criteria used in the literature, including *precision, recall* and *F-measure*. We call the execution time needed to perform the matching process including four steps of Fig. 1 the *response time*. We use it as a criterion of matching efficiency.

6.2   Experimental Results

We present results for two sets of experiments. The first set is used to answer the following question: "Which node context shall be used in computing context similarity?". To this end, we made use of five different contexts, namely children (Ch), parents (P), grandparents (GP), siblings (S), and leaves (L) in eight different combinations. The experimental results on matching quality (precision, recall, and F-measure) are reported in Fig. 3.

Figs 3(a,b) give the matching quality for matching tasks of the PO and Web directory domains, respectively. For the PO domain, all the exploited contexts, except the child and Ch+P+S contexts, produce F-measure equal to or higher than 80%. It should be noted that both P and Ch+P contexts achieves the highest F-measure (82%), as shown in Fig. 3(a). Since schemas in the Web directory domain contain more heterogeneities and a simple matcher is used, the highest F-measure merely reaches 55% using also the Ch+P context, as shown in Fig. 3(b). This motivates and verifies our selection of the Ch+P context in computing context similarity. To verify this selection, we also investigated the generated number of partitions and the response time using of the ontologies in Series 2 (Web directory).

Table 2 illustrates the average number of partitions (clusters) generated using different node contexts. The Ch+P+S, P, and Ch contexts lead to mostly higher number of partitions while the Ch+P context achieves a medium number of

Table 2. Average no. of partitions

Context	Ch	P	Ch+P	Ch+P+GP	Ch+P+S	Ch+P+S+GP	Ch+P+L	Ch+P+L+GP
Avg. partitions	38	62	22.8	19.8	112.7	28.9	25.6	19.1

partitions that is largely in the same range for different match tasks. Furthermore, using this context performs on average faster than the other contexts. Hence, we conclude that the Ch+P context is most suitable for our clustering approach and we will use this choice in the next set of experiments.

The second set of experiments is used to compare the proposed approach with two current matching strategies in COMA++ and Falcon (for the anatomy match task). For COMA++ we consider the non-partitioned strategy (*AllContext*) and different *Fragment-based* strategies [6]. We choose different techniques to select fragments, such as *inner (Fragment_inner)* and *Parent_of_Leaves (Frag_P_L)*. The first selects inner nodes as roots of fragments, while parents of leaves are selected as roots of fragments in the second strategy. The experimental results are reported in Fig. 4.

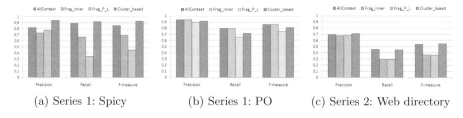

(a) Series 1: Spicy   (b) Series 1: PO   (c) Series 2: Web directory

Fig. 4. Matching quality comparison

Figs. 4(a, c) show that our proposed approach achieves, despite its reduced search space for matching, the best matching quality for the Spicy and Web directory schemas. The approach produces the highest F-measure compared with the other matching strategies. The clustering-based approach could correctly identify similar clusters which helps in achieving good recall; good precision is favored by the restricted search space reducing the risk of false positives. Fig. 4(b) also illustrates that the approach realizes a sufficient matching quality for the PO schemas.

We conducted another set of experiments to verify the matching efficiency. We measured the response time required to perform the specified match tasks of Series 2 illustrated in Table 1. We also compared the response time of the clustering-based approach to the mentioned strategies of COMA++. Results are reported in Fig. 5. The figure shows that the clustering-based approach outperforms the other strategies. The approach needs a total of 28 seconds to match the specified matching tasks. While, *AllContext*, *Fragment_inner* and *Fragement_parents_leaves* require 101, 72, and 66 seconds, respectively. We also analyzed the number of generated partitions (clusters or fragments) and we found

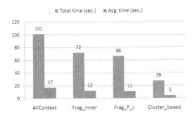

Fig. 5. Res. time comparison

Table 3. Anatomy results

	Clustering-based	Falcon
No. of partitions	84/80	139/119
Precision	0.975	0.964
Recall	0.613	0.591
F-measure	0.753	0.73
Res. time	58.8 sec	10 mins.

that COMA++ generates more partitions than the new cluster approach. We also tested with *Frag_sub* [6], we found that only few partitions are determined so that no correspondences could be found for several matching tasks. We thus do not include the detailed results produced by the *Frag_sub* strategy.

We finally evaluate our clustering approach on the anatomy match task and compare it with Falcon. For this purpose, we installed the publicly available Falcon system and run it on the same machine. Results are reported in Table 3. The table shows that our approach achieves a slight improvement in matching quality as Falcon, however, our system is about ten times faster (1 vs. 10 minutes).

In summary, the evaluation results show that the proposed approach achieves for different domains better matching response times compared to previously proposed partition-based strategies at a comparable or better match quality.

7 Conclusions

We proposed a new clustering-based matching approach for large-scale ontology matching. The proposed approach is generic and can be applied to different data models including XML schemas. It shares the same steps of other partition-based match strategies. However, it uses different techniques for partitioning and finding similar partitions. The partitioning process is based on a bottom-up clustering scheme utilizing context-based structural node similarities, while finding similar partitions to match is based on an effective and light-weight linguistic technique. To verify the performance of the proposed approach, we conducted several sets of experiments. The results show that the proposed approach presents significant and encouraging improvement, especially in runtime efficiency. In future work we want to further explore the design space of partition-based match strategies by taking further algorithms for the key steps and further application domains into account.

Acknowledgements. This work is supported by the Federal Ministry of Education and Research (BMBF), grant 03FO2152 ("Web Data Integration").

References

1. Abiteboul, S., Suciu, D., Buneman, P.: Data on the Web: From Relations to Semistructed Data and XML. Morgan Kaufmann, USA (2000)
2. Algergawy, A., Nayak, R., Saake, G.: Element similarity measures in XML schema matching. Information Sciences 180(24), 4975–4998 (2010)
3. Chiticariu, L., Hernndez, M.A., Kolaitis, P.G., Popa, L.: Semi-automatic schema integration in Clio. In: VLDB 2007, pp. 1326–1329 (2007)
4. Choi, N., Song, I.-Y., Han, H.: A survey on ontology mapping. SIGMOD Record 35(3), 34–41 (2006)
5. Cohen, W.W., Ravikumar, P., Fienberg, S.E.: A comparison of string distance metrics for name-matching tasks. In: IIWeb, pp. 73–78 (2003)
6. Do, H.H., Rahm, E.: Matching large schemas: Approaches and evaluation. Information Systems 32(6), 857–885 (2007)
7. Ehrig, M., Staab, S.: QOM- quick ontology mapping. In: International Semantic Web Conference, pp. 683–697 (2004)
8. Gal, A.: Managing uncertainty in schema matching with top-k schema mappings. Journal on Data Semantics 6, 90–114 (2006)
9. Guerrini, G., Mesiti, M., Sanz, I.: An Overview of Similarity Measures for Clustering XML Documents. Emerging Techniques and Technologies (2007)
10. Hamdi, F., Safar, B., Reynaud, C., Zargayouna, H.: Alignment-based partitioning of large-scale ontologies. In: Guillet, F., Ritschard, G., Zighed, D.A., Briand, H. (eds.) Advances in Knowledge Discovery and Management. SCI, vol. 292, pp. 251–269. Springer, Heidelberg (2010)
11. Hu, W., Qu, Y., Cheng, G.: Matching large ontologies: A divide-and-conquer approach. DKE 67, 140–160 (2008)
12. O. A. E. Initiative (2010), http://20.ontologymatching.org/
13. Massmann, S., Rahm, E.: Evaluating instance-based matching of web directories. In: 11th Workshop on Web and Databases, WebDB (2008)
14. Peukert, E., Berthold, H., Rahm, E.: Rewrite techniques for performance optimization of schema matching processes. In: EDBT, pp. 453–464 (2010)
15. Peukert, E., Massmann, S., Konig, K.: Comparing similarity combination methods for schema matching. In: GI-Workshop, pp. 692–701 (2010)
16. Rahm, E.: Towards large-scale schema and ontology matching. In: Bellahsene, Z., Bonifati, A., Rahm, E. (eds.) Schema Matching and Mapping. Data-Centric Systems and Applications series, Springer, Heidelberg (2010)
17. Rahm, E., Bernstein, P.A.: A survey of approaches to automatic schema matching. VLDB Journal 10(4), 334–350 (2001)
18. Rahm, E., Do, H.-H., Massmann, S.: Matching large XML schemas. SIGMOD Record 33(4), 26–31 (2004)
19. Seddiquia, M.H., Aono, M.: An efficient and scalable algorithm for segmented alignment of ontologies of arbitrary size. Web Semantics 7(4), 344–356 (2009)
20. Wang, Z., Wang, Y., Zhang, S., Shen, G., Du, T.: Matching large scale ontology effectively. In: Mizoguchi, R., Shi, Z.-Z., Giunchiglia, F. (eds.) ASWC 2006. LNCS, vol. 4185, pp. 99–105. Springer, Heidelberg (2006)
21. Yuruk, N., Mete, M., Xu, X., Schweiger, T.A.J.: AHSCAN: Agglomerative hierarchical structural clustering algorithm for networks. In: International Conference on Advances in Social Network Analysis and Mining, pp. 72–77 (2009)

Automatic Building of an Appropriate Global Ontology

Cheikh Niang[1,2], Béatrice Bouchou[1], Moussa Lo[2], and Yacine Sam[1]

[1] Université François Rabelais Tours - Laboratoire d'Informatique
{first.last}@univ-tours.fr
[2] Université Gaston Berger de Saint-Louis - LANI
{first.last}@ugb.sn

Abstract. Our objective is to automatically build a global ontology from several data sources, annotated with local ontologies and aiming to share their data in a specific application domain. The originality of our proposal lies in the use of a background knowledge, i.e. a reference ontology, as a mediation support for data integration. We represent ontologies using Description Logics and we combine syntactic-matching with logical-reasoning in order to build the shared global TBox from both the TBoxes of sources and that of the reference ontology.

Keywords: Background Knowledge, Data Integration, Descripion Logic.

1 Introduction

As the need for Web Data Integration is still growing, we address here the first challenge pointed out in [15]: "How to build an appropriate global schema". Indeed, many organizations hold some similar data in specific domain and want to share some parts of it. Data integration may alleviate users from knowing the structure of different sources, as well as the way they are conciliated, when making queries [15].

When the access to heterogeneous data sources is made possible using ontologies, the integration process in called *ontology-based data integration*. Ontologies offer a formal semantics which allows the automation of tasks such as heterogeneity resolution, consistency checking, inferrence, etc. There are three main ontology-based data integration architectures in the literature [21], namely *(i)* the *single-ontology*, *(ii)* the *multiple-ontologies* and *(iii)* the *hybrid* approaches. In the first one, all the data sources are related to a global ontology: this approach requires from all the sources the same view of the domain, for instance the same granularity-level, because in the presence of sources with a different view of the domain, finding a consensus in a minimal ontology commitment is a difficult task. This approach is implemented for instance in [3].

In the *multiple-ontologies* approach, for instance in the OBSERVER system [16], each data source is described with its own (local) ontology, and inter-ontology mappings must be defined for interoperability. The lack of a common vocabulary between the sources makes this task difficult. The *hybrid* approach combines the two precedent ones, allowing to overcome their drawbacks by defining a global shared vocabulary in addition to local ontologies: [11] is an example of such an integration architecture.

In this paper we build on the hybrid approach as we propose to automatically build a global ontology from local ones. As usual in data integration systems, our global

J. Eder, M. Bielikova, and A.M. Tjoa (Eds.): ADBIS 2011, LNCS 6909, pp. 429–443, 2011.

ontology must be linked to local ontologies by mappings. The two basic solutions for doing so are the LAV (local as view) and the GAV (global as view) mappings. Each of them has advantages and drawbacks: LAV approach allows to define the global ontology independently from the sources, so adding or removing a new data source is easy but query processing is harder. Query processing is less complex with GAV approach, since the global ontology is defined from the data sources, but sources must be known in advance and adding new data source is not easily supported.

We propose to overcome some drawbacks of both hybrid approach and GAV mappings *by using a background-knowledge*, represented by a reference ontology, in order to automatically build a global ontology from local sources. We call reference ontology an ontology developed independently from any specific objective by experts in knowledge engineering with the collaboration of domain experts. It is a robust conceptualization of the knowledge about a given generic domain such as medicine, tourism, agriculture, etc. AGROVOC[1] and NALT[2] in the agriculture domain and MeSH[3] in the medical field are some examples of reference ontologies. The growth of Semantic Web allows to expect that such reference ontologies will become more and more accessible and usable by machines in the coming years.

The algorithm presented in this paper follows the mediation-based process illustrated in Fig. 1: each source (S_i) involved in the sharing process is represented by its local ontology (LO_i) and the reference ontology (MO) allows to find the portion of knowledge that each source can share with others. This portion is called agreement (A in Fig. 1(a)). Then each agreement is incrementally integrated in the global ontology (GO) via MO in what we call the conciliation phase (Fig. 1(b)).

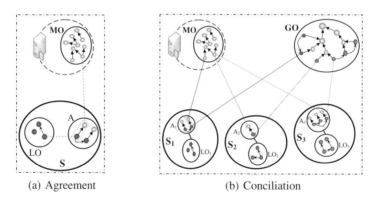

(a) Agreement   (b) Conciliation

Fig. 1. General overview of our mediation-based process

The challenge that we point out is: how to automatically build an appropriate global ontology for several data source owners that want to share parts of their data for a specific web application, but that do not want to (or can not) invest much efforts on the hard task of building a consensual appropriate shared conceptual level ? An appropriate

---

[1] http://www.fao.org/agrovoc
[2] http://agclass.nal.usda.gov/agt
[3] http://www.nlm.nih.gov/mesh/

global ontology in such a sharing context should provide an appropriate conceptualiza-tion of the application domain (maximizing relevant information for the sharing process and minimizing irrelevant one). It must allow to add easyly new data sources, and also to remove or update sources. It must allow an easy querying of sources. Finally it must be automatically built and maintained.

This is what our algorithm builds. **For an easy query processing** it lies on the GAV approach. However it generalizes existing proposals so that it is no longer necessary to have sources known in advance. An anchoring phase allows each source to participate in the global ontology to some extent, whatever it is. **For scalability**, it incrementally integrates data sources, so it is easy to add a new source involved in the sharing pro-cess. **For an appropriate conceptualization**, it selects in the reference ontology the smallest relevant information portion and, in each data source involved in the sharing process, it selects only information that is relevant to be shared in the application do-main. **For automation**, we use Description Logics to represent ontologies. Description Logics (or DL) are formalisms for conceptual representations which have already been successfully used for *(i)* linking Data to Ontologies [17] and *(ii)* building Data Integra-tion Systems [7]. Our choice of DL is mainly motivated by their capability to represent hierarchies and to automaticaly reason on these relationships. Moreover, the inference capabilities of DL are not limited to hierarchies (they are equipped with a formal *logic*-based semantics), so DL are fully justified here as a data model that allows inferences.

The rest of this article is organized as follows: in Section 2 we addres related works, in Section 3 we define the notions used in our algorithm, in Sections 4 and 5 we present our global-ontology-building process, and Section 6 concludes.

2   Related Works

Our approach deals with ontology-based data integration. Very close to our interests are the works based on (i) Description Logics and (ii) Ontology Matching.

In [8] and [19], a formalism for reasoning with multiple local ontologies connected by directional semantic mappings is presented, in other words they introduce the no-tion of distributed description logics, useful for linking different data sources. The ap-proach presented in [13] is another solution exploiting description logics, namely the $\mathcal{E}$-connections framework, to link different sources for an integration purpose. Both of these approaches have successfully shown the interest of using description logics for efficiently exploiting distributed data sources. The contribution of our proposal is the introduction of automation in the linking process. We use for this a reference ontology: on the one hand links between source ontologies are obtained from the taxonomical re-lationships of the reference ontology. On the other hand, mappings between the global ontology and sources are obtained by syntactic-matching, from source-concepts' names to reference-ontology-concepts' names.

For that reason, our algorithm depends on the performance of Ontology Matching techniques (cf. Section 4), which constitute a very active research field (see [20] for a survey). The use of reference ontologies has been investigated in this field, see for example [2], [1] and [18]. It was shown that the reference ontology can significantly improve the performance of the matching process. The contribution of our proposal is to show that the reference ontology also allows to enrich the semantics of links discovered in the matching process. As an example, we can have in a local ontology two anchored concepts, e.g. *Onion* and *Tomato*, not related: our conciliation algorithm can relate them via a common ancestor of their anchor concepts, e.g. *Vegetable* (see Fig. 7).

Finally, our work is related to data integration systems using GAV mapping ([15], [6] and [12]) and our contribution here is again the use of the reference ontology. It stands for the information about sources that allows adding easily new sources.

3   Preliminaries

In our approach, ontologies are expressed in Description Logics (DLs) [4], a family of logic-based representation formalisms. They allow representing the domain of interest in terms of *concepts*, denoting sets of objects, and *roles*, denoting binary relations between (instances of) concepts [17]. A DL ontology consists of a *TBox* (Terminological Box) and an *ABox* (Assertional Box): the former formally specifies concepts and roles and the latter represents their instances. DLs differ in constructs they allow to specify concepts and roles. In this paper, we consider the *DL-Lite$_\mathcal{A}$* description logic [17]. It is known as one of the most expressive DL in the *DL-Lite* family [9].

3.1   *DL-Lite$_\mathcal{A}$* Syntax and Semantics

The syntax of *DL-Lite$_\mathcal{A}$* expressions is defined as follows [17]:

$$B ::= A \mid \exists Q \mid \delta(U_C)$$
$$C ::= \top_C \mid B \mid \neg B \mid \exists Q.C$$
$$E ::= \rho(U_C)$$
$$F ::= \top_D \mid T_1 \mid ... \mid T_n$$

$$Q ::= P \mid P^-$$
$$R ::= Q \mid \neg Q$$
$$V_C ::= U_C \mid \neg U_C$$

Where A denotes an *atomic concept*, i.e., a concept denoted by a name, B a *basic concept*, C a *general concept*, and $\top_C$ the *universal concept*. E denotes a basic value-domain, i.e., the range of an attribute, F a *value-domain expression*, and $\top_D$ the *universal value-domain*. P denotes an *atomic role*, Q a *basic role*, and R a *general role*. U_C denotes an *atomic attribute* and V_C a *general attribute*.

The semantics of every DL expression is specified in term of its first-order interpretation. An *interpretation* is defined as a pair $I = (\Delta^I, \cdot^I)$, where Δ^I is the domain interpretation and $\cdot^I$ an *interpretation function*. In *DL-Lite$_\mathcal{A}$*, Δ^I is composed of two non-empty sets: Δ_O^I, the *domain of objects*, i.e. the set of all allowed objects in the domain, and Δ_V^I, the *domain of values*, i.e. the set of all allowed values in the domain ($\Delta^I = \Delta_O^I \cup \Delta_V^I$). The interpretation function assigns a subset of Δ^I to each concept or value domain, and a subset of $\Delta^I \times \Delta^I$ to each role or attribute, in such a way that the following conditions are satisfied:

$$\top_C^I = \Delta_O^I$$
$$\top_D^I = \Delta_V^I$$
$$A^I \subseteq \Delta_O^I$$

$$P^I \subseteq \Delta_O^I \times \Delta_O^I$$
$$U_C^I \subseteq \Delta_O^I \times \Delta_V^I$$
$$(\neg U_C)^I = (\Delta_O^I \times \Delta_V^I) \backslash U_C^I$$

$$(\rho(U_C))^I = \{v \mid \exists o.(o,v) \in U_C^I\}$$
$$(\delta(U_C))^I = \{o \mid \exists o.(o,v) \in U_C^I\}$$
$$(P^-)^I = \{(o,o') \mid (o',o) \in P^I\}$$

$$(\exists Q)^I = \{o \mid \exists o'.(o,o') \in Q^I\}$$
$$(\neg Q)^I = (\Delta_O^I \times \Delta_O^I)\backslash Q^I$$
$$(\neg B)^I = \Delta_O^I\backslash B^I$$

3.2 Our *DL-Lite$_A$* Ontologies

A *DL-Lite$_A$* ontology $O = \langle \mathcal{T}, \mathcal{A} \rangle$ specifies a given application domain in terms of a TBox $\mathcal{T}$ representing its intensional part and an ABox $\mathcal{A}$ representing the extensional one. $\mathcal{T}$ consists in a set of *intensional expressions* specified according to the following syntax: $B \sqsubseteq C \mid E \sqsubseteq F \mid Q \sqsubseteq R \mid U_C \sqsubseteq V_C \mid (funct\ Q) \mid (funct\ U_C)$

A concept (respectively, value-domain, role, and attribute) inclusion expresses that a basic concept B (respectively, basic value-domain E, basic role Q, and atomic attribute U_C) is subsumed by a general concept C (respectively, value-domain F, role R, attribute V_C). A role (attribute) functionality expresses the functionality of a role. The semantics of a *DL-Lite$_A$* TBox is defined by its interpretations. A given interpretation I satisfies:

- a concept (respectively, value-domain, role, attribute) inclusion assertion $B \sqsubseteq C$ (respectively, $E \sqsubseteq F$, $Q \sqsubseteq R$, $U_C \sqsubseteq V_C$), if $B^I \subseteq C^I$ (respectively, $E^I \subseteq F^I, Q^I \subseteq R^I, U_C^I \subseteq V_C^I$)
- a role functionality assertion $(funct\ Q)$, if for each $o_1, o_2, o_3 \in \Delta_O^I$ $(o_1, o_2) \in Q^I$ and $(o_1, o_3) \in Q^I$ implies $o_2 = o_3$
- an attribute functionality assertion $(funct\ U_C)$, if for each $o \in \Delta_O^I$ and $v_1, v_2, \in \Delta_V^I$ $(o, v_1) \in U_C^I$ and $(o, v_1) \in U_C^I$ implies $v_1 = v_2$

I is a *model* of $\mathcal{T}$ if and only if I satisfies all intensional expressions in $\mathcal{T}$. $\mathcal{T}$ is *satisfiable* (or *consistent*) if it has at least one model. In this article, we reason essentially on the structural part of an ontology, i.e., a TBox-level reasoning. Moreover, when considering the mediator ontology we restrict ourselves to specify *atomic concept inclusion* (ACI) expressions. An ACI expression is defined as an inclusion of the form $A \sqsubseteq D$, where A and D are atomic concepts. A finite set of ACI expressions is called an *atomic TBox*.

Definition 3.1. *An ACI is an expression of the form $A \sqsubseteq D$, where A and D are atomic concepts. A finite set of ACIs is called an* **atomic TBox**.

Finally, $\mathcal{A}$ consists in a finite set of *membership assertions* of the form: $A(a)$, $P(a,b)$ and $U_C(a,b)$. As said before, we don't address this part in the present article.

3.3 Inference Capabilities

One of the traditional inference services provided by DLs is computing subsumption relationships between concepts.

Definition 3.2. *Let $\mathcal{T}$ be a TBox, C and D two concept descriptions. The concept C is subsumed by D w.r.t the TBox $\mathcal{T}$ ($C \sqsubseteq_{\mathcal{T}} D$) iff $C^I \subseteq D^I$ for all models I of $\mathcal{T}$.*

In the present article, we explore subsumption reasoning in order to compute the *deductive closure* of an atomic TBox, defined as follows.

Definition 3.3. *Let $\mathcal{T}$ be an atomic TBox. The deductive closure of $\mathcal{T}$, denoted by $cl\mathcal{T}$, is the TBox inductively defined as follows:*

1. *If $A_1 \sqsubseteq_{\mathcal{T}} A_2$, then $A_1 \sqsubseteq_{cl(\mathcal{T})} A_2$.*
2. *If $A_1 \sqsubseteq_{cl(\mathcal{T})} A_2$ and $A_2 \sqsubseteq_{cl(\mathcal{T})} A_3$, then $A_1 \sqsubseteq_{cl(\mathcal{T})} A_3$.*

3.4 Building an Appropriate *DL-Lite$_{\mathcal{A}}$* Global TBox

Our objective consists in building a *global ontology*, more precisely a *DL-Lite$_{\mathcal{A}}$* TBox $\mathcal{T}_g$, based on the local-sources' TBoxes and that of a mediated (reference) ontology. Precisely, the following are the four kinds of TBoxes that we deal with here.

- **The set of local TBoxes** $\{\mathcal{T}_{li}\}$ involved in the sharing process. Each data source S_i is represented by its local TBox $\mathcal{T}_{li}$, denoted LO_i in Fig. 1.
- **The mediator TBox** $\mathcal{T}_m$. It is a *DL-Lite$_{\mathcal{A}}$* atomic TBox providing general intensional knowledge on the application domain. We consider it as a set of atomic concepts inclusions (*ACI*), *i.e.* a subsumption hierarchy. It is denoted *MO* in Fig. 1.
- **The set of agreement TBoxes** $\{\mathcal{T}_{ai}\}$, denoted A_i in Fig. 1. An agreement TBox $\mathcal{T}_{ai} = \langle \mathcal{T}'_{li}, \mathcal{M}_i \rangle$ is built for each local TBox $\mathcal{T}_{li}$. It is composed of $\mathcal{T}'_{li}$, the subset of $\mathcal{T}_{li}$ containing expressions of $\mathcal{T}_{li}$ that are relevant for the application domain, and $\mathcal{M}_i$, the set of mappings between $\mathcal{T}_{li}$ and $\mathcal{T}_m$.
- **The global TBox** $\mathcal{T}_g = \langle \{\mathcal{T}_{ai}\}, \mathcal{T}'_m \rangle$, denoted *GO* in Fig. 1. It consists in the set of agreement TBoxes $\{\mathcal{T}_{ai}\}$ together with $\mathcal{T}'_m$, which is the smallest subset of $\mathcal{T}_m$ that conciliates every $\mathcal{T}_{ai}$.

We show how to build $\mathcal{T}_g$ from $\{\mathcal{T}_{li}\}$ by using $\mathcal{T}_m$. It consists in the selection of parts of $\{\mathcal{T}_{li}\}$ to be included in $\mathcal{T}_g$ (Section 4) and then their conciliation (Section 5) in $\mathcal{T}_g$.

4   Agreement

Agreement process consists in the selection of the expressions in $\mathcal{T}_l$ to be included in the global TBox $\mathcal{T}_g$. To identify such knowledge we proceed first by applying an *anchoring* process [2] to select from the local TBox relevant concepts for the application domain. Anchoring consists in associating atomic concepts of a local Tbox, called *anchored concepts*, with concepts of the mediator TBox, called *anchor concepts*. Consider the example shown in Fig. 2, where concepts are represented by ovals, attributes by rectangles and roles by dashed arrows. Single-full arrows represent subsumption relationships between two concepts. Fig. 2(a) shows an excerpt of a local TBox $\mathcal{T}_l$ that deals with both agricultural and accommodation knowledge. We assume that the application domain in which the source $\mathcal{T}_l$ shares its data is the agricultural domain: Fig. 2(b) shows an excerpt of the agreement TBox obtained after the anchoring process. Prefix *mo* : denotes anchor concepts from the mediator TBox $\mathcal{T}_m$. We can notice that only concepts related to agriculture are anchored because no anchor is found for accommodation knowledge. Anchor concepts generalize anchored concepts and will be used for finding semantic links between concepts in different local TBoxes.

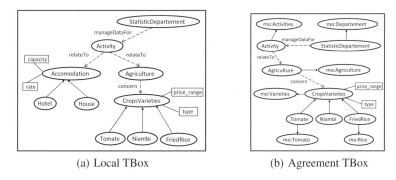

(a) Local TBox   (b) Agreement TBox

Fig. 2. Example of the agreement process

We perform two successive anchoring steps: a lexical anchoring process that selects relevant concepts to be anchored, based on syntactic-matching, followed by a semantic one (logical-inference) that selects other ones not detected in the first step.

Lexical Anchoring Process. It consists in matching a local TBox $\mathcal{T}_l$ with the mediator TBox $\mathcal{T}_m$, *i.e.* in computing a set of mappings as defined in [20].

Definition 4.1. *Let $\mathcal{T}_l$ be a local TBox and $\mathcal{T}_m$ be the mediator TBox. Lexical anchoring of $\mathcal{T}_l$ w.r.t $\mathcal{T}_m$ consists in finding a set of mappings $\mathcal{M} = \langle m_1, ..., m_n \rangle$ such that each m_i is an assertion of the form: $m_i = A_l \sqsubseteq A_m$, where $A_l \in \mathcal{T}_l, A_m \in \mathcal{T}_m, A_l$ and A_m are both atomic concepts. A_m is called the anchor of A_l.*

The key point in the lexical anchoring (or matching) process is to measure how much an atomic concept A_l in a local TBox $\mathcal{T}_l$ is related to an atomic concept A_m in the mediator TBox $\mathcal{T}_m$. This is done by syntactically comparing concepts names (labels). Many lexical similarity measures, proposed in the literature [14], [10], [20], may be used and, as noticed in [20], no similarity measure can give good results in all cases: it is still necessary to look for the best one for each specific application. However, whatever the application is, the relation between A_l and A_m is obtained as follows, considering that φ is the chosen similarity measure: $\Gamma_{N_l} \times \Gamma_{N_m} \rightarrow [0,1]$, where Γ_{N_l}, Γ_{N_m} are respectively the set of atomic concept names in $\mathcal{T}_l$ and $\mathcal{T}_m$. In general, let $n_l \in \Gamma_{N_l}$ be the name of A_l and $n_m \in \Gamma_{N_m}$ the name of A_m, the mapping $m = A_l \sqsubseteq A_m$ is established if and only if $(\varphi(n_l, n_m) \geq \alpha)$ and $(\forall n_{mi} \in \Gamma_{N_m} \varphi(n_l, n_m) \geq \varphi(n_l, n_{mi}))$ or $(\alpha \geq \varphi(n_l, n_m) \geq \beta)$ and $(n_m \succ n_l)$, where α, β are respectively the maximum and the minimum threshold similarity and $\succ$ denotes a lexical inclusion relation.

Semantic Anchoring Process. It consists in finding additional local concepts that may be relevant for the application domain and which have not been anchored during the lexical anchoring process. We assume that all atomic concepts subsumed by an anchored concept are relevant for the application domain and then must be considered as anchored concepts even if they have not been anchored during the lexical anchoring process. To identify such concepts we automatically compute what we call the *anchoring closure* of a local TBox based on subsumption relationship.

Definition 4.2. *Let $\mathcal{T}_l$ be a local TBox, $\mathcal{T}_m$ the mediator TBox, and $\mathcal{M} = \langle m_1, ..., m_n \rangle$ the result of anchoring $\mathcal{T}_l$ w.r.t $\mathcal{T}_m$. The anchoring closure of $\mathcal{T}_l$, denoted by $a\_cl(\mathcal{T}_l)$, is inductively defined as follows:*

1. *All assertions in $\mathcal{M}$ are also assertions in $a\_cl(\mathcal{T}_l)$.*
2. *All ACI assertions in $\mathcal{T}_l$ are also assertions in $a\_cl(\mathcal{T}_l)$.*
3. *If A_1, A_2, A_3 are atomic concepts and $A_1 \sqsubseteq A_2$ and $A_2 \sqsubseteq A_3$ are in $a\_cl(\mathcal{T}_l)$, then $A_1 \sqsubseteq A_3$ is in $a\_cl(\mathcal{T}_l)$.*

According to this definition we can say that an anchored concept A_l of $\mathcal{T}_l$ is a concept appearing in $a\_cl(\mathcal{T}_l)$ and is of the form: $A_l \sqsubseteq mo : A_m$, where $A_m \in \mathcal{T}_m$ and $mo :$ is a prefix used to distinguish anchor concepts from other concepts.

For example, consider the TBox in Fig. 3(a). It is composed of ACIs of the local TBox $\mathcal{T}_l$, shown graphically in Fig. 2(a), enriched with assertions of the lexical anchoring of $\mathcal{T}_l$ w.r.t $\mathcal{T}_m$. The anchoring closure of this TBox is shown in Fig. 3(b). Notice in Fig. 2(a) that we have *Niambi* that appears as an unanchored concept, because there is no assertion of the form *Niambi* $\sqsubseteq mo : A_m$, where $A_m \in \mathcal{T}_m$. But, we have the assertion *Niambi* $\sqsubseteq mo:Varieties$ in the anchoring closure in Fig. 3(b), and *Varieties* $\in \mathcal{T}_m$. So, *Niambi* becomes an anchored concept, semantically selected, and its anchor is *Varieties*.

ACI assertions	Lexical anchoring assertions		Anchoring closure assertions	
Hotel $\sqsubseteq$ *Accomodation*	*Activity* $\sqsubseteq$ *mo:Activities*		*Activity* $\sqsubseteq$ *mo:Activities*	*Tomate* $\sqsubseteq$ *CropsVarieties*
House $\sqsubseteq$ *Accomodation*	*Agriculture* $\sqsubseteq$ *mo:Agriculture*		*Agriculture* $\sqsubseteq$ *mo:Agriculture*	*FriedRice* $\sqsubseteq$ *CropsVarieties*
Tomate $\sqsubseteq$ *CropsVarieties*	*StatDepartement* $\sqsubseteq$ *mo:Departement*		*StatDepartement* $\sqsubseteq$ *mo:Departement*	*Niambi* $\sqsubseteq$ *CropsVarieties*
FriedRice $\sqsubseteq$ *CropsVarieties*	*CropsVarieties* $\sqsubseteq$ *mo:Varieties*		*CropsVarieties* $\sqsubseteq$ *mo:Varieties*	*Tomate* $\sqsubseteq$ *mo:Varieties*
Niambi $\sqsubseteq$ *CropsVarieties*	*Tomate* $\sqsubseteq$ *mo:Tomato*		*Tomate* $\sqsubseteq$ *mo:Tomato*	*FriedRice* $\sqsubseteq$ *mo:Varieties*
	FriedRice $\sqsubseteq$ *mo:Rice*		*FriedRice* $\sqsubseteq$ *mo:Rice*	*Niambi* $\sqsubseteq$ *mo:Varieties*
			Hotel $\sqsubseteq$ *Accomodation*	*House* $\sqsubseteq$ *Accomodation*
(a) ACI assertions with lexical anchoring ones			(b) Anchoring closure TBox	

Fig. 3. Semantic anchoring process

From Anchoring to Agreement. We build the agreement of $\mathcal{T}_l$ w.r.t $\mathcal{T}_m$ starting from anchored concepts, *i.e.* those in $a\_cl(\mathcal{T}_l)$. The agreement of $\mathcal{T}_l$ w.r.t $\mathcal{T}_m$ is indeed a TBox $\mathcal{T}_a$ composed by $\mathcal{T}_l'$, a subset of $\mathcal{T}_l$ containing assertions of $\mathcal{T}_l$ that are relevant for the application domain, and $\mathcal{M}$ the result of anchoring $\mathcal{T}_l$ w.r.t $\mathcal{T}_m$. We compute $\mathcal{T}_l'$ by selecting in $\mathcal{T}_l$ assertions that are related to anchored concepts.

Precisely, we aim to select assertions such that: *(i)* $\mathcal{T}_l'$ contains the maximum possible of relevant assertions w.r.t. the application domain, *(ii)* $\mathcal{T}_l'$ contains the minimum possible of irrelevant assertions w.r.t. the application domain, and *(iii)* $\mathcal{T}_l'$ is consistent if $\mathcal{T}_l$ is consistent.

Thus, in addition to anchored concepts, $\mathcal{T}_l'$ may contain unanchored concepts that we call *selected concepts*. A selected concept C is an unanchored concept that must be related to an anchored concept A in order to avoid loosing information about A and also to avoid inconsistency in $\mathcal{T}_l'$. We consider that an unanchored concept C must be a selected concept if $\mathcal{T}_l$ contains assertions of the form:

- $A \sqsubseteq C$, where A is an anchored concept.
- $\exists R \sqsubseteq A$ *and* $\exists R^- \sqsubseteq C$, where A is an anchored concept and R is a basic role.
- $C \sqsubseteq C_1$, where C_1 is a selected concept.

$\exists \, relateTo \sqsubseteq Activity$	$\exists \, concern \sqsubseteq Agriculture$
$\exists \, relateTo^- \sqsubseteq Agriculture$	$\exists \, concern^- \sqsubseteq CropsVarieties$

Fig. 4. Case of selected concept

For instance, consider the local TBox $\mathcal{T}_l$ in Fig. 2 and assume that the concept *Agriculture* is an unanchored concept. Because we have in $\mathcal{T}_l'$ the indirect relation between the two anchored concepts *Activity* and *CropsVarieties*, as illustrated with assertions in Fig. 4, it is necessary to select the concept *Agriculture* in order to keep it in $\mathcal{T}_l'$.

Definition 4.3. *Let $\mathcal{T}_l$ be a local TBox and $\mathcal{T}_m$ be the mediator TBox, the agreement $\mathcal{T}_a = \langle \mathcal{T}_l', \mathcal{M} \rangle$ for $\mathcal{T}_l$ w.r.t $\mathcal{T}_m$ is such that (i) $\mathcal{M} = \langle m_1, ..., m_n \rangle$ is the result of the anchoring of $\mathcal{T}_l$ w.r.t $\mathcal{T}_m$, and (ii) $\mathcal{T}_l'$ is inductively defined as follows:*

1. *All assertions in $\mathcal{M}$ are in $\mathcal{T}_a$.*
2. *If A is an anchored concept and $B \sqsubseteq A$ is in $\mathcal{T}_l$, then $B \sqsubseteq A$ is in $\mathcal{T}_{l'}$.*
3. *If A is an anchored or a selected concept and $A \sqsubseteq B$ is in $\mathcal{T}_l$, then $A \sqsubseteq B$ is in $\mathcal{T}_l'$.*
4. *If $Q \sqsubseteq R$ is in $\mathcal{T}_l$ and $\exists R \sqsubseteq B$ is in $\mathcal{T}_l'$, then $Q \sqsubseteq R$ is in $\mathcal{T}_l'$.*
5. *If $(funct\ Q)$ is in $\mathcal{T}_l$ and $\exists Q \sqsubseteq B$ is in $\mathcal{T}_l'$, then $(funct\ Q)$ is in $\mathcal{T}_l'$.*
6. *If $\rho(U_C) \sqsubseteq F$ is in $\mathcal{T}_l$ and $B \sqsubseteq \delta(U_C)$ is in $\mathcal{T}_l'$, then $\rho(U_C) \sqsubseteq F$ is in $\mathcal{T}_l'$.*
7. *If $U_C \sqsubseteq V_C$ is in $\mathcal{T}_l$ and $B \sqsubseteq \delta(V_C)$ is in $\mathcal{T}_l'$, then $U_C \sqsubseteq V_C$ is in $\mathcal{T}_l'$.*
8. *If $(funct\ U_C)$ is in $\mathcal{T}_l$ and $\exists Q \sqsubseteq B$ is in $\mathcal{T}_l'$, then $(funct\ U_C)$ is in $\mathcal{T}_l'$.*

Notice that all rules in Definition 4.3 are designed to keep in $\mathcal{T}_l'$ as much semantic information contained in $\mathcal{T}_l$ as possible. Fig 5 shows the agreement TBox computed from assertions of the local TBox $\mathcal{T}_l$ shown graphically in Fig. 2(a). Anchored concepts are those obtained in the example shown in Fig. 3(b).

Local TBox		Agreement TBox	
$\exists \, manageDataFor \sqsubseteq StatDeparment$	$\exists \, manageDataFor^- \sqsubseteq Activity$	$StatDeparment \sqsubseteq mo:Departement$	$Activity \sqsubseteq mo:Activities$
$\exists \, relateTo \sqsubseteq Activity$	$\exists \, relateTo^- \sqsubseteq Accomodation$	$\exists \, manageDataFor \sqsubseteq StatDeparment$	$\exists \, manageDataFor^- \sqsubseteq Activity$
$\exists \, relateTo^- \sqsubseteq Agriculture$	$Accomodation \sqsubseteq \delta(capacity)$	$Agriculture \sqsubseteq mo:Agriculture$	$\exists \, relateTo \sqsubseteq Activity$
$Accomodation \sqsubseteq \delta(rate)$	$\rho(capacity) \sqsubseteq xsd:string$	$\exists \, relateTo^- \sqsubseteq Agriculture$	$\exists \, concern \sqsubseteq Agriculture$
$\rho(rate) \sqsubseteq xsd:string$	$Hotel \sqsubseteq Accomodation$	$\exists \, concern^- \sqsubseteq CropsVarieties$	$CropsVarieties \sqsubseteq mo:Varieties$
$House \sqsubseteq Accomodation$	$House \sqsubseteq \neg Hotel$	$CropsVarieties \sqsubseteq \delta(type)$	$CropsVarieties \sqsubseteq \delta(price\_range)$
$\exists \, concern \sqsubseteq Agriculture$	$\exists \, concern^- \sqsubseteq CropsVareities$	$\rho(type) \sqsubseteq xsd:string$	$\rho(price\_range) \sqsubseteq xsd:string$
$CropsVarieties \sqsubseteq \delta(type)$	$Cropsvareities \sqsubseteq \delta(price\_range)$	$Tomate \sqsubseteq mo:Tomato$	$Tomate \sqsubseteq Cropsvarieties$
$\rho(type) \sqsubseteq xsd:string$	$\rho(price\_range) \sqsubseteq xsd:string$	$FriedRice \sqsubseteq mo:Rice$	$Niambi \sqsubseteq Cropsvarieties$
$Tomate \sqsubseteq Cropsvarieties$	$FriedRice \sqsubseteq Cropsvarieties$	$Tomate \sqsubseteq \neg Niambi$	$Tomate \sqsubseteq \neg FriedRice$
$Niambi \sqsubseteq Cropsvarieties$	$Tomate \sqsubseteq \neg Niambi$	$FriedRice \sqsubseteq \neg Niambi$	
$Tomate \sqsubseteq \neg FriedRice$	$FriedRice \sqsubseteq \neg Niambi$		

Fig. 5. Example of an agreement TBox

5   Conciliation

We can now build the global TBox $\mathcal{T}_g$ by conciliating the different agreement TBoxes $\mathcal{T}_{ai} = \langle \mathcal{T}_{li}', \mathcal{M}_i \rangle$ obtained above. The conciliation is achieved incrementally by integrating the agreement TBoxes into $\mathcal{T}_g$, one after another. Integrating an agreement TBox $\mathcal{T}_a$ in $\mathcal{T}_g$ consists in linking its concepts with the ones of other agreement TBoxes already conciliated in $\mathcal{T}_g$. Links between concepts in $\mathcal{T}_g$ are established through anchor concepts contained in $\mathcal{M}_i$ for every agreement TBox $\mathcal{T}_{ai}$. Let us recall that all anchor concepts are part of the mediator TBox $\mathcal{T}_m$. Thus, we search for links between anchor concepts in $\mathcal{T}_m$ in order to use them to conciliate concepts in $\mathcal{T}_g$. In this way, our global TBox $\mathcal{T}_g$ contains the following components:

- the set of agreement TBoxes $\mathcal{T}_{ai} = \langle \mathcal{T}'_{li}, \mathcal{M}_i \rangle$. They represent the part of local TBoxes that are shared ($\mathcal{T}'_{li}$), together with the mappings between these local concepts and the mediator ones ($\mathcal{M}_i$).
- an as small as possible subset $\mathcal{T}'_m$ of $\mathcal{T}_m$ containing only the part of the hierarchy which is usefull to link local concepts.

To illustrate this process in the context of agricultural domain, consider the example in Fig. 6. In this example the concepts *Tomate* of the agreement TBox $\mathcal{T}_{a1}$ and *FriedRice* of the agreement TBox $\mathcal{T}_{a2}$ are respectively anchored by the concepts *Tomato* and *Rice* of the mediator TBox $\mathcal{T}_m$. The structure of the mediator TBox reveals that *Tomato* and *Rice* have a common ancestor which is the concept *plan_product*. We reproduce this relation to conciliate the concepts *Tomate* and *FriedRice* in the global TBox $\mathcal{T}_g$.

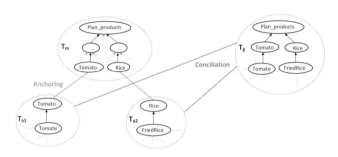

Fig. 6. General overview of the conciliation phase

Definition 5.1. *Let $\{\mathcal{T}_{li}\}$ be a set of local TBoxes and $\mathcal{T}_m$ be a mediator TBox. The corresponding Global TBox $\mathcal{T}_g$ is $\langle \{\mathcal{T}_{ai}\}, \mathcal{T}'_m \rangle$, where (i) $\{\mathcal{T}_{ai}\}$ is the set of agreement TBoxes built from local TBoxes according to Definition 4.3, and (ii) $\mathcal{T}'_m$ is the smallest subset of $\mathcal{T}_m$ that conciliates every $\mathcal{T}_{ai}$ in $\mathcal{T}_g$, built by Algorithm 1.*

As suggested by the example in Fig. 6, one particular interest in our approach is the use of the hierarchy of the mediator TBox $\mathcal{T}_m$ in order to find links between anchor concepts. These links are reproduced in the global TBox for conciliating agreements.

The relation that we are looking for within the hierarchy of $\mathcal{T}_m$ is the *least common subsumer* (*lcs*) of two anchor concepts. It is important to notice that in our first experiments we have only considered tree taxonomies, we plan to generalize this point in future work. We can follow the algorithm proposed in [5] to compute the *lcs* of two concepts C_1 and C_2 in $\mathcal{T}_m$, according to the definition of *lcs* that we recall hereafter.

Definition 5.2. *Let $\mathcal{T}_m$ be the mediator TBox, C_1 and C_2 two given atomic concepts in $\mathcal{T}_m$, the concept C of $\mathcal{T}_m$ is the lcs of C_1 and C_2 in $\mathcal{T}_m$ ($C = lcs_{\mathcal{T}_m}(C_1, C_2)$) iff (i) $C_i \sqsubseteq C$ for $i = 1, 2$, and (ii) C is the least concept with this property, i.e. if C' satisfies $C_i \sqsubseteq C'$ for $i = 1, 2$, then $C \sqsubseteq C'$.*

Based on *lcs* computation in [5], $\mathcal{T}'_m$ consists in a subsumption hierarchy between all anchor concepts of all $\mathcal{T}_{ai}$ and their *lcs* in $\mathcal{T}_m$. The algorithm that we propose to achieve this uses the hierarchical proximity measure proposed by [22], that we recall in the following definition.

Definition 5.3. *Let $\mathcal{T}_m$ be the mediator TBox, C_1 and C_2 two concepts of $\mathcal{T}_m$. The hierarchical proximity measure between C_1 and C_2 in $\mathcal{T}_m$ is such that:*

$$sim_{\mathcal{H}}(C_1,C_2) = \frac{2*depthOf(lcs_{\mathcal{T}_m}(C_1,C_2))}{depthOf(C_1)+depthOf(C_2)},$$

where $depthOf(C)$ returns the number of subsumers of C in $\mathcal{T}_m$.

Definition 5.4. *Let $C \in \mathcal{T}_m$. If $sim_{\mathcal{H}}(C,C_j) \geq sim_{\mathcal{H}}(C,C_i), \forall C_i \in \mathcal{T}_m$, then we say that C_j is the closest concept of C in $\mathcal{T}_m$ and we denote it by $closest_{\mathcal{T}_m}(C)$.*

If $sim_{\mathcal{H}}(C,C_j) \geq sim_{\mathcal{H}}(C,C_i), \forall C_i \in \mathcal{T}'_m \subseteq \mathcal{T}_m$, then we say that C_j is the closest concept of C in $\mathcal{T}'_m$ w.r.t. $\mathcal{T}_m$ and we denote it by $closest_{\mathcal{T}'_m/\mathcal{T}_m}(C)$.

The conciliation of an agreement TBox $\mathcal{T}_{ak} = \langle \mathcal{T}'_{lk}, \mathcal{M}_k \rangle$ with others agreement TBoxes already conciliated in $\mathcal{T}_g = \langle \{\mathcal{T}_{ai}\}_{i\neq k}, \mathcal{T}'_m \rangle$ consists in integrating each anchor concept A_m of $\mathcal{M}_k$ within the hierarchy $\mathcal{T}'_m$. To integrate a concept A_m within the hierarchy $\mathcal{T}'_m$ we have to compute the lcs in $\mathcal{T}_m$ between A_m and the closest concept of A_m in $\mathcal{T}_m$ among the anchor concepts already present in the hierarchy $\mathcal{T}'_m$. In order to express these features in our conciliation algorithm, we use Definitions 5.4, as it can be noticed in what follows:

Algorithm 1. *Conciliation*
Input: $\mathcal{T}_{ak} = \langle \mathcal{T}'_{lk}, \mathcal{M}_k \rangle$, $\mathcal{T}_g = \langle \{\mathcal{T}_{ai}\}_{i\neq k}, \mathcal{T}'_m \rangle$
Output: $\mathcal{T}_g = \langle \{\mathcal{T}_{ai}\} \cup \mathcal{T}_{ak}, \mathcal{T}'_m \rangle$
  1: **for each** $(m_k = A_l \sqsubseteq A_m$ in $\mathcal{M}_k)$ **do**
  2:        **if** $((\mathcal{T}'_m \neq \emptyset)$ and $(A_m \nsqsubseteq \mathcal{T}'_m))$ **then**
  3:            $A_{cl} \leftarrow closest_{\mathcal{T}'_m/\mathcal{T}_m}(A_m)$
  4:            $A_{lcs} \leftarrow lcs_{\mathcal{T}_m}(A_m, A_{cl})$
  5:            **if** $(A_{lcs} = A_{cl})$ **then**
  6:                  $\mathcal{T}'_m \leftarrow \mathcal{T}'_m \cup \{A_m \sqsubseteq A_{cl}\}$
  7:            **else if** $(A_{lcs} = A_m)$ **then**
  8:                  $\mathcal{T}'_m \leftarrow \mathcal{T}'_m \cup \{A_{cl} \sqsubseteq A_m\}$
  9:                  **if** $(\exists A \in \mathcal{T}'_m \mid A = lcs_{\mathcal{T}'_m}(A_{cl}, A))$ **then**
 10:                      $\mathcal{T}'_m \leftarrow \mathcal{T}'_m \cup \{A_m \sqsubseteq A\}$
 11:                  **end if**
 12:            **else**
 13:                  $\mathcal{T}'_m \leftarrow \mathcal{T}'_m \cup \{A_m \sqsubseteq A_{lcs}, A_{cl} \sqsubseteq A_{lcs}\}$
 14:                  **if** $(\exists A \in \mathcal{T}'_m \mid = lcs_{\mathcal{T}'_m}(A_{cl}, A))$ **then**
 15:                      $\mathcal{T}'_m \leftarrow \mathcal{T}'_m \cup \{A_m \sqsubseteq A\}$
 16:                  **end if**
 17:            **end if**
 18:        **else if** $(\mathcal{T}'_m = \emptyset)$ **then**
 19:            $\mathcal{T}'_m \leftarrow \mathcal{T}'_m \cup \{A_m \sqsubseteq \top\}$
 20:        **end if**
 21: **end for**

To illustrate our algorithm, we consider two TBoxes $\mathcal{T}_{a1} = \langle \mathcal{T}'_{l1}, \mathcal{M}_1 \rangle$ and $\mathcal{T}_{a2} = \langle \mathcal{T}'_{l2}, \mathcal{M}_2 \rangle$ such that :

- $\mathcal{M}_1 = \langle FrideRice \sqsubseteq mo : rice, Onion \sqsubseteq mo : Onion \rangle$
- $\mathcal{M}_2 = \langle Sorgho \sqsubseteq mo : Sorgho, Tomate \sqsubseteq mo : Tomato \rangle$

Results obtained by conciliating $\mathcal{T}_{a1}$ and $\mathcal{T}_{a2}$ are as follows:

1- **Integrate $\mathcal{T}_{a1}$ in $\mathcal{T}_g$**
   Input: $\mathcal{T}_{a1}$, $\mathcal{T}_g = \langle \{\}, \mathcal{T}'_m = \emptyset \rangle$

   iteration 1 $- m_{1.1} = FrideRice \sqsubseteq mo : Rice$
   $\mathcal{T}'_m = \{Rice \sqsubseteq \top\}$
   iteration 2 $- m_{1.2} = Onion \sqsubseteq mo : Onion$
   $A_{cl} = Rice$; $A_{lcs} = PlanProducts$
   $\mathcal{T}'_m = \{Rice \sqsubseteq PlanProducts, Onion \sqsubseteq PlanProducts\}$
2- **conciliate $\mathcal{T}_{a1}$ and $\mathcal{T}_{a2}$ in $\mathcal{T}_g$**
   Input: $\mathcal{T}_{a2}$, $\mathcal{T}_g = \langle \{\mathcal{T}_{a1}\}, \mathcal{T}'_m = \{Rice \sqsubseteq PlanProducts, Onion \sqsubseteq PlanProducts\} \rangle$

   iteration 1 $- m_{2.1} = Sorgho \sqsubseteq mo : Sorgho$
   $A_{cl} = Rice$; $A_{lcs} = Cereals$
   $\mathcal{T}'_m = \{Rice \sqsubseteq Cereals, Sorgho \sqsubseteq Cereals, Cereals \sqsubseteq PlanProducts, Onion \sqsubseteq PlanProducts\}$
   iteration 2 $- m_{2.2} = Tomate \sqsubseteq mo : Tomato$
   $A_{cl} = Onion$; $A_{lcs} = Vegetables$
   $\mathcal{T}'_m = \{Rice \sqsubseteq Cereals, Sorgho \sqsubseteq Cereals, Cereals \sqsubseteq PlanProducts, Onion \sqsubseteq Vegetables, Tomato \sqsubseteq Vegetables, Vegetables \sqsubseteq PlanProducts\}$

Fig. 7 illustrates graphically the global TBox $\mathcal{T}_g$ resulting from the conciliation of $\mathcal{T}_{a_1}$ and $\mathcal{T}_{a_2}$. We have distinguished the hierarchy $\mathcal{T}'_m$, composed of all anchor concepts in $\mathcal{M} = \mathcal{M}_1 \cup \mathcal{M}_2$, linked to each other by containment assertions found in $\mathcal{T}_m$. Notice that all information existing in local TBoxes also exists in $\mathcal{T}_g$. In fact, the part of $\mathcal{T}_g$ which is not in $\mathcal{T}'_m$ represents the data sources that can be accessed from the global schema, this access being supported by the mapping $\mathcal{M}$ (following a GAV approach).

It can be noticed that the global TBox $\mathcal{T}_g$ is composed of *(i)* the union of all $\mathcal{T}'_l$ (local parts of each source agreement), *(ii)* the union of all $\mathcal{M}$ (mapping parts of each source agreement) and *(iii)* the hierarchy extracted from $\mathcal{T}_m$ to relate anchors in $\mathcal{T}_g$. Thus, managing dynamic changes that occur frequently in a semantic web context requires to consider not only to add new sources but also to remove a source or to update a source's schema. We consider these three operations: adding a new source S consists of the following steps:

   – the computation of S's agreement: $\mathcal{T}_a = \langle \mathcal{T}'_l, \mathcal{M} \rangle$,
   – the union of $\mathcal{T}'_l$ with local parts of the other sources,
   – the union of $\mathcal{M}$ with mapping parts of the other sources,
   – the integration of anchors of $\mathcal{M}$ into the hierarchy $\mathcal{T}'_m$ (using Algorithm 1).

Removing a source S that became unavailable will follow two stages:

- the removal from $\mathcal{T}_g$ of $\mathcal{T}_a = \langle \mathcal{T}_l', \mathcal{M} \rangle$ corresponding to S,
- an iteration on the hierarchy $\mathcal{T}_m'$, in order to remove items that became unnecessary. We plan to design the corresponding algorithm in future works and to compare it with the simple recomputation of $\mathcal{T}_m'$ based on remaining sources.

Finally, taking into account an update performed on a source TBox $\mathcal{T}_l$ will require first to compute the corresponding new agreement. This is again one of our future works to design an incremental update algorithm, more efficient than removing the old version and adding the new one.

6   Conclusion and Future Work

Our proposition in this article brings a solution to the problem of *automatic* construction of an appropriate global ontology. This ontology will be shared between a set of loosely-interrelated partners. We have tackled this problem using a background-knowledge, i.e., a domain-reference ontology, as a mediator to build the global ontology. The global ontology offers interesting properties, especially an appropriate conceptualization and easy resource-adding and querying processes. For our solution to be automated, we use, on one hand, logical-inference techniques offered by description logics, the knowledge-representation formalism used for ontology specification. On the other hand, we make use of some classical syntactic-matching techniques for ontology matching. Our approach is that hybrid. To the best of our knowledge, no other solution has already been proposed in the literature combining these two techniques for an automatic construction of a global ontology.

We are working to go further in exploring automatic-reasoning capabilities of *DL-Lite$_\mathcal{A}$* Description Logic, in order to (i) check the global-ontology's consistency and (ii) answer queries using the global ontology. Moreover, in future work all types of relations will be considered in the reference ontology, we will thus extend our proposition

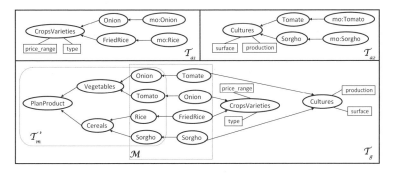

Fig. 7. Overview of a global TBox that conciliates two agreement TBoxes

that concerns here only subsumption relationship. Another important future work is to specify the complete life cycle management of the global ontology.

Finally, we will build upon our first experiments, which have been realized as a proof of concept but are not yet an actual publishable evaluation, in order to turn our proposition into a robust software for ontology-based data integration.

References

1. Aleksovski, Z., Klein, M.C.A., ten Kate, W., van Harmelen, F.: Matching Unstructured Vocabularies Using a Background Ontology. In: Staab, S., Svátek, V. (eds.) EKAW 2006. LNCS (LNAI), vol. 4248, pp. 182–197. Springer, Heidelberg (2006)
2. Aleksovski, Z., ten Kate, W., van Harmelen, F.: Exploiting the Structure of Background Knowledge Used in Ontology Matching. In: Ontology Matching (2006)
3. Amann, B., Beeri, C., Fundulaki, I., Scholl, M.: Ontology-based integration of XML web resources. In: Horrocks, I., Hendler, J. (eds.) ISWC 2002. LNCS, vol. 2342, pp. 117–131. Springer, Heidelberg (2002)
4. Baader, F., Calvanese, D., McGuinness, D., Nardi, D., Patel-Schneider, P.: The Description Logic Handbook: Theory, Implementation and Applications. Cambridge University Press, Cambridge (2003)
5. Baader, F., Sertkaya, B., Turhan, A.-Y.: Computing the least common subsumer w.r.t. a background terminology. J. Applied Logic 5(3), 392–420 (2007)
6. Bergamaschi, S., Castano, S., Vincini, M., Beneventano, D.: Semantic integration of heterogeneous information sources. Data Knowl. Eng. 36(3), 215–249 (2001)
7. Borgida, A., Serafini, L.: Distributed description logics: Directed domain correspondences in federated information sources. In: Chung, S., et al. (eds.) CoopIS 2002, DOA 2002, and ODBASE 2002. LNCS, vol. 2519, pp. 36–53. Springer, Heidelberg (2002)
8. Borgida, A., Serafini, L.: Distributed description logics: Assimilating information from peer sources. In: Spaccapietra, S., March, S., Aberer, K. (eds.) Journal on Data Semantics I. LNCS, vol. 2800, pp. 153–184. Springer, Heidelberg (2003)
9. Calvanese, D., De Giacomo, G., Lembo, D., Lenzerini, M., Rosati, R.: Tractable Reasoning and Efficient Query Answering in Description Logics: The L-Lite Family. J. Autom. Reasoning 39(3), 385–429 (2007)
10. Choi, N., Song, I.-Y., Han, H.: A survey on ontology mapping. SIGMOD Record 35(3), 34–41 (2006)
11. Cruz, I.F., Xiao, H.: Ontology driven data integration in heterogeneous networks. In: Complex Systems in Knowledge-based Environments, pp. 75–98 (2009)
12. Cruz, I.F., Xiao, H., Hsu, F.: Peer-to-peer semantic integration of XML and RDF data sources. In: Moro, G., Bergamaschi, S., Aberer, K. (eds.) AP2PC 2004. LNCS (LNAI), vol. 3601, pp. 108–119. Springer, Heidelberg (2005)
13. Grau, B.C., Parsia, B., Sirin, E.: Combining owl ontologies using epsilon-connections. J. Web Sem. 4(1), 40–59 (2006)
14. Kalfoglou, Y., Schorlemmer, M.: Ontology Mapping: The State of the Art. The Knowledge Engineering Review 18 (2003)
15. Lenzerini, M.: Data integration: A theoretical perspective. In: PODS, pp. 233–246 (2002)
16. Mena, E., Illarramendi, A., Kashyap, V., Sheth, A.P.: OBSERVER: An Approach for Query Processing in Global Information Systems Based on Interoperation Across Pre-Existing Ontologies. Distributed and Parallel Databases 8(2), 223–271 (2000)
17. Poggi, A., Lembo, D., Calvanese, D., De Giacomo, G., Lenzerini, M., Rosati, R.: Linking Data to Ontologies. In: Spaccapietra, S. (ed.) Journal on Data Semantics X. LNCS, vol. 4900, pp. 133–173. Springer, Heidelberg (2008)

18. Sabou, M., d'Aquin, M., Motta, E.: Using the Semantic Web as Background Knowledge for Ontology Mapping. In: Ontology Matching (2006)
19. Serafini, L., Borgida, A., Tamilin, A.: Aspects of distributed and modular ontology reasoning. In: IJCAI, pp. 570–575 (2005)
20. Shvaiko, P., Euzenat, J.: Ten Challenges For Ontology Matching. In: Proceedings of The 7th International Conference on Ontologies, DataBases, and Applications of Semantics, ODBASE (2008)
21. Wache, H., Vögele, T., Visser, U., Stuckenschmidt, H., Schuster, G., Neumann, H., Hübner, S.: Ontology-Based Integration of Information - A Survey of Existing Approaches. In: IJCAI 2001 Workshop on Ontologies and Informations Sharing, pp. 108–117 (2001)
22. Wu, Z., Palmer, M.S.: Verb Semantics and Lexical Selection. In: ACL, pp. 133–138 (1994)

Semantic Interoperation of Information Systems by Evolving Ontologies through Formalized Social Processes

Christophe Debruyne and Robert Meersman

Semantics Technology & Applications Research Laboratory,
Vrije Universiteit Brussel, Brussels, Belgium
{chrdebru,meersman}@vub.ac.be

Abstract. For autonomously developed information systems to interoperate in a meaningful manner, ontologies capturing the intended semantics of that interoperation have to be developed by a community of stakeholders in those information systems. As the requirements of the ontology and the ontology itself evolve, so in general will the community, and vice versa. Ontology construction should thus be viewed as a complex activity leading to formalized semantic agreement involving various social processes within the community, and that may translate into a number of ontology evolution operators to be implemented. The hybrid ontologies that emerge in this way indeed need to support both the social agreement processes in the stakeholder communities and the eventual reasoning implemented in the information systems that are governed by these ontologies. In this paper, we discuss formal aspects of the social processes involved, a so-called fact-oriented methodology and formalism to structure and describe these, as well as certain relevant aspects of the communities in which they occur. We also report on a prototypical tool set that supports such a methodology, and on examples of some early experiments.

Keywords: ontology development, methodology, social process, business semantics management, fact-orientation, natural language.

1 Introduction

Ontologies are keystone technologies for the meaningful and efficient interoperation of information systems. Information systems on the Web are in general developed and maintained *autonomously*, which necessitates agreement to be negotiated between Web services. This, in turn, requires agreement between the stakeholders and designers on the semantics of the shared concepts involved. As a consequence, ontologies in general will *evolve* while such agreements are developed and finally put in place. These ontologies are approximations of a real world; in fact to the Web services involved, ontologies *are* the world. Ontologies represent an *externalization* of the semantics *outside* of the information system. The basic techniques and architecture for semantic interoperation is based on *annotation* (of an application system) and *reasoning* (about the concepts involved, in terms of the ontology).

From above it follows that the modeling of ontologies within a community of stakeholders and designers is a critical activity for the eventual success of interoperability. In

J. Eder, M. Bielikova, and A.M. Tjoa (Eds.): ADBIS 2011, LNCS 6909, pp. 444–459, 2011.
© Springer-Verlag Berlin Heidelberg 2011

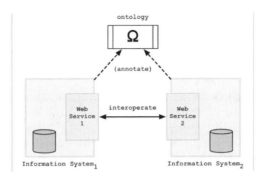

Fig. 1. Two autonomously developed information systems interoperating at runtime, exchanging messages via the application symbols annotated with the shared ontology

this paper, we discuss the social processes involved, a methodology and formalism to structure these and the communities in which they occur, and a prototypical tool set that supports such a methodology.

Fundamental to our approach is the involvement of structured natural language as a vehicle to elicit useful and relevant concepts from community communication, and the mapping of these social processes to evolutionary processes in the emerging ontology. The formalism and language presented here are therefore "upstream" from the usual ontology languages such as RDF(S) and OWL and should not be confused with those; in fact it is relatively straightforward to compile the resulting/emerging ontologies into, for example, RDF(S) and OWL at any time.

One fundamental principle of all large system design is the so-called *separation of concerns* resulting in architectures that delegate respective functionalities to the stakeholders responsible for them. Examples are modules, etc. provided by the (generic) architecture of information systems driven by a database, largely separating the concern of basic data management from that of application development, the famous paradigm of *data independence.*

We reapply this principle in our approach by the rigorous separation in conceptualizations of "fact modeling" from all application-specific interpretations. It is this interpretation process (formally, of statements shared in the application system in terms of ontology concepts) that usually is called "reasoning" in the Semantic Web literature. However, there is little or no attention to such separation of concerns in the usual reasoning formalisms of Semantic Web in terms of Description Logic and its syntactical manifestations such as OWL and its dialects. In our approach, this interpretation is *exclusively* delegated to the mapping between application system and the "lexon base"[1] of the ontology. We shall call these mappings *ontological commitments* after Guarino [10], but shall reify them in a well-defined manner suited to our formalism. Intuitively, our commitments select the facts needed, map application symbols to ontology concepts, and contain the rules and constraints, expressed in ontology terms, under which application symbols, relationships and business rules must be interpreted when they are to be shared with other autonomous systems. Those

---

[1] We shall call the facts in the ontology lexons to distinguish the terminology from application context. See Section 3 for details.

systems will share the concepts, but of course will have their own symbols, business rules, etc.

This separation of concerns now allows a natural introduction of formalized social processes in goal-oriented communities such as exist in enterprises, professional networks, standardization groups, etc. and in fact in any "human agent" context for which agreement about facts is more efficient than reasoning from axioms. Note that nearly all data models for databases and business information systems were arrived at in this manner for the last 50 or so years.

And finally, as we shall argue in the next sections, this provides for a suitable and elegant context in which such business information systems can be made truly *open* by "lifting" their data models to an ontological level by widening the scope of the social processes. It follows that data schemas (e.g. defining a relational database) should not be seen as equal to ontologies. At best they will serve, after lifting by a community into a more "agreed" and "shared" form, as first approximations of one. In fact the same is true for any "conceptual schema" [1] that was used for designing an information system within one given enterprise. For more details on the distinction between a data model and an ontology, we refer to [18].

2 Related Work

Social interactions have been studied by observation in *mediawiki* talk pages by [21], which resulted in a (fairly limited) taxonomy of possible discussion items on a Wikipedia article. [17] also noted a correlation between the number of edits and the amount of discussion in an article. [17] extended this classification by using a larger dataset and added about 5 extra types. The goal of [17] was to create subclasses of sioc:Post so different types of discussion items can be easily accessed and mined upon. The SIOC Ontology[2] focuses on the integration of online community socialization and is used in conjunction with the FOAF[3] vocabulary for expressing personal profile and social networking facts. In the context of a discussion, forum topics can range from conceptions that must be added to the ontology to meta-concept types that constitute the community meta-model itself.

Fact-oriented modeling, such as ORM [11] and NIAM [22], is a method for analyzing and creating conceptual schemas for information systems starting from (usually binary) relationships expressed as part of human-to-system communication. Using concepts and a language people are intended to readily understand, fact-oriented modeling helps ensuring the quality of a database application without caring about any implementation details of the database, including e.g. the grouping itself of linguistic concepts into records, relations, ... In fact-oriented approaches, every concept plays roles with other concepts, and those roles may be constrained. It is those constraints that allow the implementer of a database (or in fact an algorithm) to determine whether some linguistic concept becomes an entity or an attribute, or whether a role turns out to be an attribute relationship or not. This is different from other approaches such as (E)ER and UML, where these decisions are made at design time.

---

[2] http://www.sioc-project.org/
[3] http://xmlns.com/foaf/spec/

3 DOGMA

In [14, 15] a formalism and methodology for ontology development called DOGMA[4] was defined that illustrated and implemented these principles, now lifted to domain level from the mere enterprise system level. We first define a DOGMA *ontology description*. As indicated in the introduction, such descriptions must be seen as different from their eventual implementations, e.g. using RDF(S) and/or OWL. In the methodology and lifecycle of semantic systems the creation of DOGMA ontology descriptions belongs upstream from such implementation – although of course in many cases one will have to "mine" or elicit the required knowledge from existing information systems and their enterprise environments.

3.1 Towards Hybrid Ontology Descriptions

Definition 1. A DOGMA Ontology Description Ω is an ordered triple $< \Lambda, ci, K >$ where Λ is a *lexon base*, i.e. a finite set of *lexons*. A lexon is an ordered 5-tuple $< \gamma, t_1, r_1, r_2, t_2 >$ where $\gamma \in \Gamma$ is a context identifier, $t_1, t_2 \in T$ are *terms*, and $r_1, r_2 \in R$ are *role labels*. A lexon is a binary fact type that can be read in two directions: t_1 playing the role of r_1 on t_2 and t_2 playing the role of r_2 on t_1. We omit here for simplicity the usual alphabets for constructing the elements of $T \cup R$. $ci : \Gamma \times T \rightarrow C$ is a partial function mapping pairs of context identifiers and terms to (unique) elements of C, a finite given set of *concepts*. The nature of these concepts is intentionally left unspecified but intuitively it is assumed that all users of an ontology described by Ω, i.e. sharing Λ and K, agree on the nature of all concepts in C. In a concrete way, within a context γ, $ci(\gamma, t)$ *is* the definition itself of the concept agreed by all such users. To emphasize this explicit agreement, we shall avoid to label concepts as such in our formalism, and assume they are "computed" by the community from the term labels. K is a finite set of ontological *commitments*. Each commitment is an ordered triple $< \sigma, \alpha, c >$ where $\sigma \subseteq \Gamma$ is a selection of lexons, $\alpha : \Sigma \rightarrow T$ is a mapping called an *annotation* from the set Σ of application (information, system, database) symbols to terms, and c is a predicate over $T \cup R$ expressed in a suitable first-order language, not defined further in this paper but an example syntax named Ω-RIDL may be found in [19,20].

Context identifiers are pointers to a community, they can be a name, a URI to a website or even a URI to a document describing the community. To improve readability, we use names as a context identifier in the following examples. For example, the DOGMA ontology description might contain the following plausible lexons in its lexon base:

- *<VCard Community, VCard, with, of, Email Address>*
- *<Vendor Community, Offer, with, of, Title>*
- *<Vendor Community, Offer, contains, contained in, Product>*
- *<Vendor Community, Offer, made by, makes Vendor>*
- *<RFP Community, Request For Proposals, corresponds to, matches, Offer>*

---

[4] Developing Ontologies-Grounded Methodology and Applications.

The function   *ci*   maps   terms   in   those   lexons   to   concepts,   e.g.
ci(Vendor Community,Offer) points to a URL of a concept definition in which all
synonyms are centralized, e.g. *ci(RFP Community,Offer)* of a fact entered in the
lexon base by a different community with overlapping concepts in their domains.
These lexons are then used to construct commitments. **Fig. 2** depicts an example
commitment. The characters in boldface are reserved words for creating the constraints and mappings. The underlined characters represent variables. For more details
on the syntax of commitments, we refer to [19,20].

```
BEGIN SELECTION
  <Vendor Community, Offer, with, of, Title>
  <Vendor Community, Offer, contains, contained in, Product>
  <Vendor Community, Offer, made by, makes, Vendor>
  <Vendor Community, Vendor, located on, location of, Address> …
END SELECTION
BEGIN CONSTRAINTS
  Offer contains at least 1 Product.
  Vendor located on exactly 1 unique Address. …
END CONSTRAINTS
BEGIN MAPPING
  map "APP_OFF.TITLE" on Title of Offer.
  map "APP_VEN.ADDR" on Address location of Vendor. …
END MAPPINGS
```

Fig. 2. Example of a commitment for a particular application showing pieces of the three parts:
selection σ, constraints c and annotations (or application symbol mappings) α

Note that the separation of concerns mentioned in the previous section is reflected
here through the set of plausible facts in the lexon base on one side, and the constraints, rules, … on a relevant selection of those lexons on the other. In fact there are
no constraints or any other reasoning supports included in the lexon base, making for
a so-called *light* ontology.

Also note this definition imparts a well-defined *hybrid aspect* on ontologies as they
are to be resources shared among humans working in a community as well as among
networked systems such as exist in the World Wide Web. As the "unique concept"
property mentioned above informally and intuitively results from a community
agreement, for the purpose of this paper we find it useful therefore to formalize a
community precisely as such a context, and to name the resulting notion a *hybrid
ontology* (see also [16]). We also introduce a special linguistic resource, called a *glossary*, recording and supporting all the social processes.

Definition 2. A Hybrid Ontology Description (HOD) is an ordered pair
$H\Omega =< \Omega,G >$ where Ω is a DOGMA ontology description where the contexts in Γ
are labeled *communities* and G is a *glossary,* a finite set of functions either of the
form $g_1 : \Gamma \times T \rightarrow$ Gloss, the *Term Glossary* or of the form $g_2 : \Lambda \rightarrow$ Gloss, the *Lexon Glossary*. Gloss is a set of linguistically interpretable objects. We shall write
$G = G_1 \cup G_2$ if the distinction needs to be made explicit.

For example, given the DOGMA ontology description Ω from the previous example, a HOD can be constructed where G contains (among others):

- *(<VCard, Email Address>, "The address of an email (system of world-wide electronic communication in which a user can compose a message at one terminal that can be regenerated at the recipient's terminal when the recipient logs in)")*
- *(<Vendor Community, Offer>, "Represents the public announcement by a vendor to provide a certain business function for a certain product or service to a specified target audience.")*
- *(<Vendor Community, Offer, contains, contained in, Products>, "Represents the relation of a product for sale being included in an offer.")*

Note that in this paper, we shall not concern ourselves with the precise nature of the elements of Gloss. For the sake of simplicity and understanding it will be sufficient to think of Gloss as a set of natural language documents each providing an "explanation" for a term in T or a lexon in Λ adequate within a given community.

3.2 Glossaries

Glossaries turn out to require a fairly rich structure when to be deployed for the purpose of (hybrid) ontology engineering, as they are used to build agreements in communities about concepts. It is natural to associate them with concepts (in a DOGMA ontology description through the terms of lexons).

Definition 3. Given a HOD $H\Omega =< \Omega, G >$, we call a glossary *coherent* if $\forall \gamma \in \Gamma, \forall \lambda =< \gamma, t_1, r_1, r_2, t_2 > \in \Lambda : g_1(\gamma, t_1) \underset{\rightarrow}{} g_2(\gamma, \lambda) \wedge g_1(\gamma, t_2) \underset{\rightarrow}{} g_2(\gamma, \lambda)$. Where $\underset{\rightarrow}{}$ stands for "is subsumed by", which is not a logic property, but a binary (linguistic) predicate on the set Gloss, intended to express that any community agreement on its first argument is implied by a community agreement about its second. (One way to implement such a predicate may be by simply listing its extension.)

Indeed, it would be undesirable to describe a relation between two terms if one or both terms playing the roles in that relation are not described themselves, meaning that their intended meaning has not yet been made explicit.

Definition 4. The glossary consistence property. We say that a hybrid ontology satisfies this property if for every two pairs $< \gamma_1, t_1 >, < \gamma_2, t_2 > \in \Gamma \times T$, if $g_1(\gamma_1, t_1) = g_1(\gamma_2, t_2)$ then $ci(\gamma_1, t_1) = ci(\gamma_2, t_2)$. The converse does not necessarily hold. In other words, if two terms in two communities point to exactly the same gloss, then they must refer to the same concept as well. For most purposes this condition is too limiting since often glosses will express "the same thing" without being textually identical. It suffices that the communities agree on their equivalence; this lead to the following definition.

Definition 5. Two term glosses are *community-equivalent* EQ_γ if that community agrees that the described terms refer to the same (abstract) concept. A similar definition may be given for lexon glosses; it is omitted here. Two term glosses are *term-equivalent* EQ_T if any two communities agree that a given term refers to the same concept for both. It is easy to see that EQ_γ and EQ_T define equivalence relations (reflexive, symmetric and transitive) on G. Again, implementation could be by listing (or e.g. logging) its extension.

Definition 6. Given a hybrid ontology description $H\Omega = <\Omega, G>$ and communities $\gamma_1, \gamma_2 \in \Gamma$ and $t_1 \in T$, we say that community γ_2 *adopts* $<\gamma_1, t_1>$ when $gloss_1 = g_1(\gamma_1, t_1)$ and $gloss_2 = g_1(\gamma_2, t_1)$ are defined, and we have (i) $EQ_T(gloss_1, gloss_2)$, i.e. first "match" the two glosses; and (ii) $ci(\gamma_1, t_1) = ci(\gamma_2, t_1)$, i.e. agree that both concepts are equal.

In other words, γ_1 and γ_2 agree behind the concepts on their respective glosses (a symmetric condition) and γ_2 agrees to use t_1 as a term to refer to γ_1's concept behind it (an asymmetric condition). **Fig. 3** shows an example of agreeing on the equivalence of two glosses that are synonyms inside the Business Semantics Glossary (see Section 5). If their synonym relation was not already established, it will be after this action.

Fig. 3: Example of two glosses the two different communities deem to be the same. The first is a definition for the term "email address" in community interested in exchanging information about offers of parking spots. Their definition came from their domain expert. The second was the definition of a different community whose effort went to lifting vCard into a hybrid ontology description. The result of agreeing on the equality was a link stating that these two terms in those two communities were synonyms.

What the definition does *not* tell is *how* to achieve the shared understanding from those glosses that the concepts are the same. For this, one option is to let the communities involve their commitments to t_1 from their respective intended applications; in particular we need to study the reference structures for t_1 in those commitments.

Several guidelines on the construction of such glosses were given in [13]. While formally adequate to be useful in practice, we shall require more details of the *structure* (i.e. organizations) and the *processes* by which such a community achieves agreement about lexons and about the commitment of a specific information system. An immediate consequence is the requirement that a community viewed as a context must agree on unique concepts based on terms used in lexons. In this paper, we propose a formalism and methodological approach for such interactions of communities with the repositories of the knowledge they own.

The hybrid aspect is reflected in a dual perspective on the ontology Ω, and in particular on the glossary underpinning its lexons within a community: the community members agree on unique concepts based on glosses while systems interoperate (reason) based on the relationships (facts) that are deemed to exist between terms that refer to those same concepts.

Fig. 4 depicts a simplification of the iterative process involved. The Hybrid Ontology Description is used downstream (ref. **Fig. 5**) to generate a knowledge base, e.g. as RDF(S)-defined "storage structures" and constraints/rules implementing relevant commitments for the enterprise information systems to be served. The *co-evolution* of a community and its Hybrid Ontology Description is a natural consequence of this process. Externalization - identifying the key conceptual patterns that are relevant from the discussions [7]- results in a series of ontology evolution operators for the next version and (re-)internalization – by committing instance bases to the new version of the ontology [7] - changes the community's composition: members depart when their goals differ too much from the common goal, or others join.

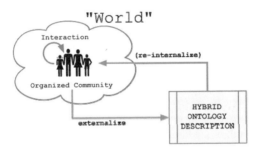

Fig. 4. Feedback loop between an organized community and an ontology. The interactions between the community result in ontology evolution operators applied to the ontology. These operators thus enact the externalization of the reflections of that community. The new ontology description, after a while, will be re-internalized as the community achieves (and discusses about) new insights.

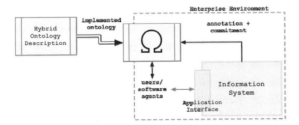

Fig. 5. "Downstream" usage of the Hybrid Ontology Description to implement the ontology, used for annotating the application symbols of an information system. Users and software agents "recognize" the kind of annotated data provided by the information system and the ontology

Before describing the procedure of the methodology and the description of social processes in the next sections, we would like to note that communities in a collaborative ontology engineering methodology are *relevant only* if there are *two or more autonomously developed information systems* that need to interoperate. When there is only one information system, the semantics resulting from that community (even if the number of people is greater than one) are of that application. This would bring us no step further from going from a closed information system to open information systems.

3.3 Procedure of the Methodology

To achieve this we first define a set of operations on a community that are intended to reflect its member interaction with the "real world" and with each other. Then we "map" those operations onto a sequence of ontology evolution operators, as defined by [4]. It is essential to observe that the ontology description evolves *only* as the result of agreements, viz. actions performed in principle by multiple community members.

Every ontology evolution operation is subject to discussion before approval during the re-internalization phase of mentioned earlier. Members request certain changes under the form of those operators with a motivation. The status of each such request is initially "candidate". Depending on the outcome of the discussion, the request is "approved", "denied" or "postponed for future iteration". The latter is useful when the community agrees that the request falls out the scope of the current iteration. Once proposed changes have been accepted and the community decides to go to a next version of the hybrid ontology description, all changes are translated into ontology evolution operators.

The next section starts from an existing collaborative ontology engineering approach built on top of DOGMA in which – for every iteration – a set of ontology engineering phases are identified. For every phase, the different social processes (e.g. the request to add a lexon and its discussion) and corresponding ontology evolution operators are identified (e.g. adding a lexon).

4 Social Processes in Ontology Engineering

These operations can be classified according to the different phases of a collaborative ontology engineering process. Business Semantics Management (BSM) [5], developed by Collibra[5], is such a collaborative ontology engineering methodology drawing from best practices in ontology management [12] and ontology evolution [6]. The representation of business semantics is based on the DOGMA approach[6]. BSM consists of two complementary cycles: semantic reconciliation and semantic application (see **Fig. 6**) that each groups a number of activities.

---

[5] Collibra nv/sa, launched in 2008, is a spin-off company of the Vrije Universiteit Brussel that validates and further develops the technology of the DOGMA research project. http://www.collibra.com/

[6] Recently, BSM adopted Semantics of Business Vocabulary and Business Rules (SBVR) [3], a recent OMG standard pushed by the business rule community and the fact-oriented modeling community (and fully compatible with DOGMA).

Semantic Reconciliation is the first cycle of the methodology. In this phase, business semantics are modeled by extracting, refining, articulating and consolidating lexons from existing sources such as natural language descriptions, existing metadata, etc. Ultimately, this results in a number of consolidated language-neutral semantic patterns that are articulated with glosses (e.g. WordNet [9] word senses). These patterns are reusable for constructing various semantic applications. This process is supported by the Business Semantics Glossary, which will serve as basis for our prototype (see Section 5).

Semantic Application is the second cycle. During this cycle, existing information sources and services are committed to a selection of lexons, as explained earlier. In other words, a commitment creates a bidirectional link between the existing data sources and services and the business semantics that describe the information assets of an organization. The existing data itself is not moved nor touched.

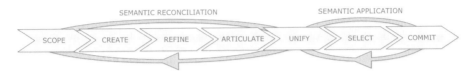

Fig. 6. Business Semantics Management consists of two complementary cycles: semantic reconciliation and semantic application. Both cycles communicate via the unify-activity

4.1 Semantic Reconciliation and Its Social Processes

Scope defines the borders of the current ontology engineering iteration and helps grounding discussions, preventing members of a community to go "off topic" on the current problem. The first iteration consists of the initial community of members representing autonomously developed information systems that need to interoperate. The discussions are grounded on the basis of a motivation and a problem scope. The motivation expresses why a HOD[7] or an incremental extension of a HOD is needed. The scope of the next iteration limits the problem that needs to be tackled, e.g. the definition and relations around one particular term, to avoid divergent discussions. During this phase, members can also propose the use of relevant sources from which inspiration can be drawn upon. Sources of inspiration can be legacy database schemas, standards, documentation, etc. Before going on to the next phase, the relevance of these sources is agreed upon by the community. The social processes here are:

- *Creating the motivation* by a member with sufficient rights to start the process. In the case of bootstrapping the ontology, the creator is the community-leader (founder) or one of the founders of the community. The motivation is discussed by all members and refined in terms of the discussion until a consensus is achieved.
- *Scoping the problem.* Similarly to the motivation, a founder or a member with sufficient rights creates the scope. It is also the subject of discussion and refinements until a consensus is achieved.

---

[7] In the case of a first iteration, the hybrid ontology description is initially empty.

- *Add (invite) member and remove member.* Members can join (or invited) to take part in refining the motivation and scope and the subsequent ontology engineering processes. When the goals of a community differ to greatly from the interest of one of the stakeholder, a member can decide to leave the community.
- *Proposing resources* that can be used to draw inspiration from. At any time, users can propose a list of resources that can be accepted or not. Examples of such resources can be the use of existing standards.

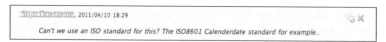

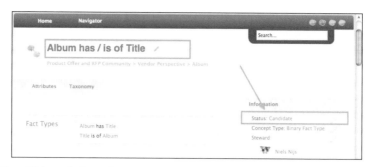

Fig. 7. A proposition to use an existing standard to create lexons concerning dates

From the creation phase onwards, every ontology operation is subject to discussion before approval as mentioned in the previous section. During the *create* process, lexons are generated from the collected sources in the scoping activity (e.g. documents or legacy database schemas). The operations in this phase are:

- *Request to add a lexon* $< \gamma, t_1, r_1, r_2, t_2 >$ on which the community members as a whole accept or *refuse* the new lexon (see **Fig. 8**).
- *Request to add a constraint* such as internal and external mandatory constraints. The role or roles on which the constraint is put on have to exist in the ontology. Constraints on lexons are modeled using a predicate language such as Ω-RIDL [19].

Fig. 8. The addition of a lexon is initially pending or "candidate". In this example, a user added a lexon around a specific type of product (see Section 5). Users can discuss this lexon before accepting, denying or postponing the decision.

The *refine* process is used to refine existing lexons and constraints in the HOD. Actions that correspond in this phase are:

- *Request to remove lexon.* All constraints involving roles in this lexon will be affected *as* well as well as the glosses around this lexon.
- *Request to change the supertype of a term.* The class hierarchy is constructed with a lexon whose roles bear a special meaning (the taxonomic relation, e.g. with *role* and co-role "is a" and "subsumes"). When no supertype was defined, a taxonomic relation is added between the two terms. When such a relation already exists between the terms and another super terms, the existing taxonomic relation is removed before the creation of the new one.

- *Request to change "super lexon" of a lexon.* Which indicates that the population of a lexon is a subset of the more general lexon. A special operation, as it corresponds with a *subset* constraint on both roles of the two lexons [11,19].
- *Request to remove a constraint.*

Articulate is used to create informal meaning description, i.e. glosses, as extra documentation that can serve as anchoring points when stakeholders have used different terms for the same terms (synonyms). When descriptions are already available, e.g. in source documents, they can already be imported to speed up this process. The operations in this phase are:

- *Request to add gloss,* for a particular term $t \in T$ or lexon $\lambda \in \Lambda$ in a community $\gamma \in \Gamma$, or request to $g_1(\gamma,t) \leftarrow gloss$ or $g_2(\lambda) \leftarrow gloss$. Lexon glosses have to follow the glossary coherence property, the glosses of those terms have to be delivered or the request is ignored.
- Request to remove a gloss. When all glosses of a term are removed, the lexons around this term that are articulated loose their glosses as well. This impact is shown to the user. All EQ_γ around this gloss will be removed.
- Request to *change a gloss.* When accepted, all EQ_γ assertions around this gloss are removed and the other communities are notified to reconsider whether the changed gloss still EQ_γ theirs.
- *Request to add synonym.* A request to link terms across different communities, such that $ci(\gamma_1,t_1) = ci(\gamma_2,t_2)$ where $t_1,t_2 \in T$ and $\gamma_1,\gamma_2 \in \Gamma$. This action will make those two terms term-equivalent EQ_T.
- *Request to remove synonym.* This action happens when the definitions of both concepts diverge in such a way that they do not handle about the same anymore. Naturally, different lexons will emerge around those terms using other operations. The EQ_T assertion is thus removed.

At any given point, in order to achieve unification, discussions between users can take place. One can compare such discussions with posts and Web forums. By linking posts with their replies, one can create threads. An item in such discussion can be a trigger for an ontology operation or a task assigned to a person. A link is therefore kept between a task and an ontology operation if the post in question was the source of this action. For example, users who do not feel comfortable with formal ontology operators or do not know how to solve a problem might request an edit.

- *Request for edit.* A general request for edit (or solving a problem). For instance used when a member feels he has not enough responsibility over the concept to propose the actual changes.
- *Request for information.* Not to be confused with a request for edit for glosses, but rather a request for clarification. Such a request might result in a request for edit or as a request for an ontology operation.
- *Request for peer review.* A invitation to review some aspects of the ontology, e.g. inviting members of the community to give comments to certain proposed changes, even though they are not immediately affected by the concepts in question.
- *Request for help*, in contributing to certain aspects of the HOD.

- *Comment.* A comment to a post or a concept, a general class of posts that are not related to the other types of posts.
- *Reply.* All posts not belonging to any category in a thread.

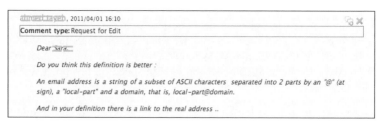

Fig. 9. Example of a request for edit, a user proposing a better definition and pointing to a problem in the existing definition

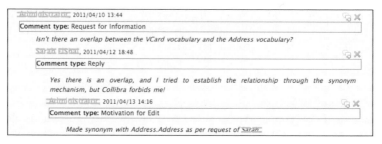

Fig. 10: Request for information. In this particular case, someone noted an overlap between two ontologies. Another member replied that an attempt was made to define synonyms, but was unable to do so. A third user then made the request instead.

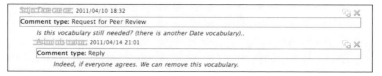

Fig. 11: Request for Peer Review. In the Business Semantics Glossary, ontologies are referred to as vocabularies, a term more "accessible" to users. In this particular case, a member of the community noticed that one of the ontologies was redundant.

4.2 Application Commitments in the Feedback Loop

Commitments provide valuable information on which terms and lexons the different members of the community representing their organization commit to. This selection is exploited by informing those members when changes are requested (and occur) in the ontology as to stimulate discussion.

The mapping α in those commitments is furthermore used to delve into the annotated data in search for support or counterexamples for certain statements made by the community, e.g. to notify the community whether proposed constraint is true for all annotated information systems currently known in the community. This process will guide the community in its dialogue to achieve agreement. This is done by generating the necessary queries using the commitments of each of the applications, populating the lexons in the conceptual schema and then reason over the data in terms of lexon populations. This tool, called the Ω-DIPPER, has been described in [8] and the results will be reported elsewhere. **Fig. 12** extends **Fig. 4** and depicts the place of Ω-DIPPER in the feedback loop.

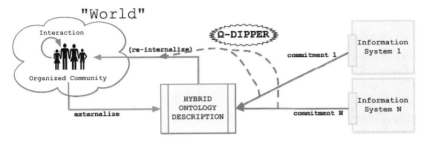

Fig. 12. Feedback loop from the ontologies to the community by not merely taking into account the lexons committed to by the application, but the data in the annotated organization information systems as well

5 Tool

These results were implemented and tested in a web application supporting BSM called Business Semantics Glossary (BSG, see **Fig. 13**), also developed by Collibra nv/sa. BSG is based on the Wiki paradigm that is a proven technique for stakeholder

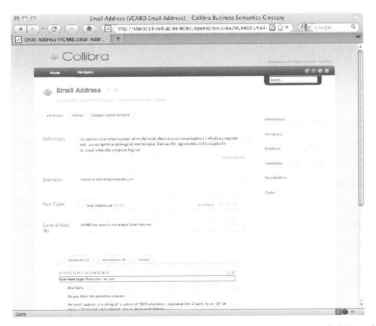

Fig. 13. Screenshot of the Business Semantics Glossary. Here we see the definition of `Email Address` in the `VCARD` ontology by the same community. Note the discussion between users around this concept.

collaboration. Governance models are built-in and user roles can be applied to distribute responsibilities and increase participation. **Fig. 13** also shows the lexons and gloss (here called "Description") of that community around the concept of Home Address.

The interactions and resulting ontology operations were implemented using the flexible architecture provided by XWiki[8], on which the BSG is based. The figures in the previous section depict some screen shots of the user interactions and resulting ontology evolution operators registered by the system. The tool was used by 45 users worked collaboratively on creating several ontologies concerning e-business (products, offers and requests for proposals). The 45 people represented 9 autonomously developed information systems: 4 request for proposals (RFP) systems and 5 vendor systems. The communities were: (i) a product community, which concerned everyone, (ii) an RFP community and (iii) a vendor community. As the experiment progressed, different communities evolved around general concepts such as Address, Dates, Contact Details, etc. in which some of the original 45 users are part of.

6 Conclusions

Ontologies are key in meaningful and efficient interoperation of autonomously developed and maintained information systems and come to be as a community effort. Those members in a community interact with each other, talking about the concepts and their relations in their natural languages. The result of those interactions is reflected in changes on the ontology. One important tool in agreeing on the meaning of concepts are glosses, natural language descriptions of those concepts.

In this paper, we presented a method for ontology engineering based on fact-oriented conceptual modeling techniques called DOGMA, in which the implementation details of an ontology has no importance at design time. This method returns ontology descriptions as an artifact, which will be used for the implementation of ontologies in languages such as RDF(S) and OWL in similar way database schema's were distilled from conceptual schemas in information systems modeling.

We extended those DOGMA ontology descriptions with hybrid aspects by giving the glosses a more prominent role in the agreement process and furthermore formalized the social processes involved in ontology engineering and how these are translated into ontology evolution operators. We proposed operators involving glosses that capture the agreement processes between members of a community and how glosses are used to identify the concepts. These describe flow from community to hybrid ontology description in the feedback loop. The use of the annotated information systems to provide feedback and steer the community in the engineering process will be reported elsewhere. These ontology evolution operators and the social interactions were implemented in a tool called Business Semantics Management, which supports the Business Semantic Management methodology built around DOGMA.

References

1. Concepts and Terminology for the Conceptual Schema, Technical Report 9007, International Organization for Standardization, Geneva, Switzerland (1987)
2. OMG MOF, version 2.0 (2009), http://www.omg.org/spec/MOF/2.0/

---

[8] http://www.xwiki.org/xwiki/bin/view/Main/

3. OMG SBVR, version 1.0 (2009), `http://www.omg.org/spec/SBVR/1.0/`
4. De Leenheer, P., de Moor, A., Meersman, R.: Context dependency management in ontology engineering: a formal approach. In: Spaccapietra, S., Atzeni, P., Fages, F., Hacid, M.-S., Kifer, M., Mylopoulos, J., Pernici, B., Shvaiko, P., Trujillo, J., Zaihrayeu, I. (eds.) Journal on Data Semantics VIII. LNCS, vol. 4380, pp. 26–56. Springer, Heidelberg (2007)
5. De Leenheer, P., Christiaens, S., Meersman, R.: Business semantics management: A case study for competency-centric HRM. Computers in Industry 61(8), 760–775 (2010)
6. De Leenheer, P., Mens, T.: Ontology evolution. In: Hepp, M., De Leenheer, P., de Moor, A., Sure, Y. (eds.) Ontology Management. Semantic Web And Beyond Computing for Human Experience, vol. 7, pp. 131–176. Springer, Heidelberg (2008)
7. De Leenheer, P., Debruyne, C.: DOGMA-MESS: A Tool for Fact-Oriented Collaborative Ontology Evolution. In: Meersman, R., Tari, Z., Herrero, P. (eds.) OTM-WS 2008. LNCS, vol. 5333, pp. 797–806. Springer, Heidelberg (2008)
8. Debruyne, C.: On the social dynamics of ontological commitments. In: Meersman, R., Dillon, T., Herrero, P. (eds.) OTM 2010. LNCS, vol. 6428, pp. 682–686. Springer, Heidelberg (2010)
9. Fellbaum, C.: WordNet: An Electronic Lexical Database. MIT Press, Cambridge (1998)
10. Guarino, N.: Formal ontology and information systems. In: Int. Conf. On Formal Ontology In Information Systems FOIS 1998, pp. 3–15. IOS Press, Amsterdam (1998)
11. Halpin, T.: Information Modeling and Relational Databases. Morgan Kaufmann, San Francisco (2008)
12. Hepp, M., De Leenheer, P., de Moor, A., Sure, Y. (eds.): Ontology Management, Semantic Web, Semantic Web Services, and Business Applications. Semantic Web And Beyond Computing for Human Experience, vol. 7. Springer, Heidelberg (2008)
13. Jarrar, M.: Towards the notion of gloss, and the adoption of linguistic resources in formal ontology engineering. In: Proc. of the 15th Int. Conf. on World Wide Web (WWW 2006), pp. 497–503. ACM Press, New York (2006)
14. Meersman, R.: Semantic ontology tools in IS design. In: Raś, Z.W., Skowron, A. (eds.) ISMIS 1999. LNCS, vol. 1609, pp. 30–45. Springer, Heidelberg (1999)
15. Meersman, R.: Reusing certain database design principles, methods and techniques for ontology theory, construction and methodology. Technical report, VUB STARLab, Brussel (2001)
16. Meersman, R., Debruyne, C.: Hybrid ontologies and social semantics. In: Proc. of 4th IEEE Int. Conf. on Digital Ecosystems and Technologies (DEST 2010). IEEE Press, Los Alamitos (2010)
17. Schneider, J., Passant, A., Breslin, J.: Enhancing mediawiki talk pages with semantics for better coordination - a proposal. In: The Fifth Workshop on Semantic Wikis: Linking Data and People Workshop at 7th Extended Semantic Web Conference, ESWC (2010)
18. Spyns, P., Meersman, R., Jarrar, M.: Data modelling versus ontology engineering. SIGMOD Record Special Issue 31(4), 12–17 (2002)
19. Trog, D., Tang, Y., Meersman, R.: Towards ontological commitments with W-RIDL markup language. In: Paschke, A., Biletskiy, Y. (eds.) RuleML 2007. LNCS, vol. 4824, pp. 92–106. Springer, Heidelberg (2007)
20. Verheyden, P., De Bo, J., Meersman, R.: Semantically unlocking database content through ontology-based mediation. In: Bussler, C., Tannen, V., Fundulaki, I. (eds.) SWDB 2004. LNCS, vol. 3372, pp. 109–126. Springer, Heidelberg (2005)
21. Viegas, F.B., Wattenberg, M., Kriss, J., van Ham, F.: Talk before you type: Coordination in wikipedia. In: Proc. of the 40th Annual Hawaii Int. Conf. on System Sciences. HICSS 2007, pp. 78–87. IEEE Computer Society, Washington, DC, USA (2007)
22. Wintraecken, J.: The NIAM Information Analysis Method, Theory and Practice. Kluwer Academic Publishers, Dordrecht (1990)

Author Index